IET SECURITY SERIES 21

Security and Privacy Schemes for Dense 6G Wireless Communication Networks

Other volumes in this series:

Volume 1	**Information Security: Foundations, technologies and applications** A. Awad and M. Fairhurst (Editors)
Volume 2	**Engineering Secure Internet of Things Systems** B. Aziz, A. Arenas and B. Crisp
Volume 3	**Mobile Biometrics** G. Guo and H. Wechsler (Editors)
Volume 4	**User-Centric Privacy and Security in Biometrics** C. Viuelhauer (Editor)
Volume 5	**Iris and Periocular Biometrics** C. Rathgeb and C. Busch (Editors)
Volume 7	**Data Security in Cloud Computing** V. Kumar, R. Ko, and S. Chaisiri (Editors)
Volume 8	**Hand-Based Biometrics: Methods and technology** M. Drahanský (Editor)
Volume 9	**Authentication Technologies for Cloud Computing, IoT and Big Data** Y.M. Alginah and M.N. Kabir (Editors)
Volume 10	**Nature-Inspired Cyber Security and Resiliency: Fundamentals, techniques and applications** E.M. El-Alfy, M. Eltoweissy, E.W. Fulp, and W. Mazurczyk (Editors)
Volume 12	**Voice Biometrics: Technology, trust and security** C. García and G. Chollet (Editors)
Volume 14	**Privacy by Design for the Internet of Things: Building accountability and security** A. Crabtree, H. Haddadi, and R. Mortier (Editors)
Volume 16	**Machine Learning, Blockchain Technologies and Big Data Analytics for IoTs: Methods, technologies and applications** A. Kumar Tyagi, A. Abraham, F. Khadeer Hussain, A. Kaklauskas, and R. Jagadeesh Kannan

Security and Privacy Schemes for Dense 6G Wireless Communication Networks

Edited by
Agbotiname Lucky Imoize, Chandrashekhar Meshram, Dinh-Thuan Do, Seifedine Kadry and Lakshmanan Muthukaruppan

The Institution of Engineering and Technology

Published by The Institution of Engineering and Technology, London, United Kingdom

The Institution of Engineering and Technology is registered as a Charity in England & Wales (no. 211014) and Scotland (no. SC038698).

First published 2023

The Institution of Engineering and Technology
Futures Place
Kings Way, Stevenage
Hertfordshire SG1 2UA, United Kingdom

www.theiet.org

British Library Cataloguing in Publication Data
A catalogue record for this product is available from the British Library

ISBN 978-1-83953-663-2 (hardback)
ISBN 978-1-83953-664-9 (PDF)

Typeset in India by MPS Limited

Cover Image: MF3d/E+ via Getty Images

Contents

About the editors

Agbotiname Lucky Imoize (senior member, IEEE) received the B.Eng. degree (Hons.) in Electrical and Electronics Engineering from Ambrose Alli University, Nigeria, in 2008 and the M.Sc. degree in Electrical and Electronics Engineering from the University of Lagos, Nigeria, in 2012. He is a lecturer in the Department of Electrical and Electronics Engineering at the University of Lagos, Nigeria. He was, until recently, a research scholar at the Ruhr University Bochum, Germany, under the Nigerian Petroleum Technology Development Fund (PTDF) and the German Academic Exchange Service (DAAD) through the Nigerian-German Postgraduate Program. He was awarded the Fulbright Fellowship as a visiting research scholar at the Wireless@VT Laboratory, Bradley Department of Electrical and Computer Engineering, Virginia Tech., USA, where he worked under the supervision of Prof. R. Michael Buehrer from 2017 to 2018. Before joining the University of Lagos, he was a Lecturer at Bells University of Technology, Nigeria. He worked as a core network products manager at ZTE Corporation, Nigeria, and as a Network Switching Subsystem Engineer at Globacom, Nigeria. His research interests cover the fields of 6G wireless communication, wireless security systems, and Artificial Intelligence. He has co-edited four books and co-authored over 150 wireless communication papers in peer review journals and conferences. Imoize is an active reviewer for over 50 international journals and conferences. He is the Vice Chair of the IEEE Communication Society, Nigeria chapter, a Registered Engineer with the Council for the Regulation of Engineering in Nigeria (COREN), and a member of the Nigerian Society of Engineers (NSE).

Chandrashekhar Meshram is an assistant professor in the Department of Post Graduate Studies and Research in Mathematics, Jaywanti Haksar Government Post Graduate College, Chhindwara University, Betul, India. His research interests include cryptography and its applications, neural networks, the IoT, wireless sensor networks, medical information systems, ad hoc networks, number theory, fuzzy theory, time series analysis, climate change, mathematical modeling, and chaos theory. He has published over 120 scientific articles in international journals and conferences on the above research fields. In addition, he is a regular reviewer for over 60 international journals and conferences. He is a member of IAENG, WASET, CSTA, IACSIT, EAI, ILAS, SCIEI, MIR Labs, and a lifetime member of the Indian Mathematical Society and Cryptology Research Society of India. He received his Ph.D. degree in Mathematics from R.T. M. Nagpur University, Nagpur, India.

Dinh-Thuan Do is an assistant professor in the School of Engineering, University of Mount Union, Alliance, USA. He was a research scientist at the Electrical Engineering Department, University of Colorado Denver, Denver, USA. Also, he was formerly a research scientist in the Department of Electrical and Computer Engineering at the University of Texas at Austin, USA. Prior to joining The University of Texas at Austin, he was an assistant professor at Asia University in Taiwan and a research assistant professor at Ton Duc Thang University in Vietnam. His research interests include signal processing in wireless communications networks, non-orthogonal multiple access, full-duplex transmission, and reconfigurable intelligent surfaces (RIS). He received the Golden Globe Award from the Vietnam Ministry of Science and Technology in 2015 (Top ten excellent scientists nationwide). He is currently serving as an editor of *Computer Communications*, associate editor of *EURASIP Journal on Wireless Communications and Networking*, associate editor of *Electronics*, associate editor of *ICT Express*, and editor of *KSII Transactions on Internet and Information Systems*. His publications include over 100 SCIE/SCI-indexed journal articles and 40 international conference papers. In addition, he is the author of two textbooks and six book chapters. He is a senior member of the IEEE. He holds a Ph.D. degree in communications engineering from the Vietnam National University (VNU-HCMC), Vietnam.

Seifedine Kadry is a professor in the Department of Applied Data Science at Noroff University College, Kristiansand, Norway, and the Department of Electrical and Computer Engineering, Lebanese American University, Byblos, Lebanon. His research focuses on data science education using technology, system prognostics, stochastic systems, and applied mathematics. He is also an ABET program evaluator for computing and for Engineering Tech. He is a fellow of the IET, IETE, and IACSIT, and a distinguished speaker of the IEEE Computer Society.

Lakshmanan Muthukaruppan is a professor and head of the Department of Electronics and Communication Engineering and dean of research at Galgotias College of Engineering and Technology, Greater Noida, Uttar Pradesh, India. His research interests include wireless communication and networks, security algorithms, digital communication, information coding theory, signal and image processing, deep learning, and machine learning. He has published over 70 papers in journals and conferences (SCI/SCIE and Scopus). He has four patents issued in India and two patents granted in Australia. He received the Best Researcher Award in International Excellence Award 2021 and the Award of Emerging Leader for his contribution and achievement in the Discipline of Engineering in Higher Education Leadership Meet – 2018 (HELA 2018). He is a senior member of the IEEE and UACEE and a fellow of the IETE and IET. He is also a life member of the ISTE, ACM, IACSIT, and IAENG. He received his Ph.D. degree in Wireless Communication and Networks from VIT University, Vellore, India.

Acknowledgments

I sincerely express my profound gratitude to God for his faithfulness and wisdom in editing this book. This book would not have been possible without the support of the Department of Electrical Engineering and Information Technology, Institute of Digital Communication, Ruhr University Bochum, Germany, and the University of Lagos, Nigeria. I acknowledge the sponsorship from the Nigerian Petroleum Technology Development Fund (PTDF) and the German Academic Exchange Service (DAAD) through the Nigerian-German Postgraduate Program. Special thanks to my beloved wife, Kelly, and our sons, Lucius, Luke, and Lucas. Also, I am indebted to the Deeper Life Bible Church, Essen Region, North Rhine-Westphalia, Germany, for their unwavering support. Last, I sincerely thank the Institution of Engineering and Technology (IET) for its editorial support.

Gelsenkirchen, North Rhine-Westphalia, Germany
Agbotiname Lucky Imoize

Preface

Fifth-generation (5G) wireless networks are now commercialized, and the research focus has shifted toward sixth-generation (6G) wireless systems. The integration of sensor nodes and massive devices (MDs) in ubiquitous 5G networks has facilitated the design of critical enabling technologies to support billions of data-hungry applications. By leveraging sensor nodes in wireless sensor networks (WSNs), sensitive users' information can be harvested and transmitted to receivers via WSN-assisted channels, which are often not secured adequately. Consequently, sensitive user information can be intercepted and used unlawfully. The measures used for the security and confidentiality of data transmitted over existing 5G WSN-assisted channels are grossly limited. 6G systems are envisaged to face fiercer security challenges. In 6G wireless networks, a new set of sensing and precise localization techniques are predicted. Thus, the need to secure the sensed user information against adversarial attacks is not negotiable.

This book highlights the proliferating security and privacy issues in commercialized 5G wireless networks. It discusses critical promising security frameworks and architectures to support the massive devices and enabling technologies for 6G wireless networks. We propose efficient, robust, and secure schemes to support the transmission of critical user data over unsecured wireless channels. We discuss the sensing traits envisioned in dense 6G devices that could be gainfully harnessed to create a new breed of context-aware security protocols to leverage the quality of the security paradigm. Additionally, we propose the implementation of low-cost security architectures to mitigate sophisticated attacks on the wireless edge. We also introduce artificial intelligence (AI) and machine learning to design cutting-edge security schemes to protect future wireless systems from malicious attacks and exploitation. Also, we present a good overview of emerging security and privacy issues in 6G wireless networks and viable solutions to address them appropriately. Additionally, we discuss extensively the prospects and societal benefits envisioned in security systems for next-generation wireless networks. Finally, we present case studies, highlight critical lessons, and provide recommendations for designing future security systems.

The key highlights of the book are as follows:

- Proposes solutions to revamp the traditional security architecture toward addressing critical security challenges in commercialized 5G and envisioned 6G wireless communication systems.
- Provides new insights into real-world scenarios of the deployment, applications, management, and associated benefits of robust, provably secure, and efficient security schemes for massive devices in 6G wireless networks.

- Discusses critical security and privacy issues affecting all parties in the wireless ecosystem and provide practical AI-based solutions to address these problems appropriately.

Specifically, the book is structured into 20 chapters outlined as follows:

Chapter 1 introduces emerging security and privacy schemes for dense 6G wireless communication networks. The chapter presents a glimpse into the future of 6G wireless networks, which promises to facilitate a higher data transmission rate incorporating space, undersea communication, and other novel technologies and applications. Additionally, it emphasized that security and privacy concerns had been raised due to the numerous data-intensive use cases and applications the 6G network is expected to support. Although there are existing reviews conducted on 6G security and privacy issues, not all considered the possible solutions to these proliferating issues. Thus, the chapter extensively reviews the emerging security and privacy schemes for dense 6G wireless networks. Specifically, the chapter presents the security and privacy challenges in the 6G enabling technologies and reviews possible solutions to show the trends and proffer probable research directions for ameliorating security and privacy concerns in 6G wireless networks.

Chapter 2 focuses on the history of security and privacy in wireless communication systems. The authors noted that wireless communication is one of the most successful technologies that have found application in various sectors of our daily lives. However, it was noted that one of the significant issues and challenges of wireless communication is security. In order to provide a holistic view of this crucial issue, the chapter presents a historical description of the evolution of cellular networks. Specifically, it provides a general overview of different attacks and vulnerabilities in the wireless network based on the open system interconnection layered protocol. The chapter further briefly reviews the security and privacy issues for each cellular network generation. The various emerging wireless communication systems and the application of artificial intelligence and machine learning to wireless security system design are broached. The security issues and challenges in the key technologies of 6G wireless networks were identified and discussed extensively. Finally, open research issues were identified, and discussions provided a path for further research on security in 6G wireless networks and beyond.

Chapter 3 discusses the potential of artificial intelligence (AI) enabling security systems for 6G wireless networks, with a laser focus on algorithms, strategies, and applications. The authors noted that considering the incredibly complex and diverse requirements, 6G is anticipated to support the extraordinary Internet of Things advances and to effectively satisfy a wide range of requirements; space-aerial-terrestrial-ocean, interconnected three-dimensional networks are envisioned in 6G wireless networks. The chapter emphasized that big data processing methods, computing capacity, and rich data availability have progressed. However, security is still a significant concern, and the application of AI to address complicated security issues in the envisioned 6G networks is only natural. The chapter comprehensively reviews AI-based security for 6G wireless networks. The concept of an AI-enabled 6G system, the motivations behind integrating AI-based security systems into 6G networks, and state of the art in AI-assisted security system models in wireless communication systems are discussed extensively.

Chapter 4 focuses on the vision of 6G security and privacy. The authors remarked that the fifth-generation (5G) network is yet to be utilized fully, especially in developing countries. 5G being at its underlying stage in commercial usage, various identified limitations have invoked research into the sixth-generation (6G) wireless networks. The deployment of 5G networks in several parts of the world has revealed the limitations of these networks, which undoubtedly supports the exploratory study of 6G networks. The fundamental privacy and security concerns envisioned in 6G technology are highlighted in the chapter. The critical security concerns with the envisioned 6G networks comprising issues with tactile communications, resources as services, variable radio access constraints, varied high-frequency bands, and other security-related issues in the 6G ecosystem are broached. Last, the potential security attacks on 6G wireless networks were highlighted, and the probable countermeasures to mitigate these attacks were suggested.

Chapter 5 explores the potential benefits of using reconfigurable intelligent surfaces (RISs) to enhance the physical layer security of promising sixth-generation (6G) wireless systems relying on non-orthogonal multiple access (NOMA) systems. RISs are a new technology that can dynamically modify the propagation environment of wireless signals, allowing for increased efficiency and security. By leveraging RISs in NOMA systems, this chapter demonstrates that it is possible to enhance the secrecy performance of the system while simultaneously improving the overall spectral efficiency. First, the chapter comprehensively overviews the physical layer security theoretical foundations for 6G systems. Then, the chapter considers a case study of RIS-aided NOMA systems, including the optimization of the RIS reflection coefficients and performance analysis of security concerns. Simulation results demonstrate that RIS-aided NOMA systems can significantly improve secrecy performance, particularly in scenarios with a high number of meta-surface of RIS. The chapter concludes that RIS-aided NOMA has the potential to be an effective solution for enhancing the physical layer security of wireless communication systems.

Chapter 6 considers dynamic optical beam transmitters of secure visible light communication systems. The chapter posits that due to the broadcast nature of visible light communications (VLC) channels, physical layer security (PLS) techniques have been considered to improve the transmission confidentiality of VLC links. However, it was remarked that almost all current schemes merely work with multiple distributed transmitters and fail to serve the scenarios with centralized transmitters, even single transmitters. In order to address this issue, the authors proposed the dynamic inclined optical beam-based PLS enhancement scheme. Unlike the conventional Lambertian beam-based technique paradigm, the projected scheme utilizes the commercially available typical non-Lambertian beams to form the secure VLC links. Numerical results show that, compared with the conventional static Lambertian configuration, up to 5.52 bps/Hz average secrecy capacity gain was derived via the proposed dynamic scheme using two inclined candidate beams.

Furthermore, it was noted that this potential gain would be further elevated to about 6.63 bps/Hz when up to four candidate beams are available at the transmitter.

Chapter 7 presents a new machine learning-based scheme for physical layer security. The work emphasized that massive MIMO is a possible physical layer security technique to meet the 6G security requirements. Massive MIMO systems are naturally immune to passive eavesdroppers, but active eavesdroppers dramatically degrade existing security architectures. The book chapter describes the use of massive MIMO enhanced through artificial intelligence to improve security on the physical layer. The authors describe several machine learning-based algorithms and a deep neural network (DNN) model capable of detecting the presence of an active eavesdropper by exploiting the particular properties and features of massive MIMO. Also, the work describes a machine learning model, and DNN applied to a realistic scenario where the Channel State Information (CSI) of the channels (i.e., legitimate users and eavesdroppers) is unknown. A set of different algorithms showing varying performances are compared. The prediction complexity of the different algorithms is discussed elaborately. The chapter explains the design issues for DNN and new machine learning-based secure transmission schemes in massive MIMO-based communication systems. Simulations prove the robustness of machine learning-based algorithms and DNN without the need for feedback overhead and when the CSI of all channels is unknown. Finally, the chapter showed that higher security of communication systems could be achieved on the physical level.

Chapter 8 covers ad hoc vehicular networks employing intelligent, reflective surfaces for physical layer security. The authors noted that a significant amount of personal data is shared by smart vehicles that are a part of vehicular ad hoc networks (VANET). As a result, security needs to be improved to stop eavesdropping and intruder attacks. The traditional encryption protocols are more complicated computationally and are intended for upper layers, and are not appropriate for applications requiring lightweight infrastructure, such as the Internet of Things (IoT). Physical layer security (PLS) was identified as the ideal solution to address these issues. Intelligent reflecting surfaces (IRS), one of several PLS solutions, are explored. Two different IRS configurations were considered: the smart reflector (SR) and the access point (AP). Analytical expressions for secrecy outage probability (SOP) and secrecy rate are developed for these arrangements. Simulations show that the IRS-assisted system outperforms the system without IRS. In addition, the results show that adding more IRS components increases the secrecy rate. Although relaying provides comparable benefits, the hardware and signal processing complexity associated with it makes IRS a better choice for PLS. The chapter concludes that one of the viable components to preserve security in vehicular applications could be IRS-assisted transmission.

Chapter 9 explores core physical layer security solutions and technologies for 6G networks. The chapter emphasized that ensuring the confidentiality of wireless communications systems requires effective encryption under computational security using hard mathematical problems to build ciphers that apparently cannot be cracked in a useful time. However, as these constructions are not agnostic to technological advances, some problems may be solved efficiently with future technologies, e.g., quantum computing. In wireless networks, physical layer security emerges as a post-quantum security solution that promises to mitigate these threats. This concept of secrecy exploits the physical properties of the wireless channel to encode information so that the eavesdropper observes a certain degree of statistical independence between the message and the ciphertext. This notion of secrecy allows building cryptosystems that are agnostic to the computing capabilities of the attacker, being widely accepted as one of the strongest notions of secrecy created so far. The objective of the chapter is to overview the concept of physical layer security and understand how its integration could be done in future wireless networks. The information-theoretical framework grounding the concept is introduced for that purpose, and some basic design approaches are presented. These include PHY key generation methods, secure beamforming techniques, and cooperative jamming constructions. The mechanisms to enable the integration of these technologies in future 6G and beyond networks are discussed elaborately.

Chapter 10 captures steganography-based secure communication via single-carrier frequency division multiple access (SC-FDMA) transceivers. The authors noted that in the past few decades, wireless communications have empowered communication between people across the globe. The communication has been achieved through various transmission systems, of which the SC-FDMA is prominently featured. It is an enhanced version of Orthogonal Frequency Division Multiple Access (OFDMA) systems with low PAPR and high-power capabilities. Despite SC-FDMA being one of the best means for transmission, it requires inalienable safety efforts from intruders. Information that is imparted remotely is less safe than wired communication. Thus, the security of SC-FDMA turns into vitality. The modified-least significant bit (MLSB) algorithm has been proposed to incorporate security in the SC-FDMA system. The data is first embedded in the proposed algorithm and transmitted through the SC-FDMA system over an Additive White Gaussian Noise (AWGN) Channel. Then using the same algorithm, the information is rooted at the receiver end of the SC-FDMA system. The performance of the proposed algorithm is analyzed through various parameters of image quality assessment and Bit Error Rate (BER). The proposed method boasts higher PSNR and SSIM values of 68.9102 and 0.9995, respectively. The MSE, AD, and NAE were observed to be lesser, with the metrics 0.0084, 0.00021, and 0.000060, respectively.

Chapter 11 presents a lightweight algorithm for detecting fake incident reports in wireless communication systems. The authors remarked that sensor devices in 6G technology are an affordable method of identifying target incidents within large wireless communication systems (WCSs). However, these devices face potential compromise or capture by compromised actors. For instance, fake incident reports in a compromised device can result in congested networks that hinder the passage of valid incident data. Nevertheless, fake incident reports can be identified. However, this approach can prove very complex because of the requirement for certification tokens, which only sometimes offer a workable solution to congested networks. One suggested strategy is creating space-efficient Bloom filters. Their creation would result from correctly combining the correct devices and placing them in each device in advance. The next stage would see an incident report featuring an XOR of the tokens (XT), with all devices confirming the information according to its Bloom filter. Illegally acquiring a device can prove costly because it would compromise the Bloom filter data and the XT allocated to the correct incident report. Thus, the study suggests using a secure algorithm to update the data. Unlike existing studies, detecting a fake incident report would only increase by approximately one hop. However, the amount of traffic a compromised device creates would decrease by around 60%. Thus, the suggested method would lessen the traffic resulting from attacks featuring file incident reports, which would, in turn, make the network less congested.

Chapter 12 presents a real-time intrusion detection system for service availability in a cloud computing environment. The authors noted that the spike in Internet usage and outsourcing of computing needs, such as databases, networking, and storage, among others, to third parties poses a significant security threat to cloud users due to the cloud deployment medium, the Internet. The Internet exposes data confidentiality, integrity, and availability of cloud users to cybercriminals, who gather cloud users' personal information for illicit activities or sometimes make the cloud service unavailable for legitimate users. The Intrusion Detection System (IDS) is a prominent second-line approach for monitoring illicit activities like distributed denial of service attacks (DDoS) over cloud communication networks. However, it faces challenges in areas of false alarm, detection time, and accuracy, primarily attributed to the enormous amount of attributes the machine learning (ML) algorithm needs to process within a short period. Feature selection (FS) using statistical and metaheuristic algorithms is a promising method to overcome the IDS challenges. The chapter explores the binarization of user information, leveraging the UNSW_NB15 network attack dataset to enhance the efficiency of the ML algorithms. The work optimizes the statistical FS method, maximum relevance, and minimum redundancy (MrMr) with a nature-inspired algorithm known as Cuckoo search. The experimental evaluation of the proposed algorithms was conducted using Python IDLE 3.7.1. Various performance metrics, like detection time, false alarm, and accuracy, using the confusion matrix obtained from four selected algorithms, K-Nearest Neighbor (KNN), Logistic Regression (LR), Decision Tree (DT), Multi-Layer perceptron (MLP), are presented. Among the contending algorithms, the DT produced the best result with an accuracy of 96%, precision of 96% and 97% (training and testing), and recall scores of 96% and 97% (training and testing). A detection time of 1.60 s was obtained, making the model the most suitable among the four algorithms for real-time IDS.

Chapter 13 proposed a study addressing the security challenges of IoT-enabled frameworks using artificial intelligence, machine learning, and blockchain technology. The Internet of Things (IoT) is one of the most trending and rapidly growing domains which is being amalgamated with lots of new technologies like machine learning, deep learning, and blockchain. Intelligent devices are improving various parameters like efficiency, complexity, and reliability. IoT facilitates the monitoring and processing of wireless communication systems to address various security challenges. Implementing these technologies comes with many heterogeneous challenges that require specific protocols to overcome. Therefore, the chapter addresses the significant challenges in implementing and deploying these IoT-enabled frameworks. Various attacks that can take place in any IoT-enabled platform were discussed. The challenges related to security have been highlighted, and the corresponding solutions are discussed.

Chapter 14 focuses on alleviating 6G security and privacy issues using artificial intelligence. The authors remarked that the emergence of sixth-generation (6G) networks has brought about new security and privacy concerns, emphasizing the need to address these issues promptly and comprehensively. This book chapter thoroughly examines the flaws and potential solutions in 6G networks, focusing on a multi-objective optimization problem that considers the deployment of mobile users and prioritizes energy efficiency, data integrity, and end-to-end encryption. A genetic algorithm scheme is employed to solve the optimization problem. Several machine learning algorithms' performance is evaluated using metrics such as mean absolute error, mean square error, root mean square, and R2 score. It is important to note that all methods for the R2 score produced exact unity results for the LR, RF, KNN, and SVM with 1.000, 0.999, 0.993, and 0.993, respectively. However, the study uses artificially generated data to overcome the lack of available data on the energy efficiency, throughput, latency, and spectral efficiency of 6G networks. Additionally, the study highlights the security and privacy challenges posed by 6G networks and draws insights from previous technological advancements to identify potential solutions and prospects. This work is a comprehensive guide for researchers, practitioners, and policymakers to navigate the emerging landscape of 6G networks while ensuring security, privacy, and optimal performance.

Chapter 15 covers interference and phase noise in millimeter wave MIMO-NOMA and OFDM systems for 5G networks. The fifth-generation mobile communication (5G) is designed to support huge connectivity, high data rates, and excellent dependability as the number of wireless devices linked to the network approaches billions. Mobile users generate most of their data traffic from video streaming, which demands a higher bandwidth and lower latency. Therefore, current mobile communication networks must be upgraded to meet these criteria. A multiuser environment will require multiple access methods, such as NOMA and OFDM. The use of mmWave spectrum and NOMA could alleviate inefficient power allocation phase noise issues and hybrid beamforming complications to meet the lowest rate needs of each user. At the transmitter base station, the user data are superimposed in the power domain NOMA, after which the user end is subjected to phase noise cancellation. Due to insufficient elimination of the unwanted interference in the multiuser downlink, the desirable user's information is vulnerable to unsatisfactory Successive Interference Cancellation (SIC). The key goal of the chapter is how to decrease interference, i.e., phase noise in millimeter wave 5G and beyond 5G systems using the parametric phase noise filtering method.

Chapter 16 presents a generative adversarial network-based approach for mitigating inference attacks in IoT wireless networks. The authors noted that the proliferation of smart, connected, always-listening devices had introduced significant privacy risks to users in home-based wireless networks. Beyond the significant risk of eavesdropping, intruders can adopt machine learning techniques to infer sensitive information from audio recordings on these devices, resulting in a new dimension of privacy concerns and attack variables for smart home users. However, sound masking and microphone jamming have effectively prevented eavesdroppers from listening to private conversations. The study investigated the problem of adversaries spying on smart home users to infer sensitive information with machine learning techniques. The role of randomness in the effectiveness of sound masking for mitigating sensitive information leakage was analyzed rigorously. Finally, a generative adversarial network (GAN)-based approach for privacy preservation in smart homes, which generates random noise to distort the unwanted machine learning-based inference, was proposed. The experimental results demonstrate that GANs can be used to generate more effective sound masking noise signals which exhibit more randomness and effectively mitigate deep learning-based inference attacks while preserving the semantics of the audio samples. The presented GANs would find useful applications in addressing the increasing privacy and security concerns in 5G and the envisioned 6G wireless networks.

Chapter 17 discusses the adversarial resilience of self-normalizing convolutional neural networks for deep learning-based intrusion detection systems. The presence of adversarial examples can easily fool deep learning-based intrusion detection systems, thus, limiting their usefulness in security-critical applications such as in 5G and 6G wireless networks. The cause for the adversarial vulnerability of the neural network is still unknown. Still, some researchers have proposed that regularization and normalization techniques applied to the neural network models play a significant role. This chapter examines the role of self-normalization in the adversarial vulnerability of neural network models within the context of intrusion detection systems for application in the envisioned 6G wireless networks. The authors propose designing and implementing a deep learning-based intrusion detection system for botnet traffic and subjecting it to various adversarial attacks. The impact of self-normalization on the adversarial resilience of the deep learning-based intrusion detection system was investigated and compared with that of image classification neural network models. Finally, the study proposes a customized convolutional neural network (CNN) model that utilizes self-normalizing activation in the fully connected layers. The results show that self-normalization of the deep learning-based intrusion detection system using scaled exponential linear unit (SELU) results in greater resilience to various adversarial examples. The projected adversarial resilience of self-normalizing convolutional neural networks for deep learning-based intrusion detection systems would be useful in future wireless communication systems such as 6G networks.

Chapter 18 discusses the legal frameworks for security schemes in wireless communication systems. The author opines that wireless communication is one of the most successful technologies that has found applications in our daily lives. It has drawn the attention of researchers, standard bodies, and organizations continuously proposing and developing different standards and regulations to advance the existing wireless communication systems. However, one of the major issues and concerns raised in wireless communication is the security aspect and its legal frameworks. Security and privacy are crucial issues in wireless communication due to the transmission of signals over unprotected media, thus exposing signals to security and privacy attacks such as eavesdropping, modification, and data theft, among others. Depending on the type of data being transmitted, security, privacy, and legal frameworks become even more critical, especially with the adoption of new technologies such as cloud computing in healthcare and other sectors. While a framework is a structure that collects different but relevant areas together in the form of a single hybrid conceptual solution, legal frameworks are simply standards that can be utilized to deal with a challenge or decide what to do. Over the years, some legal frameworks have been developed, particularly in healthcare, the Internet of Things (IoT), and Artificial Intelligence (AI). However, there need to be more legal frameworks for the security and privacy of wireless networks, particularly for the envisioned 6G networks. Thus, this chapter presents the fundamental factors of legal frameworks for security and privacy in a wireless communication network, focusing on the 6G networks.

Chapter 19 discusses the design of a quantum true random number generator using quantum gates and benchmarking its performance on an IBM quantum computer. The authors noted that random numbers are used in various domains, including quantum communication and cryptography applications such as key generation and authentication. Quantum mechanics has the intrinsic ability to generate truly random numbers, making it the ideal alternative for scientific applications that require randomness. For example, quantum wireless communication is proposed as safe, secure, and efficient for the envisioned 6G wireless networks. The chapter explored the 24-qubit random number and employed the source rotation gate for random number generation. The simplicity of the source rotation gate, combined with its independently verifiable solitary unpredictability, is essential to obtain quantum random number generators at the least cost. The worst-case entropy value for such randomly produced integers is 0.999445, with the min-entropy of such numbers being 0.0008. Additionally, steering restart tests were used to validate and verify true randomness. Finally, the True Random Number Generation (TRNG) statistical characteristics were assessed using an autocorrelation study, and the statistical test showed impressive results.

Chapter 20 presents the proliferating security challenges and prospects of 6G networks in cloud environments. The authors noted that global technological and industrial development is advancing alarmingly. The development of the 6G communication system has been facilitated by the pervasive adoption of the latest generation information and communication technologies (ICTs), such as artificial intelligence (AI), virtual reality (VR), augmented reality (AR), extended reality (XR), the Internet of things (IoT), blockchain technology, among others. Exploratory research into 6G networks as the next wave of solutions is likely motivated by the limits of 5G networks observed as more 5G wireless networks are deployed. These analyses cover the fundamental privacy and security concerns in the ubiquitous 5G technology. The chapter discusses 6G security problems as a foundation for future research directions. In addition, the prevalent security issues in cloud computing platforms were explored. Last, the chapter delves into the potential application of cloud computing in 6G wireless systems.

Gelsenkirchen, North Rhine-Westphalia, Germany
Agbotiname Lucky Imoize

Chapter 1

Introduction to emerging security and privacy schemes for dense 6G wireless communication networks

Emmanuel Alozie[1], Agbotiname Lucky Imoize[2,3], Hawau I. Olagunju[1] and Nasir Faruk[4,5]

Abstract

A glimpse into the future presents the 6G wireless network, which promises to facilitate a higher data transmission rate incorporating space, undersea communication, and other novel technologies and applications. However, due to the numerous data-intensive use cases and applications the 6G network is expected to support, security and privacy concerns have been raised. Although there are existing reviews conducted on 6G security and privacy issues, not all considered the possible solutions to these issues. Thus, this chapter aims to extensively review the emerging security and privacy schemes for the dense 6G network. Specifically, this chapter presented the security and privacy challenges in the 6G enabling technologies and reviewed their possible solutions, to show the trends and proffer further research directions for improving security and privacy in 6G networks and beyond.

Keywords: 6G security; 6G privacy; 6G enabling technologies; blockchain-based security; quantum security

1.1 Introduction

Wireless communication is one of the most utilized forms of communication today, which is the communication between two or more wireless devices without any physical connection. Over the last few decades, wireless communication networks

[1]Department of Telecommunication Science, University of Ilorin, Nigeria
[2]Department of Electrical and Electronics Engineering, Faculty of Engineering, University of Lagos, Nigeria
[3]Department of Electrical Engineering and Information Technology, Institute of Digital Communication, Ruhr University, Germany
[4]Department of Information Technology, Sule Lamido University, Nigeria
[5]Directorate of Information and Communication Technology, Sule Lamido University, Nigeria

have experienced a substantial evolution from 1G up to 5G, with each one addressing the major limitations of its predecessor in terms of speed, capacity, security, privacy, etc. The 5G implementation phase began in 2019, and further adoptions facilitated massive connections. The deployment phase of the 5G network began in the year 2019, and further adoptions facilitated massive connections, extreme reliability, and reduced latency [1]. Although the 5G network has not been fully deployed, researchers and other stakeholders are already envisioning the 6G wireless network, which is expected to be an intelligent, reliable, scalable, and secure wireless network that would also combine both space and the undersea communication to form a ubiquitous network [2]. The 6G network is expected to provide a faster data rate in Tbps, 10 times lower latency, 100 folds enhanced connection density, and significantly higher spectrum, energy, and cost efficiency [3]. These requirements can be met using several novel technologies and applications, including visible light communication (VLC), Artificial Intelligence (AI), machine learning (ML), enhanced cloud computing, molecular communication (MC), quantum computing, terahertz (THz) communication, blockchain technologies, etc. [2,4,5]. However, two of the major challenges of the 6G wireless network that has been identified are security and privacy.

Security and privacy are two critical concerns in wireless communication because wireless signals are transmitted over unprotected media from the sender to the receiver over a distance, either short or long. Security and privacy are often used interchangeably but are quite different. Security is simply preventing or restricting unauthorized access to a particular resource, while privacy is a subset of security primarily concerned with upholding confidentiality. Over the years, authors have exposed the security and privacy challenges in the earlier cellular generations, including the recently developed 5G wireless network, and also proffered some countermeasures to defend against them [6,7]. Furthermore, other research works have been conducted to review the security and privacy challenges in 6G wireless networks. However, very few reviews exist on emerging solutions to defend against these challenges. Therefore, this chapter aims to show the trends of the emerging schemes that can be utilized to defend against these vulnerabilities in a dense 6G wireless communication network.

1.1.1 Key contributions

This section presents the noteworthy contributions of this chapter which includes the following:

- An exhaustive review of previous reviews on security and privacy in the 6G network is presented.
- The security and privacy challenges in the 6G enabling technologies are presented and reviewed.
- An extensive review of the emerging solutions proffered to defend against the various threats and attacks in the 6G network is provided.
- Further research directions for security and privacy in the 6G and beyond networks are identified and elaborated.

1.1.2 Chapter organization

The remaining part of this chapter is organized as follows: Section 1.2 presents the review of previous review work on the 6G-based security and privacy challenges. Section 1.3 presents an overview of the security and privacy in the 6G network. Section 1.4 briefly introduces the 6G enabling technologies and the various security and privacy threats/attacks in these technologies. Section 1.5 presents the emerging security and privacy solutions for the 6G wireless network. Section 1.6 highlights and discusses the lessons learned from the review, and finally, Section 1.7 concludes the chapters and provides recommendations for further research directions.

1.2 Related work

This section reviews previous studies on the security and privacy challenges as well as countermeasures in the envisioned 6G wireless networks.

Insights on the critical problems and difficulties in 6G wireless networks based on security, privacy, and trust are presented in [1], where the standard technologies and security challenges are clarified and elaborated. The work examined security concerns in developed networks ranging from 1G to 5G before critically analyzing the envisaged 6G network and the security needs and project security architecture. Furthermore, the work analyzed the security challenges in the 6G enabling technologies and applications. The review emphasized that the technologies and applications supported in the envisioned 6G network are not secure. However, solutions to these several attacks were not considered in the work. Similarly, the authors in [8] also investigated five major security and privacy challenges in the 6G network, including authentication, communication, access control, malicious behaviors, and encryption. The work also reviewed the security challenges in the developed networks, 1G to 5G, and then discussed the key areas in the 6G wireless network. The security and privacy challenges in the 6G enabling technologies were elaborated, showing that all of these technologies are vulnerable to several different threats and attacks. However, solutions to these attacks were not considered in the work. In order to further the work, a systematic review of security and privacy issues in 6G wireless networks was presented in [9] based on the prospective technologies in the three different layers, including the physical, network, and application layers. The security and privacy challenges of the developed technologies, including the 5G network, were reviewed. Extensive analysis and review of the security vulnerabilities/attacks in the 6G network were also conducted based on the supported applications. The work showed that the three layers are vulnerable to different attacks as well as the impact of artificial intelligence on the 6G network.

Furthermore, the work proffered some solutions to these security and privacy issues. However, only three layers were considered. The security and privacy challenges in the 6G network and the possible challenges with different key technologies and potential solutions are presented in [10]. A brief evolution of mobile security, from 1G to 5G, and an investigation of security requirements and challenges was provided. The threat landscapes of the various 6G applications, such as

AI/ML, blockchain technology, quantum computing, VLC, and terahertz communication, were analyzed, including the possible solutions to these threats. However, only four key technologies domain that significantly impacts the security and privacy challenges in the 6G network were identified and investigated. With a focus on the potential privacy challenges and their potential countermeasures beyond 5G and 6G (B5G/6G), the authors in [11] conducted a privacy-based comprehensive review. The research provided various taxonomies defined for privacy to obtain an overview, then investigated and discussed the privacy issues in the B5G/6G network.

Furthermore, several privacy solutions were discussed, as well as 6G privacy projects and standardization. The conclusion drawn in the work indicated that these solutions had many gaps that must be resolved based on a variety of characteristics such as maturity, application, and costs. However, the research was strictly limited to privacy in B5G/6G networks. The authors in [12] surveyed the potential security and privacy challenges in 6G networks based on their requirements, network architecture, applications, and the major enabling technologies. Furthermore, the research discussed possible countermeasures for these identified threats and attacks. However, the work only considered five key enabling technologies for the envisioned 6G wireless network.

Table 1.1 presents the limitations of the existing review on security and privacy challenges for the envisioned 6G network in terms of the enabling technologies considered and the proffered countermeasures.

From this review, it is obvious that there are only a few existing reviews on the security and privacy challenges and countermeasures in the 6G network. The review also showed that the envisioned 6G network is vulnerable to several security and privacy threats and attacks; however, as seen in Table 1.1, most of these reviews did not elaborate on the potential solutions that can be utilized to defend against these security and privacy issues.

1.3 6G security and privacy

This section presents an overview of the security and privacy in the 6G network. Specifically, the section presents the key architectural components and the security and privacy requirements of the envisioned 6G network.

Each successive generation of cellular networks aims to establish or update at least one of the security architectural components, such as new authentication and key management, to address security and privacy challenges posed by new applications and business models [9]. The 6G network security architecture was proposed with transparency in perspective. Because the 6G network will be designed to be more open than 5G, the distinction between within and outside the network will become increasingly hazy. As a result, existing network security solutions such as IPsec and firewalls will be ineffective in protecting the network from intruders. The tremendous proliferation of connections in the envisioned 6G network will raise security and privacy concerns. The envisaged 6G network is expected to bring about the advent of the Internet of Everything (IoE), a network of billions of

Table 1.1 Limitations of existing works

Reference	6G enabling technologies												Privacy issues	Counter-measures
	UAV communication	**Molecular communication**	**Terahertz communication**	**Visible light communication**	**Blockchain technology**	**Reconfigurable intelligent surfaces**	**Ambient backscatter communication**	**Cell-free massive MIMO**	**Quantum communication**	**Internet of BioNano-Things**	**Internet of Nano-Things**	**Perversive artificial intelligence**		
[1]	×	✓	✓	✓	✓	×	×	×	✓	×	×	✓	×	×
[8]	×	✓	✓	✓	✓	×	×	×	✓	×	×	✓	✓	×
[9]	×	✓	✓	✓	×	✓	×	×	×	×	×	×	✓	×
[10]	×	✓	✓	✓	✓	✓	×	×	✓	×	×	✓	✓	✓
[11]	×	×	×	×	×	×	×	×	×	×	×	×	✓	×
[12]	×	×	✓	✓	✓	×	×	×	✓	×	×	✓	×	×
This chapter	✓	✓	✓	✓	✓	✓	✓	✓	✓	✓	✓	✓	✓	✓

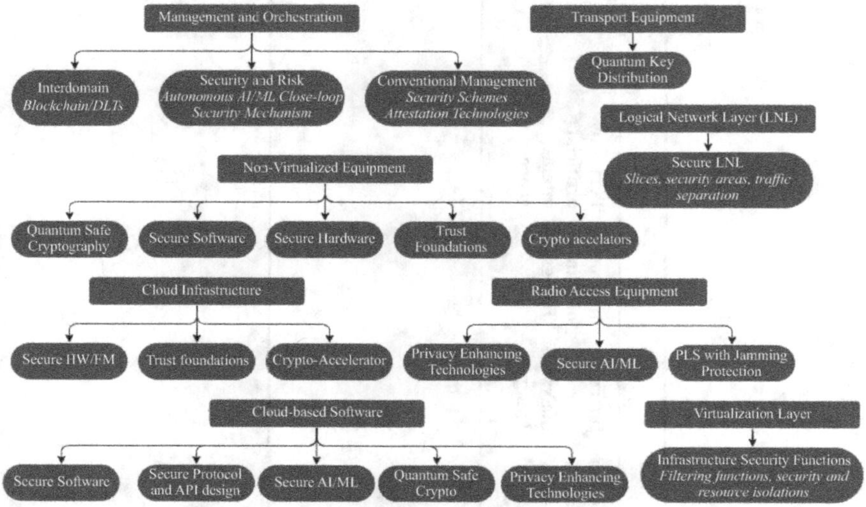

Figure 1.1 Key 6G security and privacy architectural components

different devices [1,9,13]. The core device security architecture based on SIM cards is not a realistic deployment for IoE in 6G, especially with small form factor devices like in-body sensors. In such a large network, key distribution and administration operations are exceedingly inefficient [13].

Furthermore, data collection via hyper-connected IoE to service 6G applications poses privacy concerns. Data theft through resource-constrained Internet of Things (IoT) devices will affect data privacy, location privacy, and identity privacy [13]. Security concerns must include all aspects of cyber-security, including robustness against attacks, privacy protection, and the ethical, appropriate use of automation technologies, particularly AI, to network functions and applications. Security also depends on active threat area control, which includes proactive actions such as threat prevention and protection and reactive actions such as attack detection and mitigation [14]. Figure 1.1 shows the architectural components of the 6G security and privacy [14].

Several 6G security and privacy requirements have been identified in the literature, which include the following.

1.3.1 Automated management system

The essential thing to do when addressing open-source security issues is the management of vulnerabilities caused by the updates, use, and disposal of open sources, which thus re-emphasized the urgent need for an automated management system to discover vulnerabilities and apply patches. Furthermore, an additional measure needs to be put in place using the Over-The-Air (OTA) technique to ensure the patched software is applied quickly and securely. Furthermore, a security governance framework must be established to handle the deployment of security solutions, changes in the developer perception, and open-source vulnerabilities from a long-term view [1].

1.3.2 Virtualization security solution

This necessitates using a system with a secured virtualization layer and security software that recognizes hidden malicious software. The emulator must allow total separation of processing, storage, and network services using secure protocols such as Transport Layer Security (TLS), Secure Shell (SSH), and Virtual Private Network (VPN). Virtual Machine Introspection (VMI) is a part of the emulator that analyses and identifies security risks by evaluating each virtual machine vCPU register information, communication packets, and file IO to prevent infiltration. The operating system should use containerization to appropriately set the privileges of various containers and prohibit mounting essential system directories and direct access to the host file container [1].

1.3.3 Data security using AI

AI systems require transparency on how their users and mobile communication system are safeguarded to ensure they are protected against Anti-Money Laundering (AML). First, the AI models must be developed in a reliable system. Then the digital signature technique would be utilized to verify if the AI models operating on the radio access network and user equipment and the core have been improperly updated or modified as a result of an aggressive attack. A system must perform a restoration procedure when a detrimental AI model is discovered and allow only reliable network components for AI training in the data collection for the system [1].

1.3.4 Post quantum cryptography (PQC)

Due to the known fact that asymmetric key encryption would be rendered insecure with the quantum computers making it unusable for 6G, most researchers have focused on PQC solutions such as code-based cryptography, lattice-based cryptography, hash-based signature, and multivariate polynomial cryptography. As part of its study, the US National Institute of Standards and Technology (NIST) will choose the best PQC solution between the years 2022 and 2024. However, compared to the existing Rivest-Shamir-Adleman (RSA) method, the PQC is anticipated to have a higher computational cost and longer key length. Hence, the PQC must be appropriately incorporated Hardware (HW)/Software (SW) performance and service requirements of the 6G network [1].

1.3.5 Preserving user privacy

Users' personal information should be handled and maintained according to the rules laid down between the service provider, subscriber, and mobile network operator to ensure their safety. The 6G system minimizes or encrypts the amount of publicly available information when it is used, preserving personal information securely in a trusted execution environment (TEE) and a reliable SW. Prior to sharing of personal information by the MNO, authentication, and authorization must be validated. Another alternative when dealing with user information is to utilize homomorphic encryption to encrypt the data. In addition, AI-based solutions, such as learning-based privacy-aware offloading systems, can be used to protect the user's location and usage pattern [1].

1.4 6G security and privacy challenges

This section briefly reviews the security and privacy challenges of the 6G network to comprehend the emerging schemes utilized to defend against these security difficulties. It is critical to examine the challenges based on the core enabling technologies.

The envisioned 6G network would support numerous core enabling technologies. These include, but are not limited to, unmanned aerial vehicle (UAV) or Satellite communication, pervasive AI, reconfigurable intelligent surfaces (RISs), ultra-massive MIMO, terahertz communication, molecular communication, blockchain and distributed ledger technology, quantum communication (QC), ambient backscatter communication, cell-free MIMO, etc. [4,15] as depicted in Figure 1.2.

1.4.1 UAV/satellite communication

An UAV is an autonomous as well as a remotely controlled aircraft or drone by a human pilot, and due to its high aerial movability, superior battery technology, cost-effectiveness, cameras, GPS, and gyroscopes, among other features, it has been extensively applied in both military and civilian fields for various purposes, which include weather monitoring, forest fire detections, emergency search and rescue, remote surveillance, traffic control, movie making, extensive coverage as well as for the provision of wireless communication services [4,16–18]. UAVs have been extensively reviewed in literature as part of the core enabling technologies of the envisioned 6G wireless network [16,19,20]. This technology has great usefulness and applicability such as in acquiring and disseminating data, agriculture, aerial photography, surveillance, search and rescue, and healthcare systems [21]. However, it has significant challenges in terms of security and privacy based on four levels: the sensor, hardware, software, and communication [22,23], as shown in Table 1.2.

Some privacy issues have been identified in UAV/satellite communication, such as the unsolicited taking of photos or videos and the disclosure of such photos or videos without expressed permission [23].

1.4.2 Molecular communication (MC)

MC is also one of the major 6G enabling technologies and was proposed as a solution for the limitations of the conventional electromagnetic wave-based transmission in extreme environments such as inside a human body, very small environments that require nanotechnology, as well as unsecured environments [24,25].

Figure 1.2 6G enabling technologies

Table 1.2 Various attacks in UAV/satellite communication

Levels	Threats/attacks
Sensor-based	GPS data spoofing and jamming, sensory channels attack, fake data injection attacks
Hardware-based	Physical collisions, battery depletion attacks, hijacking, hardware failure, RF module attacks
Software-based	Malicious software, system ID spoofing, fabrication of captured videos, operating system attacks
Communication-based	Eavesdropping, Man in the Middle (MITM), fabrication, Denial-of-Service (DoS), replay attacks

Although the potential of molecular communication has not been fully actualized, before transmission, information is encoded with concentration, release time, and type of molecules and then transmitted using chemical signals [24]. In essence, signals in molecular communication are biocompatible and consume very small energy particles. Various benefits and challenges of this technology have been

Table 1.3 Various attacks in molecular communication

Layers	Attacks
Molecular-based transport layer	Desynchronization, unfairness
Molecular-based network layer	Flooding, packet storage exhaustion
Molecular-based link layer	Collision, unfairness
Signaling sublayer	Jamming, replication misusing
Bio-nanomachine sublayer	Jamming, fabrication

reviewed, one of which is in terms of security and privacy. Molecular communication is vulnerable to several attacks classified based on the different layers [1,26], as shown in Table 1.3.

1.4.3 Terahertz communication (THzCom)

The 5G wireless network was developed based on frequency ranges, Frequency Range 1 (FR1) and Frequency Range 2 (FR2), enabling the network to support several novel technologies and applications. However, transmission rates in the 6G network are expected to be 100–1,000 times greater than those for the 5G network; thus, these FR1 and FR2 frequencies would be insufficient for the 6G and beyond networks. THzCom was proposed to solve this problem, also referred to as the ultra-wide THz band communication ranging from 0.1 to 10 THz band, to be used for the 6G network and beyond [27]. THzCom would be critical in realizing numerous ultra-high-throughput technologies and applications such as the Internet of Nano-Things (IoNT), extremely dense networks, Extended and Augmented Reality (XR/AR), and tactile Internet [28].

Furthermore, due to the low coverage and penetration power, THzCom is inherently secure and can defend against attacks, including jamming and eavesdropping [9,28]. However, despite its resilience, THzCom can still be attacked in special cases [9]. Furthermore, THzCom is vulnerable to access control attacks, in which an attacker can bypass access permissions to capture sensitive data or user identities to obtain unapproved access to allowed resources or modify system parameters [12]. For privacy issues, THzCom can be utilized for centimeter-level localization applications where attackers can expose and exploit users' locations for malicious intentions [9].

1.4.4 Visible light communication (VLC)

VLC is a type of optical wireless communication in which data is transmitted from source to destination using visible spectrum light waves at unlicensed frequencies ranging from 400 to 800 THz [3,29,30]. This technology has several benefits over conventional radio frequency, including, but not limited to, the provision of faster data transmission via laser beams both in free space and underwater [31], and the utilization of unlicensed and free-of-charge bandwidth [32]. VLC also has various applications, such as intelligent transport systems, smart cities, telemedicine, and underwater communications [30]. Furthermore, because VLC is limited to devices that are exposed to light, it is thought to be more secure than standard radio frequency technologies [30]. However, because of its broadcast nature, VLC is still susceptible to several attacks, such as jamming, eavesdropping, data modification, access control, and authentication attacks [1,12].

1.4.5 Blockchain/distributed ledger technology

This is another promising technology that has been considered as part of the 6G core enabling technologies. It is frequently recognized as a vital technology for establishing trust in the future wireless network, as well as providing other benefits and advancements such as (i) elimination of a single point of failure, (ii) immutability and fabrication-proof of distributed ledger's content, (iii) decentralization, that is removing the need for intermediaries or third party, (iv) authenticity and non-repudiation of completed transactions, (v) significantly reduced processing fees and delays, and (vi) transparency with anonymity [33].

Blockchain technology has several applications, such as cryptocurrency, Hyperledger, and smart contracts. It can be utilized to defend against some threats/attacks, such as distributed DoS (DDoS), because of its decentralized nature. However, blockchain technology is susceptible to different threats/attacks, including 51% attacks, forking issues, eclipse attacks, application bugs, short address attacks, timestamp dependence, regulatory issues, scalability issues, integration issues, selfish attacks, and Sybil attacks [34,35]. Other privacy issues in blockchain technology include various leakages of (i) transaction data, (ii) smart contract logic, and (iii) user privacy during smart contract execution.

1.4.6 RIS

A RIS, also known as intelligent reflecting surface (IRS) is a flat surface made up of several passive reflecting components, each of which can cause an incoming signal to achieve the desired phase shift [36]. It has also been identified as part of the 6G core enabling technologies [37] for extending coverage area, reducing power consumption, and enhancing data transmission rates [38]. Because of their passive nature, RISs can be easily placed on buildings and other structures, minimizing the interference challenge in an extremely dense network [4]. Additionally, RIS has several applications such as in mobile edge computing (MEC) networks, millimeter wave systems, multicell networks, simultaneous wireless information and power transfer (SWIPT) networks, cognitive radio networks, non-orthogonal multiple access, multicast networks, and physical layer security (PLS) networks as it can be set up to offer protection at the physical layer by concurrently enhancing the signal beam at the authorized user and dampening the beam at the unauthorized user, thereby preventing eavesdropping [4,36,39]. Although this technology can be applied as a countermeasure to some security and privacy challenges in the 6G network, it is also susceptible to several threats/attacks such as signal cancellation attacks, pilot sequence poisoning, beam management poisoning, RIS-adding jamming attack, channel equalization poisoning, PHY key generation, and PHY authentication [40].

1.4.7 Ambient backscatter communication (AmBC)

AmBC enables the communication between batteryless devices through the use of available radio frequencies, for instance, cellular communication and ambient television, among others [4,41]. The primary advantage of this technology over traditional backscatter systems is that it can achieve both information transmission and energy harvesting via pervasive ambient radio frequency (RF) sources such as cellular, Wi-Fi, broadcast, or TV signals [42]. AmBC is suitable for large interconnectivity in the 6G network and beyond due to the battery-free nodes. It has several benefits, which include (i) addressing energy efficiency in sensor networks and other low-power consumption devices, (ii) lower device costs with low-power components and less infrastructure, (iii) improving spectrum resource utilization, and (iv) improving mobile device battery life [4]. However, AmBC is vulnerable to eavesdropping attacks due to the simple coding and modulation schemes used [42].

1.4.8 Cell-free massive MIMO (CF-mMIMO) communication

CF-mMIMO communication is another cutting-edge technology that can be regarded as part of the core enabling technologies of the envisioned 6G network. The concept behind this technology is simple to deploy a significant number of access points (APs) that are networked to a central processing unit (CPU) to connect all users across a wide geographical area [43,44].

CF-mMIMO utilizes the concepts of small cells (SCs), massive MIMO, and user-based joint transmission coordinated multi-point (JT-CoMP) to deal with inter-cell interference [44,45]. Furthermore, CF-mMIMO provides a high level of macro-diversity to compensate for path loss by lowering the negative effects of spatially correlated fading and shadowing [44]. Cell-free massive MIMO communication has various advantages over traditional cellular massive MIMO communication, including high energy efficiency, flexibility and cost-effective deployment, channel hardening and optimal propagation characteristics, reliable quality of service (QoS) [46], as well as mitigation of significant pathloss variations and cell-edge performance issues common in traditional cellular networks [47]. However, CF-mMIMO is vulnerable to active eavesdropping, which can lead to other attacks, such as piloting spoofing attacks, jamming, and eavesdropping [48].

1.4.9 QC

Quantum systems are especially beneficial for addressing challenging optimization problems that require a large number of interconnected quantum bits and cannot be solved on a single quantum processor [2,49]. As a result, the requirement to interconnect numerous of these chips led to the advent of QC [2]. QC is the transfer of information with respect to the principles of quantum mechanics. It has several significant benefits, including the capacity for massively parallel computation, transfer of data in a tamper-free mode, and encoding and sending multiple data streams simultaneously [2]. QC can be applied as a high-security solution for the 6G network because it utilizes the quantum key approach based on the Heisenberg uncertainty principle and the quantum theorem of no-cloning [49]. In essence, QC can significantly enhance the detection of unauthorized eavesdropping where if an attacker tries to interfere, interrupt, or manipulate the data, the quantum state is affected; hence, the receiver is aware of any interference.

Furthermore, QC can provide tremendously high data rates as well as significant protection against various possible attacks [49]. Other potential benefits of QC include increasing the accuracy and effectiveness of AI systems that require enormous amounts of data and extensive training. However, it is faced with some attacks, such as quantum cloning and collision attacks [12].

1.4.10 Internet of BioNanoThings (IoBNT)

The IoBNT are made up of biological nanonetworks that can identify biological and chemical changes in an environment, such as the human body, and then transmit the gathered information to data centers for additional processing through the Internet [50]. It is a type of molecular communication where several molecules can communicate. This technology includes artificial cells that act as translators between distinct molecule types and a bio-cyber interface that can convert chemical signals to electrical signals and communicate them with other devices for further processing [2].

The IoBNT provides various advantages and applications, notably in the healthcare sector, including sustained health monitoring, personalized drug delivery, tissue engineering, as well as tumor detection [50]. However, this technology is susceptible to several threats/attacks, including DoS, eavesdropping, device tampering, data fabrication, Man-in-the-Middle, and resource depletion attacks [50]. Privacy risks in IoBNT include the exploitation of users' data and data theft, which can harm users' health.

Figure 1.3 shows the eavesdropping attack in UAV communication, RIS, VLC, CF-mMIMO communication, and IoBNT.

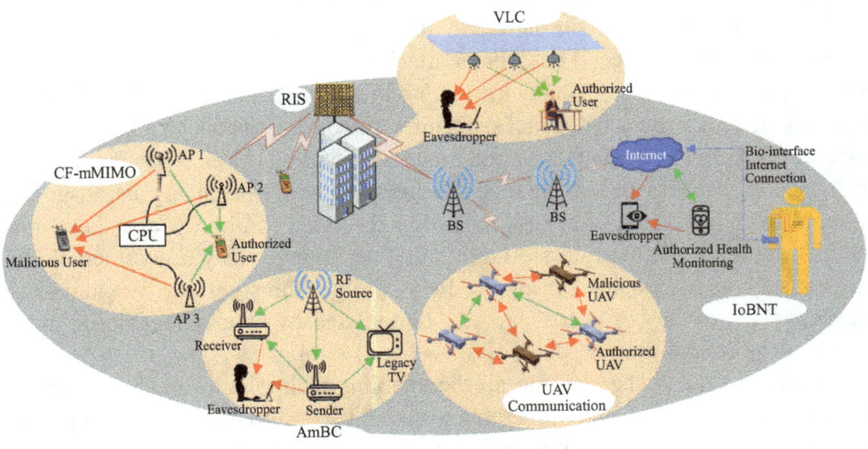

Figure 1.3 Eavesdropping attacks in 6G enabling technologies

1.4.11 Internet of NanoThings (IoNT)

A nano network is made up of numerous nanoscale devices that perform specific activities such as sensing, actuation, monitoring, and control [51]. IoNT refers to the integration of nanoscale devices with established telecommunication networks and the Internet. It adds novel dimensions to the IoT by embedding nano-sensors into the objects, allowing them to communicate and interact via nanonetworks connected to the Internet [52]. IoNT has a wide range of applications in several sectors, including Oil and Gas, Military, Agriculture, Smart Cities, Multimedia, Health monitoring, and Industry [52]. However, IoNT is susceptible to various threats/attacks, whether physical or via wireless technology, because these devices are not constantly monitored. These attacks can occur to steal private data from sensors, disrupt computer controlled applications, or manipulate communication lines in nanonetworks [53]. An invasion vector is a path or technique that allows an attacker to gain unauthorized access to a network to deliver a payload or malicious effect.

In contrast, attack vectors allow attackers to exploit system faults, install various types of malware and carry out cyber-attacks. IoNT has several attack vectors, which include Internet exposure, no encryption, DoS attacks, and wearable malware [53]. Similar to the IoBNT, privacy issues in IoNT include the misuse of users' data as well as data theft.

1.4.12 Pervasive AI

AI has expanded rapidly in recent years, resulting in its use in a diverse range of sectors in both research and business. It is a broad concept with numerous subfields that investigate related but distinct research areas such as robotics, speech recognition, ML, machine vision [2]. The 6G network is envisioned as a self-configuring, self-monitoring, self-healing, and self-optimizing network that requires minimal or no human intervention [12,54]. AI would enable seamless interconnection between the space–air–ground–underwater integrated network nodes via extensive learning and enormous multivariate data training, including network, service, and user data [55].

AI can also optimize the network, identify network state in real-time, and enhance users' experience. Thus, threat/attack on AI, particularly on the ML system, would also affect the 6G network. These threats/attacks include the infiltration of AI frameworks, physical attacks on infrastructure, evasion attacks, ML API-based attacks, and communication manipulation [55,56]. AI also has certain privacy concerns because the envisioned 6G network will extensively use user data gathered by a plethora of devices, removing the users' authority over how other systems would handle their data. Additionally, data thieves may use vulnerable IoT devices, such as inadequate sensors, to feed AI algorithms with private and sensitive data. The privacy of users may also be violated by attacks leveraging model inversion to retrieve training data [13].

The brief review showed that all the enabling technologies of the envisioned 6G network are exposed and vulnerable to several security and privacy challenges. Interestingly, eavesdropping attacks, the passive or active unauthorized listening in on the conversation between two communicating nodes without their knowledge, are one of the most common security threats/attacks in most of these technologies. In terms of privacy, it was seen that the theft of users' data is one of the most pronounced issues.

1.5 Addressing 6G security and privacy challenges

Several techniques and algorithms have been offered to solve the future requirement for security and privacy in the envisioned 6G network. As a result, this section describes and examines the emerging schemes that are being used to defend against the security and privacy vulnerabilities mentioned in the previous section.

1.5.1 PLS schemes

The 6G physical layer will be critical in supporting larger bandwidths, higher carriers, and reduced latencies while consuming less energy. The two main attacks in the physical layer are interception, jamming, and DoS attacks, and interruption, active and passive eavesdropping. Unlike traditional security techniques such as public key cryptography, PLS does not require private keys or complex encryption computation to provide security and privacy at the first layer, thereby acting as a first line of defense to make attackers' jobs more difficult [57]. It provides security protections and precautions at the most basic level of communication, which aids in overcoming bandwidth, reliability, and transmission rate constraints [58].

PLS can be considered a confidentiality enabler in the envisioned 6G network because its features, when combined with new developments in artificial intelligence algorithms and the trend of distributed computing frameworks, can either be used to improve traditional encryption methods or satisfy security requirements when interacting with simple but delicate devices that are not able to incorporate cryptography techniques, such as nano-devices of the IoT [59]. It has been proposed as a security and privacy solution scheme for most core enabling technologies in the 6G network, which are reviewed as follows.

To secure VLC against eavesdropping, replay, message injection, and modification attacks, Soderi and Nicola [60] proposed a watermark blind PLS (WB-PLS) scheme which combines an RGB LED jamming and watermarking. Results obtained suggested that it is possible to create a protected area surrounding the legitimate receiver by using jamming optical power. Similarly, Soderi *et al.* [61] also proposed WB-PLS to protect VLC at the physical layer. However, unlike in [60], a RIS was utilized to enhance the security properties of the communication. The result obtained showed that the secured communication area is broadened by integrating RIS.

Further review of the potential application as well as the challenges of PLS for the security and privacy challenges in VLC was provided in [62–64]. CF-mMIMO is currently one of the strongest candidates for PLS [59]. To secure the CF-mMIMO in the presence of a single-antenna active eavesdropper, Timilsina *et al.* [65] conducted a performance evaluation of PLS for CF-mMIMO and compared it with that of co-located MIMO. The results obtained showed that the rate leaked into the eavesdropper can be significant in the presence of active pilot attacks for CF-mMIMO. To further secure CF-mMIMO from eavesdropping, Park *et al.* [66] proposed a keyless PLS scheme known as the artificial noise (AN)-aided secure power control scheme for CF-mMIMO in the presence of passive eavesdropping to achieve a higher secrecy rate as well as guarantee security.

The efficacy of the proposed scheme was evaluated against other power control schemes, where results obtained showed the supremacy of the proposed scheme over the others. Reconfigurable intelligent surface (RIS) is another strong candidate for PLS [59] and has been utilized to improve PLS as in [67], where a novel reconfigurable intelligent system based channel randomization scheme was proposed for improving PLS for a downlink time-division duplex (TDD) cellular wiretap network made up of a base station (BS) with numerous antennas, a large number of authorized user equipment (UE), a large number of eavesdroppers, and a large number of reconfigurable intelligent systems, where each reconfigurable intelligent system generates a large number of reflection matrices pseudo-randomly and utilizes them for both pilot signal duration (PSD) and data transmission duration (DTD) in uplink and downlink, respectively.

The proposed scheme was evaluated against a conventional scheme, where results showed that the proposed scheme performed better based on achievable secrecy rates. To further show how reconfigurable intelligent system can contribute to the security improvement of the PLS for 6G network, Lipps *et al.* [68] presented and discussed five use cases which include reconfigurable intelligent system support context information addition, reconfigurable intelligent system personalized fingerprints, reconfigurable intelligent system manipulated channel profiles, reconfigurable intelligent system beamforming, and splitting, and, finally, reconfigurable intelligent system based anti-jamming. PLS for reconfigurable intelligent system aided non-orthogonal multiple access (NOMA) 6G networks were investigated in [69] to secure against eavesdropping attacks. Two scenarios were considered where a joint beamforming and power allocation scheme was proposed for the first scenario to improve PLS, which considered only internal eavesdropping. For the second scenario, an optimal power allocation scheme was proposed for both internal and external eavesdropping to enhance the system's PLS simulations further. The results showed that the proposed schemes performed better. It was also proved that a higher number of BS broadcast antennas or reconfigurable intelligent system reflecting elements improves the system's secrecy performance.

PLS has also been considered as a security solution against eavesdropping attacks for AmBC, for example, to distinguish between the reception effect of the authorized receiver and that of an eavesdropper under the condition of receiving a symmetrical signal backscattered by the tag due to the broadcast nature of the wireless channel, Hou *et al.* [70] proposed a scheme that involved injecting AN from the tag end. The proposed scheme enhanced the PLS of the AmBC system while improving the system's secrecy rate. Results from simulations and evaluations indicated that the proposed scheme could significantly enhance the PLS for the AmBC system.

Similarly, PLS can be applied to defend against active and passive eavesdropping in UAV communication as reviewed in [71] where the various security attacks in UAV systems were highlighted and discussed. Based on the review, it was noted that to defend against active and passive eavesdropping, techniques such as joint trajectory and resource allocation design, robust joint design, and AN were proposed. Furthermore, to protect against jamming attacks, the cooperative multi-point (CoMP) technique can be utilized.

Other approaches that can be utilized to enhance the PLS of the system further were also discussed in terms of NOMA, multi-antenna technology, and millimeter wave. To protect terahertz communication from eavesdropping, He *et al.* [72] investigated the secured transmission of THz waves in an enclosed area against stochastically placed eavesdroppers and developed a PLS model for this system. The investigation and results demonstrated that eavesdroppers could compromise the system's secrecy performance by employing various tactics. As an appropriate strategy to protect against numerous eavesdroppers, an AN-beamforming system with a well-designed power allocation was proposed.

Furthermore, in [10,12,58], PLS was considered a potential solution to security and privacy challenges in VLC, THzCom, RIS, and molecular communication in the 6G network. To provide privacy with ultra-low latency for emerging B5G/6G wireless networks, Yerrapragada *et al.* [73] proposed a key-based PLS to protect information against eavesdroppers by mapping to reference signals that are intelligently rotated. The research demonstrated that the shortest possible latency for two-way secret sharing and synchronization is only four symbol times. Furthermore, the simulation results obtained showed that the revised protocol achieved sub-1 ms two-way latency, including retransmissions. Further reviews of the operations and potential applications of PLS in 6G networks and beyond are provided in [57,74,75].

1.5.2 Distributed AI/ML schemes

AI/ML, as explained earlier, are the major drivers of the 6G network to achieve the envisioned network automation; remarkably, it can also be utilized to defend against some security and privacy challenges in the 6G network. However, as mentioned earlier, AI/ML is susceptible to several attacks of which a variety of countermeasures have been proposed. For instance, to defend against AI-based poisonous attacks, the proposed solutions can be categorized into two [13,56]: input validation and robust learning. Input validation is concerned with cleaning the training data of malicious and anomalous samples, such as outliers, before supplying it to the ML algorithm. The reject on negative impact (RONI) technique can be considered suitable for such a task.

On the other hand, robust learning is concerned with developing learning techniques that are resistant to training data compromises through robust statistics methodologies. Several strategies have also been proposed as potential solutions to AI-based evasive attacks, including protective generative adversarial networks (GANs), defensive distillation, ensemble methods, adversarial training, and adversarial concept drifting handling techniques [56]. Some other countermeasures recommended to defend against ML API-based attacks include adding noise to the ML model's execution time, implementing differential privacy, employing homomorphic encryption, and finally, restricting critical information offered by the ML APIs [56]. Other reviews on AI/ML being proposed as a solution for security and privacy challenges as well as to establish trust in the 6G network are as follows.

Bandi and Yalamarthi [76] conducted an extensive review of the security and privacy challenges in the various application in the envisioned 6G network, such as access control attacks, ML API attacks, denial-of-service attacks, evasion attacks, malicious behavior, jamming attacks, and access control attacks. The research further reviewed AI-enabled technologies that can be utilized as a solution scheme for these attacks. It was concluded that the explainable AI, trustworthy AI, and super-intelligent AI could enhance the deep learning-based intelligent protection mechanisms for the 6G network.

Siriwardhana *et al.* [13] reviewed the security and privacy challenges in 6G as well as the AI-based countermeasures. The research reviewed the application of AI to identify and mitigate the security and privacy challenges in pre-6G, 6G architecture, and 6G technologies, where it was mentioned that AI can predict the occurrence of attacks such as the 51% attack in blockchain technology. AI can also detect jamming attacks and provide optimal beamforming policy against beamforming in VLC systems using intelligent beamforming techniques that are based on reinforcement learning. Furthermore, AI can defend node compromise attacks again via AI-based authentication and authorization systems. To solve the AI-based privacy issues mentioned earlier, techniques such as homomorphic encryption, edge-based federated learning as well as differential privacy techniques can be utilized to preserve privacy.

1.5.3 Quantum cryptography schemes

Quantum cryptography, one of the security schemes under the umbrella of quantum security, was developed to compensate for the drawbacks of some of the conventional asymmetric cryptographic algorithms, such as RSA algorithms, particularly when it comes to providing high-rated security [77]. Quantum cryptography can provide uncompromising security regardless of the current processing capacity and has drawn attention to quantum key distribution (QKD) usage. The QKD can theoretically provide a secured key agreement between multiple communication devices [77,78]. It can be used to improve communication flexibility and efficiency since it is based on two major principles: the principle of uncertainty (Heisenberg uncertainty) and the quantum no-cloning theorem.

The principle of uncertainty precludes an eavesdropper from measuring quantum entanglement without interfering with the network, whereas the no-cloning theorem argues that it is impossible to replicate an existing quantum state [78]. Thus, this prevents an eavesdropper in a quantum channel from copying or even tampering with the information without alerting the communicating nodes of their presence [49,77]. QC can be combined with AI/ML to improve 6G network security, particularly for IoT devices with low processing power and computation ability.

Wang and Rahman [78] extensively investigated how quantum information technology (QIT) may enhance and strengthen the 6G network. The QIT was subdivided into four major areas: quantum computing, QC, quantum mechanics, and, finally, quantum sensing and metrology. Several protocols and security schemes were highlighted and discussed under each sub-division for QC protocols such as quantum key distribution (QKD), quantum secure direct communication (QSDC), and quantum secret sharing (QSS), where it was seen that unlike the QKD only secures key establishments and neglects the remaining aspect of the communication, QSDC provides direct secure communication between nodes based on quantum mechanics and does not rely on secret key sharing and management.

For the QSS, only part of the secret key would be shared among multiple communicating nodes to prevent a single node from having full knowledge of the secret key. In addition, other schemes were discussed, such as blind quantum computing (BQC), which was proposed to provide privacy in the 6G network. The research concluded that the QIT is a viable technology that can be used to guarantee security and privacy in the 6G network and can be combined with other enabling technologies; however, the technology is not without challenges.

Detailed analysis and reviews of QC, ML, and the integration of both for the potential application for the further improvement of the security and privacy in the 6G network and future wireless communication network were provided in [79,80].

1.5.4 Blockchain-based security schemes

Blockchain, one of the most prominent distributed ledger technologies (DLT), has previously been highlighted as one of the enabling technologies of the envisaged 6G network. Because of its decentralized and tamper-proof character, it has also been widely touted as a potential security solution for 6G networks and beyond. Blockchain is also a good solution for maintaining privacy in content-centric 6G networks because having a common communication channel in blockchain enables network users to be differentiated by pseudo names instead of direct and personal identities or geographical information [12]. It can help with the implementation of Internet of Everything (IoE) applications by establishing trust between connected devices, hence bypassing the demand for trusted third parties [35].

Blockchain technology has numerous uses in different sectors, particularly in the security of IoT networks. For example, Pajooh *et al.* [81] reviewed the convergence of 6G-enabled IoT and blockchain for a variety of areas, such as healthcare, smart cities, smart homes, intelligent manufacturing, and vehicle automation, where it was discovered that the blockchain could provide novel solutions to effectively address security and privacy concerns as well as advance the 6G-enabled IoT networks. Jahid *et al.* [82] presented a more in-depth examination of the combination of blockchain and IoT.

Manogaran *et al.* [83] presented a blockchain-based scheme to provide access control to resources and privacy of users in the 6G network, which, when evaluated using metrics such as true positives, access rejection and success proportion, access time and alteration, and finally, memory usage and time consumption, demonstrated that the proposed measure is credible for both access control and user privacy. Gupta *et al.* [84] presented an interplanetary file system and blockchain-based secured UAV scheme to assure data confidentiality and privacy while lowering data storage costs, hence improving UAV communication performance over the envisaged 6G network.

According to the study, combining blockchain with UAVs can protect against eavesdropping, fabrication, jamming, DDoS, and connection attacks common in UAV communication. To ensure user data privacy in UAV communication, Saraswat *et al.* [85] proposed the use of blockchain-based federated learning (FL) in UAVs for the B5G network. Blockchain technology in 6G network was reviewed in [35] in terms of security and privacy, where it was noted that although several 6G use cases can rely on blockchain's trust and security robustness, such as edge computing, federated learning, and energy trading, there are still some issues faced by the blockchain technology itself which include the security and privacy risks, scalability and quantum computing. Solutions were also provided where it was mentioned that a transaction fee could defend against DDoS attacks while to defend against the potential disclosure of users' private information due to the default transparent nature of blockchain, several solutions were proposed, including ring signatures, zero-knowledge arguments, and proofs, and coin mixers. However, these solutions are not without challenges.

1.5.5 Other security schemes

Several other methods have been proposed for security and privacy challenges in the 6G network. Zafar *et al.* [50] addressed numerous schemes and methods for providing security and privacy for IoBNT, including elliptic curve cryptography (ECC). This lightweight encryption scheme is suitable for devices with minimal resources and can be used to protect against eavesdropping attacks. Intrusion detection and external device authentication methods are proposed to protect against DoS, replay, and injection attacks. ZeroPower defense can be used to counter resource depletion attacks. Device tampering and malware attacks can be mitigated through access control, device strengthening, and system monitoring systems. Finally, firmware attacks can be mitigated by encrypting firmware, performing periodic upgrades, and detecting malicious firmware. These systems and procedures, however, are not without challenges.

1.6 Lessons learned

This section highlights and elaborates on noteworthy points that can be used for further research in 6G network security and privacy.

Lesson 1: Eavesdropping attack is the most common security attack
Eavesdropping is simply the unauthorized listening or interfering with data transmission or conversation between two or more communicating parties. It is considered the mother of all attacks because an attacker must first eavesdrop on a conversation before committing malicious intentions. It can be either passive or active. Passive eavesdropping is where the eavesdropper monitors the conversation without interfering with the communication channel, while active eavesdropping is where the eavesdropper takes control of the communication channel and starts

acting as a relay between the communicating parties. The review shows that most of the enabling technologies of the envisioned 6G network, particularly in VLC, CF-mMIMO, molecular communication, RIS, etc., are vulnerable to eavesdropping attacks. Although the proposed solution schemes discussed earlier can be utilized to potentially defend against this attack, especially the PLS schemes, further enhancement is still needed as these schemes are at the theoretical stage.

Lesson 2: Data theft attack is the most common privacy attack

The 6G network is expected to be autonomous and data-based, where a very large amount of data would be collected from several devices, particularly IoT devices. Review of the privacy issues in the enabling technologies has shown that most of the privacy issues were centered on the users' data exploitation or simply data theft. Data theft is the unauthorized exchange or retention of one's personal or confidential information. It is one of the most common forms of privacy attack, which has raised concerns, as users would not have control or knowledge of how their data have been used in the network. Users' data, in this context, include health-based data, username and password, transaction data, location data, and other confidential data, which can be exploited if there are loopholes found in the network, such as a compromised IoT device collecting data and feeding an algorithm or central server. Although some schemes have been proposed to defend against this privacy attack in the envisioned 6G network, such as homomorphic encryption, edge-based federated learning, differential privacy techniques, BQC, ring signatures, zero-knowledge arguments, and proofs, coin mixers, and so on, there are still need for further improvement and the evaluation of their efficacy.

Lesson 3: AI can be utilized to enhance other security schemes

As mentioned earlier, the 6G network is envisioned to be an autonomous network, and this can only be actualized using AI due to its robustness and versatility, among others features. Therefore, AI would be a major enabler of the 6G network and can also be used as a potential security and privacy scheme to defend against several attacks. Furthermore, from the review, it can be seen that AI can be integrated with other security and privacy schemes, for example, quantum security schemes, to speed up and further improve the efficacy of the scheme and thus improve the security and privacy of the envisioned 6G network. Furthermore, AI can also be combined with blockchain to improve security [86]. However, AI has not been integrated with the PLS schemes.

Lessons 4: Proposed security schemes are still in theoretical stage

The majority of the proposed security and privacy schemes for the 6G network are still theoretical and have not yet been implemented. This is to be expected because the 6G network is still in the works, and there are no official standards in place to demonstrate the exact frequency ranges, specifications, and other requirements, even though researchers are still providing their perspectives on how the network should look. As a result, a detailed examination and evaluation of these proposed solutions for improving the security and privacy of the 6G network are crucial.

Lesson 5: Existing security and privacy schemes cannot meet the requirement of the envisioned 6G network

Several security and privacy schemes have been proposed and utilized to defend against various types of threats/attacks in the past cellular generations, including the recently developed 5G network. However, due to the robustness and ultra-high-speed nature of the envisioned 6G network, these schemes have failed to meet these requirements. For instance, conventional cryptographic schemes such as public key cryptography (PKC) failed to defend against security and privacy threats/attacks in the 6G network due to the complex encryption and the dependency on secret keys as a result, the PLS was incorporated [57]. Other schemes as discussed in [50] can also be utilized particularly for the IoBNT. However, these are without challenges. Therefore, there is a need to improve the existing security and privacy schemes to make them efficient and effective against several emerging threats/attacks. This can be achieved by combining these with other emerging schemes, thus making the 6G network more secure.

1.7 Conclusion and recommendations

Security and privacy are the two most crucial issues that have been identified in the envisioned 6G network, particularly with the advent of novel technologies and the support of several applications. This chapter has reviewed the security and privacy challenges in the 6G core enabling technologies as well as the emerging schemes that can be used to defend against these attacks. Specifically, this chapter reviewed the PLS, distributed AI, quantum cryptography, and blockchain-based security, where it was seen that these have the potential to secure some of the enabling technologies of the envisioned 6G network. Although notable efforts have been made to propose security and privacy schemes for the envisioned 6G network, most of these schemes are still in the theoretical stage, and schemes have not been proposed for some 6G enabling technologies such as the IoNT. Also, the improvement of the PLS using AI. Therefore, for future work, the following are recommended:

- The convergence of PLS with AI for improving 6G network security and privacy at the physical layer.
- The efficacy of these security and privacy schemes should be evaluated and used as a benchmark in developing and evaluating other novel security and privacy schemes.
- Improving the efficiency and effectiveness of existing security and privacy schemes to meet the requirements of the 6G network.

Acknowledgment

The work of Agbotiname Lucky Imoize is supported in part by the Nigerian Petroleum Technology Development Fund (PTDF) and, in part, by the German Academic Exchange Service (DAAD) through the Nigerian–German Postgraduate Program under Grant 57473408.

References

[1] Abdel Hakeem, S., Hussein, H., and Kim, H. Security Requirements and Challenges of 6G Technologies and Applications. *Sensors* 2022, *22*, 1969, doi:10.3390/s22051969.

[2] Akyildiz, I.F., Kak, A., and Nie, S. 6G and Beyond: The Future of Wireless Communications Systems. *IEEE Access* 2020, *8*, 133995–134030, doi:10.1109/ACCESS.2020.3010896.

[3] You, X., Wang, C.X., Huang, J., *et al.* Towards 6G Wireless Communication Networks: Vision, Enabling Technologies, and New Paradigm Shifts. *Sci. China Inf. Sci.* 2021, *64*, 110301, doi:10.1007/s11432-020-2955-6.

[4] Imoize, A.L., Adedeji, O., Tandiya, N., and Shetty, S. 6G Enabled Smart Infrastructure for Sustainable Society: Opportunities, Challenges, and Research Roadmap. *Sensors* 2021, *21*, 1–57.

[5] Mahmoud, H.H.H., Amer, A.A., and Ismail, T. 6G: A Comprehensive Survey on Technologies, Applications, Challenges, and Research Problems. *Trans. Emerg. Telecommun. Technol.* 2021, *32*, e4233, doi:10.1002/ett.4233.

[6] Sullivan, S., Brighente, A., Kumar, S.A.P., and Conti, M. 5G Security Challenges and Solutions: A Review by OSI Layers. *IEEE Access* 2021, *9*, 116294–116314, doi:10.1109/ACCESS.2021.3105396.

[7] Ahmad, I., Kumar, T., Liyanage, M., Okwuibe, J., Ylianttila, M., and Gurtov, A. 5G Security: Analysis of Threats and Solutions. In *2017 IEEE Conference on Standards for Communications and Networking, CSCN 2017*, 2017, pp. 193–199, doi:10.1109/CSCN.2017.8088621.

[8] Wang, M., Zhu, T., Zhang, T., Zhang, J., Yu, S., and Zhou, W. Security and Privacy in 6G Networks: New Areas and New Challenges. *Digit. Commun. Networks* 2020, *6*, 281–291, doi:10.1016/j.dcan.2020.07.003.

[9] Nguyen, V.L., Lin, P.C., Cheng, B.C., Hwang, R.H., and Lin, Y.D. Security and Privacy for 6G: A Survey on Prospective Technologies and Challenges. *IEEE Commun. Surv. Tutorials* 2021, *23*, 2384–2428, doi:10.1109/COMST.2021.3108618.

[10] Porambage, P., Gur, G., Osorio, D.P.M., Liyanage, M., Gurtov, A., and Ylianttila, M. The Roadmap to 6G Security and Privacy. *IEEE Open J. Commun. Soc.* 2021, *2*, 1094–1122, doi:10.1109/OJCOMS.2021.3078081.

[11] Sandeepa, C., Siniarski, B., Kourtellis, N., Wang, S., and Liyanage, M. A Survey on Privacy for B5G/6G: New Privacy Challenges, and Research Directions. *J. Ind. Inf. Integr.* 2022, *30*, 100405, doi:10.1016/j.jii.2022.100405.

[12] Porambage, P., Gur, G., Moya Osorio, D.P., Livanage, M., and Ylianttila, M. 6G Security Challenges and Potential Solutions. In *Proceedings of the Joint European Conference on Networks and Communications 6G Summit (EuCNC/6G Summit)*, 2021. ISBN 9781665415262.

[13] Siriwardhana, Y., Porambage, P., Liyanage, M., and Ylianttila, M. AI and 6G Security: Opportunities and Challenges. In *2021 Joint European*

Conference on Networks and Communications & 6G Summit (*EuCNC/6G Summit*), 2021, pp. 616–621. ISBN 9781665415262.
[14] 5G PPP Architecture Working Group. *The 6G Architecture Landscape*, 2022.
[15] Shahjalal, M., Kim, W., Khalid, W., *et al.* Enabling Technologies for AI Empowered 6G Massive Radio Access Networks. *ICT Express* 2022, doi:10.1016/j.icte.2022.07.002.
[16] Shrestha, R., Bajracharya, R., and Kim, S. 6G Enabled Unmanned Aerial Vehicle Traffic Management: A Perspective. *IEEE Access* 2021, *9*, 91119–91136, doi:10.1109/ACCESS.2021.3092039.
[17] Van Nguyen, M.S., Do, D.T., Phan, V.D., Ullah Khan, W., Imoize, A.L., and Fouda, M.M. Ergodic Performance Analysis of Double Intelligent Reflecting Surfaces-Aided NOMA–UAV Systems with Hardware Impairment. *Drones* 2022, *6*, 408, doi:10.3390/drones6120408.
[18] Khan, A.S., Sattar, M.A., Nisar, K., *et al.* A Survey on 6G Enabled Light Weight Authentication Protocol for UAVs, Security, Open Research Issues and Future Directions. *Appl. Sci.* 2023, *13*, 277, doi:10.3390/app13010277.
[19] Khan, M.A., Kumar, N., Mohsan, S.A.H., *et al.* Swarm of UAVs for Network Management in 6G: A Technical Review. *IEEE Trans. Netw. Serv. Manag.* 2022, *19*, 188–202, doi:10.1109/TNSM.2022.3213370.
[20] Noor, F., Khan, M.A., Al-Zahrani, A., Ullah, I., and Al-Dhlan, K.A. A Review on Communications Perspective of Flying AD-HOC Networks: Key Enabling Wireless Technologies, Applications, Challenges and Open Research Topics. *Drones* 2020, *4*, 1–14, doi:10.3390/drones4040065.
[21] Aggarwal, S., Kumar, N., and Tanwar, S. Blockchain-Envisioned UAV Communication Using 6G Networks: Open Issues, Use Cases, and Future Directions. *IEEE Internet Things J.* 2021, *8*, 5416–5441, doi:10.1109/JIOT.2020.3020819.
[22] Mekdad, Y., Aris, A., Babun, L., *et al.* A Survey on Security and Privacy Issues of UAVs. CoRR abs/2109.14442, *arXiv Prepr. arXiv2109.14442* 2021.
[23] Zhi, Y., Fu, Z., Sun, X., and Yu, J. Security and Privacy Issues of UAV: A Survey. *Mob. Networks Appl.* 2020, *25*, 95–101.
[24] Haselmayr, W., Springer, A., Fischer, G., *et al.* Integration of Molecular Communications into Future Generation Wireless Networks. In *1st 6G Wireless Summit. IEEE*, Levi, Finland, 2019, pp. 1–2.
[25] Guo, W., Abbaszadeh, M., Lin, L., *et al.* Molecular Physical Layer for 6G in Wave-Denied Environments. *IEEE Commun. Mag.* 2021, *59*, 33–39, doi:10.1109/MCOM.001.2000958.
[26] Loscri, V., Marchal, C., Mitton, N., Fortino, G., and Vasilakos, A.V. Security and Privacy in Molecular Communication and Networking: Opportunities and Challenges. *IEEE Trans. Nanobiosci.* 2014, *13*, 198–207, doi:10.1109/TNB.2014.2349111.
[27] Shafie, A., Yang, G.N., Han, C., Jornet, J.M., Juntti, M., and Kurner, T. Terahertz Communications for 6G and Beyond Wireless Networks:

Challenges, Key Advancements, and Opportunities. *IEEE Netw.* 2022, 1–8, doi:10.1109/MNET.118.2200057.

[28] Singh, R. and Sicker, D. THz Communications – A Boon and/or Bane for Security, Privacy, and National Security. *SSRN Electron. J.* 2020, doi:10.2139/ssrn.3750493.

[29] Chi, N., Zhou, Y., Wei, Y., and Hu, F. Visible Light Communication in 6G: Advances, Challenges, and Prospects. *IEEE Veh. Technol. Mag.* 2020, *15*, 93–102, doi:10.1109/MVT.2020.3017153.

[30] Ariyanti, S. and Suryanegara, M. Visible Light Communication (VLC) for 6G Technology: The Potency and Research Challenges. In *Proceedings of the World Conference on Smart Trends in Systems, Security and Sustainability, WS4 2020*, 2020, pp. 490–493.

[31] Zhao, Y., Zhai, W., Zhao, J., *et al.* A Comprehensive Survey of 6G Wireless Communications, 2020. *arXiv Prepr. arXiv2101.03889*.

[32] Soderi, S. *Enhancing Security in 6G Visible Light Communications*, 2020. ISBN 9781728160474.

[33] Hewa, T., Gur, G., Kalla, A., Ylianttila, M., Bracken, A., and Liyanage, M. The Role of Blockchain in 6G: Challenges, Opportunities and Research Directions. In *2020 2nd 6G Wireless Summit*. IEEE, 2020. ISBN 9781728160474.

[34] Islam, M.R., Rahman, M.M., Mahmud, M., Rahman, M.A., Mohamad, M.H.S., and Embong, A.H. A Review on Blockchain Security Issues and Challenges. In *Proceedings of the 2021 IEEE 12th Control and System Graduate Research Colloquium (ICSGRC)*, Shah Alam, Malaysia, 2021. ISBN 9781665440110.

[35] Nguyen, T., Tran, N., Loven, L., Partala, J., Kechadi, M.T., and Pirttikangas, S. Privacy-Aware Blockchain Innovation for 6G: Challenges and Opportunities. In *6G Wireless Summit*, Levi, Finland, 2020. ISBN 9781728160474.

[36] Pan, C., Ren, H., Wang, K., *et al.* Reconfigurable Intelligent Surfaces for 6G Systems: Principles, Applications, and Research Directions. *IEEE Commun. Mag.* 2021, *59*, 14–20.

[37] Kumaravelu, V.B., Imoize, A.L., Soria, F.R.C., *et al.* Outage Probability Analysis and Transmit Power Optimization for Blind-Reconfigurable Intelligent Surface-Assisted Non-Orthogonal Multiple Access Uplink. *Sustainability* 2022, *14*, 13188, doi:10.3390/su142013188.

[38] Shen, L-H., Feng, K-T., and Hanzo, L. Five Facets of 6G: Research Challenges and Opportunities. *Comput. Surv.* 2022, *55*, 235.

[39] Basharat, S., Hassan, S.A., Pervaiz, H., Mahmood, A., Ding, Z., and Gidlund, M. Reconfigurable Intelligent Surfaces: Potentials, Applications, and Challenges for 6G Wireless Networks. *IEEE Wirel. Commun.* 2021, *28*, 184–191, doi:10.1109/MWC.011.2100016.

[40] Kibilda, J., Mahmood, N.H., Gomes, A., Latva-aho, M., and DaSilva, L.A. Reconfigurable Intelligent Surfaces: The New Frontier of Next G Security, 2022. *arXiv Prepr. arXiv2212.05101*.

[41] Chowdhury, M.Z., Shahjalal, M., Ahmed, S., and Jang, Y.M. 6G Wireless Communication Systems: Applications, Requirements, Technologies,

Challenges, and Research Directions. *IEEE Open J. Commun. Soc.* 2020, *1*, 957–975, doi:10.1109/OJCOMS.2020.3010270.

[42] Li, X., Zheng, Y., Khan, W.U., *et al.* Physical Layer Security of Cognitive Ambient Backscatter Communications for Green Internet-of-Things. *IEEE Trans. Green Commun. Netw.* 2021, *5*, 1066–1076, doi:10.1109/TGCN.2021.3062060.

[43] He, H., Yu, X., Zhang, J., Song, S., and Letaief, K.B. Cell-Free Massive MIMO for 6G Wireless Communication Networks. *J. Commun. Inf. Netw.* 2021, 6, 321–335, doi:10.23919/jcin.2021.9663100.

[44] Obakhena, H.I., Imoize, A.L., Anyasi, F.I., and Kavitha, K.V.N. Application of Cell-Free Massive MIMO in 5G and beyond 5G Wireless Networks: A Survey. *J. Eng. Appl. Sci.* 2021, *68*, 13, doi:10.1186/s44147-021-00014-y.

[45] Alamu, O., Gbenga-Ilori, A., Adelabu, M., Imoize, A., and Ladipo, O. Energy Efficiency Techniques in Ultra-Dense Wireless Heterogeneous Networks: An Overview and Outlook. *Eng. Sci. Technol. Int. J.* 2020, *23*, 1308–1326, doi:10.1016/j.jestch.2020.05.001.

[46] Zhang, J., Chen, S., Lin, Y., Zheng, J., Ai, B., and Hanzo, L. Cell-Free Massive MIMO: A New Next-Generation Paradigm. *IEEE Access* 2019, *7*, 99878–99888, doi:10.1109/ACCESS.2019.2930208.

[47] Imoize, A.L., Obakhena, H.I., Anyasi, F.I., and Sur, S.N. A Review of Energy Efficiency and Power Control Schemes in Ultra-Dense Cell-Free Massive MIMO Systems for Sustainable 6G Wireless Communication. *Sustainability* 2022, *14*, 11100, doi:10.3390/su141711100.

[48] Hoang, T.M., Ngo, H.Q., Duong, T.Q., Tuan, H.D., and Marshall, A. Cell-Free Massive MIMO Networks: Optimal Power Control against Active Eavesdropping. *IEEE Trans. Commun.* 2018, *66*, 4724–4737, doi:10.1109/TCOMM.2018.2837132.

[49] Alsabah, M., Naser, M.A., Mahmmod, B.M., *et al.* 6G Wireless Communications Networks: A Comprehensive Survey. *IEEE Access* 2021, *9*, 148191–148243, doi:10.1109/ACCESS.2021.3124812.

[50] Zafar, S., Nazir, M., Bakhshi, T., *et al.* A Systematic Review of Bio-Cyber Interface Technologies and Security Issues for Internet of Bio-Nano Things. *IEEE Access* 2021, *9*, 93529–93566, doi:10.1109/ACCESS.2021.3093442.

[51] Galal, A. and Hesselbach, X. Nano-Networks Communication Architecture: Modeling and Functions. *Nano Commun. Netw.* 2018, *17*, 45–62, doi:https://doi.org/10.1016/j.nancom.2018.07.001.

[52] Atlam, H.F., Walters, R.J., and Wills, G.B. Internet of Nano Things: Security Issues and Applications. In *ICCBDC'18: Proceedings of the 2018 2nd International Conference on Cloud and Big Data Computing*, 2018. ISBN 9781450364744.

[53] Nikhat A. and Yusuf P. The Internet of Nano Things (IoNT) Existing State and Future Prospects. *GSC Adv. Res. Rev.* 2020, *5*, 131–150, doi:10.30574/gscarr.2020.5.2.0110.

[54] Moubayed, A., Shami, A., and Al-Dulaimi, A. On End-to-End Intelligent Automation of 6G Networks. *Futur. Internet* 2022, *14*, 165.

[55] Gracia, M.B., Malele, V., Ndlovu, S.P., Mathonsi, T.E., Maaka, L., and Muchenje, T. 6G Security Challenges and Opportunities. In *2022 IEEE 13th International Conference on Mechanical and Intelligent Manufacturing Technologies (ICMIMT)*, 2022. ISBN 9781665484008.

[56] Benzaïd, C. and Taleb, T. AI for beyond 5G Networks: A Cyber-Security Defense or Offense Enabler? *IEEE Netw.* 2020, *34*, 140–147, doi:10.1109/MNET.011.2000088.

[57] Mucchi, L., Jayousi, S., Caputo, S., *et al.* Physical-Layer Security in 6G Networks. *IEEE Open J. Commun. Soc.* 2021, *2*, 1901–1914, doi:10.1109/OJCOMS.2021.3103735.

[58] Lipps, C., Baradie, S., Noushinfar, M., Herbst, J., Weinand, A., and Schotten, H.D. Towards the Sixth Generation (6G) Wireless Systems: Thoughts on Physical Layer Security. In *Mobile Communication – Technologies and Applications; 25th ITG-Symposium*, 2021. ISBN 9783800756742.

[59] Ylianttila, M., Kantola, R., Gurtov, A., *et al.* 6G White Paper: Research Challenges for Trust, Security and Privacy, 2020. *arXiv Prepr. arXiv2004.11665.*

[60] Soderi, S. and De Nicola, R. 6G Networks Physical Layer Security Using RGB Visible Light Communications. *IEEE Access* 2022, *10*, 5482–5496, doi:10.1109/ACCESS.2021.3139456.

[61] Soderi, S., Brighente, A., Turrin, F., and Conti, M. VLC Physical Layer Security through RIS-Aided Jamming Receiver for 6G Wireless Networks. In *Annual IEEE Communications Society Conference on Sensor, Mesh and Ad Hoc Communications and Networks,* 2022, September, pp. 370–378, doi:10.1109/SECON55815.2022.9918547.

[62] Arfaoui, M.A., Soltani, M.D., Tavakkolnia, I., *et al.* Physical Layer Security for Visible Light Communication Systems: A Survey. *IEEE Commun. Surv. Tutorials* 2020, *22*, 1887–1908, doi:10.1109/COMST.2020.2988615.

[63] Cho, S., Chen, G., Coon, J.P., and Xiao, P. Challenges in Physical Layer Security for Visible Light Communication Systems. *Network* 2022, *2*, 53–65, doi:10.3390/network2010004.

[64] Blinowski, G. Security of Visible Light Communication Systems—A Survey. *Phys. Commun.* 2019, *34*, 246–260, doi:10.1016/j.phycom.2019.04.003.

[65] Timilsina, S., Kudathanthirige, D., and Amarasuriya, G. Physical Layer Security in Cell-Free Massive MIMO. In *Proceedings of the 2018 IEEE Global Communications Conference, GLOBECOM 2018*, IEEE, 2018, pp. 1–7.

[66] Park, J., Yun, S., and Ha, J. Secure Power Control for Downlink Cell-Free Massive MIMO With Passive Eavesdroppers, 2022. *arXiv Prepr. arXiv2211.13920.*

[67] Youn, J., Son, W., and Jung, B.C. Physical-Layer Security Improvement with Reconfigurable Intelligent Surfaces for 6G Wireless Communication Systems. *Sensors* 2021, *21*, 1–12, doi:10.3390/s21041439.

[68] Lipps, C., Herbst, J., Reddy, R., *et al.* Reconfigurable Intelligent Surfaces: A Physical Layer Security Perspective. In *Proceedings of the 2022 4th*

International Conference on Data Intelligence and Security (ICDIS), IEEE, 2022, pp. 174–181.

[69] Zhang, Z., Zhang, C., Jiang, C., Jia, F., Ge, J., and Gong, F. Improving Physical Layer Security for Reconfigurable Intelligent Surface Aided NOMA 6G Networks. *IEEE Trans. Veh. Technol.* 2021, *70*, 4451–4463, doi:10.1109/TVT.2021.3068774.

[70] Hou, P., Gong, J., and Zhao, J. A Physical Layer Security Enhancement Scheme under the Ambient Backscatter System. *Symmetry (Basel)* 2021, *13*, 1–11, doi:10.3390/sym13010005.

[71] Sun, X., Ng, D.W.K., Ding, Z., Xu, Y., and Zhong, Z. Physical Layer Security in UAV Systems: Challenges and Opportunities. *IEEE Wirel. Commun.* 2019, *26*, 40–47, doi:10.1109/MWC.001.1900028.

[72] He, Y., Zhang, L., Liu, S., Zhang, H., and Yu, X. Secure Transmission of Terahertz Signals with Multiple Eavesdroppers. *Micromachines* 2022, *13*, 1–15, doi:10.3390/mi13081300.

[73] Yerrapragada, A.K., Eisman, T., and Kelley, B. Physical Layer Security for beyond 5G: Ultra Secure Low Latency Communications. *IEEE Open J. Commun. Soc.* 2021, *2*, 2232–2242, doi:10.1109/OJCOMS.2021.3105185.

[74] Shakiba-Herfeh, M., Chorti, A., and Vincent Poor, H. Physical Layer Security: Authentication, Integrity and Confidentiality. In *Physical Layer Security*, Springer, New York, NY, 2021, pp. 129–150.

[75] Sanenga, A., Mapunda, G., Jacob, T., Marata, L., Basutli, B., and Chuma, J. An Overview of Key Technologies in Physical Layer Security. *Entropy* 2020, *22*, 1261, doi:10.3390/e22111261.

[76] Bandi, A. and Yalamarthi, S. Towards Artificial Intelligence Empowered Security and Privacy Issues in 6G Communications. In *2022 International Conference on Sustainable Computing and Data Communication Systems (ICSCDS)*, 2022. ISBN 9781665478847.

[77] Okey, O.D., Maidin, S.S., Lopes Rosa, R., *et al.* Quantum Key Distribution Protocol Selector Based on Machine Learning for Next-Generation Networks. *Sustainability* 2022, *14*, 15901, doi:10.3390/su142315901.

[78] Wang, C. and Rahman, A. Quantum-Enabled 6G Wireless Networks: Opportunities and Challenges. *IEEE Wirel. Commun.* 2022, *29*, 58–69, doi:10.1109/MWC.006.00340.

[79] Nawaz, S.J., Sharma, S.K., Wyne, S., Patwary, M.N., and Asaduzzaman, M. Quantum Machine Learning for 6G Communication Networks: State-of-the-Art and Vision for the Future. *IEEE Access* 2019, *7*, 46317–46350, doi:10.1109/ACCESS.2019.2909490.

[80] Duong, T.Q., Ansere, J.A., Narottama, B., Sharma, V., Dobre, O.A., and Shin, H. Quantum-Inspired Machine Learning for 6G: Fundamentals, Security, Resource Allocations, Challenges, and Future Research Directions. *IEEE Open J. Veh. Technol.* 2022, *3*, 375–387, doi:10.1109/OJVT.2022.3202876.

[81] Pajooh, H.H., Demidenko, S., Aslam, S., and Harris, M. Blockchain and 6G-Enabled IoT. *Inventions* 2022, *7*, 1–18, doi:10.3390/inventions7040109.

[82] Jahid, A., Alsharif, M.H., and Hall, T.J. The Convergence of Blockchain, IoT and 6G: Potential, Opportunities, Challenges and Research Roadmap, 2021. *arXiv Prepr. arXiv2109.03184*.

[83] Manogaran, G., Rawal, B.S., Saravanan, V., *et al.* Blockchain Based Integrated Security Measure for Reliable Service Delegation in 6G Communication Environment. *Comput. Commun.* 2020, *161*, 248–256, doi:10.1016/j.comcom.2020.07.020.

[84] Gupta, R., Nair, A., Tanwar, S., and Kumar, N. Blockchain-Assisted Secure UAV Communication in 6G Environment: Architecture, Opportunities, and Challenges. *IET Commun.* 2021, *15*, 1352–1367, doi:10.1049/cmu2.12113.

[85] Saraswat, D., Verma, A., Bhattacharya, P., *et al.* Blockchain-Based Federated Learning in UAVs Beyond 5G Networks: A Solution Taxonomy and Future Directions. *IEEE Access* 2022, *10*, 33154–33182, doi:10.1109/ACCESS.2022.3161132.

[86] Taherdoost, H. Blockchain Technology and Artificial Intelligence Together: A Critical Review on Applications. *Appl. Sci.* 2022, *12*, 12948, doi:10.3390/app122412948.

Chapter 2

History of security and privacy in wireless communication systems: open research issues and future directions

Abdulwaheed Musa[1,2]

Abstract

Wireless communication is one of the most successful technologies that have found application in various sectors of our daily lives. However, one of the major issues and challenges of wireless communication is security. To provide a holistic view of this crucial issue, this chapter presents a historical description of the evolution of the cellular network and provides a general overview of different attacks and vulnerabilities in the wireless network based on the Open System Interconnection (OSI) layered protocol architecture. The chapter further presents a brief review of the security and privacy issues for each cellular generation. The various emerging wireless communication systems as well as the application of artificial intelligence (AI) and machine learning (ML) to wireless security system design. The security issues and challenges in the key technology of 6G wireless network were identified and discussed. Finally, open research issues have been identified and discussed, giving a path for further research direction for security in 5G wireless networks and beyond.

Keywords: Wireless communication; Cellular network; Open System Interconnection; Security; Privacy

2.1 Introduction

Communication is the exchange of information between people or nodes separated by a short or long distance. Wireless communication is simply the communication between one or more devices without any physical connection between them. It utilizes radio frequency transmission to transmit information from source to destination [1]. It is one of the most popular modes of communication today due to its numerous advantages such as its cost-effectiveness, minimal power consumption, flexibility, lack of physical

[1]Department of Electrical and Computer Engineering, Kwara State University, Nigeria
[2]Institute for Intelligence Systems, University of Johannesburg, South Africa

infrastructure, and ease of deployment, among others, over its wired counterpart [2]. This has led to an exponential increase in the number of wireless users, especially with the provision and implementation of wireless networks systems and telecommunication services, including cordless, cellular, and satellite phones as well as wireless local area networks (WLANs), which have found extensive applications in various sectors such as in business, government, and military. Over the past few decades, wireless communication systems have evolved progressively from the first to the fifth generation which has benefitted various stakeholders, including standard bodies, end-users, commercial service providers, academic research groups, etc. [3]. The recent 5G network included several novel technologies such as ultra-densification, network function virtualization, the adoption of small cells, referred to as heterogenous network (HetNet) as well as the utilization of new frequency bands (millimeter wave bands) which enabled the development of the different emerging applications such as machine to machine (M2M) communication, enhanced mobile broadband (eMBB), and virtual and augmented reality (VAR). This thus has demonstrated the vast potential of the 5G wireless network and beyond [3]. However, security is one of the major issues faced by these developments and applications of wireless communication systems.

Security is a critical issue in every aspect of our daily lives, such as in banking, finance, e-commerce, and medical data, especially with the increased number of hackers and hacking tools that could be used to break into networks and systems to steal users' data [4]. Similarly, security is a crucial concern in wireless communication networks due to their dynamic nature and the utilization of radio frequency for the transmission of signals over unprotected media, referred to as open-air transmission, where any attacker that is in proximity to the signal can easily attack and steal confidential information by eavesdropping, modifying, tracking or perform any other malicious attacks on the information. This has drawn the interest and attention of researchers over the years, and there have been various developed algorithms to combat these security threats and vulnerabilities. However, these techniques and algorithms, for example, cryptographic algorithms, failed to meet up with the requirements of the future wireless communication systems challenges [5]. Hence, the need for new, dynamic, and robust security algorithms that would meet the requirements of the 5G and beyond networks. This chapter aims to present the security issues and challenges in wireless communication across all the developed cellular technologies (1G to 5G), as well as to review these security challenges in the 6G wireless network showing the trends and recommendations for further research directions. The key contributions of this chapter are as follows:

- The historical description as well as a comparison between the developed cellular generations is presented.
- General security issues in wireless communication based on the OSI reference model and a tabular summary of these attacks are provided.
- The security and privacy challenges across all cellular generations are presented, including a taxonomy of these attacks.
- The security issues in 5G technology, URLLC, eMBB, and mMTC are presented and discussed.

- The contributions of artificial intelligence (AI) and machine learning (ML) to security system design in wireless communication are also reviewed.
- The security issues and challenges of future wireless communication, particularly the 6G network, are presented and reviewed.
- Critical open research issues for security in 5G networks and beyond have been identified and elaborated.

The rest of the chapter is organized as follows: Section 2.2 presents the history and evolution of wireless communication. The general security issues are provided in Section 2.3. Section 2.4 presents and discusses security and privacy in wireless communication from the first to the fifth generation. The emerging wireless communication system is discussed in Section 2.5. Section 2.6 presents the application of AI and ML to wireless security system design. The security issues and challenges in future wireless communication system are presented and discussed in Section 2.7. Finally, Section 2.8 concludes the chapter and provides some recommendations.

2.2 History and evolution of wireless communication

This section presents a historical description of wireless communication with a binocular focus on cellular technologies from the first generation (1G) to the fifth generation (5G) and compares these different technologies.

Wireless communication can be historically classified into three eras which include the following: the Pioneer era, which lasted till 1920, the pre-Cellular era which spanned between 1920 and 1979, and the Cellular era which started beyond 1979. In the Pioneer era, information was transmitted using smoke, torches, flares, mirrors, semaphore flags, etc., categorized under the first wireless networks, all of which were superseded by the first telegraph network developed by Samuel Morse in 1838 and thereafter, by telephones. A few decades after, Guglielmo Marconi, in 1895, demonstrated the first radio transmission which led to the advent of the wireless communication network [6]. The cellular system has advanced from the first to the fifth generation over the past few decades.

First generation (1G): the first generation of cellular network, launched by Nippon Telephone and Telegraph (NTT) in Tokyo, Japan in 1979, utilized analog transmission and operated at 800 MHz frequency band for speech services. Two years after the successful implementation of the network in Japan, two cellular networks were developed and deployed in Europe: Nordic Mobile Telephone (NMT) and Total Access Communication System (TACS). Afterward, another 1G network was developed and deployed in the United States of America (USA) in 1983 which was called the Advanced Mobile Phone System (AMPS). Although these systems were implemented differently, they still have similar disadvantages such as incompatibility issues, limitation to only voice communication, limited roaming services, limited coverage area, and low capacity. In addition, it is frequency modulation (FM)-based and used frequency division multiple access (FDMA) multiplexing scheme, which reduced the capacity of the network [7].

Second generation (2G): This cellular network, based on the Global System for Mobile Communication (GSM) standard, was developed in Finland in 1991 and utilized digital transmission. As an improvement over 1G, the 2G network was based on both the FDMA and time division multiple access (TDMA). This enhanced the system capacity. The GSM system employed a 25 MHz frequency spectrum in a 900 MHz range with a speed of 14.4 kbps. In 2G, circuit switching was used, and the core network was known as the public switched telephone network (PSTN). The general packet radio service (GPRS) was then developed as the demand to transmit data on air interface increased where an optimal speed of up to 150 kbps was achieved [8]. As the demand further increased, the Enhanced Data GSM Environment was developed which can be considered as 2.5G. This also increased the data rate amount to increase further up to 500 kbps. The major aim of this cellular network was to use digital signals for transmission and to provide several services such as web access facilities, digital voice calling, and short message services [9]. However, there are still some deficiencies experienced with the network such as the inability to support video communication and the dependency on strong digital signals. This means there is usually no network coverage if the digital signal is weak.

Third generation (3G): This cellular network was launched by the NTT DoCoMo in Japan in 2001 and was based on the International Mobile Telecommunication standard, precisely the IMT-2000 standard. The system utilized both the Code Division Multiple Access (CDMA) and WideBand Code Division Multiple Access (WCDMA) multiplexing schemes. Compared to the FDMA and TDMA utilized in the previous networks, the CDMA is a multiplexing scheme where each user utilizing the channel at a particular period is given a different code. This allowed for a significant number of users to utilize the network simultaneously. The 3G system utilized a 15–20 MHz frequency spectrum with 1,800–2,500 MHz, with a speed of up to 2 Mbps. WCDMA, the air interface for the Universal Mobile Telecommunication System (UMTS), utilized larger carrier frequency due to the increased number of users it can support as compared to the CDMA. The core network for the 3G system was a combination of circuit and packet switching [8]. The 3G system was known as the UMTS in Europe, CDMA2000 in America and the time division-synchronous code multiple access (TD-SCDMA) in China. As an advancement from the 3G network, the 3.5G and 3.75G networks were developed and deployed in the form of high-speed packet access (HSPA) and HSPA+, respectively to further increase the data rate. The HSPA+ enabled the first introduction of the multiple input multiple output (MIMO) technology where an optimal speed of up to 42 Mbps was achieved using a modulation scheme of 64-bit quadrature amplitude modulation (QAM) [7]. This system has numerous advantages such as supporting both voice and video communication. However, it also has some notable disadvantages such as expensive terminals, and high power consumption.

Fourth generation (4G): The cellular network, sometimes called the long-term evolution (LTE) was first launched in Oslo, Norway and Stockholm, Sweden in 2009. It was based on the ITU-defined capabilities in the IMT-advanced standard. The 4G network uses orthogonal frequency division multiplexing (OFDM) technique and utilized 5–20 MHz in a frequency band of 2,000–8,000 MHz where a speed of up to 100 Mbps and 500 Mbps was achieved for downlink and uplink, respectively. This allowed the system to support high data rate applications such as online gaming, voice as well as video over IP, and high-quality online streaming. The 4G core network was IP-based and has low latency as well as wider channel and carrier aggregation of up to 100 MHz. The 4G network operates in both frequency division duplex and time division duplex (TDD) modes and also supports the TD-SCDMA evolution. Although the LTE had numerous advantages such as high throughput and simpler architecture, it is however limited in terms of security threats, cost, battery life of mobile phones, etc. [10].

Fifth generation (5G): This is the latest developed and deployed cellular network that was first largely adopted by South Korea in 2018. Unlike 4G which depends on high-power cell towers to transfer data over long distances, 5G wireless signals are provided by a vast number of small cell stations placed on light poles or building roofs [7]. Many small cell stations are required because the 5G frequency band (millimeter wave bands) is suitable only for short-distance transmission. The 5G is completely wireless making it suitable for the World Wide Wireless Web (WWWW) and to be supported by different technologies such as large area synchronized code-division multiple access (LAS-CDMA), multi-carrier code division multiple access (MC-CDMA), OFDM, ultrawideband (UWB), network-local multipoint distribution service (NLMDS) and Internet Protocol version 6 (IPv6) [11]. The 5G wireless network would operate in two primary categories of frequencies: Frequency Range 1 (FR1) or simply sub-6 GHz and Frequency Range (FR2) or simply millimeter wave (mmWave) band. Unlike the previous, this network offers many advantageous benefits, but it is quite limited in terms of cost of deployment since small cells are required.

A summary of the evolution of cellular communication technology showing the years of the initial launch and key characteristics is given in Figure 2.1 and a comparison of the generation in terms of their features is given in Table 2.1.

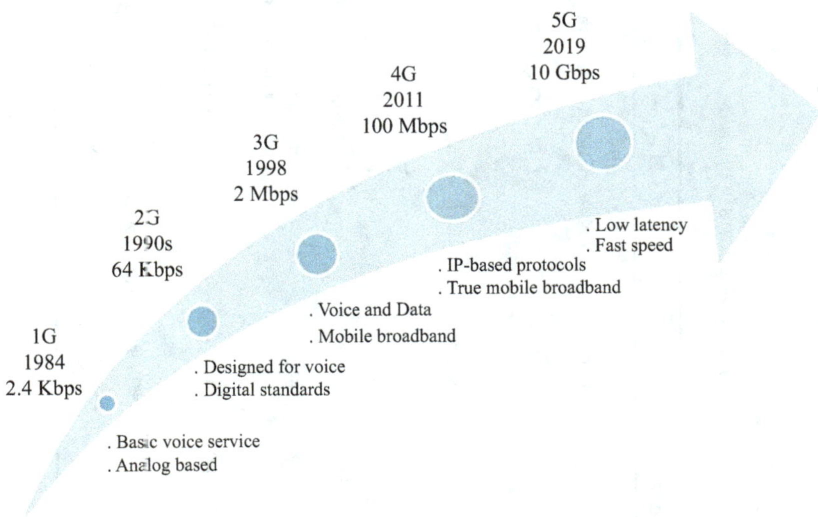

Figure 2.1 Evolution of cellular communication [12]

Table 2.1 Comparative analysis of cellular generations

Generations	1G	2G	3G	4G	5G
Deployment	1979	1991	2001	2009	2018
Switching techniques	Circuit switching	Circuit switching	Circuit and packet switching	Packet switching	Packet switching
Access techniques	FDMA	GSM, TDMA, CDMA	WCDMA, UMTS, CDMA 2000, HSUPA/HSDPA	LTEA, OFDMA, SC-FDMA, WiMAX	BDMA, NOMA FBMC
Application	Voice	Voice and data	Voice, data, and video calling	Voice, data, video calling, HD television, etc.	Voice, data, video calling, ultra HD video, VR application
Core network	PSTN	PSTN & packet	Internet	Internet	Internet
Data rate	2.4 kbps	14.4 kbps	384 kbps to 2 Mbps	100 Mbps to 500 Mbps	10 Gbps to 50 Gbps
Frequency band	800 MHz	800 MHz, 900 MHz, 1,800 MHz, 1,900 MHz	800 MHz, 900 MHz, 1,800 MHz, 1,900 MHz, 2,100 MHz	2–8 GHz	1.8 GHz, 2.6 GHz, 30–300 GHz
Advantages	Mobility	Digital voice calling, mass adoption, low power consumption	Better internet experience, support of both voice and video calling	High data rate, body area network	Low latency, higher data rate, wider coverage
Disadvantages	Poor security, low capacity, low coverage	Low data rate, inability to support video transmission	Expensive terminals and high-power consumption	Expensive and high-power consumption	Requires many small cell antennas

2.3 General security issues

This section provides a general overview of various security threats and attacks in wireless networks. Network security, either wireless or wired, refers to the prevention of unauthorized access, modification, destruction, or disclosure of information to ensure that the network's critical functions are executed properly and that there are no detrimental impacts.

Wireless and wired communication networks have some similarities, notwithstanding their differences. For instance, both utilize the open system interconnection (OSI) layered protocol architecture for their transmission, consisting of seven layers. However, this section would focus on five layers namely: the Physical, the Media Access Control (MAC), the Network, the Transport, and finally, the Application layers. Figure 2.2 shows how the data transmission from Node A to Node B occurs based on the OSI layers over a wireless medium.

Different protocols exist in the different layers of the OSI model. For instance, the application layer includes protocols such as hypertext transfer protocol (HTTP) and file transfer protocol (FTP) for providing web services and transporting large data files, respectively. The transport layer includes protocols such as the transmission control protocol (TCP) and user datagram protocol (UDP). Because each layer of the OSI model supports different protocols, they are vulnerable to various threats and attacks discussed below.

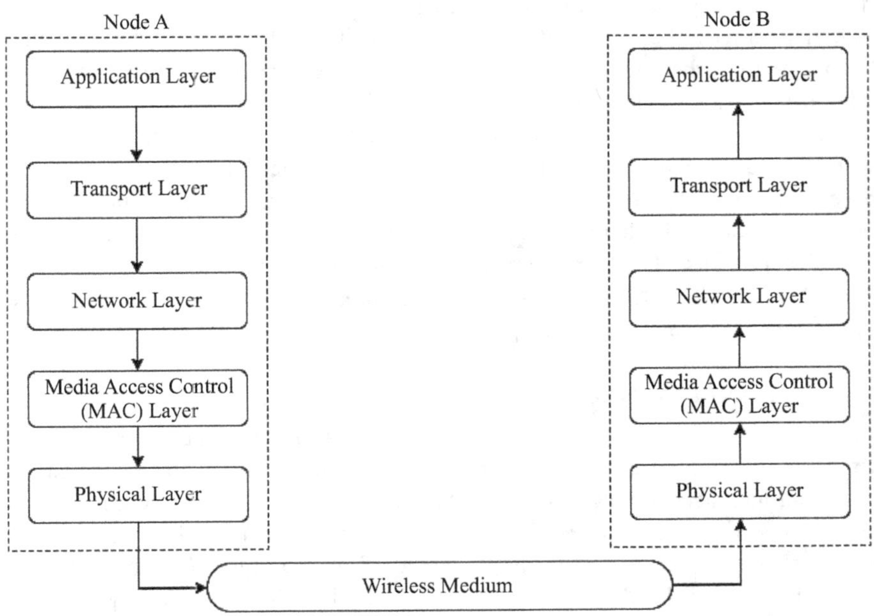

Figure 2.2 OSI-layered protocol architecture [13]

2.3.1 Physical layer attacks

The first layer of the OSI model architecture is known as the physical layer and it determines the physical characteristics of signal transmission. Signals sent through a wireless network are broadcast over the open air, making the physical layer of a wireless network susceptible to two major types of attacks which include eavesdropping attacks and jamming attacks [13]. An eavesdropping attack is a form of passive attack and it refers to an unauthorized interception of transmitted signal between the source and destination. It is a major threat to the wireless network as it is the prerequisite to all other attacks [14]. An eavesdropper situated within the coverage of the transmitter of a wireless network can overhear the communication sessions. To prevent this and maintain confidential communication, various methods have been developed, but one of the oldest of them all is known as cryptography, that is, the art and science of concealing meaning through the means of encryption. In that case, the receiver has a private key which would be utilized to decrypt the transmitted encrypted message. This will prevent the eavesdropper from understanding the message even after the interception. Jamming, on the other hand, is a type of denial of service (DoS) attack where a malicious node in the wireless network sends a large amount of radio signal to flood the channel with the intent of preventing legitimate users from utilizing the resources of the network, thus causing reduction in the network availability [15]. Jamming attacks are classified into four major types which include: deceptive, random, constant, and reactive jammers [16]. The constant jammers launch an attack by sending a continuous high-powered noise sweep from one communication channel to another based on a predetermined approach and then repeating this procedure over and over. For the random jammers, they work at random and switches from one communication channel to the other without any precise plan. The deceptive jammers transmit unauthorized packets via wireless networks making them busy. Finally, the reactive jammers consistently monitor the status of frequency channels and attack only those utilized for communication [16]. Various techniques have been developed to prevent jamming attacks. Among all of them, spread spectrum techniques have been widely utilized where the transmitted signal would be spread over a wider spectral bandwidth. The spread spectrum techniques include the direct-sequence spread spectrum (DSSS), frequency-hopping spread spectrum (FHSS), and time-hopping spread spectrum (THSS) [17].

2.3.2 MAC layer attacks

This is the second layer after the physical layer that allows the intelligent allocation of the shared medium using intelligent control mechanisms which include carrier sense multiple access/congestion detection (CSMA/CD), carrier sense multiple access/congestion avoidance (CSMA/CA), orthogonal FDMA (OFDMA), CDMA, and so on. In the wireless-MAC layer, there are four main types of attacks that can occur. They include MAC spoofing, man-in-the-middle (MITM), identity theft, and network injection [18]. Every node in the network is typically equipped with a network interface card (NIC) and a distinct MAC address, which is used for user identification. A MAC spoofing attack happens when an attacker alters its assigned MAC address to conceal its identity or impersonates another node on the network to eavesdrop traffic and carry out illegal activities such as stealing private information; this is referred to as identity theft. The MITM attack involves the attackers first sniffing the network to capture the MAC addresses of the authorized communicating nodes, after which the attacker impersonates the two nodes and establishes a connection with them to act as a relay between the victims, giving them the impression that they are communicating directly with one another over a private connection. Network injection, on the other hand, attempts to prevent networking equipment such as routers, switches, and access points from functioning by the injection of fraudulent re-configurable network commands. This is done to shut down the entire network, necessitating resetting or reprogramming of all networking equipment.

2.3.3 Network layer attacks

The Internet protocol (IP) is the major protocol utilized in this layer for the delivery of packets from the sender to the receiver through a relay router based on IP address. In this layer, the attacks are focused on exploiting the IP weaknesses and can be classified into three types namely: IP spoofing, IP hijacking, and Smurf attacks [13,19]. Similar to MAC address spoofing, IP spoofing is when an attacker generates a fabricated IP address to conceal their identity or impersonates another node on the network to conduct illicit activities. This can also waste network capacity and potentially bring the network down since the network node that receives packets with a faked source IP address will interact with the same faked address. IP hijacking is another attack conducted by hijackers to obtain the IP address of an authorized user to disconnect the authorized user and then impersonate them on the network to access and steal confidential information. Prefix, route, and border gateway protocol hijacking are the three major types of IP hijacking attack [20]. A Smurf attack is a type of DoS attack where a significant number of Internet Control Message Protocol (ICMP) packets are sent to a target or group of targets from a faked source IP broadcast address with the intention of bringing down the target's network upon the receipt of the ICMP packets. The target would be required to respond which would lead to a substantial amount of traffic in the network. Firewalls are widely utilized to prevent this attack. It can be utilized to reject any malicious packets from a fabricated IP address. Another solution, although not scalable, is to configure each node in the network to not respond to ICMP requests.

2.3.4 Transport layer attacks

This is the fourth layer and the heart of the OSI model. As mentioned earlier, TCP and UDP are the two major protocols in this layer. The TCP is a connection-based protocol that supports the seamless and reliable delivery of segments from one node to another, such as email delivery and file transfer. UDP is the opposite of the TCP as it is not a connection-based protocol and does not guarantee the reliable delivery of segments from one node to the other, but it has lower overhead and latency. It is typically used for time-dependent applications which do not demand reliable delivery of segments, such as IP television, Voice and Video over IP (VoIP). Attacks such as TCP flooding, UDP flooding, and TCP sequence number predictions are the major attacks faced in this layer [19]. TCP flooding or Ping flooding is another type of DoS attack where the attacker sends a significant number of ping requests to a target or group of targets such as Address Resolution Protocol (ARP) echo requests, and the target responds by sending ping replies. This will eventually cause flooding for the target node's input and output buffers, as well as connection delay, especially if the ping requests are of a substantial amount. To perform a TCP sequence number prediction attack, the attacker tries to estimate the TCP segment sequence index of the source node, fabricates the segment, and then sends it to the target receiver. This has thus infringed on the integrity of the packet. UDP flooding is similar to TCP flooding; however, instead of sending ping queries, malicious UDP packets are delivered to the target node, forcing it to respond to these multiple packets, rendering the node unreachable to other authorized nodes in the network. A firewall can be used to defend against this attack as well as to limit the response rate of UDP segments.

2.3.5 Application layer attacks

This is the last layer of the OSI model, but the first layer that the users interact with. It comprises various protocols such as HTTP, FTP, and SMTP. However, these protocols expose this layer to several security vulnerabilities. There are five major application-layer attacks which include malware attacks, SQL injection, SMTP attack, FTP bounce, Cross-site scripting that can be categorized under HTTP attacks, FTP attacks, and SMTP attacks [17]. The malware attack is an example of attacks categorized under the HTTP attacks that include key loggers, backdoors, worms, viruses, and Trojan horse. A malware refers to any damaging software in the form of code or script developed by an attacker to interrupt the authorized transmission or steal confidential data. SQL injection is another type of HTTP attack commonly used to exploit data-driven systems by inserting malicious SQL statements to gain unauthorized access to a website. Cross-site scripting is also another type of HTTP attack where the attacker inserts client-side script into a web page causing the browser to automatically download malicious scripts when the user loads the web page, thus allowing the attacker to easily bypass security restrictions and steal sensitive information such as cookies information and perform other series of malicious activity [21]. An FTP bounce attack is when an attacker exploits the vulnerabilities of the FTP protocol using the PORT command to request port access using a victim's IP address, acting as a middleman. The SMTP is the protocol utilized to send emails over the Internet, but lacks encryption for sensitive information such as user credentials or the original message contained in the email, which has raised severe privacy issues. The SMTP attack includes password sniffers, SMTP viruses, worms, and email spoofing [18].

Table 2.2 presents the summary and taxonomy of the major attacks in each layer of the OSI reference model.

Table 2.2 Attacks in the OSI layers

OSI layers	Major attacks
Physical layer	Eavesdropping and Jamming attacks
MAC layer	MAC spoofing, MITM, identity theft, and network injection
Network layer	IP spoofing, IP hijacking, and Smurf attacks
Transport layer	TCP flooding, UDP flooding, and TCP sequence number
Application layer	Malware attacks, SQL injection, SMTP attack, FTP bounce, cross-site scripting

2.4 Security and privacy in wireless communication

This section presents the various security and privacy issues and vulnerabilities of the different cellular generations from 1G to 5G.

2.4.1 1G

There are several limitations in the 1G network as mentioned earlier, one of which is in terms of security. Both the AMPS and TACS did not have authentication and data encryption because 1G was analog, making it relatively simple for the interception of calls and the retrieval of the two unique identities, which include the mobile identity number (MIN) and the electronic serial number (ESN), which can be utilized to impersonate a device on the network. This can thus lead to an impersonation attack [22]. However, the first security measure implemented in 1G to defend against eavesdropping is known as voice scrambling which was used in Nordic Mobile Telephone (NMT) for both the mobile station (MS) and the base station (BS). Although, it was not a particularly strong encryption technique, it did prevent call interception by attackers [23].

2.4.2 2G – GSM

The digital nature of the 2G wireless network enabled the security limitations of the 1G network to be adequately addressed, particularly with the introduction of cryptography for authentication and the subscriber identity module (SIM) for verifying user's identity. However, the 2G network also has security and privacy issues. For example, spamming is one of the major security issues where attackers send unwanted messages to victims. Furthermore, the Abis interface, which is the interface between the BS and BS controller (BSC), is not encrypted. As a result, audio and signaling signals can be easily intercepted by attackers who can intercept E1-based communication. Another vulnerability is that the only method operators can authenticate the MS is through a unilateral process, which leaves the MS with no way to verify the operator, allowing an attacker to pose as the actual operator and launch an MITM attack [24]. A fabricated BS can be erected with a higher transmitting power than the other, making it the most preferred BS for several MSs to interact with. This can result in session hijacking, which is an attack on network authentication in 2G, where an attacker may be able to obtain the international mobile subscriber identity (IMSI) information from the MS [22]. Furthermore, some of the security solutions deployed in the 2G network were shown to have serious weaknesses, such as the COMP128 and A5 cryptographic algorithms, which were used to provide radio-path encryption and user authentication, respectively [25].

2.4.3 3G – UMTS

The 3G mobile system was developed to overcome the limitations in the 2G mobile systems including the security issues identified. In the 3G network, security measure implemented can be classified into five as follows: Network Access Security, User Domain Security, Application Security, Network Domain Security, and, finally, Visibility and Configurability Security. This thus provided flexibility and possible scalability to accommodate new threats. However, with the significant number of users, the network vulnerabilities were exposed and easily exploited. They include eavesdropping, access to confidential information, MITM, DoS, location update spoofing, as well as identity theft attacks [23].

2.4.4 4G – LTE

The 4G network provided security solutions to the security flaws identified in the previous generation as well as new cryptographic algorithms. The evolved packet system (EPS) encryption and EPS integrity algorithms as well as the use of 256 bits were the major security measure that was introduced and implemented in 4G network. Further significant improvement seen in the 4G network as regard to security includes the utilization of different algorithms and key sizes for the user and control plane traffic. However, the 4G network is IP-based and utilizes diameter signaling which made it less secure and very vulnerable against various attacks which include eavesdropping attacks, TCP flooding attacks, user identity theft, DoS attack, intrusion attack, address spoofing, GPRS tunneling protocol (GTP)-based attacks, particularly the backhaul traffic in 4G network. It also has many entry points including compromised smart mobile devices and access networks [26]. Furthermore, authentication in the 4G network was provided by the EPS-Authentication and Key Agreement (EPS-AKA). However, it was still susceptible to various vulnerabilities such as user identity theft and impersonation, computational overhead, MITM attack, malware spreading, and authentication delay [27].

2.4.5 5G

The 5G network addresses most of the crucial limitations of the earlier generations while facilitating requirements such as increased data rate and capacity, enhanced device connectivity, reduced cost and latency, and guaranteed quality of service (QoS) [28]. However, the 5G network is also faced with security issues due to the numerous applications and technologies it supports. To evaluate these security issues, it is imperative to analyze the threats and vulnerabilities of the enabling technologies.

2.4.5.1 Device-to-device (D2D) communication

D2D communication allows users to connect directly without the usage of BSs or access points (APs) [29]. It offers various benefits in terms of spectrum efficiency, expansion of coverage, increase in capacity, radio resource reusing, management of power, etc. It is regarded to be the fundamental technology for developing point-to-point communications to obtain improved coverage and higher data speeds for 5G user devices [30]. D2D communication is similar to vehicle to everything (V2X) communication and M2M communication which are also enabling technologies in 5G wireless network. Some of the characteristics of D2D communication include the following: the utilization of the open-air transmission which is freely accessible, the network comprising mobile phones, exposure of the network architecture, and the limited resources (bands and channels). These characteristics make the 5G technology vulnerable to several attacks including eavesdropping, jamming, DoS attacks, and injection attacks. Although different solutions have been utilized to combat these attacks. For instance, movement from the eavesdropper was a proposed solution for an eavesdropping attack [29].

2.4.5.2 Network function virtualization (NFV)

NFV is another essential enabling technology in 5G wireless networks, allowing for customized network slicing across distributed clouds to build programmable networks for 5G applications. In other words, it is one of the solutions for 5G automation in the service provider production environment because it enables service providers to design network functions such as firewall, Intrusion Detection, Network Address Translation (NAT), Domain Name Service (DNS), dynamic that meets Service Level Agreements (SLAs) and service models by implementation in software running on clouds accessible from a centralized repository [31,32]. Although the NFV technology has numerous benefits, it has security issues which makes it vulnerable to attacks such as configuration attack, DoS attack, hijacking attack, penetration attacks, and resource (network slice) theft attacks [33].

2.4.5.3 Software-defined network (SDN)

The SDN, similar to the NFV, allows for a centralized and programmable network management function in a software-oriented network management platform while the infrastructure of the network deals with the applications and network service [34]. This simplifies network control, management, and operation while also accelerating the development and deployment of network features. However, due to the fact that the network control is centralized, it makes the controller a bottleneck for the entire network which can be vulnerable to several attacks including eavesdropping, DoS, hijacking, configuration, MITM, and saturation attacks. Saturation attack also known as malware spreading attack, where if a malicious application is to be granted access, it can easily spread throughout the network [33].

2.4.5.4 Massive MIMO

Massive MIMO (mMIMO) is also known as an extensive antenna system where a base station is equipped with more than 100 antennas' elements that would be utilized to service a significant number of users simultaneously with a frequency band. It is one of the core enabling technologies of the 5G wireless network, improving its data speed and bandwidth efficiency while servicing a significant number of users [35]. Some advantages of this technology include lower latency and energy usage, simplified MAC layer, and increased capacity because of spatial multiplexing. mMIMO is vulnerable to majorly eavesdropping attack which can be either passive or active [24].

2.4.5.5 HetNet

HetNet is one of the solutions proposed to enhance the 5G network and it encompasses key elements such as small cells, mMIMO, mmWaves, and D2D communication [36]. The deployment of low-power base stations or small cells was proposed as a method to densify the network due to the increasing need for high data rates to operate applications. This would result in higher spectral efficiency as well as reduce the power consumption of mobile phones due to its communication with nearby small cells. Modern methods of data transmission used by small cells in 5G wireless network include beamforming, millimeter wave, and MIMO. HetNet is vulnerable to several attacks including eavesdropping, DoS, jamming, MITM, and hacking attacks [37].

From this brief review of the key enabling technologies in 5G wireless network, it can be seen that they are all vulnerable to different attacks, particularly eavesdropping attacks, based on their features, thus making the 5G network itself to be vulnerable. The taxonomy of these various attacks in cellular communication across the generations are presented in Table 2.3.

Table 2.3 Taxonomy of attacks in cellular communication

Attacks	1G	2G	3G	4G	5G
Eavesdropping	✓	✓	✓	✓	✓
Jamming	×	×	×	✓	✓
Identity theft & impersonation	✓	×	✓	✓	×
Spamming	×	✓	×	×	×
Fabricated BS	×	✓	✓	×	×
MITM	✓	×	✓	×	✓
DoS/DDoS	×	×	✓	✓	✓
Spoofing	×	×	✓	×	✓
TCP or UDP flooding	×	×	×	✓	✓
Malware	×	×	×	✓	✓
Injection	×	×	×	×	✓
Sniffing	×	×	×	×	✓
Hijacking	×	×	×	×	✓
Configuration	×	×	×	×	✓
Network slice theft	×	×	×	×	✓
Penetration	×	×	×	×	✓

2.5 Emerging wireless communication systems

This section presents and discusses the different emerging wireless communication system as well as provides a brief review of different research works that have considered these technologies and systems.

2.5.1 Low-cost IoT devices

Internet of Things (IoT) simply refers to a connection of physical objects, such as microwaves, doors, fans, and lighting integrated with electronic sensors and software to allow collection and sharing of data using a communication protocol. The collected data are sent to the server or microcontroller for further analysis which then processes the information to derive knowledge and translate them into a particular action [38]. One of the requirements of an IoT system is low cost which can be met using boards based on Alf and Vegard's RISC (AVR) processors including the Arduino as well as the ESP8266 board. These boards widely utilized several applications and systems including GPS trackers, home assistants, inventory management system, and can be configured with the Arduino Integrated Design Environment (IDE) [39]. The Arduino microcontroller, however, was seen to be the least expensive when compared with other latest existing microcontrollers [40]. The different types of Arduino boards as well as the working principle have been reviewed in [41]. To fully comprehend the potential of IoT, it is imperative to examine its convergence with the 5G wireless network which among other things, offers to manage the complexities of IoT such as scalability, wider coverage, backhaul connectivity, cost, and installation, accomplished using the 5G enabling technologies [42]. It is estimated that there would be a significant number of connected IoT devices that would produce enormous amounts of data and this has led to the development of an architecture known as the 5G I-IoT model, that is, the convergence of 5G, AI, and IoT [43]. This has found application in critical services such as remote healthcare, VAR, drones, security surveillance, and autonomous cars as shown in Figure 2.3. However, security is the major challenge faced by the 5G I-IoT environment due to the openness of the network and the lack of robust security schemes [43]. As a result, the 5G I-IoT environment is vulnerable to several attacks including eavesdropping, replay, MITM, impersonating, and malware attacks [43].

2.5.2 Ultra-reliable and low latency communications (URLLC)

The URLLC is one of the key application scenarios in the 5G New Radio (NR) defined by the International Telecommunication Union (ITU-R) along with eMBB and massive machine-type communication (mMTC) to satisfy the requirements of several mission-critical applications, such as remote healthcare, drone-based logistics, factory automation, smart grid, autonomous cars, VAR, and AI-based personal assistants, among others, that require low latency, better reliability, high connection density, higher throughput, and scalability [44,45]. Figure 2.4 shows the various applications as well as the key requirements of URLLC.

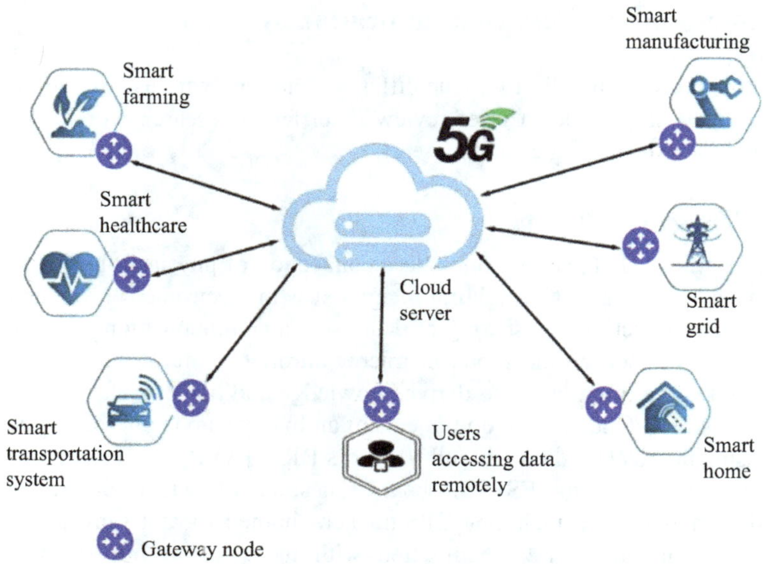

Figure 2.3 5G I-IoT environment [43]

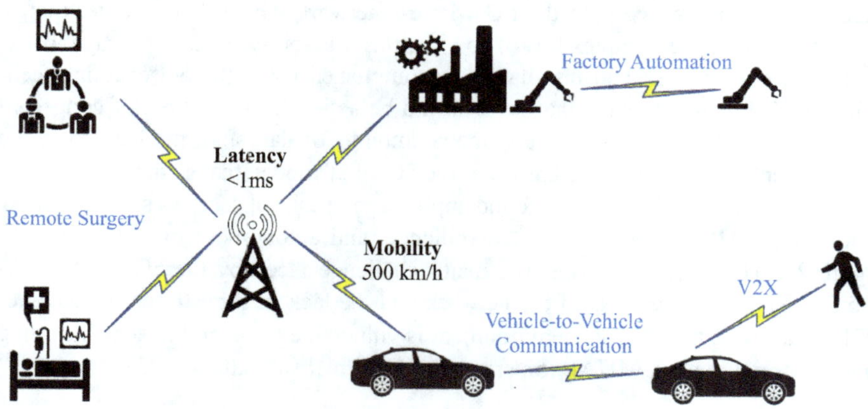

Figure 2.4 Applications and key requirements of URLLC [46]

Ultra-reliability and reduced latency are the two key requirements of the URLLC. Where reliability is the capability to transmit a specific volume of packet with high success rate within a particular period, latency refers to the delay in an accurate transmission from the source to the destination [47]. In other words, latency is the time required for a packet to make a round-trip across the network. It is classified into User Plane Latency (UPL) and Control Plane latency (CPL). The

UPL is the required time taken between when the sender transmits the packet to when the receiver receives it and according to the ITU-R, the minimum requirement of UPL in URLLC is 1 ms, assuming unloaded conditions with small packets for both uplink and downlink [47]. The CPL, on the other hand, is the time taken between when the user equipment (UE) changes status from idle state to active state. According to the ITU-R, the minimum requirement of the CPL is 20 ms, but less than 10 ms is most preferred [47].

Security challenges associated with most URLLC applications are also critical, as an attack can have disastrous consequences. For example, if a remote surgeon's communication is hijacked by an attacker, the entire communication will not only slow down but the patient's life is also endangered. The major attack in the URLLC application scenario is the eavesdropping attack which is an attack on confidentiality and privacy. To ensure confidentiality and protection against eavesdropping attacks, cryptography algorithms have been widely used in traditional networks. For URLLC, however, these cryptography algorithms are not efficient and may affect the two key requirements of URLLC due to the high-intricacy signal processing used for encrypting and decrypting. Key distribution may cause an additional delay in some application scenarios [5]. The physical layer security (PLS) has been proposed and demonstrated to be a potential solution for protecting against attacks in 5G URLLC network such as eavesdropping attacks because unlike cryptography, PLS does not require a key exchange for encryption and decryption [5].

2.5.3 eMBB

The eMBB is another key application scenario in 5G NR and an extension of the 4G broadband service. eMBB is concerned with the provision of increased data rate, enhanced connectivity and mobility. High throughput is the key requirement of eMBB which is to increase data rate while ensuring reasonable accuracy with a packet error rate (PER) of around 10^{-3} [48,49]. Throughput is generally referred to as the rate at which information is successfully delivered over a channel. It is affected by various factors such as channel type, network protocol, user density, packet loss, power, latency, and bandwidth [48]. Different techniques have been utilized to improve mobile broadband [50]; however, three strategies – broadband expansion, cell density expansion, and spectral efficiency increase – were utilized to increase the throughput of the 5G network [48]. To increase the bandwidth, the 5G network utilized the millimeter wave band and adopted flexible TDD as well as full duplex schemes. To increase the cell density, increased cell density was adopted as well as a heterogeneous network which include small cells. To enhance the efficiency of the spectrum, low-density parity-check (LDPC) codes as well as mMIMO were adopted. Figure 2.5 shows the various applications as well as the key requirement of eMBB.

For eMBB, two major security issues are associated with it. They lack adequate monitoring system and user privacy leakage [51]. The eMBB applications such as 4K/8K high-definition video streaming, and VAR-based mobile roaming

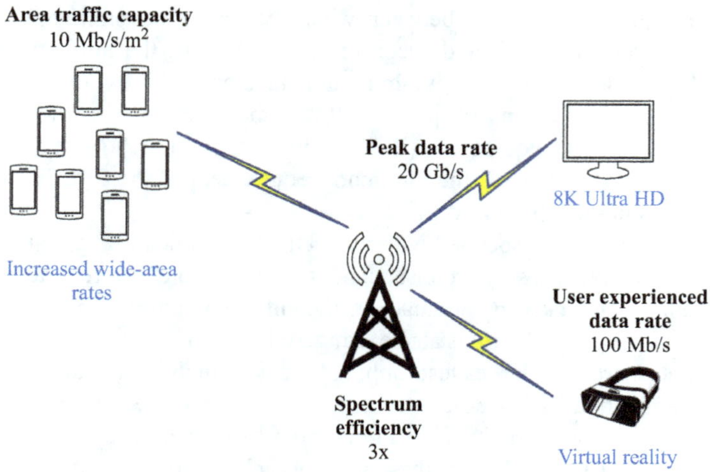

Figure 2.5 Applications and key requirements of eMBB [46]

immersive services produce an enormous amount of traffic which makes it difficult to monitor using security nodes including Firewalls and Intrusion Detection systems [51]. Furthermore, user privacy leakage can occur due to the openness of the network as these services and applications supported contain users-sensitive information including device identification and address identification.

2.5.4 Massive machine-type communication (mMTC)

mMTC, another key application scenario defined by the ITU-R, is also known as massive M2M (mM2M) communication that supports a significant number of devices to directly communicate with one another or to access the Internet with or without any human inputs [48]. Connection density and network energy efficiency are the two key requirements of the mMTC with the aim of connecting a significant number of low-rate energy-saving devices, also known as machine-type devices, to cellular network [46,52]. Figure 2.6 shows the various applications as well as the key requirements of mMTC.

Connection density, in this context, is concerned with the capability of the 5G network to support a very large number of device connections to the Internet, particularly IoT devices while network energy efficiency is the capability of the radio access technology (RAT) to reduce the consumption of energy of the radio access network (RAN) in respect to the generated amount of traffic. Various techniques have been proposed and developed to satisfy these requirements. However, similar to the other scenarios, the mMTC is also vulnerable to various threats and attacks which include fake terminals, modification attacks, eavesdropping attacks, and remote control due to the simplicity of the IoT devices and weak security protection capabilities [51].

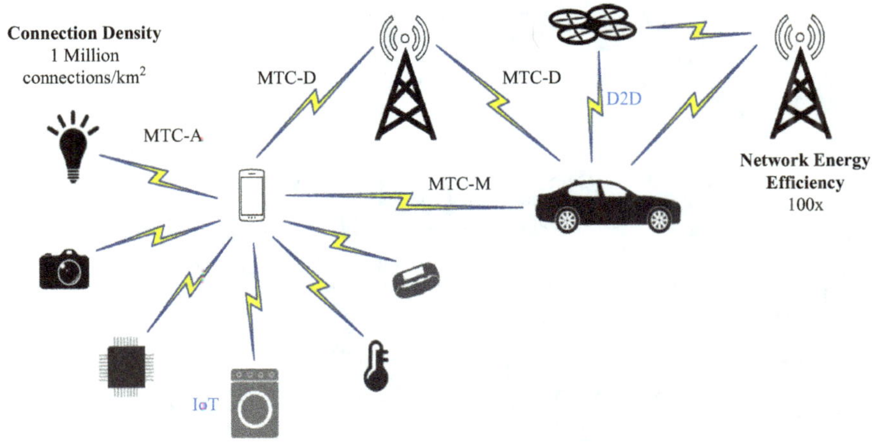

Figure 2.6 Applications and key requirements of mMTC [46]

2.6 Application of AI and ML to wireless security system design

In this section, AI and ML are introduced and a brief review of the application of AI and ML to the security aspect of wireless communication is carried out. As illustrated in Figure 2.7, AI is a vast field that covers both ML and deep learning (DL). AI is a general phrase that refers to computer intelligence that may replicate human intelligence and abilities including learning, reasoning, decision-making, perception, and problem-solving [48].

ML, on the other hand, is a subset of AI with a focus on extracting knowledge from a significant amount of data. It refers to the capability of machines to learn without being explicitly programmed. AI and ML have proven their effectiveness in various fields including medicine, education, engineering, transportation, industrial robotics, agriculture, etc., for their high accuracy in terms of classification, identification, and automation. Similarly, AI and ML have been utilized to develop, simulate, and implement several efficient and effective security algorithms to defend against various threats and attacks [53].

ML was applied for detecting jamming attack in [54] for wireless IoT networks using the received signal strength indication (RSSI) data from both simulation and real networks where the results obtained after analysis showed that the proposed model achieved high accuracy with no communication cost or overhead for nodes in the network. Similarly, the work [55] used ML for the detection of distributed DoS (DDoS) for consumer IoT devices where results showed that routers or other intermediary nodes can detect sources of the DDoS attack automatically using the proposed low-cost ML algorithm.

Furthermore, the work [56] proposed an ML-based antenna design scheme to enhance the security of IoT network where the simulation results obtained showed

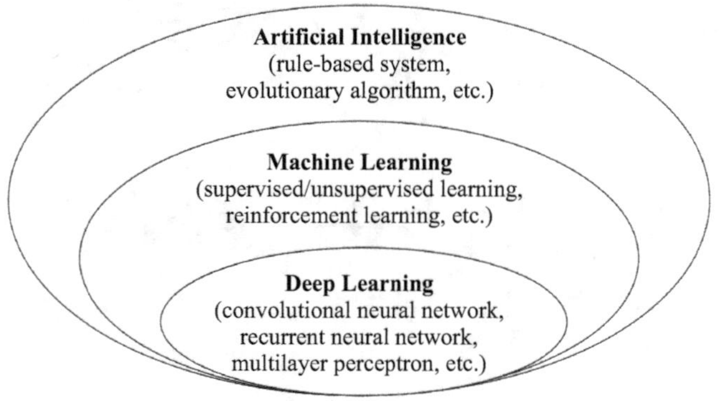

Figure 2.7 Relationship between AI, ML, and DL [48]

that the proposed scheme is suitable to be applied in PLS communication in which signal-to-noise ratio (SNR) of wiretap channels is decreased while ensuring reliable communication on the major channel thus preventing the leakage of information. The work [57] proposed a model that can detect abnormalities in an IoT network using deep neural network (DNN) with chicken swarm optimization (CSO) where results showed that the proposed model is highly accurate with better detection rate compared to other models. An ML-based security approach was proposed in [58] for autonomous IoT systems to achieve optimal energy efficiency and better reliable transmission. Results showed that the proposed model optimized the network performance, achieved fault–tolerant data transmission, and accomplished network confidentiality against adversaries using a cryptography-based deterministic algorithm. Further reviews of AI- and ML-based techniques for the security of IoT systems were presented in [59,60] against various threats and attacks including the future challenges when employing these techniques.

A generative adversarial network (GAN), an unsupervised learning scheme, was utilized in [61] to secure wireless sensor network middleware. Results obtained showed that the algorithm improves the data accuracy and ensures protection from adversaries with less energy, and throughput compared to other conventional schemes. Similarly, the work [62] reviewed the AI applications for intrusions detection in software-defined wireless sensor network (SDWSN) where it was shown that AI techniques are effective techniques that can be used in SDWSNs to defend against malicious attacks. Further reviews of the security of IoT and WSNs against various threats and attacks using ML techniques are presented in [63].

In [64], the constraints of traditional cryptographic and physical layer authentication solutions, such as low dependability, increased latencies, and security overheads, led to the application of ML for intelligent authentication in 5G and beyond wireless networks. Similarly, the work in [46] investigated the ML contributions to satisfying each target requirement of the 5G network, particularly the 5G use cases. The research showed that ML can be utilized as a solution mechanism to solve some of the impending issues in the 5G-and-beyond network.

Furthermore, the work [53] presented the AI- and ML-driven applications for the security of the 5G network, their challenges as well as future research recommendations.

From the brief review, it can be seen that both AI and ML have been proven to be effective solutions to the security issues in wireless communication, particularly in IoT, WSNs, and mobile communication systems, against adversaries such as eavesdropping attacks, DDoS attacks, intrusion attacks, and jamming attacks. It was also seen that these learning-based techniques are energy-efficient while ensuring confidentiality without security overheads and can be applied to the security of future wireless communication systems. However, there are gaps to fill in this research area.

2.7 Security issues and challenges in future wireless communication systems

The future wireless communication system, that is, the 5G-and-beyond network or simply 6G wireless network, is faced with different security threats and attacks. This section presents these issues and challenges as well as proffers further research directions.

The 5G wireless network addressed the limitations of the previous network generations by facilitating increased reliability, ultra-low latency, connection density, increased throughput, network energy efficiency, and low IoT devices using the key application scenarios as discussed in Section 2.2. However, the 5G wireless network is also insecure due to the numerous applications and enabling technologies such as NFV, SDN, HetNet, M2M, and D2D communication, that it supports. They are vulnerable to different attacks including eavesdropping, MITM, impersonation, hijacking, malware, and sniffing attacks to mention a few. This motivated the interest of researchers and other stakeholders who have tried to propose algorithms to defend against these attacks. This include the use of the PLS technique to protect against eavesdropping attacks instead of cryptographic algorithms which are not efficient. However, there are still gaps to fill in this research area.

A glimpse into the future presents the 6G wireless network which promises to have a radio latency as low as 0.1 ms, one-tenth of the 5G network, and supports both underwater and space communication [65]. There is a plethora of advantages that the 6G wireless network has over the existing 5G network. Figure 2.8 shows the emerging use cases for the 6G wireless network.

The 6G network has numerous security issues and challenges as a result of the key-supported technologies and applications that are discussed as follows.

2.7.1 AI

AI, as discussed earlier, is a very broad field that has so many applications in various aspects of our daily lives. Specifically, AI is commonly regarded as one of the important technologies in the future wireless networks [65]. For a complete autonomous network, the 6G network depends on AI. Therefore, threats and attacks

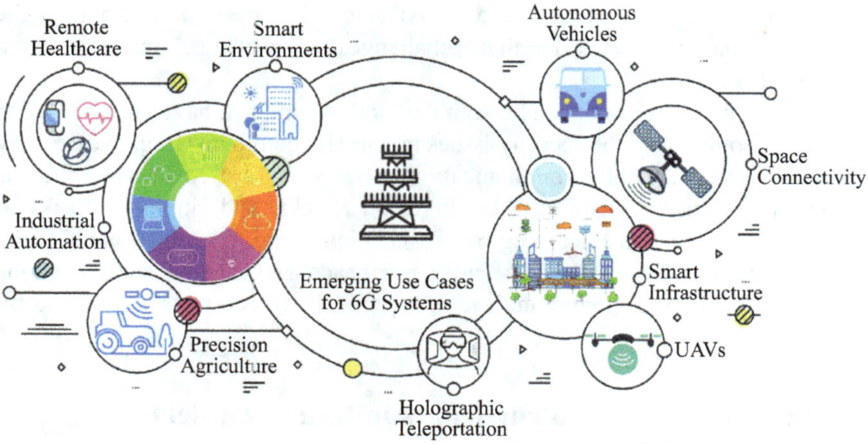

Figure 2.8 6G emerging use cases [3]

against AI systems, especially ML-based systems such as a poisonous attack, model manipulation, data manipulation, data injection, membership inference, and logic corruption attacks, would also affect 6G [66].

2.7.2 Molecular communication (MC)

MC is one of the key technologies in 6G network that enables information transmission using biochemical signals which is a solution to allow introduction of human in-body nano-networks under the scope of Internet of nano-things [67,68]. In MC, information is encoded using the concentration, release time, and type of molecules. Also, the propagation from the source to the destination can be passive, involving only molecular diffusion, or active, involving diffusion–advection and molecular motors [68]. Various threats and attacks have been identified at several levels of MC which include Jamming, flooding, and desynchronization attacks [66].

2.7.3 Quantum communication (QC)

QC is one of the technologies with promising applications in the 6G wireless network. Enhancement of security as well as the accurate delivery of transmitted data is one of most significant advantages of QC where if an attacker tries to interfere, interrupt, or modify the data, the quantum state would be affected; thus the receiver would be aware if interference has occurred [66]. QC offers a lot of solutions and is suitable for long-distance communication. However, it is not a universal remedy to all security threats and attacks as it is vulnerable to quantum cloning attacks and quantum collision attacks [69].

2.7.4 Blockchain

Blockchain technology is another technology that will unveil the potential of 6G network. It has a wide range of possible applications, some of which include spectrum sharing, distributed ledger technology, and network decentralization [66]. The blockchain technology is regarded as an essential technology for establishing trust in future wireless networks, and it has found applications in smart contracts, national ID systems, censorship resistance systems, bitcoins, healthcare systems, and other areas [70]. However, the blockchain technology also has several security issues, which include 51% attacks, forking issues, eclipse attacks, application bugs, short address, timestamp dependence, scalability issues, regulatory issues, and integration issue related to network access management, communication procedures, and authentication [71].

2.7.5 Terahertz (THz) technology

mmWave bands have been extensively used to satisfy the demands of the 5G network, but they are insufficient for the 6G network due to increased transmission rate demand. Additionally, the radio frequency is nearly exhausted and cannot be used for future wireless networks. As a result, the terahertz communication, which operates in the 0.1–10 THz band, has been proposed to be used for future wireless communication. Due to numerous benefits of the 6G wireless network which include supporting data rate of up to 100 Gbps or higher, minimized attenuation through certain materials and because of its narrow beam and short pulse duration, THz can be utilized to minimize the probability of eavesdropping. However, there are other security challenges in THz wireless communication as the technology is susceptible to access management attacks [69].

2.7.6 Visible light communication (VLC)

The VLC, a technology that utilizes visible spectrum light waves to transmit signals with wavelength range of 380–750 nm, is another emerging technology that can be employed to address the ever-increasing demand for wireless connectivity in future wireless communication systems [72]. The benefits of the technology include the use of unlicensed frequency with free-of-charge optical bandwidth, cost effectiveness, longevity, and energy-saving. VLC is thought to be more secure than other wireless technologies since its coverage is restricted to devices that are open to light, implying that it has fewer security weaknesses than radio frequency [72]. The VLC is also facing some security and privacy challenges which include eavesdropping attacks and jamming or data modification attacks [69]. Further reviews of the security issues and challenges in the 6G wireless network can be seen in [65,69].

2.8 Conclusions and recommendations

Wireless communication is one of the most successful technologies that has impacted our daily lives. This chapter has presented the historical description of wireless communication focusing on the mobile generations from 1G to 5G as well as the various security and privacy issues and vulnerabilities. The emerging wireless communication systems in the 5G wireless network have also been discussed and the application of AI and ML to wireless security design has been reviewed.

The various security issues and challenges in the 6G wireless network based on the key technologies have been briefly reviewed. The review on the security issues and challenges has shown that both 5G and 6G are vulnerable to threats and attacks. Although some key highlighted technologies such as AI, THz, and QC have been utilized to defend against some security threats and attacks such as eavesdropping attacks, however, these technologies are also vulnerable to several attacks.

In the future work, further research issues which can assist the global communities to achieve improved security for the 5G wireless network and beyond have been identified and elaborated. Some efforts on potential solutions to these several attacks are yet to meet the requirements of future wireless communication networks as the current security protocols are limited in terms of efficiency, overhead information, robustness, and would be unable to handle massive connections and large amount of data as the authentication mechanisms requires a longer period to complete preliminary authentication. Further research especially in the following areas is therefore recommended:

1. The development of novel delay-sensitive and robust protocols for security and privacy against different attacks across all the OSI layers of the 5G-and-beyond network.
2. The modification and optimization of existing security protocols to meet the fast pace requirement of the 5G-and-beyond wireless network as well as dynamic nature of the attacks across all the OSI layers.

References

[1] C. Azad, S. Agrawal, T. Medininagar, and V. K. Jha, "Security, privacy and accountability in wireless network: a review," *Int. J. Eng. Res. Technol.*, vol. 2, no. 7, pp. 399–408, 2013.

[2] S. B. Sadkhan and N. A. Abbas, "Privacy and security of wireless communication networks," in *Mobile Networks and Cloud Computing Convergence for Progressive Services and Applications*, IGI Global, Pennsylvania, PA, 2013, pp. 58–78. doi: 10.4018/978-1-4666-4781-7.ch004.

[3] I. F. Akyildiz, A. Kak, and S. Nie, "6G and beyond: the future of wireless communications systems," *IEEE Access*, vol. 8, pp. 133995–134030, 2020, doi: 10.1109/ACCESS.2020.3010896.

[4] K. R. Rao, S. N. T. Rao, and P. C. Reddy, "Wireless communication security and privacy issues and challenges," *Int. J. Comput. Sci. Inf. Secur.*, vol. 15, no. 7, pp. 202–209, 2017.

[5] R. Chen, C. Li, S. Yan, R. Malaney, and J. Yuan, "Physical layer security for ultra-reliable and low-latency communications," *IEEE Wirel. Commun.*, vol. 26, no. 5, pp. 6–11, 2019, doi: 10.1109/MWC.001.1900051.

[6] A. Maitra, "Early history of wireless communications," in *Souvenir of 33rd Annual Convention of Radio Physics and Electronics Association, University of Calcutta*, 2003.

[7] M. H. Mahmud, "Cellular mobile technologies (1G to 5G) and massive MIMO," *Int. J. Sci. Res.*, vol. 8, no. 7, pp. 929–937, 2019. Available: https://www.researchgate.net/profile/Md-Mahmud-22/publication/349493734_Cellular_Mobile_Technologies_1G_to_5G_and_Massive_MIMO/links/608c112192851c490fa9c763/Cellular-Mobile-Technologies-1G-to-5G-and-Massive-MIMO.pdf

[8] S. Patel, V. Shah, and M. Kansara, "Comparative study of 2G, 3G and 4G," *Int. J. Sci. Res. Comput. Sci. Eng. Inf. Technol.*, vol. 3, no. 3, pp. 1962–1964, 2018.

[9] M. Arshad, A. Farooq, and A. Shah, "Evolution and development towards 4th generation (4G) mobile communication systems," *J. Am. Sci.*, vol. 6, no. 12, pp. 6, 2010.

[10] N. Rawat, "Future and challenges of 4G wireless technology," *Int. J. Sci. Eng. Res.*, vol. 3, no. 12, pp. 1–7, 2012. Available: http://www.ijser.org/researchpaper%5CFuture-and-Challenges-of-4G-Wireless-Technology.pdf

[11] B. S. Usmonov, R. O. Sattorivich, and U. A. Rustamov, "5g Technology evolution," in *International Conference on Information Science and Communications Technologies: Applications, Trends and Opportunities, ICISCT 2019*, 2019, pp. 1–5. doi: 10.1109/ICISCT47635.2019.9011957.

[12] M. Attaran, "The impact of 5G on the evolution of intelligent automation and industry digitization," *J. Ambient Intell. Humaniz. Comput.*, vol. 2021, pp. 1-17, 2021, doi: 10.1007/s12652-020-02521-x.

[13] Y. Zou, J. Zhu, X. Wang, and L. Hanzo, "A survey on wireless security: technical challenges, recent advances, and future trends," *Proc. IEEE*, vol. 104, no. 9, pp. 1727–1765, 2016, doi: 10.1109/JPROC.2016.2558521.

[14] H. N. Dai, H. Wang, H. Xiao, X. Li, and Q. Wang, "On eavesdropping attacks in wireless networks," in *Proceedings – 19th IEEE International Conference on Computational Science and Engineering, 14th IEEE International Conference on Embedded and Ubiquitous Computing and 15th International Symposium on Distributed Computing and Applications to Business*, Engi, 2017, pp. 138–141, doi: 10.1109/CSE-EUC-DCABES.2016.173.

[15] A. Mpitziopoulos, D. Gavalas, C. Konstantopoulos, and G. Pantziou, "A survey on jamming attacks and countermeasures in WSNs," *IEEE Commun. Surv. Tutorials*, vol. 11, no. 4, pp. 42–56, 2009, doi: 10.1109/SURV.2009.090404.

[16] Y. Arjoune, F. Salahdine, M. S. Islam, E. Ghribi, and N. Kaabouch, "A novel jamming attacks detection approach based on machine learning for wireless communication," in *International Conference on Information Networking*, January 2020, vol. 2020, pp. 459–464, doi: 10.1109/ICOIN48656.2020.9016462.

[17] P. Sinha, V. K. Jha, A. K. Rai, and B. Bhushan, "Security vulnerabilities, attacks and countermeasures in wireless sensor networks at various layers of OSI reference model: a survey," in *Proceedings of IEEE International Conference on Signal Processing and Communication, ICSPC 2017*, January 2018, vol. 2018, pp. 288–293, doi: 10.1109/CSPC.2017.8305855.

[18] G. K. Ijemaru, I. A. Adeyanju, K. O. Olusuyi, T. J. Ofusori, E. T. Ngharamike, and A. A. Sobowale, "Security challenges of wireless

communications networks: a survey," *Int. J. Appl. Eng. Res.*, vol. 13, no. 8, pp. 5680–5692, 2018, [Online]. Available: http://www.ripublication.com

[19] E. Sula, "A review of network layer and transport layer attacks on wireless networks.," *Int. J. Mod. Eng. Res.*, vol. 8, no. 12, pp. 23–27, 2018.

[20] X. Hu and Z. M. Mao, "Accurate real-time identification of IP prefix hijacking," *Proceedings – IEEE Symposium on Security and Privacy*, 2007, pp. 3–17, doi: 10.1109/SP.2007.7.

[21] X. Wang and W. Zhang, "Cross-site scripting attacks procedure and prevention strategies," in *MATEC Web Conference*, 2016, vol. 61, p. 3001, doi: 10.1051/matecconf/20166103001.

[22] F. Njoroge and L. Kamau, "A survey of cryptographic methods in mobile network technologies from 1G to 4G," *Jomo Kenyatta Univ. Agric. Technol.*, vol. 10, no. 4, pp. 1–6, 2018.

[23] S. Sullivan, A. Brighente, S. A. P. Kumar, and M. Conti, "5G security challenges and solutions: a review by OSI layers," *IEEE Access*, vol. 9, pp. 116294–116314, 2021, doi: 10.1109/ACCESS.2021.3105396.

[24] I. Ahmad, S. Shahabuddin, T. Kumar, J. Okwuibe, A. Gurtov, and M. Ylianttila, Security for 5G and beyond, *IEEE Commun. Surv. Tut.*, vol. 21, no. 4, pp. 3682–3722, 2019, doi:10.1109/COMST.2019.2916180.

[25] S. Gindraux, "From 2 G to 3G: a guide to mobile security," in *Third International Conference on 3G Mobile Communication Technologies*, no. 489. 2002, doi:10.1049/cp:20020410.

[26] S. Mavoungou, G. Kaddoum, M. Taha, and G. Matar, "Survey on threats and attacks on mobile networks," *IEEE Access*, vol. 4, pp. 4543–4572, 2016, doi: 10.1109/ACCESS.2016.2601009.

[27] K. A. Alezabi, F. Hashim, S. J. Hashim, and B. M. Ali, "An efficient authentication and key agreement protocol for 4G (LTE) networks," in *IEEE TENSYMP 2014 – 2014 IEEE Region 10 Symposium*, 2014, pp. 502–507, doi: 10.1109/tenconspring.2014.6863085.

[28] A. K. Jain, R. Acharya, S. Jakhar, and T. Mishra, "Fifth generation (5G) wireless technology 'Revolution in Telecommunication,'" in *Proceedings of the International Conference on Inventive Communication and Computational Technologies, ICICCT 2018*, 2018, pp. 1867–1872, doi: 10.1109/ICICCT.2018.8473011.

[29] A. Celik, J. Tetzner, K. Sinha, and J. Matta, "5G device-to-device communication security and multipath routing solutions," *Appl. Netw. Sci.*, vol. 4, no. 1, pp. 102, 2019, doi: 10.1007/s41109-019-0220-6.

[30] N. T. Le, M. A. Hossain, A. Islam, D. Y. Kim, Y. J. Choi, and Y. M. Jang, "Survey of promising technologies for 5g networks," *Mob. Inf. Syst.*, vol. 2016, p. 25, 2016, doi: 10.1155/2016/2676589.

[31] B. Han, V. Gopalakrishnan, L. Ji, and S. Lee, "Network function virtualization: challenges and opportunities for innovations," *IEEE Commun. Mag.*, vol. 53, no. 2, pp. 90–97, 2015, doi: 10.1109/MCOM.2015.7045396.

[32] A. A. Barakabitze, A. Ahmad, R. Mijumbi, and A. Hines, "5G network slicing using SDN and NFV: a survey of taxonomy, architectures and future

challenges," *Comput. Netw.*, vol. 167, p. 106984, 2020, doi: 10.1016/j.comnet.2019.106984.

[33] I. Ahmad, T. Kumar, M. Liyanage, J. Okwuibe, M. Ylianttila, and A. Gurtov, "5G security: analysis of threats and solutions," in *2017 IEEE Conference on Standards for Communications and Networking, CSCN 2017*, pp. 193–199, 2017, doi: 10.1109/CSCN.2017.8088621.

[34] N. Panwar, S. Sharma, and A. K. Singh, "A survey on 5G: the next generation of mobile communication," *Phys. Commun.*, vol. 18, pp. 64–84, 2016, doi: 10.1016/j.phycom.2015.10.006.

[35] S. A. Khwandah, J. P. Cosmas, P. I. Lazaridis, Z. D. Zaharis, and I. P. Chochliouros, "Massive MIMO systems for 5G communications," *Wirel. Pers. Commun.*, vol. 120, no. 3, pp. 2101–2115, 2021, doi: 10.1007/s11277-021-08550-9.

[36] S. I. Al-Qasrawi, "Proposed technologies for solving future 5G heterogeneous networks challenges," *Netw. Comput. Syst.*, vol. 7, no. 1, pp. 11–19, 2017.

[37] A. Sharma, V. Balasubramanian, and A. Jolfaei, "Security challenges and solutions for 5G HetNet," in *2020 IEEE 19th International Conference on Trust, Security and Privacy in Computing and Communications (TrustCom)*, 2020, doi:10.1109/TrustCom50675.2020.00177.

[38] M. M. Alsulami and N. Akkari, "The role of 5G wireless networks in the internet-of- things (IoT)," in *2018 1st International Conference on Computer Applications & Information Security (ICCAIS)*, 2018, pp. 1–8, doi: 10.1109/cais.2018.8471687.

[39] A. Ciuffoletti, "Low-cost IoT: a holistic approach," *J. Sens. Actuator Netw.*, vol. 7, no. 2, 2018, doi: 10.3390/jsan7020019.

[40] Vidhyotma and J. Singh, *Comparative Analysis of Existing Latest Microcontroller Development Boards*, vol. 545. Springer, Singapore, 2019, doi:10.1007/978-981-13-5802-9_88.

[41] A. S. Ismailov and Z. B. Jo'rayev, "Study of arduino microcontroller board," *Sci. J.*, vol. 3, no. 3, pp. 172–179, 2022.

[42] M. Agiwal, N. Saxena, and A. Roy, "Towards connected living: 5G enabled Internet of Things (IoT)," *IETE Tech. Rev. (Institution Electron. Telecommun. Eng. India)*, vol. 36, no. 2, pp. 190–202, 2019, doi: 10.1080/02564602.2018.1444516.

[43] M. Wazid, A. K. Das, S. Shetty, P. Gope, and J. J. P. C. Rodrigues, "Security in 5G-enabled Internet of Things communication: issues, challenges and future research roadmap," *IEEE Access*, vol. 9, pp. 4466–4489, 2021, doi: 10.1109/ACCESS.2020.3047895.

[44] H. Ji, S. Park, J. Yeo, Y. Kim, J. Lee, and B. Shim, "Ultra-reliable and low-latency communications in 5G downlink: physical layer aspects," *IEEE Wirel. Commun.*, vol. 25, no. 3, pp. 124–130, 2018, doi: 10.1109/MWC.2018.1700294.

[45] D. Feng, L. Lai, J. Luo, Y. Zhong, C. Zheng, and K. Ying, "Ultra-reliable and low-latency communications: applications, opportunities and

challenges," *Sci. China Inf. Sci.*, vol. 64, no. 2, p. 120301, 2021, doi: 10.1007/s11432-020-2852-1.

[46] M. E. Morocho-Cayamcela, H. Lee, and W. Lim, "Machine learning for 5G/B5G mobile and wireless communications: potential, limitations, and future directions," *IEEE Access*, vol. 7, pp. 137184–137206, 2019, doi: 10.1109/ACCESS.2019.2942390.

[47] A. Slalmi, H. Chaibi, A. Chehri, R. Saadane, G. Jeon, and N. Hakem, "On the ultra-reliable and low-latency communications for tactile internet in 5G era," *Procedia Comput. Sci.*, vol. 176, pp. 3853–3862, 2020, doi: https://doi.org/10.1016/j.procs.2020.09.003.

[48] H. Kim, *Design and Optimization for 5G Wireless Communications*, John Wiley & Sons, New York, NY, 2020, doi: 10.1002/9781119494492.

[49] P. Popovski, K. F. Trillingsgaard, O. Simeone, and G. Durisi, "5G wireless network slicing for eMBB, URLLC, and mMTC: a communication-theoretic view," *IEEE Access*, vol. 6, pp. 55765–55779, 2018, doi: 10.1109/ACCESS.2018.2872781.

[50] D. Abdullah and S. Ameen, "Enhanced mobile broadband (EMBB): a review," *J. Inf. Technol. Informatics*, vol. 1, no. 1, pp. 13–19, 2021, [Online]. Available: https://qabasjournals.com/index.php/jiti/article/download/24/36

[51] Q. Qiu, S. Liu, S. Xu, and S. Yu, "Study on security and privacy in 5G-enabled applications," *Wirel. Commun. Mob. Comput.*, vol. 2020, p. 8856683, 2020, doi: 10.1155/2020/8856683.

[52] C. Bockelmann, N. K. Pratas, G. Wunder, *et al*, "Towards massive connectivity support for scalable mMTC communications in 5G networks," *IEEE Access*, vol. 6, pp. 28969–28992, 2018, doi: 10.1109/ACCESS.2018.2837382.

[53] N. Haider, M. Z. Baig, and M. Imran, "Artificial intelligence and machine learning in 5G network security: opportunities, advantages, and future research trends," *CoRR*, vol. abs/2007.0, 2020. Available: http://arxiv.org/abs/2007.04490

[54] B. Upadhyaya, S. Sun, and B. Sikdar, "Machine learning-based jamming detection in wireless IoT networks," in *Proceedings – 2019 IEEE VTS Asia Pacific Wireless Communications Symposium, APWCS 2019*, 2019, pp. 1–5, doi: 10.1109/VTS-APWCS.2019.8851633.

[55] R. Doshi, N. Apthorpe, and N. Feamster, "Machine learning DDoS detection for consumer internet of things devices," in *Proceedings – 2018 IEEE Symposium on Security and Privacy Workshops, SPW 2018*, 2018, pp. 29–35, doi: 10.1109/SPW.2018.00013.

[56] T. Hong, C. Liu, and M. Kadoch, "Machine learning based antenna design for physical layer security in ambient backscatter communications," *Wirel. Commun. Mob. Comput.*, vol. 2019, p. 4870656, 2019, doi: 10.1155/2019/4870656.

[57] R. Khilar, K. Mariyappan, M. S. Christo, *et al*, "Artificial intelligence-based security protocols to resist attacks in internet of things," *Wirel. Commun. Mob. Comput.*, vol. 2022, pp. 1–10, 2022, doi: 10.1155/2022/1440538.

[58] T. Saba, K. Haseeb, A. A. Shah, A. Rehman, U. Tariq, and Z. Mehmood, "A machine-learning-based approach for autonomous IoT security," *IT Prof.*, vol. 23, no. 3, pp. 69–75, 2021, doi: 10.1109/MITP.2020.3031358.

[59] S. M. Tahsien, H. Karimipour, and P. Spachos, "Machine learning based solutions for security of Internet of Things (IoT): a survey," *J. Netw. Comput. Appl.*, vol. 161, pp. 102630, 2020, doi: 10.1016/j.jnca.2020.102630.

[60] S. Zaman, K. Alhazmi, M. A. Aseeri, *et al.*, "Security threats and artificial intelligence based countermeasures for internet of things networks: a comprehensive survey," *IEEE Access*, vol. 9, pp. 94668–94690, 2021, doi: 10.1109/ACCESS.2021.3089681.

[61] R. A. Alshinina and K. M. Elleithy, "A highly accurate deep learning based approach for developing wireless sensor network middleware," *IEEE Access*, vol. 6, pp. 29885–29898, 2018, doi: 10.1109/ACCESS.2018.2844255.

[62] S. M. Wa Umba, A. M. Abu-Mahfouz, T. D. Ramotsoela, and G. P. Hancke, "A review of artificial intelligence based intrusion detection for software-defined wireless sensor networks," in *IEEE International Symposium on Industrial Electronics*, June 2019, vol. 2019, pp. 1277–1282, doi: 10.1109/ISIE.2019.8781458.

[63] M. Mamdouh, M. A. I. Elrukhsi, and A. Khattab, "Securing the internet of things and wireless sensor networks via machine learning: a survey," in *2018 International Conference on Computer and Applications, ICCA 2018*, 2018, pp. 215–218, doi: 10.1109/COMAPP.2018.8460440.

[64] H. Fang, X. Wang, and S. Tomasin, "Machine learning for intelligent authentication in 5G and beyond wireless networks," *IEEE Wirel. Commun.*, vol. 26, no. 5, pp. 55–61, 2019, doi: 10.1109/MWC.001.1900054.

[65] M. Wang, T. Zhu, T. Zhang, J. Zhang, S. Yu, and W. Zhou, "Security and privacy in 6G networks: new areas and new challenges," *Digit. Commun. Netw.*, vol. 6, no. 3, pp. 281–291, 2020, doi: 10.1016/j.dcan.2020.07.003.

[66] Y. Siriwardhana, P. Porambage, M. Liyanage, and M. Ylianttila, "AI and 6G security: opportunities and challenges," in *2021 Joint European Conference on Networks and Communications & 6G Summit (EuCNC/6G Summit)*, 2021, doi:10.1109/EuCNC/6GSummit51104.2021.9482503.

[67] W. Guo, W. Guo, M. Abbaszadeh, L. Lin, *et al.*, "Molecular physical layer for 6G in wave-denied environments," *IEEE Commun. Mag.*, vol. 59, no. 5, pp. 33–39, 2021, doi: 10.1109/MCOM.001.2000958.

[68] W. Haselmayr, A. Springer, G. Fischer, *et al.*, "Integration of molecular communications into future generation wireless networks," in *1st 6G Wireless Summit. IEEE, Levi, Finland*, pp. 1–2, 2019, [Online]. Available: https://www.gov.uk/government/news/robots-to-fix-

[69] P. Porambage, G. Gur, D. P. Moya Osorio, M. Livanage, and M. Ylianttila, "6G security challenges and potential solutions," in *2021 Joint European Conference on Networks and Communications & 6G Summit (EuCNC/6G Summit)*, 2021, doi:10.1109/EuCNC/6GSummit51104.2021.9482609.

[70] G. Bindu, I. T. Joseph S, V. R. Kanakala, G. L. K. Niharika, and B. E. Raj, "Impact of blockchain technology in 6G network: a comprehensive survey,"

in *2022 International Conference on Inventive Computation Technologies (ICICT)*, 2022, doi:10.1109/ICICT54344.2022.9850463.

[71] M. R. Islam, M. M. Rahman, M. Mahmud, M. A. Rahman, M. H. S. Mohamad, and A. H. Embong, "A review on blockchain security issues and challenges," in *2021 IEEE 12th Control and System Graduate Research Colloquium (ICSGRC)*, 2021, doi:10.1109/ICSGRC53186.2021.9515276.

[72] S. Ariyanti and M. Suryanegara, "Visible light communication (VLC) for 6G technology: the potency and research challenges," in *Proceedings of the World Conference on Smart Trends in Systems, Security and Sustainability, WS4 2020*, 2020, pp. 490–493, doi: 10.1109/WorldS450073.2020.9210383.

Chapter 3

Artificial intelligence-enabled security systems for 6G wireless networks: algorithms, strategies, and applications

Joseph Bamidele Awotunde[1], Agbotiname Lucky Imoize[2], Emmanuel Abidemi Adeniyi[3], Muyideen AbdulRaheem[1], Idowu Dauda Oladipo[1], Rasheed Gbenga Jimoh[1] and Peace Busola Falola[3]

Abstract

With incredibly complex and diverse requirements, 6G is anticipated to support the extraordinary Internet of Things advances. To effectively satisfy a wide range of requirements, space–aerial–terrestrial–ocean, interconnected three-dimension networks are what 6G intended to utilize various slices. This made it possible by new technologies and paradigms to increase the system's flexibility and intelligence. Because 6G networks are getting more sophisticated, varied, and dynamic, it is incredibly challenging to accomplish good resource utilization, a seamless user experience, autonomous administration, and orchestration. In addition, big data processing methods, computing capacity, and rich data availability have all progressed. Hence, applying artificial intelligence (AI) to address complicated 6G network difficulties is natural. Therefore, this chapter comprehensively reviews AI-based enabled security for 6G wireless networks. The chapter begins by outlining the concept of an AI-enabled 6G system, the motivations behind integrating AI-enabled security systems into 6G, and state of the art in AI-based security systems models in the 6G network. It then fully explored how to apply AI-enabled security solutions to important 6G network concerns such as smart traffic regulation, cybersecurity, administration and orchestration, and network optimization are all examples of network optimization. Lastly, we highlight major outstanding issues to

[1]Department of Computer Science, Faculty of Communication and Information Sciences, University of Ilorin, Nigeria
[2]Department of Electrical and Electronics Engineering, Faculty of Engineering, University of Lagos, Nigeria
[3]Department of Computer Science, Precious Cornerstone University, Nigeria

spur continued research toward a 6G network that is sophisticated, efficient, and secured.

Keywords: 6G wireless communication technology; Artificial intelligence; Internet of Things; 3-dimensional network; Security; Privacy

3.1 Introduction

Worldwide technology and industrial transformation are advancing in the modern world. The increasing use of the latest generation current expertise, such as the Internet of Things (IoT), artificial intelligence (AI), blockchain technology, extended reality (XR), augmented reality (AR), and virtual reality (VR), has resulted in the creation of the 6G technological system [1]. The growth of 6G will greatly influence communications cleverness progress, building on the foundation of 5G [2]. This consists of deep connectedness, holographic connectivity, intelligent connectivity, and universal connectivity.

Without being constrained by current network models or technology, 6G will investigate novel communication techniques. It incorporates compatible new ideas, architectures, protocols, and fresh approaches to support current and next generation challenges. The 6G system specifically entails smart connectedness, deep interconnection, holographic, and pervasive connectivity [3]. All communication system intelligence is reflected in intelligent connectivity: the intelligence of the connected object, the intelligence of the network components and architecture (the terminal device), and the data transmitted to enable the smart facility. Deep sensing, knowledge, and thinking are all correlated with deep connection. Holographic communication is one of the traits of holographic connectedness (and whenever), continuous, high-fidelity coverage of AR, VR, and XR. A link with ubiquitous connectivity covers all surfaces and the space in several dimensions [4,5]. Local, mobile cellular, ocean, satellite, and other undefinable networks will all be merged to form the 6G network [6].

Even though 5G commercialization remains in its early phases, relevant technological elements still need to be improved. Additionally, issues involving the implementation of the IoT and vertical industries should be considered. It must also proactively anticipate the communication requirements of the coming information society and begin researching the concept and technology for the upcoming mobile communication system [7]. A 6G wireless service infrastructure must give incredibly high speed, increasing capacity, non-proximity, computer disaster forecasting, VR, and persuasive medication to enable new applications. Given the prior control of mobile networks primarily based on the present 5G structure, the first 6G links leverage the features of 5G. For example, the increase in permissible frequency ranges improved the layout of a decentralized system and altered how we play and do our activities [8,9]. Furthermore, most audiences will likely be impacted by data around 2030, enabling instantaneous and unrestricted wireless connectivity [10]. Hence, 6G should advance present wireless expertise and achieve scheme enforcement.

Conventional AI methods, particularly machine learning (ML) and deep learning (DL), have gained traction as a result of successful deployments in sectors such as computer vision, automated speech synthesis, and natural-language processing (NLP) have recently attracted significant attention from academia and business [11]. Numerous academics aim to integrate AI into mobile network systems in response to the enormous success. As an indication, ML/DL has been applied to radio interfaces, such as the physical layer and upper layers of cognitive radio, resource management [12], link adaptation [13], and modulation mode identification [14,15].

Implementing ML at the physical layer can make radio parameterization, resource management, and interference reduction easier. A thorough examination of AI-based models in wireless networks was conducted, highlighting how well suited conventional mathematical model-based design methodologies are too data-driven AI-based methods. Additionally, AI-based models have increased network intelligence and can lessen the difficulty of managing and optimizing networks [16,17]. Contrary to the widely used traditional optimization techniques in message systems, AI-based solutions can automatically get better at what they do through data-based training. According to earlier research, ML technologies can strengthen mobile networks from a variety of angles, including network optimization [18], reasoning radio systems [19], and sensor networks [20,21].

The works recently made clear the idea of AI-based models allowing a 6G system [22]. However, neither of the two studies comprehensively analyzes the common network problems that AI can solve. Furthermore, most available research does not provide precise directions on how, when, and where to use various AI approaches to tackle typical mobile system challenges. No comprehensive study explains how to apply AI-based frameworks to boost mobile networks and attain 6G ambitions. Moreover, various review studies have not comprehensively explored the problems of AI-based models enabled by 6G technology. Therefore, this study examines how AI is being applied to mobile carriers as they progress toward 6G. The study attempt to evaluate the importance of launching concept and technology development for the 6G mobile communication technology straightaway.

3.1.1 Contribution

The study has the following contributions:

(i) The present study provides the latest and cutting-edge information on AI integration-based models enabled with 6G technology.
(ii) The study examines the benefits and drawbacks of common AI-based strategies to enhance network performance by analyzing the most recent literature and resolving numerous complicated network issues in dynamic situations.
(iii) The study concludes by identifying AI-based security and privacy challenges in 6G wireless communication technology.
(iv) The future directions in AI-based models enabled 6G mobile communication was also discussed.

3.1.2 Chapter organization

Section 3.2 presents an overview of the 6G of the wireless communication network. Section 3.3 discusses the security and privacy issues with 6G wireless communication and prospective attacks. Section 3.4 presents AI-based security and privacy for 6G wireless communication technology. Section 3.5 discusses the future directions of AI-based security and privacy for 6G wireless communication technology, and Section 3.6 finishes the chapter by discussing research gaps.

3.2 Overview of 6G technology

The primary forces behind 6G come from more than just the difficulties and performance constraints that 5G brings but also from the fundamental change brought on by technology and the ongoing development of wireless networks. Fundamental 6G necessities created by industry revolutions and intelligent transportation networks include mobile ultra-broadband (uMUB) is prevalent, as are ultrahigh-speed with low latency connectivity (uHSLLC) and extreme high traffic capacity carriers (uHDD). The existing 5G network is facing significant difficulties as a result of the network features of virtual world system services, such as the mixing of virtual and real worlds and real-time interaction. Research on generic information theory and tailored transmission technologies must be done to fulfill these 6G application requirements and idea-driven network technology founded on underlying concepts and enabling tools.

Emerging telecommunication, sensing, and computing end-to-end codesign are necessary for uMUB, uHSLLC, and uHDD services. They stimulate the fusion of photonics and AI, giving rise to two 6G-enabling technology contenders: cognitive radio based on photonics and computational holographic radio. There are now two potential 6G system architectures: (1) the full-spectral, all-photonic, integrated, and multipurpose radio access networks (RAN) and (2) laser-millimeter wave (mm-wave) converging at 100 Gb/s, hyperspectral, space-terrestrial integrated network. Although 5G was launched commercially around the end of 2019, several nations and organizations are conducting 6G research at the moment. For example, the Telecommunication Standardization Union launched the Network 2030 focus group in July 2018 to explore the advancement of techniques and plans for 2030 and beyond. New holographic multimedia, operations, system architectures, and Internet Protocol (IP) are being developed for some of the 6G concepts proposed by Network 2030 [23,24].

5G has issues in a variety of areas. While URLLC is supported by the 5G wireless technology, short packet, sensing-based URLLC functions, which are a disadvantage, restrict the provision of elevated, reduced solutions at huge data volumes speeds like in AR, XR, and VR. Applications for the IoT will need to converge on functions for communication, sensing, control, and computing, which 5G has generally failed to consider. A recent study by Rethink Technology Research of 74 mobile providers found that the unpredictability of the pricing and management of the virtual network function is one of the barriers to the commercial implementation of superior technologies. The price of fresh development in RAN hardware resources, a lack of faith in the platform's durability, and so on.

More so than many industry analysts, the implementation of virtualized RAN (vRAN), as anticipated a few years ago, is significantly later than other networking devices. The primary challenges to widespread vRAN deployment have been vendor hesitation and the fronthaul problem. Beginning with 4G, supervisors sought to link the baseline and distant radio units. However, the transition from the traditional upgraded radio broadcast platform appears to fall well short of this expectation goal [25]. Therefore, a new 6G network architecture is necessary to satisfy the aims of low latency with high data rates, connectivity, sensor convergence, and open engagements.

From 2030 forward, the health business will be dominated by the promised 6G communication technology. It will rule a variety of industries in addition to the health industry. Healthcare is one of the many industries that 6G is predicted to transform. The future of healthcare will be AI-powered and relies on 6G connectivity technology, transforming how we view lifestyle. The main obstacles to health care today are time and distance, which 6G can remove. Additionally, 6G will demonstrate its potential as a game-changing medical technology.

The authors in [26,27] reveal the problems and difficulties with 6G connectivity technology. 6G communication technology has already been launched in several nations for timely implementation. First, the 6G initiative began in Finland in 2018 [28]. Second, the United States announced the 6G plan in 2019 in South Korea and China [29]. Third, Japan started a 2020 6G innovation effort [30]. Numerous algorithms have also been created for 6G [31]. Launching the 6G project is now crucial to prevent slipping below competitors' nations [32]. B5G, on the other hand, has yet to be created, and 5G technological advancement is yet to be developed and completely adopted internationally. To alter current lifestyles, society, and industry, 5G and B5G will have several downsides. For instance, the decreased data rate prevents it from allowing holographic communication. Therefore, now is the greatest time to think about the possible applications of 6G communications technologies in the future.

Sixth-sense communication is the foundation of 6G communication. The technology will be three-dimensional, especially in terms of time, space, and frequency. A communication system powered by AI-based models will be 6G. High data rate (about 1 Tbps), high working recurrence (around 1 THz), short start-to-finish delay (around 1 ms), and low latency are requirements for 6G communication technology, and consistently good quality (10–9), high portability (1,000 km/h), and a frequency of 300 μm [33]. Additionally, enhanced augmented simulation and holographic communication will benefit intelligent communications networks.

With the aid of developing innovation, 6G will provide 3D forms of assistance; blockchain, AI, distributed computing, and edge innovation, for instance [34]. The 6G communications system will be all-encompassing and interconnected. Using device-to-device, LEO, and satellite technology connectivity, 6G will provide further and more widespread integration [35]. The five major designing future 6G networks technology under spectrum management is carrier aggregation (CA), CR, small cell, high-spectrum access, and M-MIMO [7], as shown in Figure 3.1.

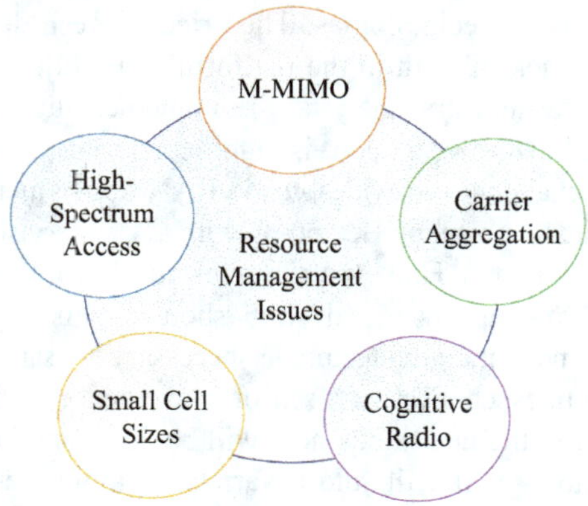

Figure 3.1 Challenges with spectrum management

The communication corporation will receive a combined calculation, route, and detection from 6G. 6G will handle the privacy, secrecy, and preservation of the huge amounts of data created in the safety domain by billions of smart technologies. The phrase "smart devices" will be replaced with "smart devices." Intelligent gadgets require rapid URLLC connections. The operational speed is 1 THz, the data rate is 1 Tbps, the frequency is 300 m, and the flexibility range is 1,000 km are essential components of 6G technology [32]. For time, frequency, space, and oscillation, the 6G configuration is 3D. For 6G communication, the end-to-end latency, radio-only delay, and processing delay are each ≤ 1 ms, ≤ 10 ns, and ≤ 10 ns, respectively [36]. Since AI drives 6G communication technology, it necessitates massively broad bandwidth machines (mBBMT), massively low latency machines (mLLMT), and broad mobile bandwidth and low latency (MBBLL) [37]. Ten developments in 6G communication are the main emphasis of the authors in [38]. The authors of [39] presented their concept for AI in 6G communication. The authors of [24] similarly concentrate on further-enhanced portable broadband (FeMBB), exceptionally trustworthy and low-latency connections (ERLLC), ultra-massive machine-type connectivity (umMTC), lengthy and elevated communication systems, and very minimal connectivity.

3.2.1 6G technology requirements

The peak data speed, data transfer rate received by users, region traffic availability, and frequency and energy efficiency are the main performance parameters for assessing 6G wireless networks (or traffic capacity in space), mobility, latency, and connectivity density [24]. The specific technological objectives comprise the following:

- An ultimate data throughput that is at least 1 Tb/s [40], or one hundred times faster than 5G. For some unique situations, such as THz wireless front- and backhaul (x-haul) [40], up to 10 Tb/s of peak data rate is anticipated.

- 1 Gb/s user-experienced data rate is 10 times quicker than 5G. A data rate of up to 10 Gb/s as perceived by the user is also anticipated in specific circumstances, for example, indoor hotspots.
- Over-the-air latency of 10–100 s and rapid mobility ($1,000 km/h). This will offer a satisfactory QoE in the circumstances such as hyper-HSR and aircraft systems.
- Ten times the connection rate of 5G. For instances such as hotspots, this will reach up to 10 devices/k 7 2 m and an area traffic ability of up to 1/Gb/s m^2.
- 5–10 times the measures of the effectiveness of 5G and 10–100 times the energy performance.

6G will provide expanded network features to address common scenarios and services in the advanced information environment of 2030. The Terahertz (THz) band, AI, 3D connectivity, unmanned aerial vehicles (UAV), and wireless transmission are all examples of optical wireless communication (OWC) and are the most critical innovations that will power 6G. Many fascinating technologies have the potential to make 6G possible. Here, the emphasis is on a few 6G technological solutions that could materially improve mobile carriers' businesses in terms of revenue generation and business expansion. To fulfill all of these demands and the long-term goal, the combination of societal needs and mobile communication systems would be the primary drivers surpassing present mobile communication as a result of the revolutionary change technologies and the advancements in technology that enable those demands. When taken as a whole, these elements present a compelling case for future wireless network borders. Cognition, reliability, flexibility, and protection are all attributes of terrestrial mobile networks and are not the only benefits of 6G mobile technology. They will enable the incorporation of satellite telecommunications to create a global mobile network, in keeping with the requirement for a genuinely international wireless network presence [41]. Therefore, several important solutions have been researched in terms of the 6G technology vision, specifications, criteria, and anticipated technology [42,43].

To achieve the 6G goals, move past 5G restrictions, and support new challenges, mobile communication systems need to be improved with cutting-edge features for 5G and beyond. The 6G system will reduce the 5G scheme lag by adding a unique set of technologies. Future mobile networks are also anticipated to evolve as a result of three planned characteristics. However, they would not be prepared for 6G. Examples of these characteristics include the Internet of Nano Things (IoNT), the Internet of Bio-Nano Things (IoBNT), and quantum connectivity [5,44]. Unfortunately, most recent articles also lack consideration of technologies exceeding 6G (such as IoNT, IoBNT, and quantum connectivity). The most recent research that examines advancements in several 6G domains is shown in Table 3.1.

Although various studies have thoroughly evaluated the domain 6G, despite the limited investigations that have been undertaken, little or no former study has comprehensively surveyed the security and privacy of advances in 6G technology research. As a complement to earlier research, this study systematically reviews the

Table 3.1 Research that examines advancements in 6G technology

Authors	Research area	Significant contributions and future directions for research
Nayak and Patgiri (2020) [27]	Key problems and constraints	The problems and difficulties that 6G technology may confront in the future are highlighted.
Chowdhury *et al.* (2020) [45]		A few challenging problems are introduced together with the anticipated 6G coming technology. Moreover, the prerequisites, prospective technologies, and projected applications are discussed.
Yaacoub and Alouini (2020) [46]	Networks for access, fronthaul, and backhaul	It conducted a survey of technology for connecting rural communities. Techniques for access/fronthaul and backhaul are also explored. The evaluated technologies' energy needs and cost effectiveness are also looked at.
Tomkos *et al.* (2020) [47]	Vision of 6G for edge computing	Investigations are conducted on the development in addition to the integration of 5G and IoT advancements toward 6G networks. Attempts to make the case that 6G must be human-centric, with safety, confidentiality, and privacy as top priorities components. A structured framework, necessary technology, and difficulties are described to accomplish this vision.
Giordani *et al.* (2020) [4]	Use-cases	Outlines several possible use case scenarios and important technological developments that have been identified as allowing 6G usage scenarios.
Chen *et al.* (2020) [48]	Literature review	This paper aims to emphasize the ways of solving 6G communications network range, scalability, data rate, and mobility challenges through a complete 6G debate based on an examination of SG advances.
Liu *et al.* (2020) [49]	Literature review	The article offers a logical mobile network architecture while defining vision, novel application situations, and critical functionality criteria.
Gui *et al.* (2020) [37]	Literature review	An outline of the main 6G concerns, such as essential services, use cases, demands, novel approaches, designs, and typical application situations, obstacles, and future paths.
Saad *et al.* (2019) [5]	Literature review	Regarding use applications, technological breakthroughs, service kinds, requirements, and the naming of specific supporting technologies and open

(Continues)

Table 3.1 (Continued)

Authors	Research area	Significant contributions and future directions for research
		research, the 6G vision is detailed below problems.
Tariq *et al.* (2020) [50]	Literature Review	The article speculates on potential innovative solutions that might offer the enormous breakthroughs required to enable 6G, expanding the 5G concept to more ambitious potential trends.
Bariah *et al.* (2020) [51]	Systematic review	A detailed analysis of the 6G requirements, visions, obstacles, and unresolved research questions, highlighting seven (7) disruptive technologies: the Tactile Internet, back-scatter optical wireless connectivity, customizable conceptual, drone-based connectivity, millimeter wave (mmWave) communications, and others.
Wang *et al.* (2020) [52]	Channel	The article examines measurements and models for 6G wireless channels that span all signal frequencies, applications, and global coverage.
Chi *et al.* (2020) [53]	VLC	The authors talk about the potential and difficulties of VLC in 6G, its improvements in communications at high speeds, and research findings pursuits like novel components and tools, sophisticated modulation, and undersea VLC.
Strinati *et al.*, (2019) [44]; Chi *et al.* (2020) [53]; Kishk *et al.* (2020) [54]; Jiang *et al.* (2021) [55]	UAV	This study offers a network architecture centered on tethering UAVs and focuses on the benefits of UAVs in increasing network capacity and coverage.
Jiang *et al.* (2021) [55]	Literature review	An in-depth examination of the drivers, use cases, use situations, needs, and key performance factors (KPIs), architectural and infrastructure networks for the 6G system is offered.
De Alwis *et al.* (2021) [56]	Systematic review	A thorough analysis of current 6G developments is conducted.
Zhao *et al.* (2020) [57]	Systematic review	This survey offers a thorough 6G wireless technology overview by presenting needs, features, important technologies, difficulties, and applications.
De Lima *et al.* (2021) [58]	Systematic review	This research engrossed convergent 6G communication, navigation, and sensor technology, finding key scientific

(Continues)

Table 3.1 (*Continued*)

Authors	Research area	Significant contributions and future directions for research
		drivers, attempting to understand fundamental concerns and probable impending issues, and proposing viable solutions.
Mahmoud *et al.* (2021) [59]	Systematic review	This article covers the system needs, possible trends, innovations, applications, deployments, and recent research developments.
Abdel Hakeem *et al.* (2022) [60]	Review	The examination of the significant issues and critical circumstances with 6G information security, confidentiality, and trust issues is presented in this paper.
El Mettiti and Oumsis (2022) [61]	Systematic review	This report undertakes a relevant study in terms of the idea, needs, and anticipated application situations for the 6G network. Additionally, it defines the combination of networks in space, the air, the ground, and the sea. Additionally, it highlights and assesses the most significant prospective critical technologies needed for 6G in the future.
Alraih *et al.* (2022) [62]	Review	This paper discusses the 6G mobile technology vision, objectives, networking technologies, and obstacles. It also contrasts essential technology, allowing communication methods and 5G and 6G services.
Banafaa *et al.* (2022) [63]	Systematic review	This broadly classified into the following analyses: the recent news in 6G investigation offers a wide conversation in regard to necessities, goals, evaluated ideas, key performance indicators (KPIs), structures, software, difficulties, and prospects for facilitating technological advances that 6G systems will convey into our lives.
Imoize *et al.* (2021) [3]	Systematic review	A thorough examination of UAVs and CubeSats and how they will support 6G requirements is offered. Clear descriptions of 6G applications are provided, along with a comparison of the restrictions of the current 5G network. Finally, there is a clear outline of Open Research Issues and Future Research Directions.

latest 6G safety and privacy research, delivers AI-based 6G security and wireless confidentiality communication, and gives future directions of AI-based models for supporting innovative technologies for 6G networks.

3.3 The security and privacy issues with 6G wireless communication and prospective attacks

A few fundamental innovations have already been demonstrated to be effective in vital parts of the 6G technology. They support 6G systems that offer high security, low latency resilience, and effective communication services. However, the vulnerabilities to privacy and security are higher with the majority of new 6G technologies. This segment analyses the basic refuge and secrecy in 6G wireless technologies and the specifications for these technologies' security and privacy. Business and society's reliance on networks and Information Technology (IT) will grow during the 6G era. A continuation of what we see in 5G, the significance of channels and information technology in national security is expanding. In 6G networks, the transition to edge and cloud-native architectures is anticipated to continue.

Additionally, the preparation for the design of the 6G network security is necessary. Security automation raises new issues: Can AI be utilized to create safer systems? But with potentially more lethal assaults. Techniques for physical layer security can also be effective first lines of defense for safeguarding less explored network parts.

There is no clear means to tell when connected, deidentified datasets have crossed the line into becoming individually identifiable. This is a significant, unresolved issue for many digital technologies in various industries. Without explicit measurements of the amount of personal information, courts in various world regions decide if privacy is being violated. At the same time, businesses are looking for new methods to profit from the use of private data. We could consider solutions for distributed ledger technology, blockchain, and various privacy strategies.

Since the introduction of 2G for digital mobile communication, mobile networks rely on a SIM card, which stands for Subscriber Identification Module, to physically store symmetric keys. International standards for encryption techniques were adopted, and additional cryptographic mechanisms for mutual authentication were implemented. However, the SIM card remains the foundation of the 5G security model [16]. Although SIM cards shrank (currently at "nano" size), they still need to be connected to devices, which restricts their suitability for IoT. The introduction of eSIMs somewhat resolves this problem, but physical size problems remain. Future gadgets may include the currently under development iSIM as part of their System-on-Chip, despite the resistance from operators who fear a loss of control.

Traditional SIM cards use symmetric key encryption that has been successfully scaled to billions of users. The IoT, privacy, bogus base stations, and Internet verification are a few of its drawbacks. Will the transition transitioning from symmetrical crypto to asymmetrical shared key be fundamental? Such a large-scale deployment has never before occurred. 5G intends to provide authentication via a public-key infrastructure in addition to SIM (PKI). A collection of microservices interacting over HTTPS make up the central component of 5G. TLS, which uses Elliptic Curve Encryption (ECC), enables such communications' authentication, secrecy, and consistency. However, as this has not been implemented, it can wait until 6G. Figure 3.2 displays the analysis of the 6G networks security threat background.

THz technology and molecular communication both support intelligent radio. As it relates to communication, authentication, and encryption, molecular communication technology raises security and privacy concerns, whereas malevolent activity and weak authentication security are particularly problematic for THz technology. Blockchain and quantum communication technologies are related to distributed AI and intelligent radio. The major areas of privacy and protection include authorization, access control, data transport, and cryptography concerns in this situation. Additionally, 6G applications have certain weaknesses. AI is generally used by linked robotics and autonomous systems, as well as visible range transmission (VLC) technologies, where issues with data transmission, encryption, and criminal activity can arise. Molecular communication technology is used in multimodal XR applications, and quantum information technologies using THz equipment imply that they are data transmission leaks, malicious behavior, and access control risks. The multisensory XR application techniques are also used in wireless brain–computer

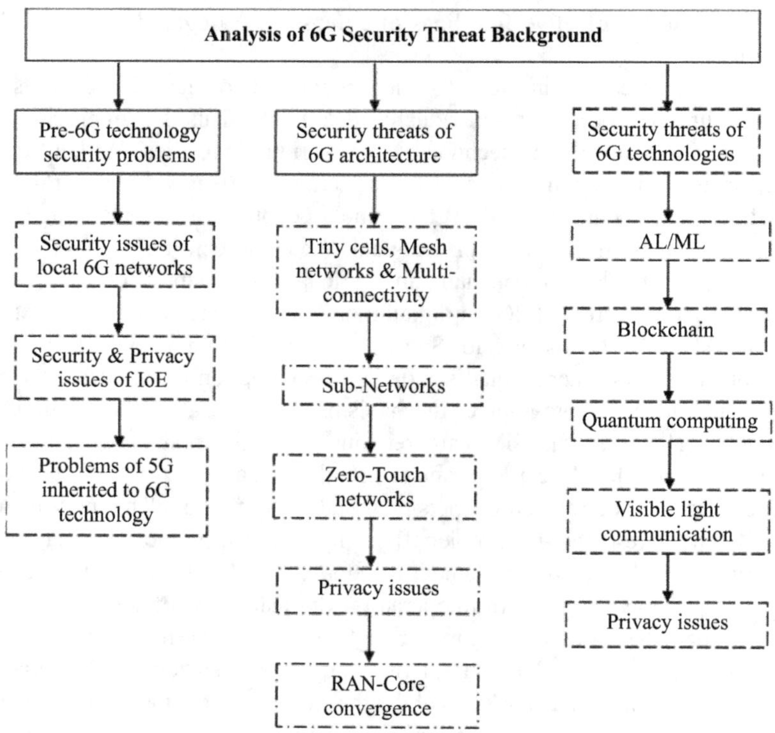

Figure 3.2 Overview of the security threat scenario for 6G

communications but have particular security and privacy problems. The most serious issues are malevolent behavior and cryptography. Distributed ledger technology and blockchain are the final 6G network applications (built primarily on the blockchain) and are generally secure. However, they might still be the object of bad actions. The five key security and privacy issues that can arise in these new domains are authentication, access control, malevolent behavior, encryption, and data transit.

As indicated in the preceding section, some critical innovations have already been shown to be beneficial in critical sections of the 6G network. They provide high dependability, low latency, secure, and effective transmission services for 6G networks [64]. But as was also said in the preceding section, these technologies raise fresh privacy and security issues. Table 3.2 provides a synopsis of the privacy and security issues related to important 6G wireless communication technology.

Table 3.2 An overview of the main 6G technology-related security and privacy issues

Technology	**Authors**	**Security and privacy concerns**	**Important technological contribution**
AI	Lovén *et al.* (2019) [65]	Access control	Processes for fine-grained control
	Dang *et al.* (2020) [29]	Malicious behavior	Find network abnormalities and send out early alerts
	Sattiraju *et al.* (2019) [66]	Authentication	A technique for unsupervised learning that could be applied to the verification procedure to strengthen the protection of the different layers
	Hong *et al.* (2019) [67]	Communication	An ML-based antenna design that might be applied to PHY layer communication to stop data leaks
	Nawaz *et al.* (2019) [42]	Encryption	Systems for quantum encryption and ML based
Molecular communication	Farsad *et al.* (2016) [68]	Malicious behavior	An enemy that interferes with molecular communication or its procedures
	Lu *et al.* (2015) [69]	Encryption	An encoding scheme capable of increasing data confidentiality and security
	Loscri *et al.* (2014) [70]	Authentication	Gives guidance for creating new authentication methods
Quantum communication	Nawaz *et al.* (2019) [42]	Encryption	Protection techniques for quantum encryption keys
	Hu *et al.* (2016) [71]	Communication	Various quantum communication techniques
Blockchain	Kiyomoto *et al.* (2017) [72]	Authentication	A novel theoretical framework for authorization of mobile services
	Kotobi and Bilen (2018) [73]	Access control	A technique for enhancing access protocols
	Ferraro *et al.* (2018) [74]	Communication	Hashing power is used to verify transactions.

(Continues)

Table 3.2 (*Continued*)

Technology	Authors	Security and privacy concerns	Important technological contribution
THz	Akyildiz *et al.* (2014) [75]	Authentication	The use of magnetic signatures for authentication
	Ma *et al.* (2018) [76]	Malicious behavior	A signal can still be intercepted by an observer even if it is sent via a narrow beam
VLC	Ucar *et al.* (2016) [77]	Communication	A communication method that uses a secure protocol
	Cho *et al.* (2019) [78]	Malicious behavior	Security can be compromised by compliant eavesdroppers

In conjunction with the network infrastructure and physical layers, AI is beneficial in various ways, such as network optimization, decentralized AI, resource management, and big data analysis. According to authors in [29], AI may be able to identify network abnormalities and offers early warning systems to boost 6G network security. According to authors in [47], dispersed and regulated AI can help a 6G network. Network security is further increased by the edge devices' lack of data-sharing requirements. Additionally, authors in [79] highlight that some ML based algorithms' use of data correlation may increase privacy leaks. In [80], the authors proposed various differential privacy-based techniques that might be useful for resolving some of the 6G privacy problems.

A common occurrence among living things with nanoscale structures is molecular communication [81]. Microscale and nanoscale technologies are becoming a common place due to the advancements made in the last 10 years in nanotechnology, bioengineering, and synthetic biology [68]. Additionally, the synthesis and transmission of a molecular transmission medium consume extremely little energy. Although these phenomena have been studied for a long time in biology, communication researchers have only begun to explore them. A particularly promising technique for 6G communications is molecular communication. Since it is still in its infancy, it is a multidisciplinary technique. The essential standard of molecular communication is communication transfer via biological signals. A portable molecular message method permits the source, according to [82], the linked nodes, and the receiver to communicate while traveling. But some communication, identification, and encryption methods already have security and privacy flaws [83].

Theoretically, the quantum transmission may provide total security, and given the necessary technological advancements, it ought to be ideal for long-distance correspondence. It provides a wide range of novel solutions and raises communication to a level above what is possible with conventional communications technologies [84]. However, the quantum message is not yet a solution to every safekeeping and secrecy problem. Long-distance quantum transmission is a promising technology substantial barrier despite significant advancements in the development of quantum cryptography for the quantum message. These issues include fiber weakening and procedure mistakes. Authors in [71] hypothesize that multiple alternative forms of quantum encryption and other approaches may be necessary to guarantee entirely secure quantum communications. The quantum public key, quantum information exchange, quantum protection through interaction, quantum teleportation, and quantum dense coding are all examples of quantum technologies are only a few examples.

3.4 AI-based security and privacy for 6G wireless communication technology

Due to the development of DL-based models and their application's simplicity, advancements in AI have accelerated over the past 10 years [85,86]. These developments, combined with AI's ability to analyze data and make decisions, have made it a common tool in a variety of industries, including wireless communication. The ability to link billions of heterogeneous systems to the network is one of the main goals of 6G. This requirement necessitates using sophisticated, ML-based techniques based on data in place of conventional mathematical models and algorithms [87]. It is possible to use transfer learning [88] and generating networks [89] in the absence of a large amount of training data.

Additionally, classical theories are less able to adapt optimally to non-linearities, defects, and real-time operating state variations [90], which are better addressed by DL and reinforcement learning techniques [91]. Hence, AI is considered one of the most important scientific solutions to address the unresolved issues with the 6G communication infrastructure. In this section, we summarize the research on the use of AI-based assisted 6G wireless communication security and privacy.

The 6G wireless transmission network is anticipated to provide interconnected global intelligent and autonomous solutions by utilizing AI expertise. FL stands for federated learning, recommended in particular to encourage the adoption of pervasive AI applications in 6G, where huge, distributed data is used to jointly train AI models, safeguard privacy, and conserve resources. However, given the volume of contacts between infrastructures or mobile devices in FL, there are numerous possible privacy and security hazards associated with AI. Authors in [92] investigate the security and privacy of AI-enabled 6G. The paper specifically presents the design of the space-air-ground-ocean integrated network (SAGOIN) and two distinct FL-based AI model building environments in 6G as the first AI-enabled 6G architecture. Additionally, we detail the challenges to privacy and security, including attacks that steal privacy, use low-quality local models, and cheat.

In contrast to every other technology anticipated to be utilized in 6G networks, AI is commonly believed to be one of the indispensable mechanisms of the upcoming network design. To say that AI has received a lot of responsiveness in the network space would be an understatement. And as a result of this focus, more and more fresh security and privacy concerns are surfacing [93]. Although AI in the 5G network is purportedly run in remote locations with access to enormous training data and potent but private computer hubs, the 6G network will increasingly prioritize AI [94]. Once more, the architectural layers that AI technology provides can be separated [95]: the physical layers, which comprise elements like network infrastructure and data lines, and the computing layers, among other things, software-defined networking, system purpose virtualization, and cloud computing, edge, and fog computing, among other.

THz spectrum, which has a substantially larger bandwidth, is being adopted by 6G by connecting billions of devices and nodes with a worldwide reach for terrestrial, oceanic, and space. Further densification and cloudification are needed for a hyperconnected society. Automated security using virtualization, ML, and security function softwarization will become necessary [96]. To remove obstacles to the security of 5G networks, both current and future, and to meet the requirements of 6G security, security solutions utilizing the current ideas of SDN and NFV must be improved dynamically with smart sensors. To enforce policies, and identify, contain, mitigate, and prevent threats or active assaults, intelligent security functions in containerized VNF boxes will watch over 6G traffic residing in gateways. They will scan the data using continuous DL on a packet/byte level.

Utilizing container technology for security operations results in higher usage rates, reduced storage needs, improved security, and quicker reboot times. Pods will be created from groups of containers containing many containers on a single system with security service features. By scaling up or down, availability is provided. To maintain and implement security functions, cloud computing innovations like edge and fog computing will be deployed (safety VNFs) as needed in various network perimeters by proactive decision-making utilizing ML. Using SDN, global resource visibility, event monitoring as a foundation, configurable APIs, and coordinated network security policies amongst several parties, end-to-end network security will be achieved using network abstractions. In an integrated context, 6G networks will reconcile the ideas of SDN, NFV, and AI to secure end-to-end network security as well as essential service [97].

With the use of ML and DL models, 6G is expected to be a very intelligent network that will enable intelligent and effective data monitoring, allocation of resources, and network monitoring [98]. The intelligent 6G can enable various AI uses, including unmanned factories, self-directed cars, intelligent healthcare, and smart homes. AI applications strive to create fully autonomous services to enhance industrial productivity and convenience for people. To complete these applications, a high-precision AI algorithm must be built in advance, which needs to be trained using a lot of data about the particular duty and goal. Conversely, the training of conventional AI models typically relies on a base station, for example, a cloud server, which raises many privacy and security issues. This is unsolvable to enable widespread and protected AI for 6G. Antagonists who gain access to the central node can specifically evaluate learning data from users' portable devices to infer their anonymity for prospective uses [92]. Additionally, because the central node must receive a substantial volume of raw data for processing and aggregation, the standard centralized AI model training takes a lot of resources, including processing power, bandwidth, and electricity. It makes it more difficult for 6G to integrate AI applications on a broad scale. 6G can make AI omnipresent if the personal details may be securely kept and utilized to train AI models. FL is recommended to be used as a distributed training usually forming for emerging AI in 6G [99].

In FL, distributed mobile devices are used to locally and cooperatively train the AI model, and the model is updated (such as weight, gradient, and others) and subsequently uploaded for convergence to the network nodes in place of raw data [100]. Smartphone anonymity may be securely protected, and the overhead for training AI systems is considerably reduced because only the AI parameter estimates need to be supplied. FL still faces some sophisticated privacy and security issues despite being able to take advantage of 6G AI service provisions [92].

First, the third-party-created shown is fascinating and might be taken advantage of by enemies who want to use AI model upgrades to infer mobile device private information illegally. Second, malicious actors may take control of nearby mobile devices and use them to deliver the central node updates for fake and forged AI models. As a result, FL regular operation is compromised, which causes the AI model to converge slowly or possibly not at all. Third, using local training models impacts scarce resources like electricity, processing, storage, and so on. Mobile gadgets are self-centered and could offer better training services (less training data samples, for instance) with adequate compensation, significantly reducing the AI model accuracy.

3.5 The future directions of AI-based security and privacy for 6G wireless communication technology

The use of AI in wireless networks has provided several issues. First, obtaining training data takes time and effort. Therefore, training data is seldom available, unlike in other disciplines like computer vision and NLP. These data demand processing and calculation resources, which raise the cost of communication [101]. The various data characteristics, with fluctuating values, pose a hurdle because it is difficult to operate with increased dimensionality [102]. Fourth, high computational complexity is required for DL-based systems, which may not be compatible with modern mobile phones [103,104]. Except for the intricacy, AI-based algorithms must be carefully constructed to minimize the number of computational resources needed on these devices [105]. A potential solution to the problem of constrained computational power and energy efficiency is quantum communication [106]. Accuracy and computational/energy needs present a trade-off dilemma when applying AI to the IoT [107]. Therefore, selecting the appropriate use cases for AI-based applications should take precedence over fine-tuning neural networks and network operations [108].

Cloud-based VR/AR deployment makes it more portable and accessible, but images must be compressed for 5G bandwidths. Therefore, real-time transmission of massive volumes of uncompressed images or films will have to delay till the 6G system is ready. In 6G technology, the realistic VR/AR experiences will be much greater. Numerous devices will be employed to assemble sensual data and offer an opinion to customers. Therefore, the XR in 6G networks is anticipated to combine conventional URLLC with improved Mobile BroadBand (eMBB). This is also denoted as Mobile Broad Bandwidth and Low Latency (MBBLL) [37]. Further research in this area will enhance the 6G wireless communication technology.

The second use of application for the 6G system is automation and automated vehicles. In 5G networks, self-sufficient driving is a crucial application. However, 6G networks and autonomous cars are insufficient; a more inclusive driverless system is necessary. A multi-dimensional network should be included in the autonomous system in addition to the driving procedure. Additionally, this system should include intelligence throughout the entire network and the integration of AI logic into the network architecture [109]. This would allow us to connect and control all internal components using AI automatically. Future work can look in this direction to explore how 6G technology can increase the potential of autonomous systems. Due to the limits of 5G networks, installing a self-directed drone scheme completely has not yet been conceivable. The full potential of these technologies might be realized with 6G networks. However, these systems are equally subject to some attacks.

The UAV communicates global AI model parameters to smartphones on the landscape in FL for the AI-enabled 6G for local model training. However, model parameters and testing datasets may be disclosed via portable device enemies when interacting with other mobile devices or infrastructural facilities. The opponents can deduce the AI model applications to get the prospective benefits. As a result, it is difficult to implement a trusted confined ideal exercise to stop the disclosure of AI model knowledge.

The aggregating UAV for FL AI-enabled 6G is not always accessible; mobile gadgets, for instance, are not inside any UAV range of operation. Hence, a network of mobile devices can work together to disseminate training of an AI model. However, it is simple for privacy to be compromised when local AI model updates are shared around mobile devices, where the adversary can determine how other mobile devices' training data is distributed by comparing their local model updates with their own. Therefore, it is important to discuss safeguarding mobile device privacy for FL in distributed networks.

Discussing how to protect mobile device privacy in distributed networks is crucial for FL. However, as the malicious attackers' actions are typically documented on a central server, replacing and removing the misconduct records is simple, causing the FL to hire the malicious attackers once more. To prevent hostile attackers from entering FL, it is crucial to record the inappropriate activities for finding immutably.

3.6 Conclusion and future directions

Many scholars are currently focused on 6G networks because the study time for 5G technology is coming to an end, and it will be adopted soon. Using the 6G network, network service will undoubtedly increase over prior generations. Therefore, this study looks into the security and privacy challenges of AI-based enabled 6G technology architectures. The study presents the most recent and cutting-edge research on integrating AI-based models enabled with 6G technology. The main benefits and drawbacks of common AI-based approaches to enhance 6G network technology were discussed. The future directions in AI-based models enabled 6G mobile communication were also discussed. Finally, the study identifies AI-based security and privacy challenges in 6G wireless communication technology. We anticipate that the literature reviews of this study will pique researchers' interest and stimulate more research on AI-powered 6G network security and privacy challenges.

Acknowledgment

The work of Agbotiname Lucky Imoize is supported in part by the Nigerian Petroleum Technology Development Fund (PTDF) and, in part, by the German Academic Exchange Service (DAAD) through the Nigerian–German Postgraduate Program under Grant 57473408.

References

[1] Lu, Y. and Zheng, X. (2020). 6G: a survey on technologies, scenarios, challenges, and related issues. *Journal of Industrial Information Integration*, 19, 100158.

[2] Obakhena, H. I., Imoize, A. L., Anyasi, F. I., and Kavitha, K. V. N. (2021). Application of cell-free massive MIMO in 5G and beyond 5G wireless networks: a survey. *Journal of Engineering and Applied Science*, 68(1), 1–41.

[3] Imoize, A. L., Adedeji, O., Tandiya, N., and Shetty, S. (2021). 6G enabled smart infrastructure for sustainable society: opportunities, challenges, and research roadmap. *Sensors*, 21(5), 1709.

[4] Giordani, M., Polese, M., Mezzavilla, M., Rangan, S., and Zorzi, M. (2020). Toward 6G networks: use cases and technologies. *IEEE Communications Magazine*, 58(3), 55–61.

[5] Saad, W., Bennis, M., and Chen, M. (2019). A vision of 6G wireless systems: applications, trends, technologies, and open research problems. *IEEE Network*, 34(3), 134–142.

[6] Zhao, Y., Zhao, J., Zhai, W., Sun, S., Niyato, D., and Lam, K. Y. (2021). A survey of 6G wireless communications: emerging technologies. In *Advances in Information and Communication: Proceedings of the 2021 Future of Information and Communication Conference (FICC)*, vol. 1 (pp. 150–170). Springer International Publishing, New York, NY.

[7] Qamar, F., Siddiqui, M. U. A., Hindia, M. N., Hassan, R., and Nguyen, Q. N. (2020). Issues, challenges, and research trends in spectrum management: a comprehensive overview and new vision for designing 6G networks. *Electronics*, 9(9), 1416.

[8] David, K. and Berndt, H. (2018). 6G vision and requirements: is there any need for beyond 5G? *IEEE Vehicular Technology Magazine*, 13(3), 72–80.

[9] Adeniyi, A. E., Abiodun, K. M., Awotunde, J. B., Olagunju, M., Ojo, O. S., and Edet, N. P. (2023). Implementation of a block cipher algorithm for

medical information security on cloud environment: using modified advanced encryption standard approach. In *Multimedia Tools and Applications* (pp. 1–15). Springer, New York, NY.
[10] Nawaz, F., Ibrahim, J., Muhammad, A. A., Junaid, M., Kousar, S., and Parveen, T. (2020). A review of vision and challenges of 6G technology. *International Journal of Advanced Computer Science and Applications*, 11(2), 643–649.
[11] Awotunde, J. B., Jimoh, R. G., Imoize, A. L., Abdulrazaq, A. T., Li, C. T., and Lee, C. C. (2022). An enhanced deep learning-based deepfake video detection and classification system. *Electronics*, 12(1), 87.
[12] Challita, U., Dong, L., and Saad, W. (2018). Proactive resource management for LTE in unlicensed spectrum: a deep learning perspective. *IEEE Transactions on Wireless Communications*, 17(7), 4674–4689.
[13] Mahdi, A. H. and Kalil, M. A. (2015). Cross-layer optimization and link adaptation in cognitive radios. In *Handbook of Research on Software Defined and Cognitive Radio Technologies for Dynamic Spectrum Management* (pp. 680–710). IGI Global, Pennsylvania, PA.
[14] Surya, S., Gupta, S., Mehbodniya, A., *et al.* (2022). Addressing the real world problem of managing wireless communication systems using explainable AI-based models through correlation analysis. *Mathematical Problems in Engineering*, 1–6.
[15] Zhang, Z. and Jianping, A. N. (2023). Applications and prospects of artificial intelligence in covert satellite communication: a review. *Information Sciences*, 66, 121301.
[16] Awotunde, J. B., Misra, S., and Pham, Q. T. (2022). A secure framework for internet of medical things security based system using lightweight cryptography enabled blockchain. In *Future Data and Security Engineering. Big Data, Security and Privacy, Smart City and Industry 4.0 Applications: 9th International Conference, FDSE 2022*, Ho Chi Minh City, Vietnam, November 23–25, 2022, Proceedings (pp. 258–272). Springer Nature, Singapore.
[17] Klaine, P. V., Imran, M. A., Onireti, O., and Souza, R. D. (2017). A survey of machine learning techniques applied to self-organizing cellular networks. *IEEE Communications Surveys & Tutorials*, 19(4), 2392–2431.
[18] Zorzi, M., Zanella, A., Testolin, A., De Grazia, M. D. F., and Zorzi, M. (2015). Cognition-based networks: a new perspective on network optimization using learning and distributed intelligence. *IEEE Access*, 3, 1512–1530.
[19] Zhou, X., Sun, M., Li, G. Y., and Juang, B. H. F. (2018). Intelligent wireless communications enabled by cognitive radio and machine learning. *China Communications*, 15(12), 16–48.
[20] Alsheikh, M. A., Lin, S., Niyato, D., and Tan, H. P. (2014). Machine learning in wireless sensor networks: algorithms, strategies, and applications. *IEEE Communications Surveys & Tutorials*, 16(4), 1996–2018.
[21] Awotunde, J. B., Imoize, A. L., Ayoade, O. B., *et al.* (2022). An enhanced hyper-parameter optimization of a convolutional neural network model for

leukemia cancer diagnosis in a smart healthcare system. *Sensors*, 22(24), 9689.

[22] Zhang, S. and Zhu, D. (2020). Towards artificial intelligence enabled 6G: state of the art, challenges, and opportunities. *Computer Networks*, 183, 107556.

[23] Littleboy, A., Keenan, J., Ordens, C. M., *et al.* (2019). A sustainable future for mining by 2030? Insights from an expert focus group. *The Extractive Industries and Society*, 6(4), 1086–1090.

[24] Zhang, Z., Xiao, Y., Ma, Z., *et al.* (2019). 6G wireless networks: vision, requirements, architecture, and key technologies. *IEEE Vehicular Technology Magazine*, 14(3), 28–41.

[25] AbdulRaheem, M., Oladipo, I. D., González-Briones, A., Awotunde, J. B., Tomori, A. R., and Jimoh, R. G. (2022). An efficient lightweight speck technique for edge-IoT-based smart healthcare systems. In *5G IoT and Edge Computing for Smart Healthcare* (pp. 139–162). Academic Press, London.

[26] Ahammed, T. B., Patgiri, R., and Nayak, S. (2023). A vision on the artificial intelligence for 6G communication. *ICT Express*, 9(2), 197–210.

[27] Nayak, S. and Patgiri, R. (2020). 6G communication: envisioning the key issues and challenges. arXiv preprint arXiv:2004.04024.

[28] Illa, P. K. and Padhi, N. (2018). Practical guide to smart factory transition using IoT, big data and edge analytics. *IEEE Access*, 6, 55162–55170.

[29] Dang, S., Amin, O., Shihada, B., and Alouini, M. S. (2020). What should 6G be? *Nature Electronics*, 3(1), 20–29.

[30] Nakamura, T. (2020). 5G evolution and 6G. In *2020 IEEE Symposium on VLSI Technology* (pp. 1–5). IEEE.

[31] Dong, W., Xu, Z. H., Li, X. X., and Xiao, S. P. (2020). Low-cost subarrayed sensor array design strategy for IoT and future 6G applications. *IEEE Internet of Things Journal*, 7(6), 4816–4826.

[32] Nayak, S. and Patgiri, R. (2021). 6G communication technology: a vision on intelligent healthcare. *Health Informatics: a Computational Perspective in Healthcare*, 1–18.

[33] Nayak, S. and Patgiri, R. (2022, April). 6G communication: a vision on the potential applications. In *Edge Analytics: Select Proceedings of 26th International Conference—ADCOM 2020* (pp. 203–218). Springer, Singapore.

[34] McMahan, B., Moore, E., Ramage, D., Hampson, S., and y Arcas, B. A. (2017). Communication-efficient learning of deep networks from decentralized data. In *Artificial Intelligence and Statistics* (pp. 1273–1282). PMLR, London.

[35] Zappone, A., Di Renzo, M., and Debbah, M. (2019). Wireless networks design in the era of deep learning: model-based, AI-based, or both? *IEEE Transactions on Communications*, 67(10), 7331–7376.

[36] Luong, N. C., Hoang, D. T., Gong, S., *et al.* (2019). Applications of deep reinforcement learning in communications and networking: a survey. *IEEE Communications Surveys & Tutorials*, 21(4), 3133–3174.

[37] Gui, G., Liu, M., Tang, F., Kato, N., and Adachi, F. (2020). 6G: opening new horizons for integration of comfort, security, and intelligence. *IEEE Wireless Communications*, 27(5), 126–132.

[38] Bi, Q. (2019). Ten trends in the cellular industry and an outlook on 6G. *IEEE Communications Magazine*, 57(12), 31–36.

[39] Letaief, K. B., Chen, W., Shi, Y., Zhang, J., and Zhang, Y. J. A. (2019). The roadmap to 6G: AI empowered wireless networks. *IEEE Communications Magazine*, 57(8), 84–90.

[40] Ji, B., Han, Y., Liu, S., *et al.* (2021). Several key technologies for 6G: challenges and opportunities. *IEEE Communications Standards Magazine*, 5 (2), 44–51.

[41] Abiodun, M. K., Awotunde, J. B., Adeniyi, A. E., Ademuagun, D., and Aremu, D. R. (2022). Securing digital transaction using a three-level authentication system. In *Computational Science and Its Applications–ICCSA 2022 Workshops*, Malaga, Spain, July 4–7, 2022, Proceedings, Part IV (pp. 135–148). Springer International Publishing, Cham.

[42] Nawaz, S. J., Sharma, S. K., Wyne, S., Patwary, M. N., and Asaduzzaman, M. (2019). Quantum machine learning for 6G communication networks: state-of-the-art and vision for the future. *IEEE Access*, 7, 46317–46350.

[43] Van Nguyen, M. S., Do, D. T., Phan, V. D., Ullah Khan, W., Imoize, A. L., and Fouda, M. M. (2022). Ergodic performance analysis of double intelligent reflecting surfaces-aided NOMA–UAV systems with hardware impairment. *Drones*, 6(12), 408.

[44] Strinati, E. C., Barbarossa, S., Gonzalez-Jimenez, J. L., *et al.* (2019). 6G: the next frontier: from holographic messaging to artificial intelligence using subterahertz and visible light communication. *IEEE Vehicular Technology Magazine*, 14(3), 42–50.

[45] Chowdhury, M. Z., Shahjalal, M., Ahmed, S., and Jang, Y. M. (2020). 6G wireless communication systems: applications, requirements, technologies, challenges, and research directions. *IEEE Open Journal of the Communications Society*, 1, 957–975.

[46] Yaacoub, E. and Alouini, M. S. (2020). A key 6G challenge and opportunity—connecting the base of the pyramid: a survey on rural connectivity. *Proceedings of the IEEE*, 108(4), 533–582.

[47] Tomkos, I., Klonidis, D., Pikasis, E., and Theodoridis, S. (2020). Toward the 6G network era: opportunities and challenges. *IT Professional*, 22(1), 34–38.

[48] Chen, S., Liang, Y. C., Sun, S., Kang, S., Cheng, W., and Peng, M. (2020). Vision, requirements, and technology trend of 6G: how to tackle the challenges of system coverage, capacity, user data-rate and movement speed. *IEEE Wireless Communications*, 27(2), 218–228.

[49] Liu, G., Huang, Y., Li, N., *et al.* (2020). Vision, requirements and network architecture of 6G mobile network beyond 2030. *China Communications*, 17(9), 92–104.

[50] Tariq, F., Khandaker, M. R., Wong, K. K., Imran, M. A., Bennis, M., and Debbah, M. (2020). A speculative study on 6G. *IEEE Wireless Communications*, 27(4), 118–125.

[51] Bariah, L., Mohjazi, L., Muhaidat, S., *et al.* (2020). A prospective look: key enabling technologies, applications and open research topics in 6G networks. *IEEE Access*, 8, 174792–174820.

[52] Wang, C. X., Huang, J., Wang, H., Gao, X., You, X., and Hao, Y. (2020). 6G wireless channel measurements and models: trends and challenges. *IEEE Vehicular Technology Magazine*, 15(4), 22–32.

[53] Chi, N., Zhou, Y., Wei, Y., and Hu, F. (2020). Visible light communication in 6G: advances, challenges, and prospects. *IEEE Vehicular Technology Magazine*, 15(4), 93–102.

[54] Kishk, M., Bader, A., and Alouini, M. S. (2020). Aerial base station deployment in 6G cellular networks using tethered drones: the mobility and endurance trade-off. *IEEE Vehicular Technology Magazine*, 15(4), 103–111.

[55] Jiang, W., Han, B., Habibi, M. A., and Schotten, H. D. (2021). The road towards 6G: a comprehensive survey. *IEEE Open Journal of the Communications Society*, 2, 334–366.

[56] De Alwis, C., Kalla, A., Pham, Q. V., *et al.* (2021). Survey on 6G frontiers: trends, applications, requirements, technologies and future research. *IEEE Open Journal of the Communications Society*, 2, 836–886.

[57] Zhao, Y., Zhai, W., Zhao, J., *et al.* (2020). A comprehensive survey of 6G wireless communications. arXiv preprint arXiv:2101.03889.

[58] De Lima, C., Belot, D., Berkvens, R., *et al.* (2021). Convergent communication, sensing and localization in 6G systems: an overview of technologies, opportunities and challenges. *IEEE Access*, 9, 26902–26925.

[59] Mahmoud, H. H. H., Amer, A. A., and Ismail, T. (2021). 6G: a comprehensive survey on technologies, applications, challenges, and research problems. *Transactions on Emerging Telecommunications Technologies*, 32(4), e4233.

[60] Abdel Hakeem, S. A., Hussein, H. H., and Kim, H. (2022). Security requirements and challenges of 6G technologies and applications. *Sensors*, 22(5), 1969.

[61] El Mettiti, A. and Oumsis, M. (2022). A survey on 6G networks: vision, requirements, architecture, technologies and challenges. *Networks*, 3, 4.

[62] Alraih, S., Shayea, I., Behjati, M., *et al.* (2022). Revolution or evolution? Technical requirements and considerations towards 6G mobile communications. *Sensors*, 22(3), 762.

[63] Banafaa, M., Shayea, I., Din, J., *et al.* (2022). 6G mobile communication technology: requirements, targets, applications, challenges, advantages, and opportunities. *Alexandria Engineering Journal*, 64, 245–274.

[64] Wang, M., Zhu, T., Zhang, T., Zhang, J., Yu, S., and Zhou, W. (2020). Security and privacy in 6G networks: new areas and new challenges. *Digital Communications and Networks*, 6(3), 281–291.

[65] Lovén, L., Leppänen, T., Peltonen, E., *et al.* (2019). EdgeAI: a vision for distributed, edge-native artificial intelligence in future 6G networks. In *The 1st 6G Wireless Summit* (pp. 1–2).

[66] Sattiraju, R., Weinand, A., and Schotten, H. D. (2019). AI-assisted PHY technologies for 6G and beyond wireless networks. arXiv preprint arXiv:1908.09523.

[67] Hong, T., Liu, C., and Kadoch, M. (2019). Machine learning based antenna design for physical layer security in ambient backscatter communications. *Wireless Communications and Mobile Computing*, 2019, 1–10.

[68] Farsad, N., Yilmaz, H. B., Eckford, A., Chae, C. B., and Guo, W. (2016). A comprehensive survey of recent advancements in molecular communication. *IEEE Communications Surveys & Tutorials*, 18(3), 1887–1919.

[69] Lu, Y., Higgins, M. D., and Leeson, M. S. (2015). Comparison of channel coding schemes for molecular communications systems. *IEEE Transactions on Communications*, 63(11), 3991–4001.

[70] Loscri, V., Marchal, C., Mitton, N., Fortino, G., and Vasilakos, A. V. (2014). Security and privacy in molecular communication and networking: opportunities and challenges. *IEEE Transactions on Nanobioscience*, 13(3), 198–207.

[71] Hu, J. Y., Yu, B., Jing, M. Y., *et al.* (2016). Experimental quantum secure direct communication with single photons. *Light: Science & Applications*, 5 (9), e16144–e16144.

[72] Kiyomoto, S., Basu, A., Rahman, M. S., and Ruj, S. (2017). On blockchain-based authorization architecture for beyond-5G mobile services. In *2017 12th International Conference for Internet Technology and Secured Transactions (ICITST)* (pp. 136–141). IEEE, New York, NY.

[73] Kotobi, K. and Bilen, S. G. (2018). Secure blockchains for dynamic spectrum access: a decentralized database in moving cognitive radio networks enhances security and user access. *IEEE Vehicular Technology Magazine*, 13(1), 32–39.

[74] Ferraro, P., King, C., and Shorten, R. (2018). Distributed ledger technology for smart cities, the sharing economy, and social compliance. *IEEE Access*, 6, 62728–62746.

[75] Akyildiz, I. F., Jornet, J. M., and Han, C. (2014). Terahertz band: next frontier for wireless communications. *Physical Communication*, 12, 16–32.

[76] Ma, J., Shrestha, R., Adelberg, J., *et al.* (2018). Security and eavesdropping in terahertz wireless links. *Nature*, 563(7729), 89–93.

[77] Ucar, S., Coleri Ergen, S., Ozkasap, O., Tsonev, D., and Burchardt, H. (2016). Secvlc: secure visible light communication for military vehicular networks. In *Proceedings of the 14th ACM International Symposium on Mobility Management and Wireless Access* (pp. 123–129).

[78] Cho, S., Chen, G., and Coon, J. P. (2019). Enhancement of physical layer security with simultaneous beamforming and jamming for visible light communication systems. *IEEE Transactions on Information Forensics and Security*, 14(10), 2633–2648.

[79] Zhang, T., Zhu, T., Xiong, P., Huo, H., Tari, Z., and Zhou, W. (2019). Correlated differential privacy: feature selection in machine learning. *IEEE Transactions on Industrial Informatics*, 16(3), 2115–2124.

[80] Zhu, T., Li, G., Zhou, W., and Philip, S. Y. (2017). Differentially private data publishing and analysis: a survey. *IEEE Transactions on Knowledge and Data Engineering*, 29(8), 1619–1638.

[81] Akan, O. B., Ramezani, H., Khan, T., Abbasi, N. A., and Kuscu, M. (2016). Fundamentals of molecular information and communication science. *Proceedings of the IEEE*, 105(2), 306–318.

[82] Nakano, T., Okaie, Y., Kobayashi, S., Hara, T., Hiraoka, Y., and Haraguchi, T. (2019). Methods and applications of mobile molecular communication. *Proceedings of the IEEE*, 107(7), 1442–1456.

[83] Ajagbe, S. A., Florez, H., and Awotunde, J. B. (2022). AESRSA: a new cryptography key for electronic health record security. In *Applied Informatics: 5th International Conference, ICAI 2022, Arequipa, Peru, October 27–29, 2022, Proceedings* (pp. 237–251). Springer International Publishing, Cham.

[84] Gyongyosi, L., Imre, S., and Nguyen, H. V. (2018). A survey on quantum channel capacities. *IEEE Communications Surveys & Tutorials*, 20(2), 1149–1205.

[85] Nie, S., Jornet, J. M., and Akyildiz, I. F. (2019). Intelligent environments based on ultra-massive MIMO platforms for wireless communication in millimeter wave and terahertz bands. In *ICASSP 2019—2019 IEEE International Conference on Acoustics, Speech and Signal Processing (ICASSP)* (pp. 7849–7853). IEEE.

[86] Zhang, C., Patras, P., and Haddadi, H. (2019). Deep learning in mobile and wireless networking: a survey. *IEEE Communications Surveys & Tutorials*, 21(3), 2224–2287.

[87] Akyildiz, I. F., Kak, A., and Nie, S. (2020). 6G and beyond: the future of wireless communications systems. *IEEE Access*, 8, 133995–134030.

[88] Weiss, K., Khoshgoftaar, T. M., and Wang, D. (2016). A survey of transfer learning. *Journal of Big Data*, 3(1), 1–40.

[89] Celard, P., Iglesias, E. L., Sorribes-Fdez, J. M., Romero, R., Vieira, A. S., and Borrajo, L. (2023). A survey on deep learning applied to medical images: From simple artificial neural networks to generative models. *Neural Computing and Applications*, 35(3), 2291–2323.

[90] Bjornson, E. and Giselsson, P. (2020). Two applications of deep learning in the physical layer of communication systems [lecture notes]. *IEEE Signal Processing Magazine*, 37(5), 134–140.

[91] O'Shea, T. and Hoydis, J. (2017). An introduction to deep learning for the physical layer. *IEEE Transactions on Cognitive Communications and Networking*, 3(4), 563–575.

[92] Xu, Q., Su, Z., and Li, R. (2022). Security and privacy in artificial intelligence-enabled 6G. *IEEE Network*, 36(5), 188–196.

[93] Ughegbe, G. U., Adelabu, M. A., and Imoize, A. L. (2021). Experimental data on radio frequency interference in microwave links using frequency scan measurements at 6 GHz, 7 GHz, and 8 GHz. *Data in Brief*, 35, 106916.

[94] Awotunde, J. B., Ajagbe, S. A., and Florez, H. (2022). Internet of things with wearable devices and artificial intelligence for elderly uninterrupted healthcare monitoring systems. In *Applied Informatics: 5th International Conference, ICAI 2022, Arequipa, Peru, October 27–29, 2022,* Proceedings (pp. 278–291). Springer International Publishing, Cham.

[95] Awotunde, J. B., Adeniyi, E. A., Ajamu, G. J., Balogun, G. B., and Taofeek-Ibrahim, F. A. (2022). Explainable artificial intelligence in genomic sequence for healthcare systems prediction. In *Connected e-Health: Integrated IoT and Cloud Computing* (pp. 417–437). Springer International Publishing, Cham.

[96] Ylianttila, M., Kantola, R., Gurtov, A., *et al.* (2020). 6G white paper: research challenges for trust, security and privacy. arXiv preprint arXiv:2004.11665.

[97] Ahmad, I., Shahabuddin, S., Sauter, T., *et al.* (2020). The challenges of artificial intelligence in wireless networks for the Internet of Things: exploring opportunities for growth. *IEEE Industrial Electronics Magazine*, 15(1), 16–29.

[98] Yang, H., Alphones, A., Xiong, Z., Niyato, D., Zhao, J., and Wu, K. (2020). Artificial-intelligence-enabled intelligent 6G networks. *IEEE Network*, 34 (6), 272–280.

[99] Wahab, O. A., Mourad, A., Otrok, H., and Taleb, T. (2021). Federated machine learning: survey, multi-level classification, desirable criteria and future directions in communication and networking systems. *IEEE Communications Surveys & Tutorials*, 23(2), 1342–1397.

[100] Nie, Y., Zhao, J., Gao, F., and Yu, F. R. (2021). Semi-distributed resource management in UAV-aided MEC systems: a multi-agent federated reinforcement learning approach. *IEEE Transactions on Vehicular Technology*, 70(12), 13162–13173.

[101] Awotunde, J. B., Oluwabukonla, S., Chakraborty, C., Bhoi, A. K., and Ajamu, G. J. (2022). Application of artificial intelligence and big data for fighting COVID-19 pandemic. In *Decision Sciences for COVID-19: Learning Through Case Studies* (pp. 3–26). Springer, New York, NY.

[102] L'heureux, A., Grolinger, K., Elyamany, H. F., and Capretz, M. A. (2017). Machine learning with big data: challenges and approaches. *IEEE Access*, 5, 7776–7797.

[103] Kato, N., Mao, B., Tang, F., Kawamoto, Y., and Liu, J. (2020). Ten challenges in advancing machine learning technologies toward 6G. *IEEE Wireless Communications*, 27(3), 96–103.

[104] Huang, H., Guo, S., Gui, G., *et al.* (2019). Deep learning for physical-layer 5G wireless techniques: opportunities, challenges and solutions. *IEEE Wireless Communications*, 27(1), 214–222.

[105] Chen, M., Challita, U., Saad, W., Yin, C., and Debbah, M. (2019). Artificial neural networks-based machine learning for wireless networks: a tutorial. *IEEE Communications Surveys & Tutorials*, 21(4), 3039–3071.

[106] Manzalini, A. (2020). Quantum communications in future networks and services. *Quantum Reports*, 2(1), 221–232.

[107] Awotunde, J. B., Folorunso, S. O., Jimoh, R. G., Adeniyi, E. A., Abiodun, K. M., and Ajamu, G. J. (2021). Application of artificial intelligence for COVID-19 epidemic: an exploratory study, opportunities, challenges, and future prospects. In *Artificial Intelligence for COVID-19* (pp. 47–61). Springer, New York, NY.

[108] Björnson, E., Sanguinetti, L., Wymeersch, H., Hoydis, J., and Marzetta, T. L. (2019). Massive MIMO is a reality—What is next? Five promising research directions for antenna arrays. *Digital Signal Processing*, 94, 3–20.

[109] Huang, T., Yang, W., Wu, J., Ma, J., Zhang, X., and Zhang, D. (2019). A survey on green 6G network: architecture and technologies. *IEEE Access*, 7, 175758–175768.

Chapter 4

The vision of 6G security and privacy

Promise Elechi[1], Robinson Tombari Sibe[2], Kingsley Eyiogwu Onu[1] and Agbotiname Lucky Imoize[3,4]

Abstract

While the fifth-generation (5G) network has not been fully utilized globally, being at its underlying stage in terms of commercial stage, various identified limitations have invoked research into the next-generation network called the sixth-generation (6G) network. The deployment of more and more 5G networks has revealed the limitations of these networks, which undoubtedly supports the exploratory study of 6G networks as the next-generation solutions. The fundamental privacy and security concerns raised by 6G technology are covered in these evaluations. The following concerns with the projected 6G network were explored in this work: issues with tactile communications, resources as services, variable radio access constraints, and varied high-frequency bands are all security-related problems with the network. The 6G spectrum was analyzed, which ranged between 0.1 and 10 THz and a wavelength of 30–3,000 μm. Also, the various attacks on 6G were highlighted, the countermeasures to mitigate the attacks were also proposed, and the recent trends and future direction for 6G.

Keywords: 6G; Security and privacy; Network availability; Connectivity; Artificial Intelligence; Edge intelligence

4.1 Introduction

While the fifth-generation (5G) network has not been fully utilized globally, being at its underlying stage in terms of commercial stage, various identified limitations have invoked research into the next-generation network called the sixth-generation (6G)

[1]Department of Electrical/Electronic Engineering, Rivers State University, Nigeria
[2]Department of Computer Engineering, Rivers State University, Nigeria
[3]Department of Electrical and Electronics Engineering, Faculty of Engineering, University of Lagos, Nigeria
[4]Department of Electrical Engineering and Information Technology, Institute of Digital Communication, Ruhr University, Germany

network. Championing this research are those in academia and in the industry. The research is intended to find basic requirements for adequate system performance, basic drivers, the innovations related to 6G network in terms of technology, etc.; therefore, in Finland, a summit was held in 2019 by telecommunications experts across the world where a 6G white paper was drafted. Following this summit in Finland on the 6G network, many countries have encouraged research into the 6G network.

Currently, the 5G network for cellular communication is the latest standard in use in some countries of the world. The next-generation that is expected to succeed in the 5G network is the 6G network. The 6G network, when finally deployed, is expected to have more capacity, speed, and better latency compared to the 5G network [1]. Many experts are of the view that if the 6G network is finally deployed, it will create connectivity expansion for applications in conventional coverage zones within the 5G network as well as space-to-air-to-ground-to-sea [2]. Services like applications of telepresence are made possible by network capability and coverage. Other services enabled by 6G include autonomous driving [3,4], mixed reality, implantable devices with artificially made joints, intraocular lenses, etc.

As research into the possibility of the 6G network began, the 6G network only has many possibilities but no specifications and standard functions. The improvement required for the 5G network that will give rise to the current 6G technology that is being canvassed must go beyond the speed but an improvement of the 5G technology in all ramifications. This higher speed means that coverage as we have in the 5G network should not be limited to the ground level, which is the current situation with the 5G network but even beneath the sea surface coverage. Another area where improvement can be made is in the area of the capabilities of artificial intelligence (AI). This means that if it is practically possible, the 6G network that will succeed the 5G should have AI as the driver and most key feature, being a network-empowered AI. Unlike the 5G, which merely uses AI as part of its architecture, the 6G network should be driven by AI.

In this research, the following will be undertaken:

(a) Reasons for migration from 5G to 6G.
(b) Review of emerging issues in the 6G network.
(c) Evolution of security and private schemes in wireless systems from 1G to 5G.
(d) Technical overview of the 6G network.

4.1.1 Why the migration from 5G to 6G

The 5G network standard is a new technology that began to be deployed worldwide by companies dealing with cellular phones barely 3 years ago, in 2019. It is planned to succeed in the fourth-generation (4G) network. Despite being deployed a few years ago, 5G networks have been predicted to attract beyond 1.7 billion users worldwide by 2025 [5,6].

Being cellular networks, the 5G networks have the service areas split into smaller geographical areas. Like other cellular networks, these smaller geographical areas are called cells. Some basic technologies of the 5G network are getting outdated despite the fact that the technology was deployed only recently [7]. Examining these technologies and the issues peculiar to each will be undertaken to provide clear reasons for migration from 5G to 6G.

4.1.2 Carrier aggregation

Carrier aggregation is the technology in 5G that enables the use of one lone component carrier to serve different subscribers to provide much higher bandwidth, as was pointed out in [8]. This technology has some serious effects on the side of subscribers' hardware to allow the use of different frequency bands. Though 5G has the C-RAN, that is, the Cloud Radio Access Network, that is designed to address any limitations of the hardware end devices, and this is not possible again when the network size grows exponentially [9]. Cloud alone becomes incapable of mitigating such limitations unless edge and fog node computations are integrated [9].

4.1.3 Security

Fifth-generation (5G) network does not have highly advanced technology that guarantees security when deployed on very large scales [9]. When employing software-defined networks, 5G does not have the mechanisms required to ascertain trust between the system controller and management apps [9]. Besides, when software-level parts like the virtual infrastructure manager are targeted by attackers such that there is a production of fake fogs that affect the operation of network function visualization, 5G lacks the security to address such a situation [9].

4.1.4 Heterogeneity

At the core of the 5G, networks are network heterogeneity. However, outside terrestrial networks, heterogeneous networks in the generation system do not work unless a 3-dimensional network includes space and aerial networks as part of the main network [9].

4.1.5 Latency of links

5G networks have real-time services that are part of the overall network. One example of these real-time services is creating smart cities like autonomous factories and vehicles. Most of these services are very time-sensitive and have stringent latency requirements [9]. The latency requirement should be 10 ms and below to ensure the operation mode is effective. It is also very clear that latency can be degraded due to factors related to the technology. An example is the cyclic prefix length in an orthogonal frequency division modulation system or the use of dedicated channels in machine communication.

Several applications, like vehicle control, have latency requirements from 0.1 ms to 1 ms [10]. However, to realize autonomous operation as fully as possible within the industrial circle, some applications need support for URLCC simultaneously [9]. Noted that latencies of 250 μs are required for operating factories when virtual presence is used. As pointed out in [10], the latency improvement over the present 5G network requires 10-fold and 50-fold. However, 5G does not fully consider the customization of latency and data rate for different applications. Hence, in the next generation, improvements have to be made.

4.1.6 Network availability

5G network has no immunity against denial-of-service attacks, even any other attacks that adversely affect network availability. This was the point made in [11]. The fact is that since the capability of modern computers for data processing is yet to be enhanced as 5G develops, the system becomes vulnerable to attacks designed to compromise the server. This is usually done by transmitting huge data flow [11]. And because it is very difficult to retrieve certain features between benign and malicious huge data flows, the difficulties associated with the detection of denial of service (DoS) in a 5G network get amplified. This significant deficiency of the 5G network can be improved.

4.1.7 Scalability and communication speed

The deployment of the 5G network brought about mobile broadband that is highly enhanced, capable of offering a network speed of about 20 gigabytes per second (20 Gbps), according to [12]. In [13], it was noted that because of the machine–machine (M–M) communication, the projection is that by 2030, mobile traffic globally will rise to well above 670 times what it used to be back in 2010. However, it is very much expected that the 5G network will get to its limit by the same 2030 [14]. Besides, the 5G network is designed to make use of the 20–100 GHz range, which is the range of a millimeter wave, but realizing that level of speeds within this millimeter wave range is practically impossible because of the limitations in various digital modulation schemes as well as transceiver design limitations, as was argued in [7]. In view of the above, it becomes appropriate for frequencies well above 100 GHz to be considered in the next-generation network. Besides, there is a strong expectation that certain services, like connected robotics, augmented reality, autonomous systems, etc., will be adopted widely as soon as possible [7]. If this is so, then there is a real justification for high data rates. It is also expected that because of M–M communications, devices to be connected to the Internet will be hundreds of billions [7]. In this instance, the 5G is deficient since its best performance is within the range of a billion devices, as can be seen in [7]. Hence, there is a need for future upgrades.

4.1.8 Link reliability

According to [7], the 5G network supports link reliability of about 0.00001. meanwhile, some applications today require high-level connection reliability to keep incident rates as low as possible in factory automation or other applications. It was mentioned in [7] that the link reliability requirement in certain applications is up to 0.000000001 in terms of frame error rate (FER). However, since the 5G network cannot provide this level of link reliability, the current network needs to be upgraded to enable the implementation of the smart cities concept and machine operations that are fully dependable.

The advanced points explain why there is an urgent need to shift from the 5G to the 6G.

4.2 Review of emerging issues in 6G

Though the 6G network is yet to be deployed, some key issues have been identified that could affect the smooth operation of the network when it is finally deployed. Therefore, there is every reason to ensure that the issues are carefully weighed, and the possible solutions to the issues are found so subscribers can easily enjoy the 6G network when deployed. In this work, some such issues will be discussed:

(a) Security issues associated with the network
(b) Tactile communications issue
(c) Resources as a service issued
(d) Flexible radio access limits
(e) Heterogeneous high-frequency bands

4.2.1 Quantum communication issue

Quantum communication has very great potential to be used in 6G networks. It is believed to significantly improve the reliability and security of data transmission in communication systems [15]. Theoretically, quantum communication can provide high-level security and can be applied to long-distance communication. However, as [15] pointed out, quantum communication, as it is currently seen, cannot be used as a panacea for every privacy and security issue. Although true, notable progress has been made in developing what is called quantum cryptography, which is applied to quantum communication. Operation errors and fiber attenuation still exist, and they make long-distance quantum communication a very serious challenge [15].

4.2.2 Molecular communication issue

It is well known that molecular communication occurs within living things, a natural phenomenon. The development of certain technologies, such as synthetic biology, bioengineering, and nanotechnology saw the appearance of nanoscale and microscale devices in many parts of the world. The major idea in this molecular communication is to rely on signals of biochemical nature in transmitting the information. Ref. [16] gave a process of molecular communication that permeates the receiver, the sender, and the relevant nodes while they are moving. However, as was argued in [17], encryption processes, authentication, and even communication in molecular communication have many serious privacy and security issues that must not be ignored. The works [15,18] reported attack methods at different molecular communication stages.

4.2.3 Visible light communication

The technology associated with visible light communication can be relied upon in tackling the problem of increasing wireless connectivity demands. Compared with radio frequency RF, which is known to have high latency and interference, visible light communication (VLC) is capable of resisting electromagnetic interference and has much higher bandwidth. Visible light communication technology has further advanced since solid-state lighting was developed. A system of visible light communication that has the potential to give a huge number of mobile subscribers needed high-speed services were devised by [19], and it was called LiFi. It is believed that visible light communication technology can be fully utilized in the 6G network. VLC has many security issues. These security issues include communication processes, malicious behavior, etc.

4.2.4 Distributed ledger technology issue

In distributed ledger technologies, blockchain technology has dragged the most attention from the industrial sector of telecommunication [20]. Because of the advantages that blockchain offers, such as immutability and disintermediation, differing services in a secure and trusted manner can be enabled in the 6G network. This was the point made by [21]. AI/ML, in addition to other technologies for data analysis that can be used in the future 6G network, has the possibility of being a source of certain attacks vectors [20], and these attacks can either be at the training phase or testing phase or even at both phases. Because of the anticipated alliance of the 6G and distributed ledger technology (DLT), the security of the 6G network may be impacted due to the blockchain and smart contracts vulnerabilities [22]. Most of these attacks happen because of issues related to errors in software programming, programming language restrictions, etc., and they occur both in private and public blockchain platforms [23,24].

Among the most serious attacks in blockchain networks are majority attacks, privacy leakages, double spending attacks, Sybil attacks, and reentrancy attacks.

4.2.4.1 Majority attacks

Most attacks occur when malicious users have captured 51% of the network modes. In this case, the malicious users may begin to control the blockchain [20]. Maliciousness sometimes stops real transactions from being confirmed and even changes the transaction history. The vulnerability of blockchain systems to majority attacks is usually high [20]. The majority attack is often called a 51% attack since 51% of the nodes or more in the blockchain are often attacked.

4.2.4.2 Privacy leakages

This is a serious threat. Organizations have some information that is not for the public, very sensitive information. But because of privacy leakages in blockchains and smart contracts, this sensitive information may be revealed to unauthorized individuals [25]. Some authors identified these possible threats as leaking smart contract logic privacy [25], leaking transaction data privacy [26], and leakage of privacy in the process of executing smart contracts [27].

4.2.4.3 Double spending attack

This very attack occurs in blockchain platforms as one of the main aspects of blockchain is to spend cryptographic tokens [20]. Now, there is a greater possibility that a user may end up using just one token many times, as physical notes are lacking [28].

4.2.4.4 Sybil attacks

This attack in blockchain occurs when a group of attackers or a single attacker attempts to take over the blockchain peer network. The attackers try to hide their real identity [29]. This type of attack usually occurs in the blockchain system [30].

4.2.4.5 Reentrancy attacks

Sometimes a smart contract invokes some other smart contracts iteratively. When this happens, the vulnerability of reentrancy may occur. However, the smart contract that invoked the other is secondary to the malicious one. Regarding the 6G network, security is a key issue, particularly with the technique of terrestrial space integrated network being employed, as argued by [31]. Tackling the security issue requires a method or way of comprehensively identifying the issues, providing minimal difficulty and a much higher level of security. However, this is not so easy and simple.

Despite security issues, other issues also exist for the future 6G network. Issues like flexible radio access limits, heterogenous high-frequency bands, resources as a service issues, tactile communication issues, etc.

4.2.5 Flexible radio access limits

The orthogonal frequency division multiplexing numerology option can be limited by the carrier frequency and the cell size. That was the case made by the authors in [32]. Using numerology with a large subcarrier spacing is far better for small cell sizes. A much larger subcarrier space is most appropriate when digital cells are relatively large, though the performance may be low [31]. As pointed out in [33], in the case of high-frequency carriers and high mobility, cell size is restricted due to issues that border the route and Doppler propagation.

4.2.6 Heterogeneous high-frequency band (HHFB)

The future use of millimeter-wave and the terahertz frequency in the 6G networks brings a number of issues. For instance, supporting higher movement at millimeter-wave frequencies is a problem that [31] describes as 'an open central problem'.

4.2.7 Tactile communication

Physical communication can be exchanged remotely via an Internet connection in real time. Services that allow the use of random control throughout communication networks are telecommunication services, interpersonal communication, etc. The very strict requirements here mean that designing a communication system for greater efficiency is necessary. An instance of this was cited by [31] as the establishment of new physical layer diagrams, which help design congestion waveforms, signaling systems, and so forth to improve motivated protocol and transfer.

4.3 Evolution of security and privacy schemes in wireless systems: 1G to 5G

In this section, beginning with the first-generation (1G) network through the 5G network, we will consider the evolution of these different networks with special attention to the security and privacy schemes used by each of the five generations of wireless systems. A comparison of these network generations will also be included to provide an on-the-spot and clear understanding of the various wireless network generations (1G, 2G, 3G, 4G, and 5G).

4.3.1 1G network

The 1G system was brought into existence in the 1980s and was purely an analogue system designed specifically for voice services. Being an analogue, the 1G network had a lot of disadvantages, including a clear lack of privacy and security guarantees. There was no encryption of phone services. Hence, conversations over the phone and data transmission were not secure and private. Ref. [16] gave examples of the challenges, security, and privacy-wise, including eavesdropping, cloning, and illegal access. As a result of these serious and worrisome challenges, experts in academia and the industry worked hard to improve the 1G network. This effort gave birth to the second-generation (2G) wireless network.

4.3.2 2G network

The 2G network was introduced barely 10 years after the launch of the 1G network. Unlike 1G, which is an analogue system, 2G is a digital one that uses digital modulation schemes like Code Division Multiple Access [16]. Still, unlike the 1G system, the 2G system supports SMS services and voice calls. An example of a 2G system is the widely used Global System for Mobile Communications, called GSM.

The 2G improved the security and privacy issues in the 1G network. To improve the security issues in 1G, a few services were combined in 2G. These services include authentication, anonymity, data protection for users, and signaling protection. Unlike the 1G where there was no encryption of phone services, in 2G, encryption plays a major part in providing security for the system. Encryption of radio paths and the use of temporary mobile subscriber identity are methods that protect users' privacy.

Undoubtedly, the 2G network greatly improved the first-generation (1G) wireless system. However, despite these improvements, 2G has its own limitations and shortcomings regarding security and privacy issues [16]. For example, while there is authentication in the 2G system, the authentication is one-way: only the network authenticates the users, while the users are not authorized to authenticate the network. Therefore, malicious devices can easily disguise themselves as real network devices and steal user information. Furthermore, because of the limitations of TMSI and the encryption of the radio path, privacy issues are not guaranteed since the system can still experience attacks like eavesdropping attacks [16]. Besides, there is no end-to-end encryption; the fixed part of the network has no encryption.

4.3.3 3G network

The 3G network was introduced in the 2000s. 3G was launched to provide web applications with high-level speed for data transmission. Despite using the security system based on the 2G technology, the 3G network provides better security and privacy than the 2G because of the addition of other security features to the 2G network technology. Therefore, security is improved in 3G, where we have a two-way authentication unlike the 2G. In addition to this, the Third Generation Partnership Project (3GPP) provided a system that ensures the security of the air interface and subscribers' physical reliability.

Notwithstanding the improvements made by the 3G in terms of security and privacy of the network and the users, 3G is not a perfect system with no issues. Problems like Denial of Service (DoS), access to data without authorization, access to services without authorization, integrity threats, etc. These threats are peculiar or related to the wireless interface. Furthermore, privacy in 3G is also an issue because of some attacks that destroy subscribers' identities, including some sensitive information. One such attack is the authentication and key agreement error messages [16,34].

4.3.4 4G network

The 4G network is known as the LTE network. The 4G network data rate in the downlink and uplink is 1 Gb/s and 500 Mb/s, respectively. 4G network has low latency compared to 3G network and enjoys far better spectrum efficiency than 3G. The technology used in 4G system includes multiple input multiple outputs (MIMO), multipoint transmission and reception, and orthogonal frequency division multiplexing. True, the 4G network is far better than 3G in terms of services and application areas. However, there are still some security and privacy issues that are of serious concern, emanating from eavesdropping, data deletion, security issues in the wireless link, authentication problems in network entity, etc. It could be expected that in terms of security and privacy, the 4G is better than the 3G but quite on the contrary [16] noted that the 4G network is more vulnerable to privacy and security threats than all the other generations before. This is because subscribers of 4G interact with mobile terminals more closely. In addition, Denial of Service (DoS) attacks, tracking of MAC layer location, attacks on data integrity, and illegal use of subscribers and mobile devices are very common with 4G networks. There are also some defects in the basic management protocols, which make the MAC layer of WiMAX (Worldwide Interoperability for Microwave Access) have some vulnerabilities.

4.3.5 5G network

The 5G network is the latest network launched so far. It can provide connection to far more devices than the previous generations. 5G supports different devices like smartphones, Internet of Things (IoT) equipment, etc. In 5G technology, the key techniques are the wireless network in the urban area, known as WLAN for short, and the 802.11 wireless networks in local areas. In addition, artificial intelligent capabilities are provided by 5G portable devices.

Table 4.1 Security issues from 1G through 5G

	1G	2G	3G	4G	5G
Year Launched	1980s	1990s	2000s	2010s	2020
Main Services provided	Voice calls	SMS and voice	Web applications, video services, TV streaming	Mobile application, HDTV, DVB	IoT, smart city, etc.
Security issues	No guarantee of security and privacy	Authentication is one-way, illegal devices	The encryption key, vulnerabilities in IP traffic	Vulnerabilities in the MAC layer, threats from new devices	Vulnerabilities in SDN, NFV, and cloud techniques, the openness of the network

Security and privacy issues in 5G can be split into three parts: the backhaul network, the access network, and the core network. The backhaul network has some issues that emanate from the elements of the core network that have been adjusted, like the eNB, core networks MME, E-UTRAN, etc. On the other hand, when it comes to access networks, the nodes' diversity coupled with access mechanisms make new security issues, such as handover between various access technologies, add to attack risk.

The techniques adopted to enhance 5G's network performance introduce security holes. Farsad *et al.* [16] said that massive multiple input multiple output help disguises active and passive eavesdropping. Also, there are new privacy issues to battle with because of application scenarios and business types of diversity [16] in 5G. Table 4.1 shows the security from 1G through 5G.

4.4 Technical overview of 6G network

There are high expectations for the 6G network, one of such is to provide the required support for a network that connects about five hundred billion devices [35], with a total capacity 1,000 times more than 5G's capacity. In addition, since the frequency spectrum to be used in 6G is much higher, subscribers can be sure of higher speed and higher data transfer of up to 100–1,000 times more than 5G currently offers [35]. 6G also has low latency compared to any previous generation, and it can provide a data transfer rate of Gb/s to Tb/s.

The year in which the 6G network is expected to be launched is 2030, and that will be the very first 6G commercializing system. As the 6G network is launched, one particular feature is the frequency spectrum for transmission, as shown in Figure 4.1. For bit rates that go up to Tb/s, it should be expected that operations at much higher frequencies for the bandwidth and available spectrum will exist. The visible light spectrum and terahertz spectrum are the two key sixth-generation spectrum technologies.

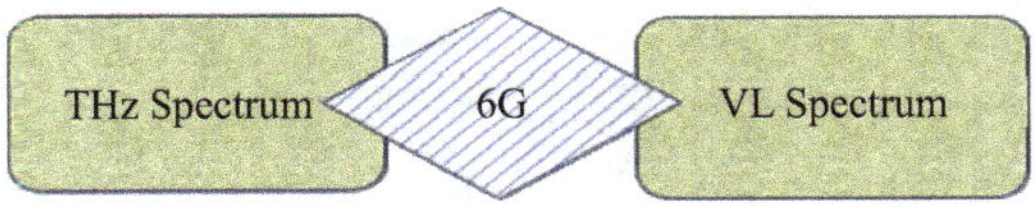

Figure 4.1 6G spectrum technology

Terahertz waves are purely electromagnetic (EM) waves, the wavelength range being between 30 and 3,000 μm, while the spectrum range is between 0.1 THz and 10 THz. As mentioned in [35], the transmission region existing between microphonics and macro-electrons is where the terahertz spectrum falls. There is a high transmission rate and good spectrum resources that terahertz communication shows, which makes the terahertz spectrum an excellent broadband wireless access for future use. Terahertz communication makes interactions of objects at the microscale possible because of the scale-down antennas that have a wavelength of about one millimeter for 300 GHz [35].

For visible light, the spectrum has a range of 430 THz–790 THz. In optical wireless communication (OWC), visible light shows the greatest potential because of technological advancements and the wide use of light emitting diodes (LEDs). Without any form of electromagnetic interference, visible light provides frequency reuse, an ultra-high bandwidth in terahertz, and clear or unrestricted access to the spectrum. Realizing good optical fiber performance necessary for the 6G network requires using VLC and visible light technologies.

The technical aspects of 6G to be considered include intelligent reflecting surfaces, AI, cell-free mMIMO, edge intelligence, holographic beamforming, and terahertz communication.

4.4.1 Intelligent reflecting surface

The intelligent reflecting surface is anticipated to be one of the key technologies of the 6G network when it is finally deployed. Benn [35] said that IRS has the capacity to minimize the utilization of energy in wireless networks. But what really is an intelligent reflecting surface (IRS)? IRS is a material with a reasonably large surface area mounted against any flat surface. It has a very reasonable quantity of reflecting electronic parts with variables that can be reconfigured. It uses low energy, and it is quite simple to install. Hence, intelligent-reflecting-surface-assisted communications can be used in 6G networks to significantly increase network coverage [34,36,37]. In cities with many telecommunication infrastructures, much electromagnetic radiation is produced, but the use of IRS significantly reduces such electromagnetic radiation [38]. Figure 4.2 shows the illustration of an intelligent reflecting surface.

4.4.2 AI

AI is one of the key driving forces of the 6G network. Though this AI was never part of the previous four generations, 1G to 4G, it became part of the 5G network following the Release 18 specifications by 3GPP [2,39]. Therefore, AI will be a basic part of the sixth generation (6G) network, bringing several interesting new revolutions into the 6G network. At least AI will play a key role in network selection, resource allocation, handover, etc., in the 6G network, leading to better network performance [37,39].

4.4.3 Cell-free mMIMO

Broadband connectivity has been a serious problem when a better quality of service is to be provided within a coverage area. To overcome this problem and ensure far better network coverage, implementing cell-free mMIMO technology, as shown in Figure 4.3, is seriously considered. In cell-free mMIMO, all the possible mMIMOs

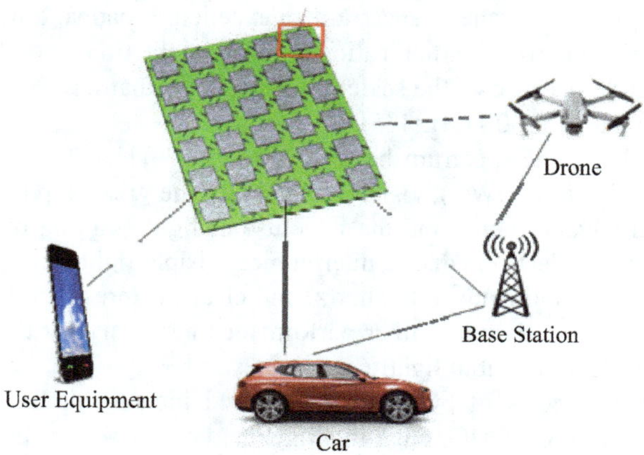

Figure 4.2 IRS

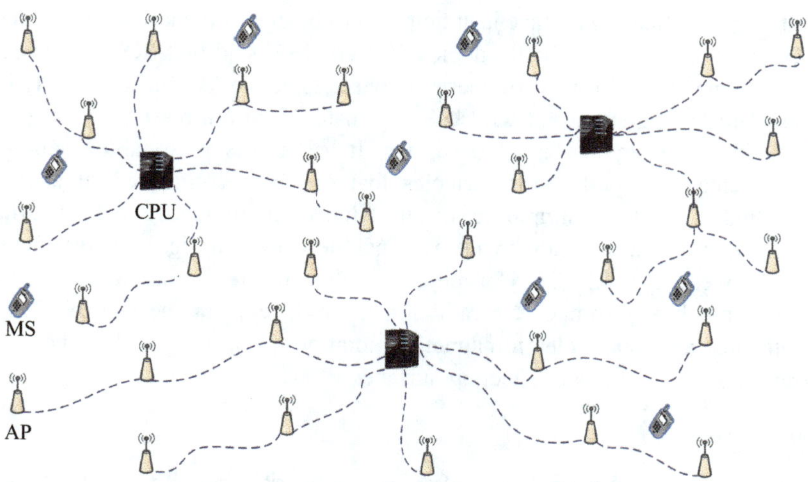

Figure 4.3 Cell-free mMIMO technology

are distributed in such a way that they appear to be just a single physically collocated mMIMO forming an ultra-mMIMO. Indeed, mMIMO technology is yet a very new technology that is still being deployed in markets, as [37] noted. Still, it can be a very important part of the 6G network because of the applications it can provide in 6G [37]. The advantages of cell-free mMIMO technology include its energy efficiency, the capability of serving many subscribers simultaneously over the same radio resources, efficient time synchronization, interference mitigation, reduction of radiation power consumption, etc.

4.4.4 Edge intelligence

Edge computing is a part of the 5G network, and it brings great benefits to the end users of the network. Because of the capability of edge computing to have systems that work at closer edges without the need to go global, end users of the network can enjoy a more secure and private connection to the edge server. According to many experts, using edge computing technology helps when data analysis and AI workloads must be run [40]. But since edge computing ensures the provision of connectivity to numerous end users at high communication speed through, the millimeter wave used by the 5G network is only applicable to short-range communication. Now that it is expected that in the 6G network, terahertz (THz) communications will be employed to provide a far greater speed than the current 5G, using the technology of edge computing used in 5G network will create more limitations to the range of communication. Also, the use of edge computing

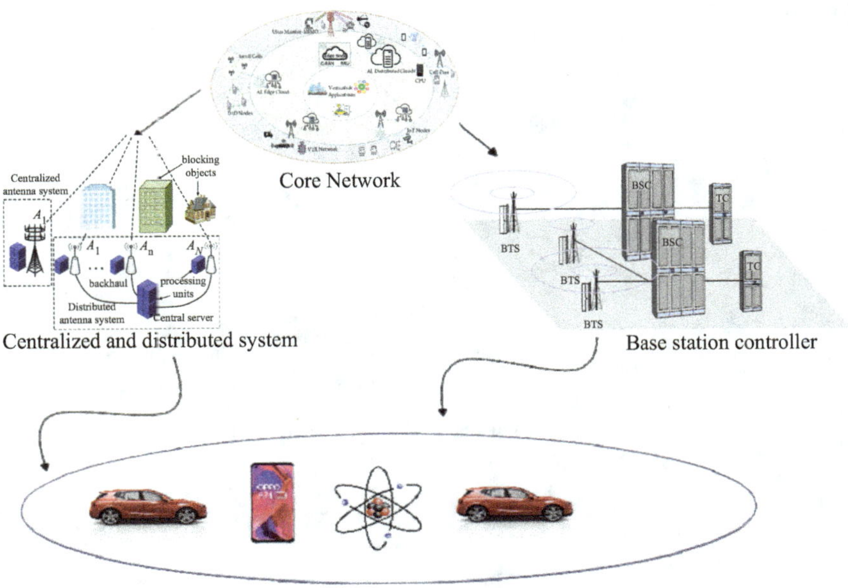

Figure 4.4 Intelligence native 6G network

technology in a 6G network will bring about speed mismatch due to the connectivity involved in providing services to all devices under the access point [37]. To overcome all these challenges, edge computing technology currently in use is expected to be improved upon to get another technology: edge intelligence technology. Intelligence native 6G network is given in Figure 4.4.

4.4.5 Holographic beamforming

Beamforming is one of the key technologies used in a communication system to provide better network coverage as well as throughput. In beamforming, antenna arrays are used at both transmitting and receiving points to provide much better gain. Holographic beamforming uses a software-defined antenna, thereby taking beamforming to an advanced level. If the holographic beamforming technique is used in a 6G network, IoT devices and 6G systems can effectively and efficiently use radio frequencies.

4.4.6 Terahertz communication

Globally, there is greater demand for wireless communication, and that demand keeps increasing daily. Hence, the radio frequency band lacks the required capacity to provide what is needed to satisfy network users. Di Renae *et al.* [37] have mentioned that a frequency band that goes beyond 0.1 THz is needed to provide a higher data transfer rate. This is one of the reasons why many experts are researching the use of terahertz communication in the future 6G network. No doubt employing terahertz communication will bring about improvement in the speed, security of the network, capacity, and even bandwidth. 6G with satellite network is given in Figure 4.5.

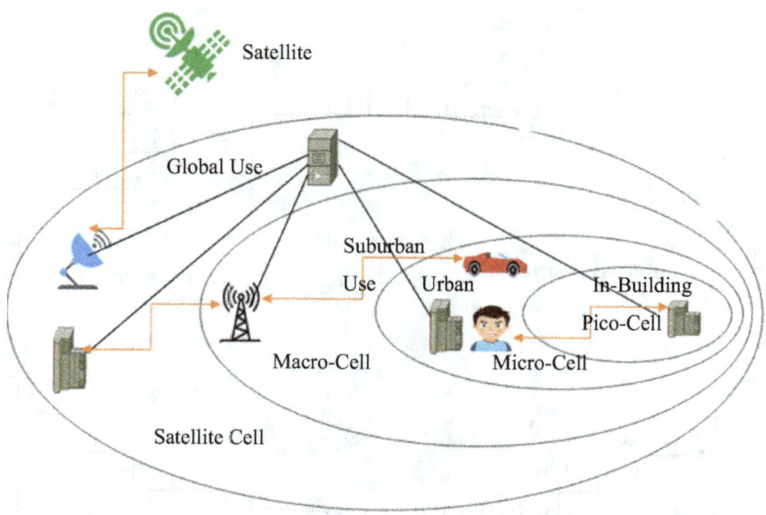

Figure 4.5 6G with satellite network

4.5 Security concerns in 6G

Interestingly, 5G is not even fully exhaustively implemented, yet discussions have started for 6G. The pace of technology has been blistering; therefore, researchers are already looking ahead to 6G. Like every innovation, 6G is coming with a lot of promises. Expectedly, 6G will also come with emerging challenges. 6G is still largely envisioned, with no fixed standard – just concepts, prospects, and possibilities. It is still largely driven by the possibilities to overcome some of the shortcomings of 5G. Due to this fact, there is limited literature on the 6G technology [19,40,41]. However, to understand the potential threat landscape of 6G, it will be well first to have an outlook of the potential 6G network. This section will present the security challenges of 6G, based on these prospective possibilities, in terms of 6G technologies.

4.5.1 An overview of 6G specification

The term 6G is the 6G standard for telecommunications [40]. 6G mobile network is expected to deliver extremely fast speeds, at data throughput up to 100–1,000 times faster than 5G. Advancements in virtual reality, autonomous vehicles, medicine, computer-based disaster forecasting, the IoT, and other new applications suggest even the extremely fast speed of 5G [42,43]. Therefore, researchers have begun to look beyond 5G to an envisioned world of blistering possibilities.

For instance, Viswanathan and Mogensen [44] outlined the complex technical and performance requirements expected for 6G communication. Some of these include data speeds greater than or equal to 1 Tbps, high operating frequency at 1 THz and above, low end-to-end delays ($\leq$ 1 ms), high reliability (10−9), high mobility ($\geq$ 1,000 km/h) and wavelength of $\leq$ 300 μm. 5G cannot deliver this successfully, and the proposed 6GB is expected to meet these performance expectations for intelligent healthcare.

Authors [45] outlined some of the future technological trends that will catalyze the development of 6G. Wearable devices, skin patchable and bio-implants, and textile-integrated devices may become more commonplace. Talking and gesturing to control context-aware devices will be quite common. Self-driving cars will litter busy roads. This and more complexities will mean that processing might be delegated to remote devices, which will need speeds beyond what 5G can guarantee. 6G can be that enabler for the future man–machine interface.

4.6 6G architecture

The envisioned 6G architecture is expected to be largely characterized by intelligent radios, edge intelligence, and intelligent network management. These concepts are discussed briefly below.

4.6.1 Intelligent radio

1G to 5G of networks had devices and transceivers integrated into one design. The integrated design meant that the hardware capabilities, such as the decoder's computational strength, the number of antennas, RF chains, phase shifters, and others, were quasi-static through the generational evolutions [46]. The rapid evolution in hardware design has catalyzed the possibility of separating the hardware from transceiver algorithms. This possibility of having an intelligent radio that operates as a unified framework and could allow for a dynamic and autonomous upgrade and configuration holds great prospects for 6G. It is expected that transceiver algorithms will be able to calculate the transceiver capabilities dynamically and use this to configure itself automatically based on the hardware capabilities [41,47].

6G is expected to leverage this algorithm-hardware separation, allowing for agile adaptation to diversified hardware that allows for upgrades. Huang *et al.* [46] noted that to maximize the possibilities of this separated design, an operating system will coordinate the handshake between the separated algorithm-hardware architecture, where the transceiver algorithm is seen as software running on the operating system. Plastiras *et al.* [48] highlighted the prospect of combining software-defined radio (SDR) and networking techniques to leverage high frequency bands and intelligently take advantage of different frequencies. Thus, providing support for intelligent radios. 6G technology can dynamically use different frequencies to maximize the benefits of the shared architecture.

4.6.2 Real-time intelligent edge (RTIE)

The use of AI/ML to acquire, store, and process data at the network edge is referred to as edge intelligence. In Edge Intelligence implementation, data generated from multiple devices are aggregated by an edge server [49]. Furthermore, aggregated data is shared among the edge servers associated with it for training models, which is used for analysis and prediction [50]. All this happens at the edge and, expectedly, achieves faster feedback for devices and reduced latency.

Specifically, key advantages of edge intelligence are the improvement of time-to-action, a significant reduction in latency, lower implementation cost, and minimize bandwidth usage. In addition, it arguably offers greater privacy over cloud computing since you have greater control over the data sent to the cloud [49]. That is, some of the processing can be achieved at the edge rather than the cloud, which could be resident in a different jurisdiction.

In its current implementation, 5G already offers some measure of support for autonomous vehicles and unmanned aerial vehicles (UAVs). However, this is limited and inefficient, given the obvious technical limitations. Achieving an efficient implementation of driverless cars and UAVs requires very low network latency, which the current network implementation cannot achieve [41]. Furthermore, it does not efficiently support network entities that are self-aware, self-adaptive, and with predictive capabilities [51]. 6G network is expected to handle complexities using real-time edge intelligence. For instance, the RTIE could be leveraged to achieve autonomous driving even in an unfamiliar environment in real-time [41]. Therefore, RTIE in 6G networks can be achieved for intelligent prediction, inference, and decision using live data [47].

However, from a security point of view, there are valid concerns. In RTIE implementation, data is aggregated from multiple sources, influencing the outcome of AI/ML algorithms. With a widened attack surface, there is a propensity for myriad attacks. Malicious attackers could target these widened attack surfaces to launch attacks such as data poisoning, data evasion, and others that could affect the AI/ML application and the expected result [50].

4.6.3 Intelligence network management

AI and ML already have several applications in the 5G cellular networks [52]. However, these applications are limited to the optimization threshold of the traditional network architecture of the 5G network. Note that at the time of the design of the 5G architecture design, AI was not necessarily taken into consideration. Therefore, these limitations prevent it from fully realizing the potential of AI in the 5G architecture [47]. The complexity of the 6G requirements is such that it will need a whole new level of network service orchestration and management [47]. For instance, 6G will require increased capacity, very low latency, high reliability, and the need for end-to-end (E2E) automation of network and service management using AI security in 6G.

Security has always been an important consideration in mobile communications. This is even more so given the possibilities of the envisioned scope and application of 6G. High-security considerations, secrecy, and privacy are expected features of 6G [53].

4.6.4 The 6G threat landscape

6G will undoubtedly revolutionize the technological landscape and have a wide range of use cases. Key highlights of the expected applications areas to be driven by 6G include industry 5.0, UAVs, CAVs, Smart Grid 2.0, collaborative robots, hyper-intelligent healthcare, digital twin, and extended reality [50]. While 5G was designed to support the IoT, 6G is envisioned to support the Internet of Everything (IoE), which is a collection of billions of heterogenous devices [54]. Therefore, 6G implementation will involve fundamental design and architectural shifts. The technological, architectural, and design changes and application range of the envisioned 6G would no doubt experience emerging risks and new security challenges. Some of the security challenges of 5G will be inherited, while new security challenges introduced by new technology and design will also be expected. This section will present the potential threats in the emerging 6G landscape.

4.6.5 Legacy design security (pre-6G)

As noted earlier, 6G will inherit some of the security challenges of the pre-6G era, given that some of the 5G technologies, such as software-defined networking (SDN), network function virtualization (NFV), multi-access edge computing (MEC), and network slicing will still form part of the 6G technology. Zhu and

Philip [54] identified the security challenges associated with the network softwarization technologies above, and they are summarized in Table 4.2.

4.6.6 AI-related security challenges

6G is expected to be largely AI-driven. AI has increasingly grown in popularity and in range of applicability. These growing use cases come a widened attack surface with emerging threats and privacy issues [55]. In the case of 5G, AI exists in isolated areas with large amounts of training data; in 6G, AI will become a core part of the system at almost all levels. With such enormous data generated from the IoE, AI will be at the core of 6G [13,59,64]. Expectedly, there will be emerging risks seeking to exploit possible vulnerabilities in the AI use case. The following potential security and privacy issues have been identified, as shown in Table 4.3.

Table 4.2 Security issues associated with the network softwarization

S. no.	Network softwarization technologies	Security issues	Threat mitigation and countermeasures
1	SDN	1. Attacks on SDN controller 2. Attacks on the northbound and southbound interface 3. Vulnerabilities of platforms used in deploying SDN controllers [55]	Automating Malware detection and mitigation for SDN controller [56].
2	NFV	Attacks targeting: 1. Virtual machines 2. Virtual network functions (VNF) 3. Hypervisor 4. VNF manager 5. NFV orchestrator [57]	1. Automating and mitigating the adversarial attack 2. Detect and mitigate at the origin 3. Improved hypervisor security 4. Use of inline solutions and scrubbing centres [58]
3	MEC	1. Physical security threats 2. Distributed Denial of Service (DDoS) 3. Man-in-the-middle attacks [59]	1. Improve physical security 2. AI-based anomaly detection system in 6G MEC [60] 3. Use a graph neural network (GNN)-based collaborative deep reinforcement learning (GCDRL) model for resource provisioning and mitigation [61]
4	Network slicing	1. DoS attacks 2. Information theft via compromised slices	Use of slice isolation to proactively mitigate DDoS [62] Enable security controls at different network layers [63]

Table 4.3 Forms of AI-related attacks

S. no.	AI attacks	Description
1	Poisonous attacks	Attackers may launch poisonous attacks. For example, AI relies on training data to influence outcomes, and attackers could maliciously inject manipulated samples. This would influence learning outcomes [19]
2	Evasion attacks	Here, the attacker targets the test phase and bypasses the learning model, by tampering with test instances. That is, by tampering with the input to the AI algorithm, the attacker can bypass the right classification mechanism, which will definitely affect the outcome [19,47]
3	ML API-based attacks	In this attack type, the adversary attacks the API of a ML model to obtain predictions on input feature vectors [19]. API-based attacks could include any of the following: 1. Parameter attacks – poorly validated parameters leading to cross-domain injection attacks. Data injection, data manipulation and logic corruption. 2. Identity attacks – exploitation of authentication and authorization flaws 3. Man-n-the-middle attacks – attackers can take advantage of API vulnerabilities to sit in the middle and obtain data from the unencrypted transmission 4. DoS/DDoS attacks – distorting API service by flooding it with massive requests [19]
	Infrastructure physical attacks	Adversaries may decide to intentionally attack the physical network to disrupt communication. ML algorithms rely on data to make decisions and once there is a disruption in flow, can obviously affect processing and decision-making [19]
5	Compromise of AI framework	There are existing vulnerabilities in AL/ML frameworks which can be exploited by adversaries

4.7 Threat mitigation and countermeasures

4.7.1 Poisonous attacks on ML systems

Traditional network management systems are largely manual in nature. The advent of AI has automated network management and made networking systems more intelligent [65]. Expectedly, 6G will introduce a deluge of data; with more accurate data, the learning outcomes of AI systems produce better results. However, this strength may be exploited by adversaries that introduce poisonous attacks to distort the training data [64]. With the share volume of data available in the 6G landscape, this could be an easy attack method by adversaries.

There are several scenarios this could be achieved. For instance, the adversary may manipulate the ML system's training sensor data to distort the learning outcomes [66,67]. Another scenario is in an autonomous vehicle, where an attacker may introduce stickers or specific background colors to the stop sign during training. The ML model would be "deceived" to think they are speed limits and cause the vehicle to keep moving, even at an intersection [65]. 6G is expected to catalyze mass adoption of driverless cars, and with potential threats such as this, it calls for a better approach to handling poisonous attacks on AI systems.

To mitigate poisonous attacks, intelligent systems need to be able to detect malicious data or distortions in training data. One way this is achieved is the intuitive use of anomaly detectors to filter out suspicious points. A recommended countermeasure for poisonous attacks that rely on statistical knowledge of training data is to learn the normal pattern or distribution of training data rather than remove all the malicious data. In conclusion, Wang *et al.* [65] summarizes the key defensive techniques against poisonous attacks as:

(i) Training data filtering – this basically detects malicious distortion of training data and filters them out. This can be achieved in different ways. One way this can be achieved is through input manipulation detection. Detecting changes in feature characteristics or labels can filter poisoned samples from training data [68]. An advantage of training data filtering is that it effectively removes outliers. A disadvantage is an inability to detect inliers [65].

(ii) Robust learning – to defend against poisonous attacks requires robust learning. This approach is based on statistical methods for detecting noise and outliers in training datasets. Wang *et al.* [65] classified robust learning into model robustifying and model verification. In model robustifying, the architecture of the trained model is modified to make the system more resilient and reduce the impact of the poisoned dataset [69]. A key advantage of model robustifying is that it is adversarially resistant. The disadvantage is that it is not strong enough to detect inliers [65]. In model verification, the characteristics of the model or training data are utilized to mitigate poisonous attacks. For instance, Subedar *et al.* [70] modeled the prediction of each layer of a deep neural network with parametric distributions and utilized model uncertainty to detect and filter out poisoned samples from clean ones. An advantage of model verification is that it can effectively detect backdoor attacks. The disadvantage is that its application is limited to detecting backdoor attacks [65].

(iii) Use of auxiliary tool – in the case of neural networks, the countermeasure adopted is to rely on an auxiliary tool to protect the neural network model. There are several implementations of this: for instance, in [71,72]. The advantages of this are that it improves the ability of the deep neural network, and the learned representations amplify signals important to classification. However, the key disadvantage is that it is limited to deep neural networks [65].

These methods are already in the 5G landscape but would have to be improved to greater effect.

4.7.2 Evasion attacks

While poisonous attacks target the training phase [73], evasion attacks target test time [74]. With the massive data explosion that will characterize the 6G landscape, adversaries may take advantage of this to target the test process. There are several ways of mitigating evasion attacks. For instance, Goodfellow *et al.* [75] proposed adversarial training where the training dataset is augmented with adversarial examples. Papernot *et al.* [76] proposed defensive distillation as a mitigation against evasion attacks. Cao and Gong [77] proposed a region-based classification method to mitigate evasion attacks in deep neural networks. In the 6G landscape, there will be a massive deluge of data, and these methods can be improved upon to cope with the emerging threat vectors.

4.7.3 ML API-based attacks

The 6G landscape will see many handshakes between devices. Adversaries will look to take advantage of the vulnerabilities in the API input vectors. There are many ways to mitigate this. This would include improved API security, the use of machine learning, and a combination of other methods.

4.7.4 Infrastructure physical attacks

The 6G landscape will involve a large interconnection of sensors and devices. As a result, adversaries can target the physical infrastructure itself to disrupt the system. Improving physical security is one way of mitigating this risk.

4.7.5 Compromise of AI framework

Adversaries may exploit existing vulnerabilities in AI/ML frameworks. Therefore, AI systems must be secure and periodically fixed. It is essential to do a comprehensive security risk assessment of the AI system and fix all vulnerabilities.

4.8 Recent trends and future directions

4.8.1 Recent trends

6G network architecture will usher in unprecedented growth in technology in terms of security and privacy. These include driverless buses and taxis on British roads by 2021 as autonomous driving research advances with highly coded security systems [78]. The populace's psychological trust is desired for this technology's smooth adoption, as stated in [79]. Spatial modulation [79] has been suggested considering Terahertz transmission to reduce hardware limitations. A growing trend is the development of Wireless Network on Chip (WNoC), made possible by transceivers and antennae that have been downsized [80]. El-Fatyany *et al.* [81] brought the trends and advancements of 6G wireless communication systems closer to reality. Also, the authors suggested a privacy system based on biocyber interfaces [81]. In the IoBNT model, the privacy method operated on top of the biocyber interface and delivered positive results with little adverse impacts. Imoize *et al.* [82] proposed the application of 6D holographic optical field technology which is constantly evolving. The strict requirements for faster data throughput and reduced latency, made possible by 6G, will make this possible. This technology provides a novel and appealing user experience in media and entertainment. The Internet of Bio-Nano Things adoption faces difficulties related to security.

4.8.2 *Future directions*

More investigations into quantum communication are required. As mentioned before in this paper, quantum communication and the Internet of Bio-Nano Things may not be developed enough to support 6G. Nevertheless, they will be important for future wireless networks. Quantum communication increases computational performance while also enhancing security. In this chapter, some security threats and attacks have been highlighted, and their proposed countermeasures have been given. There are promising areas for future research in software-defined networking, the Internet of Space Things, and resource allocation for CubeSats [82,83]. The open research questions are underwater communication, security, and channel estimation. Because of their low cost and low orbital altitude, Saeed *et al.* [84,85] envision the CubeSats as facilitating future wireless communication in space.

4.9 Conclusion

The deployment of more and more 5G networks has revealed the limitations of these networks, which undoubtedly supports the exploratory study of 6G networks as the next-generation solutions. The fundamental privacy and security concerns raised by 6G technology are covered in these evaluations. The following concerns with the projected 6G network were explored in this work: Issues with tactile communications, resources as services, variable radio access constraints, and varied high-frequency bands are all security-related problems with the network. The 6G spectrum, which had a wavelength of 30–3,000 μm and a frequency range of 0.1–10 THz, was examined. Additionally, the numerous attacks on 6G were outlined, and mitigation strategies, current trends, and future directions were suggested.

Acknowledgment

The work of Agbotiname Lucky Imoize is supported in part by the Nigerian Petroleum Technology Development Fund (PTDF) and, in part, by the German Academic Exchange Service (DAAD) through the Nigerian–German Postgraduate Program under Grant 57473408.

References

[1] NTT Docomo (2022). White paper: 5G evolution and 6G, *Technical Report, version 4.0*, 11–48.

[2] Nguyen, V. L., Lin, P. C., Cheng, B. C., Hwang, R. H., and Lin, Y. D. (2021). Security and privacy for 6G: a survey on prospective technologies and challenges. *IEEE Communications Surveys & Tutorials*, 23(4), 2384–2428.

[3] Zhang, S. and Zhu, D. (2020). Towards artificial intelligence-enabled 6G: state of the art, challenges, and opportunities. *Computer Networks*, 183, 107556.

[4] Tang, F., Kawamoto, Y., Kato, N., and Liu, J. (2020). Future intelligent and secure vehicular network toward 6G: machine-learning approaches. *Proceedings of the IEEE*, 108, 292–307.

[5] Positive 5G Outlook post-Covid-19: What Does it Mean for Avid Games? June 29, 2022, visited on 14 August 2022

[6] Market Share of Mobile Telecommunication Technologies worldwide from 2016 to 2025, by Generation. Accessed Oct. 15, 2022

[7] Salameh, A. I. and El Tarhuni, M. (2022). From 5G to 6G – challenges, technologies, and application. *Future Internet 2022*, 14, 117. Accessed from https:// doi.org/10.3390/fi140117.

[8] Mihovska, N. A. and Prasad, R. (2020). Overview of 5G new radio and carrier aggregation: 5G and beyond the network. In *Proceedings of the 2020 23rd International Symposium on Wireless Personal Multimedia Communications (WPMC)*, Okayama, Japan, 1–6.

[9] Berardinelli, G., Mahmood, N. H., Rodriguez, I., and Mogensen, P. (2018). Beyond 5G wireless IRT from industry 4.0: Design principles and spectrum aspects. In *Proceedings of the 2018 IEEE Globe Com Workshop*, Abu Dhabi, United Arab Emirate, 1–6.

[10] Fang, I., Zang, B., Li, Y., Lu, Z., Ge, C., and Meng, W. (2020). Countermeasure based on smart contract and A1 against Dos/DDoS attack in 5G circumstances. *IEEE Networks,* 34, 54–61.

[11] David, K. and Berndy, H. (2018). 6G vision and requirements: is there any for beyond 5G? *IEEE Vehicular Technology Magazine,* 13, 72–80.

[12] ITU-R M.2370-0. IMT Traffic Estimates for the year 2020 to 2030. ITU Publications 2015. https://www.itu.int/dms-pub/itu-r/op6/R-REP-M.2370-2015-PDE-E.pdf. Accessed October 15, 2022.

[13] Tariq, F., Khandaker, M. R., Wong, K. K., Imran, M. A., Bennis, M., and Debbah, M. (2020). A speculative study on 6G. *IEEE Wireless Communications*, 27(4), 118–125.

[14] Minghao, W., Tianqing, Z., Tao, Z., Jun, Z., Shui, Y., and Wanlei, Z. (2020). Security and privacy in 6G network: new area and new challenges. *Digital Communications and Networks*, 6, 281–291. https://doi.org/1016/j.dcan.2020 07.003. Accessed October 15, 2022.

[15] Nakano, T. Y., Okaie, Y., Kobayashi, S., Hara, T., Hiraoka, Y., and Haraguchi, T. (2019). Methods and application of mobile molecular communications science. *Proceedings of IEEE,* 107(7), 1442–1456.

[16] Farsad, N., Yilmaz, H. B., Eckford, A., Chae, C. B., and Guo, W. (2016). A comprehensive survey of recent advancements in molecular communication, *IEEE Communications Survey Tutorials*, 18(3), 1887–1919.

[17] Khan, L. U. (2017). Visible light communications: applications, architecture, standardization and research challenges. *Digital Communications Network,* 3(2), 78–88.

[18] Chen, C., Bian, R., and Haas, H. (2018). Omnidirectional transmitter and receiver design for wireless infrared uplink transmission in lifi. In *IEEE International Conference on Communication Workshop*, 1–6.

[19] Porambage, P., Gür, G., Osorio, D. P. M., Liyanage, M., Gurtov, A., and Ylianttila, M. (2021). The roadmap to 6G security and privacy. *IEEE Open Journal of the Communications Society*, 2, 1094–1122.

[20] Hewa, T., Ylianttila, M., and Liyanage, M. (2021). Survey on blockchain based smart contract: application, opportunities and challenges. *Journal of Network Computer Applications*, 177, 101–113.

[21] Liu, Z. (2019). A survey on blockchain: a game theoretical perspective. *IEEE Access*, 7, 47615–47643.

[22] Hewa, T., Hu-Yu Liyanase, M., Kanbare, S., and Yilanttila, M. (2021). Survey on blockchain based smart contract: technical aspect and future research. *IEEE Access,* 9, 87643–87662. doi:10.1109/ACCESS.2021.3068178

[23] Bunz, B., Agrwwal, S., Zamani, M., and Bonch, D. (2020). Zether: towards privacy in a smart contract world. In *Proceedings of International Conference on Finance Cryptography & Data Security*, pp. 423–443.

[24] Feng, Q., He, D., Zeadally, S., Khan Mok, and Kumar, N. (2019). A survey on privacy protection in blockchain system. *Journal of Network Computer Application*, 26, 45–58.

[25] Bao, Z., Wang, Q., Shi, W., Wang, L., Lei, H., and Chen, B. (2020). When blockchain meets SGX: an overview, challenges and open issues. *IEEE Access*, 8, 170404–170420.

[26] Cohan, V. W. (2021). The double spending problem and cryptocurrencies, *SSRN Electronic Journal*, 11, 3090174. https://ssrn.com/abstract=3090174. Accessed October 23, 2022.

[27] Otte, P., De Bos, M., and Pouwelse, J. (2020). Trustchain: a Sybil resistance scalable blockchain. *Future Generation Computer System*, 107, 770–780.

[28] Cai, Y. and Zhu, D. (2016). Fraud detections for online business: a perspective from blockchain technology. *Finance Innovation*, 2(1), 20.

[29] Nwaza, F., Ibrahim, J., Junaid, M., Kousar, S., and Ali, M. A. (2020). A review of vision and challenges of 6G technology. *International Journal of Advanced Computer Science and Application*, 11(2), 643–649.

[30] Zaidi, A. A., Baldemair, R., Moles-Case, V., He, N., Werner, K., and Cedergren, A. (2018). OFDM numerology design for 5G new radio to support IoT, eMBB, and MBSFN. *IEEE Communications Standards Magazine*, 2(2), 78–83.

[31] Lee, Y. L., Qin, D., Wang, L. C., and Hong, G. (2019). 6G massive radio access networks: key issues, technologies, and future challenges. arXiv preprint arXiv:1910.10416.

[32] Hao, Y. M (2021). Investigation and technological comparison of 4G and 5G networks. *Journal of Computer and Communications*, 9, 36–43.

[33] Stuber, G. L. (2020). *Principles of Mobile Communication.* Springer International, Berlin, pp. 24–47.

[34] Long, W., Chen, R., Moretti, M., Zhang, W., and Li, J. (2021). A promising technology for 6G wireless networks: intelligent reflecting surface. *Journal of Communication Information Network*, 6, 1–16.
[35] Benn, H. (2020). Hyper Connectivity with 6G; TelecomTV. https://www.telecomtv.com/content/what-next-for-wireless-infrastructure-summit/hyper-connectivity-with-6g-40147/. Accessed October 24, 2022.
[36] Alraih, S., Shayea, I., Behjati, M., *et al.* (2022). Revolution or evolution? Technical requirements and considerations towards 6G mobile communications. Sensors, 22, 1–29. https://doi.org/10.3390/s22030762. Accessed October 28, 2022.
[37] Di Renae, M., Zappone, A., Debbah, M., *et al.* (2021). Smart radio environments powered by reconfigurable intelligent surfaces: how it works, state of research and the road ahead. *IEEE Journal of Selected Areas of Communication*, 38, 2450–2525.
[38] 3GPP. Release 18. https://www.3gpp.org/release18. Accessed October 24, 2022.
[39] Ojo, S., Imoize, A. L., and Alienyi, D. (2021). Radial basis function neural network path loss prediction model for LTE networks in multi-transmitter signal propagation environments. *International Journal of Communication Systems*, 34, 1–26.
[40] Wang, M., Zhu, T., Zhang, T., Zhang, J., Yu, S., and Zhou, W. (2020). Security and privacy in 6G networks: new areas and new challenges. *Digital Communications and Networks*, 6(3), 281–291.
[41] Nawaz, F., Ibrahim, J., Muhammad, A. A., Junaid, M., Kousar, S., and Parveen, T. (2020). A review of vision and challenges of 6G technology. *International Journal of Advanced Computer Science and Applications*, 11(2), 643–649.
[42] Cacciapucti, A. S., Sankhe, K., Caleffi, M., and Chowdhury, K. R. (2018). Beyond 5G: THz-based medium access protocol for mobile heterogeneous networks. *IEEE Communications Magazine*, 56(6), 110–115.
[43] Nayak, S. and Patgiri, R. (2021). 6G communication technology: a vision on intelligent healthcare. In *Health Informatics: A Computational Perspective in Healthcare* (pp. 1–18). Springer, Singapore.
[44] Viswanathan, H. and Mogensen, P. E. (2020). Communications in the 6G era. *IEEE Access*, 8, 57063–57074.
[45] Letaief, K. B., Chen, W., Shi, Y., Zhang, J., and Zhang, Y. J. A. (2019). The roadmap to 6G: AI empowered wireless networks. *IEEE Communications Magazine*, 57(8), 84–90.
[46] Huang, T., Yang, W., Wu, J., Ma, J., Zhang, X., and Zhang, D. (2019). A survey on green 6G network: architecture and technologies. *IEEE Access*, 7, 175758–175768.
[47] Yang, P., Xiao, Y., Xiao, M., and Li, S. (2019). 6G wireless communications: vision and potential techniques. *IEEE Network*, 33(4), 70–75.
[48] Plastiras, G., Terzi, M., Kyrkou, C., and Theocharidcs, T. (2018). Edge intelligence: challenges and opportunities of near-sensor machine learning

applications. In *2018 IEEE 29th International Conference on Application-Specific Systems, Architectures and Processors (asap)* (pp. 1–7). IEEE.

[49] Porambage, P., Gür, G., Osorio, D. P. M., Livanage, M., and Ylianttila, M. (2021). 6G security challenges and potential solutions. In *2021 Joint European Conference on Networks and Communications & 6G Summit (EuCNC/6G Summit)* (pp. 622–627). IEEE.

[50] Kibria, M. G., Nguyen, K., Villardi, G. P., Zhao, O., Ishizu, K., and Kojima, F. (2018). Big data analytics, machine learning, and artificial intelligence in next-generation wireless networks. *IEEE Access*, 6, 32328–32338.

[51] Li, R., Zhao, Z., Zhou, X., *et al.* (2017). Intelligent 5G: when cellular networks meet artificial intelligence. *IEEE Wireless Communications*, 24(5), 175–183.

[52] Dang, S., Amin, O., Shihada, B., and Alouini, M. S. (2020). What should 6G be? *Nature Electronics*, 3(1), 20–29.

[53] Siriwardhana, Y., Porambage, P., Liyanage, M., and Ylianttila, M. (2021). AI and 6G security: opportunities and challenges. In *2021 Joint European Conference on Networks and Communications & 6G Summit (EuCNC/6G Summit)* (pp. 616–621). IEEE.

[54] Zhu, T. and Philip, S. Y. (2019). Applying differential privacy mechanism in artificial intelligence. In *2019 IEEE 39th International Conference on Distributed Computing Systems (ICDCS)* (pp. 1601–1609). IEEE.

[55] Ahmad, I., Kumar, T., Liyanage, M., Okwuibe, J., Ylianttila, M., and Gurtov, A. (2018). Overview of 5G security challenges and solutions. *IEEE Communications Standards Magazine*, 2(1), 36–43.

[56] Dharma, N. G., Muthohar, M. F., Prayuda, J. A., Priagung, K., and Choi, D. (2015). Time-based DDoS detection and mitigation for SDN controller. In *2015 17th Asia-Pacific Network Operations and Management Symposium (APNOMS)* (pp. 550–553). IEEE.

[57] Khan, R., Kumar, P., Jayakody, D. N. K., and Liyanage, M. (2019). A survey on security and privacy of 5G technologies: potential solutions, recent advancements, and future directions. *IEEE Communications Surveys & Tutorials*, 22(1), 196–248.

[58] Repetto, M., Bruno, G., Yusupov, J., Lamanna, G., Ertl, B., and Carrega, A. (2022). Automating mitigation of amplification attacks in NFV services. *IEEE Transactions on Network and Service Management*, 19(3), 2382–2396.

[59] Ranaweera, P., Jurcut, A. D., and Liyanage, M. (2021). Survey on multi-access edge computing security and privacy. *IEEE Communications Surveys & Tutorials*, 23(2), 1078–1124.

[60] Guşatu, M. and Olimid, R. F. (2022). Improved security solutions for DDoS mitigation in 5G multi-access edge computing. In *International Conference on Information Technology and Communications Security* (pp. 286–295). Springer, Cham.

[61] Deng, Y., Jiang, H., Cai, P., *et al.* (2022). Resource provisioning for mitigating edge DDoS attacks in MEC-enabled SDVN. *IEEE Internet of Things Journal*, 9(23), 24264–24280.

[62] Sattar, D. and Matrawy, A. (2019). Towards secure slicing: using slice isolation to mitigate DDoS attacks on 5G core network slices. In *2019 IEEE Conference on Communications and Network Security (CNS)* (pp. 82–90). IEEE.

[63] Wichary, T., Mongay Batalla, J., Mavromoustakis, C. X., Żurek, J., and Mastorakis, G. (2022). Network slicing security controls and assurance for verticals. *Electronics*, 11(2), 222.

[64] Wijethilaka, S. and Liyanage, M. (2021). Survey on network slicing for Internet of Things realization in 5G networks. *IEEE Communications Surveys & Tutorials*, 23(2), 957–994.

[65] Wang, C., Chen, J., Yang, Y., Ma, X., and Liu, J. (2022). Poisoning attacks and countermeasures in intelligent networks: status quo and prospects. *Digital Communications and Networks*, 8(2), 225–234. https://doi.org/10.1016/j.dcan.2021.07.009.

[66] Qayyum, A., Usama, M., Qadir, J., and Al-Fuqaha, A. (2020). Securing connected & autonomous vehicles: challenges posed by adversarial machine learning and the way forward. *IEEE Communications Surveys & Tutorials*, 22(2), 998–1026.

[67] Dibaei, M., Zheng, X., Jiang, K., *et al.* (2020). Attacks and defences on intelligent connected vehicles: a survey. *Digital Communications and Networks*, 6(4), 399–421.

[68] Saha, A., Subramanya, A., and Pirsiavash, H. (2020). Hidden trigger backdoor attacks. In *Proceedings of the AAAI Conference on Artificial Intelligence* (Vol. 34, No. 07, pp. 11957–11965).

[69] Jagielski, M., Oprea, A., Biggio, B., Liu, C., Nita-Rotaru, C., and Li, B. (2018). Manipulating machine learning: poisoning attacks and countermeasures for regression learning. In *2018 IEEE Symposium on Security and Privacy (SP)* (pp. 19–35). IEEE.

[70] Subedar, M., Ahuja, N., Krishnan, R., Ndiour, I. J., and Tickoo, O. (2019). Deep probabilistic models to detect data poisoning attacks. arXiv preprint arXiv:1912.01206.

[71] Tran, B., Li, J., and Madry, A. (2018). Spectral signatures in backdoor attacks. In *32nd Conference on Neural Information Processing Systems (NeurIPS 2018)* (pp. 1–16), Montréal, Canada.

[72] Zhao, Y., Chen, J., Zhang, J., Wu, D., Teng, J., and Yu, S. (2020). PDGAN: a novel poisoning defence method in federated learning using generative adversarial network. In *International Conference on Algorithms and Architectures for Parallel Processing* (pp. 595–609). Springer, Cham.

[73] Biggio, B., Nelson, B., and Laskov, P. (2012). Poisoning attacks against support vector machines. arXiv preprint arXiv:1206.6389.

[74] Biggio, B., Corona, I., Maiorca, D., *et al.* (2013). Evasion attacks against machine learning at test time. In *Joint European Conference on Machine Learning and Knowledge Discovery in Databases* (pp. 387–402). Springer, Berlin, Heidelberg.

[75] Goodfellow, I. J., Shlens, J., and Szegedy, C. (2014). Explaining and Harnessing Adversarial Examples. CoRR, abs/1412.6572.

[76] Papernot, N., McDaniel, P., Wu, X., Jha, S., and Swami, A. (2016). Distillation as a defence to adversarial perturbations against deep neural networks. In *2016 IEEE Symposium on Security and Privacy (SP)* (pp. 582–597). IEEE.

[77] Cao, X. and Gong, N. Z. (2017). Mitigating evasion attacks to deep neural networks via region-based classification. In *Proceedings of the 33rd Annual Computer Security Applications Conference* (pp. 278–287).

[78] Independent. Driverless Buses and Taxis to Be Launched in Britain by 2021 (2018). https://www.independent.co.uk/travel/news-and-advice/self-driving-buses-driverless-cars-edinburgh-fife-forth-bridge-london-greenwich-a8647926.html. Accessed December 18, 2022.

[79] Alafia, A., Ajose, S., and Imoize, A. (2018). A study on low-complexity transmits antenna selection for generalized spatial modulation. *IIUM Engineering Journal*, 19, 105–117.

[80] Imoize, A. L., Ibhaze, A. E., Atayero, A. A., and Kavitha, K. V. N. (2021). Standard propagation channel models for MIMO communication systems. *Wireless Communications and Mobile Computing*, 2021, 36.

[81] El-Fatyany, A., Wang, H., Abd El-Atty, S. M., Khan, M. (2020). Biocyber interface-based privacy for internet of bio-nano things. *Wireless Personal Communications*, 114, 1465–1483.

[82] Imoize, A. L., Adedeji, O., Tandiya, N., and Shetty, S. (2021). 6G enabled smart infrastructure for sustainable society: opportunities, challenges, and research roadmap. *Sensors*, 21, 1709. https://doi.org/10.3390/s21051709.

[83] Ndjiongue, A. R., Ngatched, T. M. N., Dobre, O. A., and Armada, A. G. (2020). VLC-based networking: feasibility and challenges. *IEEE Networks*, 34, 158–165.

[84] Saeed, N., Elzanaty, A., Almorad, H., Dahrouj, H., Al-Naffouri, T. Y., and Alouini, M. S. (2020). CubeSat communications: recent advances and future challenges. *IEEE Communications Surveys and Tutorials*, 22, 1839–1862.

[85] Yamin, M. M., Ullah, M., Ullah, H., Katt, B. (2021). Weaponized AI for cyber-attacks. *Journal of Information Security and Applications*, 57, 102722.

Chapter 5

Security threat landscape for 6G architecture

Gabe Obrist[1], Oscar Okechukwu[1], Andrew Cross[1], Dinh-Thuan Do[1] and Agbotiname Lucky Imoize[2,3]

Abstract

This chapter explores the potential benefits of using reconfigurable intelligent surfaces (RISs) to enhance the physical layer security (PLS) of promising sixth-generation (6G) wireless systems relying on non-orthogonal multiple access (NOMA) systems. RISs are a new technology that can dynamically modify the propagation environment of wireless signals, allowing for increased efficiency and security. By leveraging RISs in NOMA systems, this chapter demonstrates that it is possible to enhance the secrecy performance of the system while simultaneously improving the overall spectral efficiency. The chapter provides a comprehensive overview of the theoretical foundations of PLS for 6G systems. We aim to consider a case study of RIS-aided NOMA systems, including the optimization of the RIS reflection coefficients and performance analysis of security concerns. Simulation results demonstrate that RIS-aided NOMA systems can achieve significant improvements in secrecy performance, particularly in scenarios with a high number of meta-surfaces of RIS. The chapter concludes that RIS-aided NOMA has the potential to be an effective solution for enhancing the PLS of wireless communication systems.

Keywords: 6G; Physical security; Reconfigurable intelligent surface; Non-orthogonal multiple access

5.1 Introduction

The next generation of cellular communications, known as sixth generation (6G), is currently in development and is set to succeed fifth generation (5G). With an ambitious vision of truly autonomous networks, 6G is expected to be commercially deployed in next decade. This new standard will be able to support speeds of

[1]School of Engineering, University of Mount Union, USA
[2]Department of Electrical and Electronics Engineering, Faculty of Engineering, University of Lagos, Nigeria
[3]Department of Electrical Engineering and Information Technology, Institute of Digital Communication, Ruhr University, Germany

over 1 Tbps, which is 50 times faster than 5G, and latency is projected at 10–100 µs. Scientists predict that this criterion will increase connectivity not only for traditional coverage areas in 5G but also for applications related to communications at many environments such as space, air, ground, and sea. By providing improved coverage and network functionality, 6G technology will facilitate the development of various digital services, such as implantable devices, tactile Internet, wearable displays, and autonomous vehicles [1].

Despite the ongoing progress in building 6G networks, there is limited research conducted on security and privacy matters. The majority of existing research in this field has centered on Internet of Things (IoT) networks, with only brief examinations of particular technologies, such as Artificial Intelligence (AI) and machine learning (ML) [2,3]. This is because many fundamental components of 6G networks are yet to be defined, which has resulted in a lack of related work. Therefore, further research is required to understand and address the potential security and privacy challenges that may arise in 6G networks.

The objective of this chapter is to take cues from the security improvements made in prior cellular network generations and conduct a systematic review of existing research on security and privacy in 6G wireless networks. While some recent studies have investigated the three-layer architecture of 6G, consisting of the physical layer, network layer, and application layer, we only focus on assessing security performance at the physical layer. The results of this chapter can be used to detect possible attacks and corresponding solutions for IoT networks powered by 6G. This chapter presents a narrow subject matter that can be highly beneficial to operators and developers in guarding against particular attacks that might threaten essential protocols that govern a broad range of 6G applications.

5.2 Designing 6G wireless systems with reconfigurable intelligent surfaces

Reconfigurable intelligent surfaces (RISs) are a type of artificial material that can be used to control and manipulate radio waves. They consist of a large number of small, passive elements that can be electronically controlled to reflect, refract, or absorb electromagnetic waves in a way that is tailored to a specific application or environment. Its applications have been studied regarding how this approach may assist with wireless coverage due to their reflection ability. Primarily, the placement of this surface can maximize the coverage, due to aiming for great reflection. This can also be helpful for applications regarding advanced networks such as non-orthogonal multiple access (NOMA) networks [4–6]. In regards to eavesdropper (Eve) distributions, the distribution of the Eves location in the wiretap link and the RIS-aided communication systems are the main factors that impact the secrecy performance and showcase a tradeoff between the efficiency of the system and the RIS elements present [7]. The physical layer security (PLS) of RIS-aided transmission for a random user with a multi-antenna Eve, however, is preferable to those without it due to the probability density function (PDF) and cumulative distribution function (CDF) of the signal-to-interference-plus-noise ratio, both metrics being used to evaluate security performance such as secrecy outage and nonzero secrecy capacity probability. A RIS with a small number of reflective elements, however, does not improve systems with small path loss of non-line-of-sight [8]. Device-to-device networks can rely on both RIS and a full duplex (FD) jamming receiver for added security. This is achieved through a beamforming design for an FD receiver that helps suppress and inject artificial noise signals in the direction of Eves to prevent information leakage [9]. In terms of wireless transmission, this can also be made more secure through RIS. Originally RIS' effectiveness is limited due to a 'double fading' effect on the reflecting channel link between the user and the transmitter. However, through a new design with reflective elements that adjust phase shift and now signal amplitude, we can relieve the impact of double fading and achieve significantly higher secrecy performance gain compared to the original RIS design [10]. Spectrum efficiency (SE) and security can also be improved through RIS, especially for multi-user cellular networks that are prone to jamming and eavesdropping attacks. The re-occurring issue of worst-case sum rate maximization is tackled through imperfect angular channel state information (CSI). A heuristic scheme is proposed that converts the imperfect CSI into a more robust one, and through simulations of realistic RIS and communication models, the SE and security are improved [11].

5.3 PLS for 6G systems

In [12], PLS is a fairly new concept when discussing communications. PLS allows confidential information to be exchanged between sender and user via wireless communications without any 'eavesdroppers' intercepting said information. There are two ways to carry out PLS without relaying a higher-encryption: secret keys are generated or there is no need for a generated key. The authors in [13] studied PLS. The key-generated PLS is one form that hides information from Eves by mapping to planned rotated signals. This way, the two-way exchange between sender and receiver includes six OFDM symbol times. The goal is to ultimately have the exchange and synchronization occur in four symbols. Beyond 5G and 6G, securing more applications at a low latency will be more successful by extending PLS algorithms using a key-based PLS. In [14], visible light communications (VLC) are presented as a key supporting technology for 6G wireless communication. This technology uses artificial environmental lighting as a transfer channel. Benefits of using a VLC system include resistance to interference and less likely security breaches that occur. A key feature of using a VLC system includes the Watermark Blind PLS (WBPLSec). This feature uses RGB LEDs that offer wavelength division multiplexing as a useful support for the spread-spectrum watermarking. What is now being discussed is to combine WBPLSec with RGB LED jamming to provide improvements on secrecy capacity and the probability of outage.

Heterogeneous IoTs and multi-access mobile edge computing (MA-MEC) are seen as supporting technologies for building a smart city which is another application using PLS. MA-MEC is known for supporting resource-limited and computational-sensitive services by computation offloading and distributed content delivery/caching. In [15], a solution to this would be using PLS technologies to investigate wiretap coding, resource allocation, signal processing, multi-node cooperation, etc.

The end-end secure communication of N-pair cellular network is also headed towards NOMA. It also includes physical layer network coding (PLNC) and, in [16], both can improve spectral and temporal efficiency. Networking coding is a very big aspect of improving the throughput, efficiency and reliability of the two-way relay networks (TWRNs). This essentially brings new security problems, however, when considering the physical layer. To enhance security performances, network coding schemes are incorporated with key diagram generations in [17].

If 6G wants to remain in concept, then two key areas will need to be addressed: threat vectors from new radio technologies and protection mechanisms will need to be very efficient in preventing Eves of gaining access to information, for example, according to [18]. In [19], new security threats will need to be addressed in addition to the ones discussed previously: sensing capabilities will be a huge piece to this becoming a reality. In addition to edge and device-embedded intelligence, these technologies can bring new protocols that will be even more secure than ever. This is a fairly new concept and there is going to be difficulties that must have solutions for this concept to become a reality in the near future. In addition, the authors of [20] also state that reinforcement learning (RL) algorithms that can boost wireless devices optimize security performances against smart attacks, which is another kind of threat. RL algorithms can be applied in 6G systems focusing on jammers, eavesdropping, spoofers, and interference attackers.

5.4 The related works considering performance analysis of RIS-NOMA

This chapter deals with the security of 6G architecture by focusing on the physical layer. The importance of signal processing at the physical layer motivates us to look for advanced techniques to enlarge the coverage and spectrum sharing among users. The authors in [21] presented that the emergence of 5G and 6G wireless networks has created a substantial need for extensive device connectivity and efficient use of the radio spectrum. Multiple-input-multiple-output (MIMO) antennas have played a significant role in achieving this connectivity and enabling high-quality signal transmission among devices. However, the deployment complexity and hardware costs of MIMO technology are major drawbacks. To address these concerns, RISs have been introduced. RIS can perform almost as well as MIMO-equipped devices while using low-cost meta-surfaces to control the propagation of radio waves and do not rely on external energy sources. RIS can also work in full-duplex mode, making them an ideal choice for enabling massive device connectivity. Despite being a relatively new technology, there is a significant amount of research and publications available on integrating RIS with emerging technologies such as NOMA, CR, VLC, and PLS. Growing concerns regarding privacy and PLS have resulted in significant research efforts in this area.

In a recent study, the authors of [22] explored RISs, which are artificial surfaces made of electromagnetic material that can be electronically controlled using integrated electronics. RISs have unique wireless communication capabilities, as they can control the transmission environment and improve signal quality at the receiver. They can transform propagation environments into smart ones and have several advantages over other wireless relaying technologies. RISs devices are energy efficient and can shape the transmission signal through soft programming. They are not easily affected by external electromagnetic interference and can operate in FD transmission mode with a full-band response. While RISs are a new technology and have only been studied in a few papers, their potential applications in wireless communication systems have been recognized. However, their use to enhance the secrecy performance of wireless systems is still relatively unexplored. Previous research has focused on optimizing designs such as beamforming and jamming. In contrast, this paper examines the secrecy performance of systems using RISs in a general model with direct links between the source and the legitimate user or Eve. The study aims to quantify the benefits of using RISs in such a setup on the PLS of the system, as measured by the secrecy outage probability (SOP) metric. Additionally, the paper provides an asymptotic SOP analysis for high signal-to-noise ratios (SNRs) and a large number of RIS elements to gain more insights. Overall, RISs are expected to become a critical technology in future wireless systems.

The authors in [23] reviewed the evolution of wireless communication technologies that began in the 1980s with first-generation (1G) cellular networks and have since seen significant advancements with 2G, 3G, and 4G networks. The deployment of 5G wireless technologies began in 2020, and it is expected to continue developing until full coverage is achieved by 2025, primarily on a software-based level. The key feature of 5G is the cloudification of networks with microservice-based architecture that abstracts physical resources to virtual and logical environments, enabling on-demand automated learning management functions. Despite the incomplete rollout of 5G, researchers are already envisioning the 6G of mobile communication, with standardization expected to begin around 2026. The 6G wireless networks will be based on intelligent network orchestration and management, utilizing multiple technologies such as RISs, VLC, electromagnetic-orbital angular momentum, cell-free communications, and quantum computing. The 6G architecture will incorporate a heterogeneous cloud infrastructure, requiring the capability to discover multiple cloud services and dynamic function placement. Security and privacy concerns will also be addressed, particularly with regards to novel technologies like blockchain, VLC, TeraHertz (THz), and quantum computing. The security considerations will cover PLS, network information security, application security, and deep learning-related security.

In [24], the study focused on NOMA which is expected to play a crucial role in 5G and beyond due to its superior SE compared to conventional orthogonal multiple access (OMA). It enhances SE and user connectivity by broadcasting signals in wireless communication networks. However, ensuring communication confidentiality between the base station (BS) and legitimate users (LUs) is equally important. RIS is a new technology that can manage the reflection properties of radio waves by adjusting amplitude-reflection and phase coefficients, thereby enhancing or reducing received signals by users. This provides more possibilities for PLS issues. Previous studies have investigated the secrecy outage probability (SOP) and PLS of RIS-aided networks, but few have considered RIS-aided NOMA networks with both direct and reflected links. In this paper, the authors propose a new RIS-aided NOMA network in which a BS communicates with LUs and an Eve via the assistance of the RIS. Unlike previous literature that focuses on enhancing LU performance, the authors propose a new RIS design that eliminates signals received by the Eve, thereby improving secrecy performance and opening new directions for PLS design of communication networks.

In [25], the authors explored PLS-aided RIS system which is expected to play a critical role in the development of 6G wireless networks, especially in terms of achieving secure communications. RISs can be integrated with other emerging communication technologies to enhance their performance gains. In particular, RISs can be utilized to improve the PLS of wireless communication systems by deploying them near eavesdroppers to cancel out their signal and decrease information leakage. Mobile wireless communication has seen rapid growth in data traffic due to the increased usage of smart devices. This has led to an enormous demand for radio spectrum resources, such as bandwidth and energy, which requires the consideration of spectral and energy efficiency when designing future wireless networks. Cognitive radio (CR) has emerged as an efficient technique to improve spectral efficiency. RISs can be deployed to improve the performance of the secondary network (SN) while enhancing the secrecy rate of the primary network (PN) in CR networks. The RIS technology is used as a friendly jammer to ensure a high-secrecy performance for the PN, enabling a win–win situation between the two networks. The paper derives closed-form expressions for the secrecy outage probability (SOP) of the PN and provides asymptotic analysis for the SOP of the PN. Numerical and simulation results confirm the benefits of the proposed system model.

In [26], the advent of 5G wireless communications has brought about several benefits, including inflated data rates, highly capable base stations, and low latency, which are pivotal to the development of billions of devices. To meet the demand for larger data traffic in mobile communications, researchers have come up with several solutions, such as mmWaves, m-MIMO, small cell techniques, RIS, NOMA, ultra-dense heterogeneous networks, and free space and underwater optical wireless communication networks. Of these, RIS is gaining immense attention as an efficient mechanism for future secure wireless communication systems, with its passive beamforming and the ability to customize the propagation medium by means of adjustable phase and amplitude. RIS has several promising applications, such as enhancing m-MIMO systems, maximizing SNR, promoting signal coverage, and optimizing the beamforming of multi-user channels. Several studies have investigated RIS-assisted systems, including MISO networks, NOMA networks, optical wireless communication networks, and UOWC systems. RIS has been proposed as a promising solution for PLS, and joint beamforming and jamming techniques have been analyzed to determine the secrecy performance without Eve CSI.

In [27], a new metamaterial, called the simultaneous transmitting and reflecting RIS (STAR-RIS), has been proposed. This metamaterial supports both electric-polarization and magnetization currents, which allows it to provide full-space service coverage and more flexible deployment. NOMA is a technique that can enhance spectral efficiency by serving multiple users within one spectrum resource using successive interference cancellation. However, it is sensitive to wireless channels and introduces additional co-channel interference. Integrating NOMA and STAR-RIS can naturally overcome this randomness and boost spectral performance. Previous studies have investigated coverage characterization, proposed NOMA-integrated over-the-air federated learning (AirFL), studied resource allocation, and investigated STAR-RIS-assisted NOMA-enhanced coordinated multi-point (CoMP) transmission. A general analytical framework for STAR-RIS-assisted NOMA networks has also been proposed. However, wireless transmissions can expose confidential or sensitive data to vulnerable communication environments due to their broadcast characteristic. To provide secure and private information transmission, PLS has been proposed, which exploits the inherent characteristics of wireless channels to degrade legitimate information leakage. The integration of RIS into PLS has received much attention due to its ability to adjust wireless channels. Studies have investigated the secrecy capacity maximization problem, proposed an artificial noise (AN)-based jamming protocol via RIS, and a robust AN-aided transmission scheme. For NOMA networks, the secure transmission problems of RIS-assisted NOMA networks have been studied under imperfect and statistical eavesdropping CSI. Recent research has shown that compared with conventional RIS, STAR-RIS can provide higher secrecy performance.

5.5 A case study: PLS for RIS-NOMA

5.5.1 System model

In situations where there is not existence of a direct link between the access point (AP) and IoT devices that results in a strong signal received and processed at the destination, and the RIS link can strengthen the signal at the destination, shown in Figure 5.1. RIS is equipped with M metasurface. We assume a flat fading channel model that varies slowly for all channels.

Then, the received signal reflected by the RIS at D_1 can be expressed as [21]

$$y_{D_1}^{no} = \upsilon \sum_{t=1}^{T} \frac{h_{sr}h_{r1}}{\sqrt{d_{sr}^{\omega} d_{r1}^{\omega}}} e^{j\chi_t} \left(\sqrt{P\beta_1}x_1 + \sqrt{Pa_2}x_2\right) + \eta_1 \tag{5.1}$$

In this model, $\upsilon \in (0,1]$ represents the amplitude reflection coefficient and χ_t is the adjustable phase applied by the tth reflecting element of the RIS. The signal for D_i is denoted as x_i, where $i = 1, 2$, and P is the normalized transmission power at the AP. The power allocation coefficients α_i are defined such that $\beta_1 + \alpha_2 = 1$, and

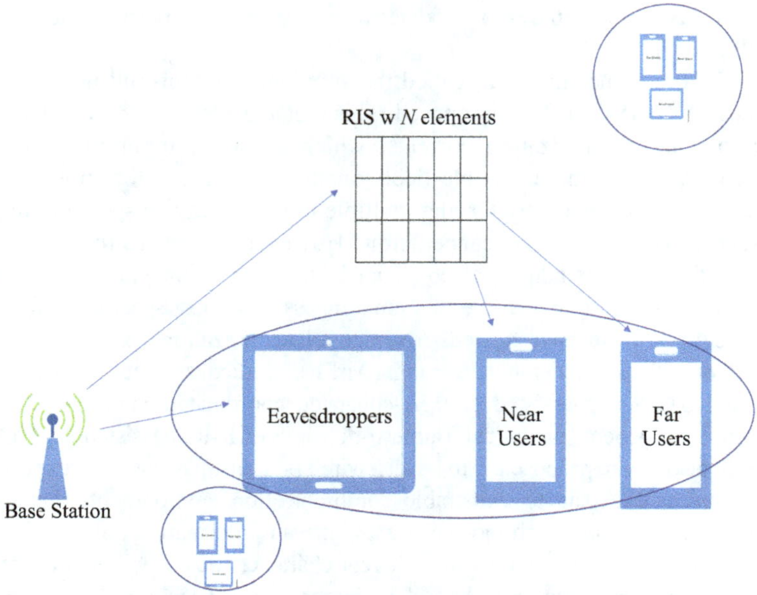

Figure 5.1 PLS-aided two-user RIS-NOMA for 6G wireless systems

to ensure better fairness between users, we assume that $\alpha_2 > \beta_1$ [21]. The additive white Gaussian noise (AWGN) at D_i is denoted by $\eta_i \sim CN(0, N_0)$, which is modeled as a zero-mean complex Gaussian distribution with variance N_0. Additionally, h_{s_i}, h_{sr}, h_{r_i}, h_{se}, and h_{re} are complex Gaussian random variables with zero mean and unit variance, where d_{s_i}, d_{sr}, d_{r_i}, d_{se}, and d_{re} are the distances for the BS-D_i, BS-RIS, RIS-D_i, BS-E, and RIS-E links, respectively. We assume a slowly varying and flat fading channel model for all the channels. The path loss exponent is denoted by ω, and all small-scale fading channel coefficients are modeled as independent and identically distributed $CN(0, 1)$ variables. As M becomes large, we can use the central limit theorem to find that $\sum_{t=1}^{M} h_{sr} h_{r_i} \sim CN(0,M)$ and $h_{s_i} \sim CN(0, 1)$. When the radio frequency (RF) source transmits its data signal x_i to the receiver, the RIS receives the same signal and then adjusts the phase $\chi_t \in [0, 2\pi)$ of reflector $t \in t, \ldots, M$ based on channel state information (CSI) [21].

The signal-to-noise ratio (SNR) at the legitimate user D_1 to decode x_2 can be expressed as

$$\gamma_{D_1}^{no,x_{21}} = \frac{\tau_{B_1} \rho \alpha_2 A_1^2}{\tau_{A_1} \rho \beta_1 A_1^2 + 1} \tag{5.2}$$

where $\tau_{A_1} = \upsilon^2 d_{sr}^{-\omega} d_{r_1}^{-\omega}$, $\rho = \frac{P}{N_0}$, $A_1 = |h_{s_1}|$, $\Phi_1 = d_{s_1}^{-\omega} A_1{}^2 + \tau_{A_1} A_1^2$. It is noted that $A_1 = \left| \sum_{t=1}^{T} h_{sr} h_{r_1} e^{j\chi_t} \right| = \sum_{t=1}^{T} |h_{sr}||h_{r_1}|$ in the case of perfect CSI.

After SIC, the resulting SNR at the legitimate user D_1 to decode x_1 can be formulated as $\gamma_{D_1}^{no,x_1} = \tau_{A_1}\rho\beta_1 A_1^2$.

Then, the received signal reflected by the RIS at D_2 can be expressed as

$$y_{D_2}^{no} = \upsilon \sum_{t=1}^{T} \frac{h_{sr}h_{r_2}}{\sqrt{d_{sr}^{\omega}d_{r_2}^{\omega}}} e^{j\chi_t}(\sqrt{P\beta_1}x_1 + \sqrt{Pa_2}x_2) + \eta_2. \tag{5.3}$$

The resulting SNR at the legitimate user D_2 to decode x_2 can be formulated as

$$\gamma_{D_2}^{no,x_2} = \frac{\tau_{B_2}\rho\alpha_2 A_2^2}{\tau_{B_2}\rho\beta_1 A_2^2 + 1}, \tag{5.4}$$

where $\tau_{A_2} = \upsilon^2 d_{sr}^{-\omega} d_{r_2}^{-\omega}, A_2 = \sum_{r=1}^{R} |h_{sr}||h_{r_2}|$.

The received signal reflected by the RIS at E can be expressed as

$$y_E^{no} = \upsilon \sum_{t=1}^{T} \frac{h_{sr}h_{re}}{\sqrt{d_{sr}^{\omega}d_{re}^{\omega}}} e^{j\chi_t}(\sqrt{P\beta_1}x_1 + \sqrt{Pa_2}x_2) + \eta_e, \tag{5.5}$$

where $\eta_e \sim CN(0, N_e)$ are the AWGN at E that is modeled as a zero-mean complex Gaussian distribution with variance N_e.

Using PIC, the resulting SNR at the legitimate E to decode x_i can be formulated as $\gamma_E^{no,x_i} = \tau_{A_e}\rho_e\alpha_i A_e^2$, where $\tau_{A_e} = \upsilon^2 d_{re}^{-\omega}$, $\rho_e = \frac{P}{N_e}$, $A_e = \sum_{r=1}^{R} |h_{sr}||h_{re}|$, $\Phi_e = d_{se}^{-\omega} {A_e}^2 + \tau_{A_e} A_e^2$. A_e can be approximated with an exponential RV parameter λ_{Φ_e}.

The instantaneous secrecy rate at D_1 can be expressed as

$$S_{D_1}^{no} = \max\left\{\frac{1}{2}\log_2(1 + \min(\gamma_{D_1}^{no,x_{21}}, \gamma_{D_1}^{no,x1})) + \frac{1}{2}\log_2(1 + \gamma_E^{no,x_1}), 0\right\}. \tag{5.6}$$

The instantaneous secrecy rate at D_2 can be expressed as

$$S_{D_2}^{no} = \max\left\{\frac{1}{2}\log_2(1 + \min(\gamma_{D_2}^{no,x_2}, \gamma_{D_2}^{no,x2})) + \frac{1}{2}\log_2(1 + \gamma_E^{no,x_2}), 0\right\}. \tag{5.7}$$

5.5.2 *Secrecy outage probability analysis*

In NOMA systems, the source transmits two signals to D_1 and D_2 with the aid of a RIS. Outage event occurs when either S_{D_1} or S_{D_2} falls below its own target rate. Using this definition, the SOP can be represented as [21]

$$SOP = 1 - \underbrace{\Pr\left(\frac{1+\gamma_{D_1}^{no,x_{21}}}{1+\gamma_E^{no,x_2}} \geq S_{th_1}, \frac{1+\gamma_{D_1}^{no,x_1}}{1+\gamma_E^{no,x_1}} \geq S_{th_1}\right)}_{\theta_1} \times \underbrace{\Pr\left(\frac{1+\gamma_{D_1}^{no,x_{21}}}{1+\gamma_E^{no,x_{21}}} \geq S_{th1}, \frac{1+\gamma_{D_1}^{no,x_1}}{1+\gamma_E^{no,x_1}} \geq S_{th1}\right)}_{\theta_2}, \tag{5.8}$$

where $S_{th_i} = 2^{2R_i}$, R_i is the target data rate for D_i.

The approximate closed-form expression for SOP_{no} is given by [21]

$$\begin{aligned} SOP_{no} \approx 1 &- \frac{\tau_{A_1}\rho M \varsigma_1 \varsigma_2}{(S_{th_1}\tau_{A_e}\rho_e\lambda_{A_e} + \tau_{A_1}\rho T)\lambda_{A_e}\lambda_{A_e}} \exp\left(-\frac{\mu_1}{\tau_{A_1}\rho\beta_1 T}\right) \\ &\times \int_0^1 \exp\left(-\frac{S_{th_1}\tau_{A_e}\rho_e\beta_1\xi_1 t + \mu_1}{(\beta_1 - \alpha_2(S_{th_1}\tau_{A_e}\rho_e\beta_1\xi_1 t + \mu_1))\tau_{A_1}\rho T} - \frac{\varsigma_1 t}{\lambda_{A_e}}\right) dt \\ &\times \int_0^1 \exp\left(-\frac{S_{th_2}\tau_{A_e}\rho_e\alpha_2\varsigma_2 q + \mu_2}{(\beta_1 - \alpha_2(S_{th_2}\tau_{A_e}\rho_e\alpha_2\varsigma_2 Q + \mu_2))\tau_{A_2}\rho T} - \frac{\varsigma_2 q}{\lambda_{A_e}}\right) dq, \end{aligned} \tag{5.9}$$

where $\varsigma_1 = \frac{\alpha_2 - \beta_1\mu_1}{\beta_1\beta_1 S_{th_1}\tau_{A_e}\rho_e}$, $\varsigma_2 = \frac{\alpha_2 - \beta_1\mu_2}{\beta_1\alpha_2 S_{th_2}\tau_{A_e}\rho_e}$, $\mu_i = S_{th_i} - 1$.

5.6 Numerical results and discussions

The parameters set for simulations are similar to recent work [21].

In Figures 5.2–5.4, we can see the secure performance of RIS-NOMA system focusing on two users with existence of Eve. The higher the SNR provided at the access point, the system can achieve better secure performance. It is worth noting that two users need different power allocation factors α which lead to different secure performance for each user, and hence the total secure outage performance can be affected as well. The 6G architecture will be led by advanced technologies including RIS and NOMA. In particular, this chapter demonstrated that RIS and NOMA are key enablers to improve security and reliability. It is noted that the SOP could be improved significantly if the IoT devices work with both direct link AP-RIS-device and AP-RIS-device link. We can see from Figure 5.2 that the SOP when the IoT devices get signals from the direct link, AP-device is better than the case without a direct link. In Figures 5.3 and 5.4, we try to change the path loss exponent, the SOP performance just changes a bit.

In Figure 5.5, the design of RIS with more number of metasurface shows the better secure performance. In particular, we compare SOP of such NOMA-RIS system and the gap among two cases of $M = 100$, $M = 500$ can be observed when the SNR at the AP is less than 5 dB. This result confirms the benefits of RIS to improve secure performance when higher number of metasurface enables that RIS is flexible to provide good signals to IoT devices.

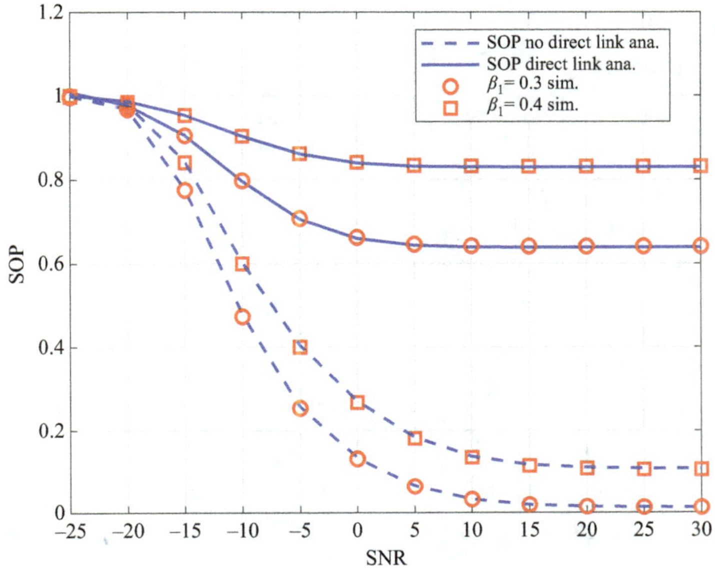

Figure 5.2 Secure outage probability of two users in PLS-aided RIS-NOMA for 6G wireless systems with a path loss exponent of 4 and a target rate of 0.1

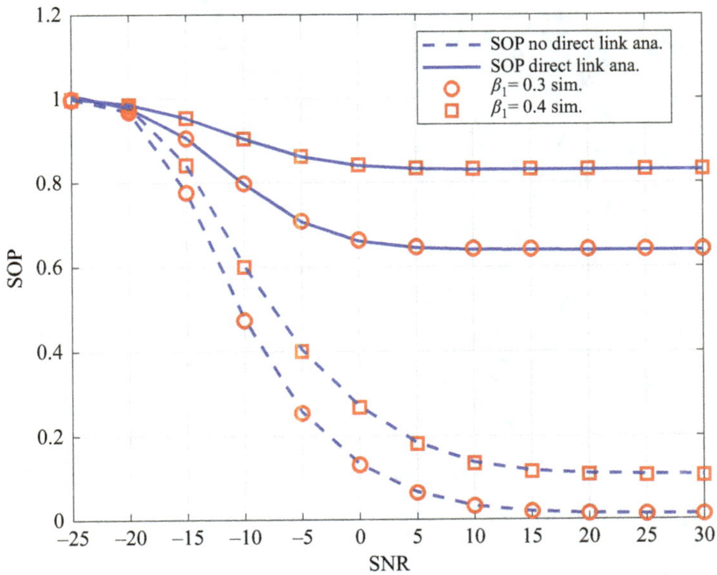

Figure 5.3 Secure outage probability of two users in PLS-aided RIS for 6G wireless systems with a path loss exponent changed to a value of 1

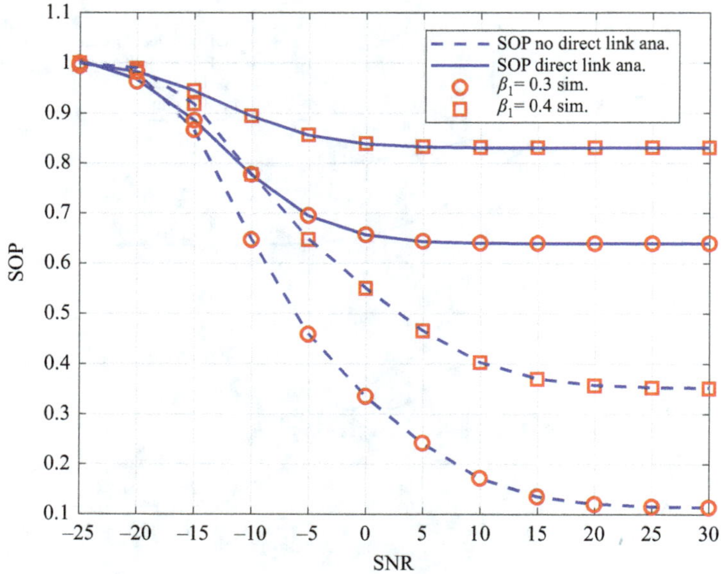

Figure 5.4 Secure outage probability of two users in PLS-aided RIS for 6G wireless systems with a path loss exponent changed to a value of 2

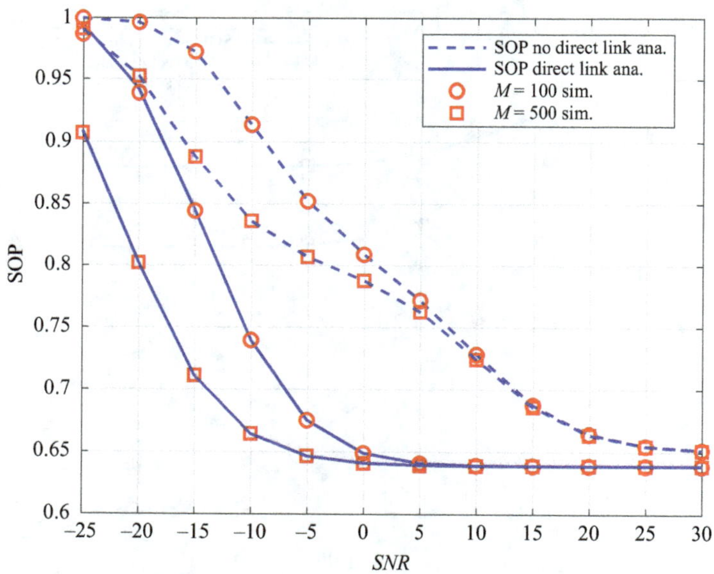

Figure 5.5 Secure outage probability with the number of metasurface changed from 100 to 500

5.7 Conclusion

Through a case study, this chapter has highlighted the potential of RIS to enhance the PLS of 6G architecture using NOMA. By leveraging the dynamic nature of RISs, it is possible to enhance the secrecy performance of the system while improving spectral efficiency. The authors have provided a detailed theoretical framework for RIS-aided NOMA systems, and presented how the system performance depends on the RIS reflection coefficients and power allocation. The simulation results demonstrate that RIS-aided NOMA can significantly improve the secrecy performance of wireless communication systems, particularly in high SNR region at the base station (access point). These findings suggest that RIS-aided NOMA has the potential to be an effective solution for enhancing the security of wireless communication systems, paving the way for the development of more secure and efficient wireless networks. Further research and development in this field may ultimately lead to the implementation of RIS-aided NOMA in practical secure 6G wireless systems, providing a more secure and reliable communication platform for users.

References

[1] W. Saad, M. Bennis, and M. Chen, “A vision of 6G wireless systems: applications, trends, technologies, and open research problems,” *IEEE Netw.*, vol. 34, no. 3, pp. 134–142, 2020.

[2] Y. Sun, J. Liu, J. Wang, Y. Cao, and N. Kato, “When machine learning meets privacy in 6G: a survey,” *IEEE Commun. Surveys Tuts.*, vol. 22, no. 4, pp. 2694–2724, 2020.

[3] M. A. Ferrag, L. Maglaras, and A. Derhab, “Authentication and authorization for mobile IoT devices using biofeatures: recent advances and future trends,” *Secur. Commun. Netw.*, vol. 2019, 2019, Art. no. 5452870.

[4] S. Zeng, H. Zhang, B. Di, Z. Han, and L. Song, “Reconfigurable intelligent surface (RIS) assisted wireless coverage extension: RIS orientation and location optimization,” *IEEE Commun. Lett.*, vol. 25, no. 1, pp. 269–273, 2021, doi:10.1109/LCOMM.2020.3025345.

[5] Z. Tang, T. Hou, Y. Liu, J. Zhang, and C. Zhong, “A novel design of RIS for enhancing the physical layer security for RIS-aided NOMA networks,” *IEEE Wirel. Commun. Lett.*, vol. 10, no. 11, pp. 2398–2401, 2021, doi:10.1109/LWC.2021.3101806.

[6] N. Ye, X. Zhuo, J. Li, B. Di, and J. An, “Secure directional modulation in RIS-aided networks: a low-sidelobe hybrid beamforming approach,” *IEEE Wirel. Commun. Lett.*, vol. 11, no. 8, pp. 1753–1757, 2022, doi:10.1109/LWC.2022.3180931.

[7] X. Gu, W. Duan, G. Zhang, Q. Sun, M. Wen, and P. -H. Ho, “Physical layer security for RIS-aided wireless communications with uncertain eavesdropper distributions,” *IEEE Syst. J.*, vol. 17, no. 1, pp. 848–859, 2023, doi:10.1109/JSYST.2022.3153932.

[8] J. Zhang, H. Du, Q. Sun, B. Ai, and D. W. K. Ng, "Physical layer security enhancement with reconfigurable intelligent surface-aided networks," *IEEE Trans. Inf. Forensics Secur.*, vol. 16, pp. 3480–3495, 2021, doi:10.1109/TIFS.2021.3083409.

[9] W. Khalid, H. Yu, D. -T. Do, Z. Kaleem, and S. Noh, "RIS-aided physical layer security with full-duplex jamming in underlay D2D networks," *IEEE Access*, vol. 9, pp. 99667–99679, 2021, doi:10.1109/ACCESS.2021.3095852.

[10] L. Dong, H.-M. Wang, and J. Bai, "Active reconfigurable intelligent surface aided secure transmission," *IEEE Trans. Veh. Technol.*, vol. 71, no. 2, pp. 2181–2186, 2022, doi:10.1109/TVT.2021.3135498.

[11] Y. Sun, K. An, Y. Zhu, *et al.*, "RIS-assisted robust hybrid beamforming against simultaneous jamming and eavesdropping attacks," *IEEE Trans. Wirel. Commun.*, vol. 21, no. 11, pp. 9212–9231, 2022, doi:10.1109/TWC.2022.3174629.

[12] A. Mukherjee, S. A. A. Fakoorian, J. Huang, and A. L. Swindlehurst, "Principles of physical layer security in multiuser wireless networks: a survey," *IEEE Commun. Surv. Tutor.*, vol. 16, no. 3, pp. 1550–1573, Third Quarter 2014, doi: 10.1109/SURV.2014.012314.00178.

[13] A. K. Yerrapragada, T. Eisman, and B. Kelley, "Physical layer security for beyond 5G: ultra secure low latency communications," *IEEE Open J. Commun. Soc.*, vol. 2, pp. 2232–2242, 2021, doi:10.1109/OJCOMS.2021.3105185.

[14] S. Soderi and R. De Nicola, "6G networks physical layer security using RGB visible light communications," *IEEE Access*, vol. 10, pp. 5482–5496, 2022, doi:10.1109/ACCESS.2021.3139456.

[15] D. Wang, B. Bai, K. Lei, W. Zhao, Y. Yang, and Z. Han, "Enhancing information security via physical layer approaches in heterogeneous IoT with multiple access mobile edge computing in smart city," *IEEE Access*, vol. 7, pp. 54508–54521, 2019, doi:10.1109/ACCESS.2019.2913438.

[16] S. Rajkumar, D. N. K. Jayakody, R. Alkanhel, and A. Muthanna, "Physical layer security in N-pair NOMA-PLNC wireless networks," *IEEE Access*, vol. 10, pp. 91356–91371, 2022, doi:10.1109/ACCESS.2022.3198978.

[17] H. Wang, L. Xu, W. Lin, P. Xiao, and R. Wen, "Physical layer security performance of wireless mobile sensor networks in smart city," *IEEE Access*, vol. 7, pp. 15436–15443, 2019, doi:10.1109/ACCESS.2019.2895338.

[18] V.-L. Nguyen, P.-C. Lin, B.-C. Cheng, R.-H. Hwang, and Y.-D. Lin, "Security and privacy for 6G: a survey on prospective technologies and challenges," *IEEE Commun. Surv. Tutor.*, vol. 23, no. 4, pp. 2384–2428, 2021, https://doi.org/10.1109/comst.2021.3108618.

[19] A. Chorti, A. N. Barreto, S. Köpsell, *et al.*, "Context-aware security for 6G wireless: the role of physical layer security," *IEEE Commun. Standards Mag.*, vol. 6, no. 1, pp. 102–108, 2022, doi: https://doi.org/10.1109/mcomstd.0001.2000082.

[20] X. Lu, L. Xiao, P. Li, *et al.*, "Reinforcement learning based physical cross-layer security and privacy in 6G," *IEEE Commun. Surv. Tutor.*, vol. 25, no. 1, pp. 425–466, 2022, doi: https://doi.org/10.1109/comst.2022.3224279.

[21] H.-P. Dang, M.-S. Van Nguyen, D.-T. Do, M.-H. Nguyen, M.-T. Pham, and A.-T. Kim, "Secure performance analysis of aerial RIS-NOMA-aided systems: deep neural network approach," *Electronics*, vol. 11, no. 16, pp. 2588, 2022, doi: 10.3390/electronics11162588.

[22] L. Yang, J. Yang, W. Xie, M. O. Hasna, T. Tsiftsis, and M. D. Renzo, "Secrecy performance analysis of RIS-aided wireless communication systems," *IEEE Trans. Veh. Technol.*, vol. 69, no. 10, pp. 12296–12300, 2020.

[23] P. Porambage, G. Gur, D. P. Moya Osorio, M. Liyanage, A. Gurtov, M. Ylianttila, and A. Karuppiah, "The roadmap to 6G security and privacy," *IEEE Commun. Mag.*, vol. 59, no. 3, pp. 126–133, 2021, doi:10.1109/MCOM.001.2000527.

[24] Z. Tang, T. Hou, Y. Liu, J. Zhang, and C. Zhong, "A novel design of RIS for enhancing the physical layer security for RIS-aided NOMA networks," *IEEE Trans. Veh. Technol.*, vol. 70, no. 6, pp. 6016–6020, 2021, doi:10.1109/TVT.2021.3064348.

[25] M. H. Khoshafa, T. M. N. Ngatchede, and M. H. Ahmed, "RIS-aided physical layer security improvement in underlay cognitive radio networks," *IEEE Access*, vol. 9, pp. 21796–21806, 2021, doi:10.1109/ACCESS.2021.3060685.

[26] T. Hossain, S. Shabab, A. S. M. Badrudduza, M. K. Kundu, and I. S. Ansari, "On the physical layer security performance over RIS-aided dual-hop RF-UOWC mixed network," *IEEE Trans. Veh. Technol.*, vol. 70, no. 7, pp. 6837–6850, 2021, doi:10.1109/TVT.2021.3087077.

[27] Z. Zhang, J. Chen, Y. Liu, Q. Wu, B. He, and L. Yang, "On the secrecy design of STAR-RIS assisted uplink NOMA networks," *IEEE Trans. Commun.*, vol. 69, no. 11, pp. 6938–6951, 2021.

Chapter 6

Dynamic optical beam transmitter of secure visible light communication systems

Jupeng Ding[1] *and Chih-Lin I*[2]

Abstract

Due to the broadcast nature of visible light communications (VLC) channel, physical layer security (PLS) techniques have been considered to improve the transmission confidentiality of VLC links. However, almost all current schemes merely work with multiple distributed transmitters and fail to serve the scenarios with centralized transmitters, even single transmitter. For addressing this issue, in this work, the dynamic inclined optical beam-based PLS enhancement scheme is proposed. Unlike conventional Lambertian beam-based technique paradigm, the above scheme utilizes the commercially available typical non-Lambertian beams to form the secure VLC links. Numerical results show that, compared with the conventional static Lambertian configuration, up to 5.52 bps/Hz average secrecy capacity gain could be derived via the proposed dynamic scheme using two inclined candidate beams. Moreover, this potential gain will be further elevated to about 6.63 bps/Hz when up to four candidate beams are available at the transmitter.

Keywords: Dynamic optical beam; MIMO; Physical layer security; Secrecy capacity; Visible light communications

6.1 Introduction

Visible light communication (VLC) is being considered as the complementary technology for radio frequency-based technology, thanks to its low cost, license free and immunity to the conventional electromagnetic interference. Moreover, the abundant optical spectrum could be explored by VLC to address the frequency spectrum crisis [1–4]. At the same time, due to the broadcast nature of VLC channel, the relevant data privacy and confidentiality must be sufficiently

[1]Key Laboratory of Signal Detection and Processing in Xinjiang Uygur Autonomous Region, School of Information Science and Engineering, Xinjiang University, China
[2]China Mobile Research Institute, China

considered. Upper layers security, such as encryption and access control mechanism, could not always provide enough wireless like safeguarding, which is disadvantageous in the required storage and computational cost. On the other side, physical layer security (PLS) techniques have been actively introduced and investigated in the secure VLC link design [5–8].

Recently, impressive secure transmission gain has been identified for typical PLS technique including null steering, artificial noise transmission in medium-size indoor scenario. However, almost all these schemes merely work with multiple distributed VLC transmitters. Nevertheless, these above schemes apparently fail to adapt to the small-size scenario with centralized transmitters, even single transmitter. It should be noted that, to a large extent, the current secure optical wireless designs remain in the well-known Lambertian beam research paradigm [6–8]. As a matter of fact, there are many commercially available non-Lambertian beam waiting for consideration and discussion in secure VLC [9–13]. Objectively, the optical beam dimension is attractive to provide novel research paradigm. To a large extent, the secure VLC channel gain and coverage characteristic are dominated by the optical beam pattern of light emitting diodes (LEDs) source in transmitter. Actually, the distinct beam patterns objectively provide one novel design and optimization dimension for secure VLC performance enhancement.

Based on the above discussion, the typical non-Lambertian beams can be utilized to form the secure VLC transmitter. The basic idea of this design is to dynamically select the candidate non-Lambertian beam to derive better PLS performance. This scheme is capable of matching various indoor scenarios with transmitter shortage issue.

In this paper, the Lambertian and typical non-Lambertian beams radiation characteristics are given in Section 6.2. And the secure VLC transmitters with fixed and dynamic beam configuration are investigated in Section 6.3. Numerical results are compared and discussed in Section 6.4. Finally, Section 6.5 concludes this chapter.

6.2 Optical beams characteristics

The following section describes the radiation characteristics of conventional Lambertian optical beam and that of the typical commercially available non-Lambertian optical beam.

6.2.1 Lambertian optical beams

Radiation intensity is the key metric to measure the angular intensity distribution of optical beams. If the LED optical source matches with Lambertian beam, the radiation intensity analytical representation is given by (6.1):

$$I_{\mathrm{Lam}}(\phi) = \frac{m_{\mathrm{Lam}} + 1}{2\pi} \cos^{m_{\mathrm{Lam}}}(\phi), \tag{6.1}$$

where ϕ denotes the emission elevation angle, and m_{Lam} denotes the Lambertian order. And this order is set as 1 for the generalized Lambertian beam, then the respective indoor application scenario and 3D radiation pattern is described in

Figure 6.1(a) and (b), separately, where the red arrow indicates the optical beam normal direction.

In a typical indoor scenario, the propagation contribution of non-line of path components is much weaker than that of line of sight (LOS) path components. For convenience of analysis, this letter only considers the effects of line of sight paths. Therefore, when the radiation characteristic of the optical source follows the Lambertian beam, the direct current channel gain at the receiver R is given as (6.2):

$$H(S^{\text{Lam}}, R) = \begin{cases} \dfrac{A_R}{d_0^2} I_{\text{Lam}}(\phi)\cos(\theta)\dfrac{n^2}{\sin^2(\theta_{FOV})}r, & 0 \le \theta_0 \le \theta_{FOV} \\ 0, & \theta_0 > \theta_{FOV} \end{cases}, \tag{6.2}$$

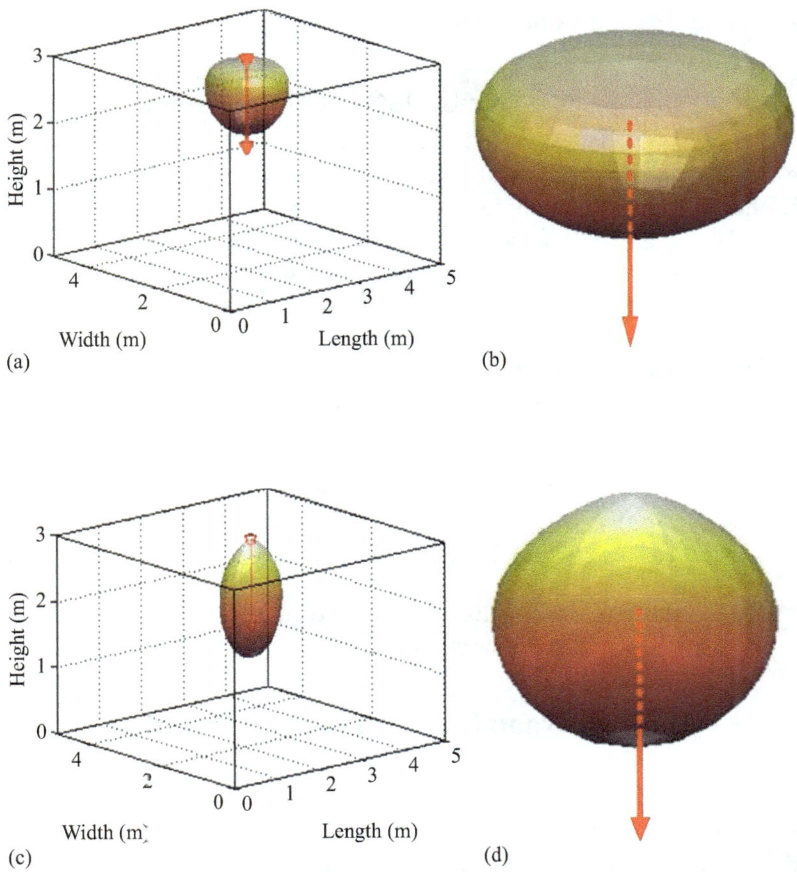

Figure 6.1 (a) Indoor application scenario and (b) 3D radiation pattern of conventional Lambertian optical beam; (c) indoor application scenario and (d) 3D radiation pattern of the typical non-Lambertian optical beam

where A_R is the detection area of receiver; d_0 is the distance between the optical source and the receiver; θ denotes the angle of incidence at receiver; n denotes the refractive index of the optical concentrator, and θ_{FOV} is the field of view (FOV) of receiver.

6.2.2 Non-Lambertian optical beams

Distinct from the well-discussed Lambertian optical beams, there are commercially available LEDs render asymmetric non-Lambertian spatial radiation characteristic, due to the production package and the secondary optics added by the manufacture procedure. Without loss of generality, the beam pattern of NSPW345CS type LED from Nichia is considered and discussed as a representative example.

Thanks to mentioned beam asymmetry, the relevant spatial radiation intensity is simultaneously dependent on the emission elevation and azimuth angle of the emitted optical signal and can be calculated as (6.3).

$$I_{\mathrm{NSPW}}(\phi,) = \sum\nolimits_{i=1}^{2} g1_i \exp\left[-(\ln 2)(|\phi| - g2_i)^2\left(\frac{\cos^2\alpha}{(g3_i)^2} + \frac{\sin^2\alpha}{(g4_i)^2}\right)\right], \tag{6.3}$$

where α is the azimuth angle within the source plane. Specifically, the values of coefficients in this expression are: $g1_1 = 0.13$, $g2_1 = 45°$, $g3_1 = g4_1 = 18°$, $g1_2 = 1$, $g2_2 = 0$, $g3_2 = 38°$, and $g4_2 = 22°$. Similar to the above Lambertian case, the indoor scenario with this typical non-Lambertian beam and the respective 3D beam patterns from the wide cross-section view angle is illustrated in Figure 6.1(c) and (d), accordingly.

Under this beam configuration, the renewed channel gain expression could be given as (6.4):

$$H(S^{\mathrm{NSPW}}, R) = \begin{cases} \dfrac{A_R I_{\mathrm{NSPW}}(\phi, \alpha)}{P_{\mathrm{norm}} d_0^2} \cos\theta_0 \dfrac{n^2}{\sin^2(\theta_{FOV})} r, & 0 \le \theta_0 \le \theta_{FOV} \\ 0, & \theta_0 > \theta_{FOV} \end{cases}, \tag{6.4}$$

where P_{norm} is the power normalization factor of this non-Lambertian beam, to ensure that the whole radiated power equal 1 W.

6.3 The static and dynamic optical beam transmitter

6.3.1 Static optical beam transmitter

At the receiver end, if the conventional fixed beam configuration is applied, the secrecy capacity (SC) of the legitimate receiver (also called Bob in some literature) can be calculated as follows (6.5):

$$C_{\mathrm{fixed}} = [\log_2(1 + \gamma_{\mathrm{Legi}}) - \log_2(1 + \gamma_{\mathrm{Eve}})]^+, \tag{6.5}$$

where γ_{Legi} is the signal-to-noise ratio (SNR) of the legitimate user; γ_{Eve} is the SNR of the eavesdropper; and [b]+ is the max{0, b}. The SNR for any receiver could be given by (6.6):

$$\gamma = \frac{r^2 a^2 P_{DC}^2 \|H(S,R)\|_1^2}{\delta^2}, \tag{6.6}$$

where r is the PD responsivity, a is the modulation index, P_{DC} is the emitted direct current (DC) optical power, and δ^2 denotes the additive noise variance at the receiver.

6.3.2 Dynamic optical beam transmitter

In the proposed dynamic configuration scheme, the transmitter includes two or four inclined non-Lambertian candidate beams with distinct original azimuth offset. For the jth candidate inclined beam, the related radiation intensity is given by (6.7):

$$I_{\text{NSPW}}(\phi + \Delta\phi, \alpha + \Delta\alpha_j) = \sum_{i=1}^{2} g1_i \exp\left[-(\ln 2)(|\phi + \Delta\phi| - g2_i)^2 \left(\frac{\cos^2(\alpha_0 + \Delta\alpha_j)}{(g3_i)^2} + \frac{\sin^2(\alpha_0 + \Delta\alpha_j)}{(g4_i)^2}\right)\right] \tag{6.7}$$

where $\Delta\phi = 45°$ is the emission elevation offset angle, $\Delta\alpha_j$ is the azimuth offset of the ith candidate beam. For dynamic configuration 1 with two candidate beams, the specific offsets are set as 45° and 225°, respectively while for dynamic configuration 2 with four candidate beams, the specific offsets are set as 45°, 135°, 225°, and 315°. The channel gain for jth candidate inclined beam is given by (6.8):

$$H(S_{\text{Beam}}^{\text{NSPW}}, R, \Delta\alpha_j) = \begin{cases} \dfrac{A_R I_{\text{NSPW}}(\phi + \Delta\phi,\ \alpha + \Delta\alpha_j)}{d_0^2} \cos\theta_0 \dfrac{n^2}{\sin^2(\theta_{FOV})} r, & 0 \le \theta_0 \le \theta_{FOV} \\ 0, & \theta_0 > \theta_{FOV} \end{cases}, \tag{6.8}$$

For each legitimate receiver and eavesdropper pair, assume the channel state information of this pair is perfect known to the transmitter, the candidate beam that provides the best SC is adopted to emit the information optical signal. Then the SC for the proposed dynamic configurations is expressed as (6.9):

$$C_{\text{dyna}} = \max_j \left[\log_2\left(1 + \gamma_{\text{Legi}}^{\text{dyna}}(\Delta\alpha_j)\right) - \log_2\left(1 + \gamma_{\text{Eve}}^{\text{dyna}}(\Delta\alpha_j)\right)\right], \tag{6.9}$$

Such that, the related SNR for the dynamic beam configurations can be given by (6.10):

$$\gamma^{\text{dyna}} = \frac{r^2 a^2 P_{DC}^2 \|H(S_{\text{Beam}}^{\text{NSPW}}, R, \Delta\alpha_j)\|_1^2}{\delta^2}, \tag{6.10}$$

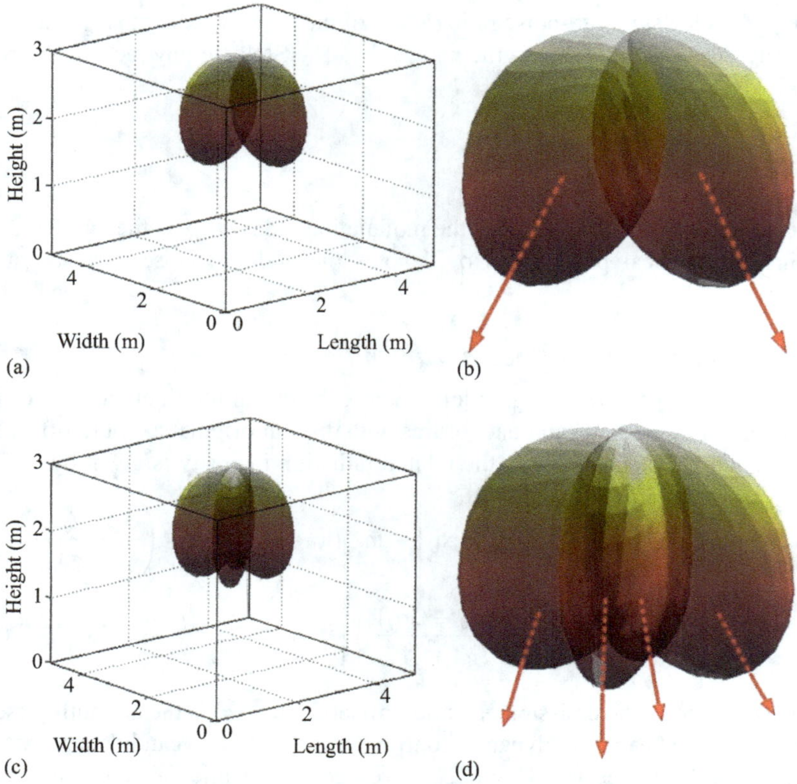

Figure 6.2 *(a) Indoor application scenario and (b) potential 3D radiation pattern of typical two candidate inclined optical beams; (c) Indoor application scenario and (d) potential 3D radiation pattern of typical four candidate inclined optical beams*

The envisioned indoor application scenario and potential 3D radiation pattern of the above two dynamic configurations is shown in Figure 6.2.

6.4 Numerical evaluation

In this section, the numerical comparison is made between the conventional static optical beam configurations and dynamic optical beam configurations in PLS performance.

In Figure 6.3, when the legitimate receiver (Bob) is located in working plane center i.e. (2.5 m, 2.5 m, 0.85 m), the eavesdropper SNR spatial distribution is illustrated for conventional Lambertian beam, static non-Lambertian beam and considered two dynamic non-Lambertian beam configurations, respectively. Intuitively, the eavesdropper SNR fluctuation extent is mitigated, thanks to the flexibility provided by the dynamic beam configurations.

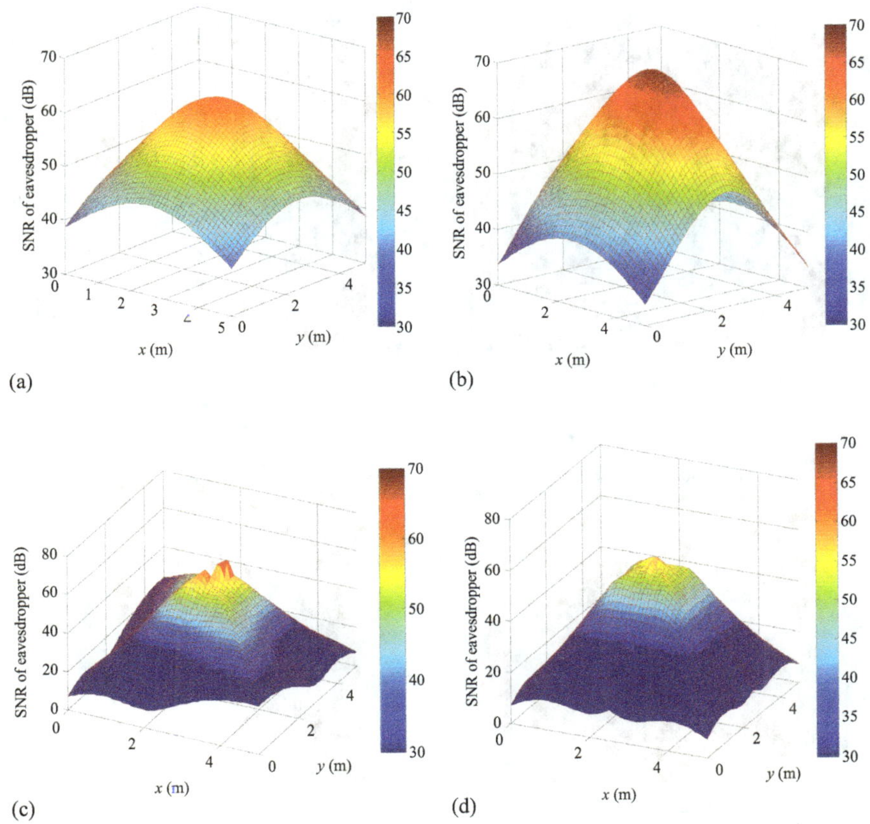

Figure 6.3 Eavesdropper SNR spatial distribution for the center legitimate receiver (Bob): (a) conventional Lambertian beam case; (b) typical non-Lambertian beam case; (c) dynamic non-Lambertian beam case 1 and (d) dynamic non-Lambertian beam case 2

Consistent with Figures 6.1 and 6.2, all configurations are bounded by a 5×5×3 m room where the single transmitter is placed on the ceiling center.

In Figure 6.4, the respective secrecy capacity spatial distributions are presented. In the first case, the SC range between 0 and 7.56 bps/Hz while the average SC is just 3.54 bps/Hz. Once the Lambertian beam is replaced by the non-Lambertian beam, the average SC is lightly increased to 5.87 bps/Hz while the SC range is adjusted to between 0 and 11.69 bps/Hz. Impressively, when the dynamic non-Lambertian beam configuration 1 is adopted, the average SC is dramatically enhanced to 9.06 bps/Hz. And the respective range is between 0 and 16.54 bps/Hz. Furthermore, once four candidate beams are available, in other words, the dynamic beam configuration 2 is applied, the

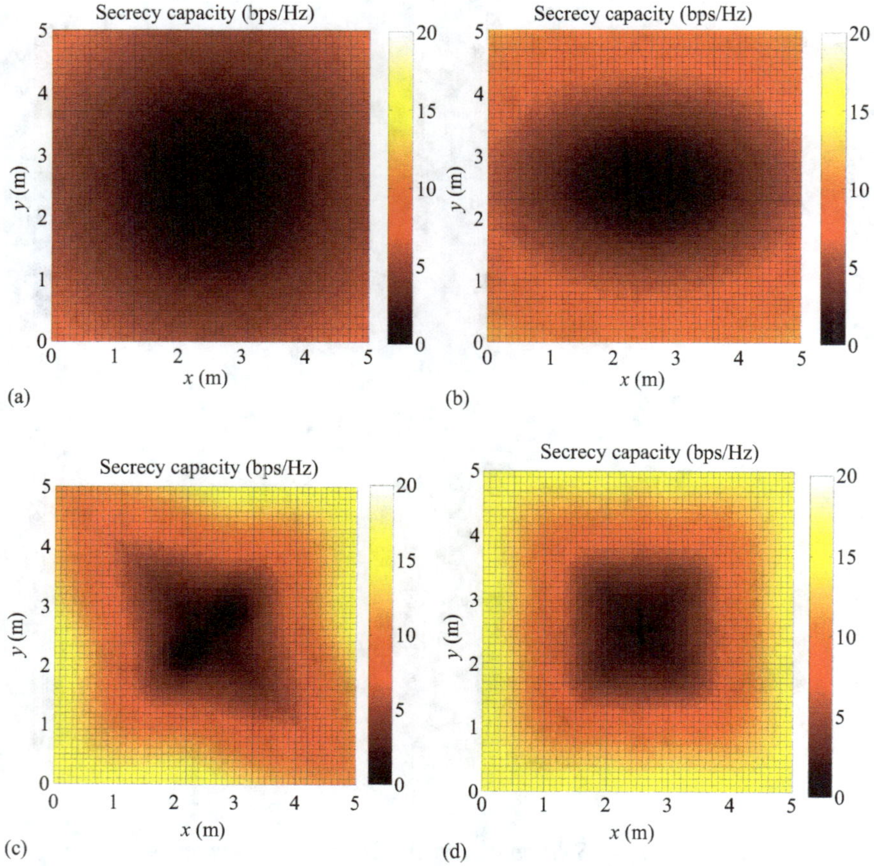

Figure 6.4 Secrecy capacity spatial distribution for the center legitimate receiver (Bob): (a) conventional Lambertian beam case; (b) typical non-Lambertian beam case; (c) dynamic non-Lambertian beam case 1 and (d) dynamic non-Lambertian beam case 2

average SC is in-creased to 10.17 bps/Hz. For clarity, for the same legitimate receiver, the cumulative distribution function (CDF) of the secrecy capacity spatial distribution is shown in Figure 6.5 as well.

Similarly, as shown in Figure 6.6, for the corner legitimate receiver located at (0.5 m, 0.5 m, 0.85 m), the CDF of the SC spatial distribution is illustrated. For fixed Lambertian and non-Lambertian link configuration, the average SC is just 0.066 bps/Hz and 0.099 bps/Hz, while that of two dynamic configurations is tremendously improved to 5.09 bps/Hz and 5.41 bps/Hz, respectively.

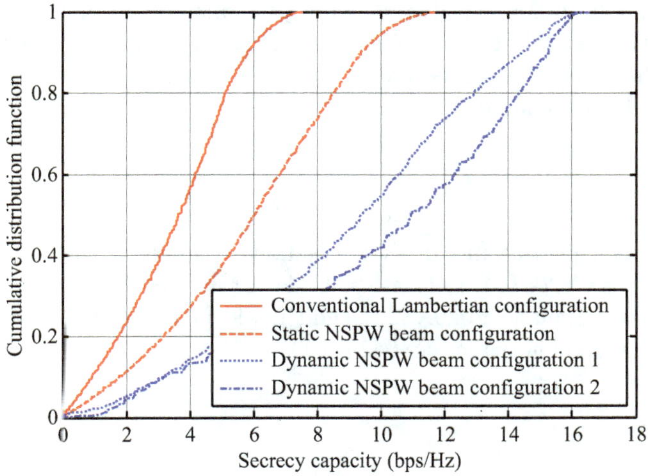

Figure 6.5 CDF of the secrecy capacity spatial distribution for the center legitimate receiver (Bob)

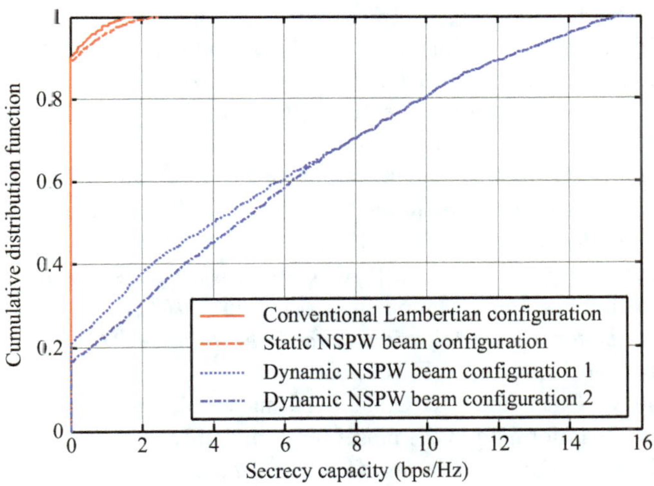

Figure 6.6 CDF of the secrecy capacity spatial distribution for the corner legitimate receiver (Bob)

6.5 Conclusion

In the proposed scheme, the distinct radiation pattern of the typical non-Lambertian beam is utilized to configure the secure VLC links in transmitter shortage scenario. For the corner legitimate receiver, the average SC is up to 5.09 bps/Hz with two candidate inclined beams, while the counterpart of benchmark Lambertian configuration is just 0.066 bps/Hz. Moreover, this metric will be further enhanced to 5.41 bps/Hz once four candidate inclined beams are available at the transmitter. Therefore, the potential benefits offered by dynamic configuration is identified for the PLS of VLC techniques.

Funding

This work was supported in part by the National Natural Science Foundation of China (Grant no. 62061043), Tianshan Cedar Project of Xinjiang Uygur Autonomous Region (Grant no. 2020XS27), and High-level Talents Introduction Project in Autonomous Region (Grant no. 042419004).

References

[1] A. Mukherjee, S. A. A. Fakoorian, J. Huang, and A. L. Swindlehurst, Principles of physical layer security in multiuser wireless networks: a survey, *IEEE Communications Surveys & Tutorials*, 2014, 16, 1550–1573.

[2] A. Mostafa and L. Lampe, Physical-layer security for MISO visible light communication channels, *IEEE Journal on Selected Areas in Communications*, 2015, 33, 1806–1818.

[3] A. Mostafa and L. Lampe, Physical-layer security for indoor visible light communications, in *2014 IEEE International Conference on Communications (ICC)*, Sydney, Australia, June 10–14; New York, NY: IEEE, 2014.

[4] A. Mostafa and L. Lampe, Pattern synthesis of massive LED arrays for secure visible light communication links, in *2015 IEEE International Conference on Communication Workshop (ICCW)*, London, England, June 8–12; Piscataway, NJ: IEEE, 2015.

[5] Z. Chen and X. Wang, A method for improving physical layer security in visible light communication networks, in *2018 IEEE Conference on Standards for Communications and Networking (CSCN)*, Paris, France, October 29–31; New York, NY: IEEE, 2018.

[6] S. Cho, G. Chen, and J. P. Coon, Securing visible light communication systems by beamforming in the presence of randomly distributed eavesdroppers, *IEEE Transactions on Wireless Communications*, 2018, 17, 2918–2931.

[7] S. Cho, G. Chen, and J. P. Coon, Physical layer security in visible light communication systems with randomly located colluding eavesdroppers, *IEEE Wireless Communications Letters*, 2018, 7, 768–771.

[8] S. Cho, G. Chen, and J. P. Coon, Enhancement of physical layer security with simultaneous beamforming and jamming for visible light communication systems, *IEEE Transactions on Information Forensics and Security*, 2019, 14, 2633–2648.

[9] T. Komine and M. Nakagawa, Fundamental analysis for visible-light communication system using LED lights, *IEEE Transactions on Consumer Electronics*, 2004, 50, 100–107.

[10] I. Moreno and C.-C. Sun, Modeling the radiation pattern of LEDs, *Optics Express*, 2008, 16, 1808–1819.
[11] J. Ding and C.-L. I, Z. Xu, Indoor optical wireless channel characteristics with distinct source radiation patterns, *IEEE Photonics Journal*, 2016, 8, 1–15.
[12] J. Ding, C.-L. I, R. Xie, H. Lai, and C. Zhang, Actual radiation patterns oriented non-deterministic optical wireless channel characterization, in *Springer Chinese Conference on Biometric Recognition*, Urumchi, China, August 11–12; Berlin, Germany: Springer, 2018.
[13] J. Ding, I. Chih-Lin, H. Zhang, X. Chen, B. Yu, and H. Lai, Cells planning of VLC networks using non-circular symmetric optical beam, in *ICC 2019 - 2019 IEEE International Conference on Communications (ICC)*, Shanghai, China, May 20–24; New York, NY: IEEE, 2019.

Chapter 7

A new machine learning-based scheme for physical layer security

Hefdhallah Sakran[1] *and Klaus Moessner*[2]

Abstract

Massive MIMO is seen as a possible physical layer security technique to meet the sixth-generation (6G) security requirements. Massive multiple-input and multiple-output (MIMO) systems are naturally immune to passive eavesdroppers (Eves), but this is dramatically degraded by active Eves. In this chapter, we describe the use of massive MIMO enhanced through artificial intelligence (AI) to improve security on the physical layer. We describe a number of machine learning-based algorithms and a deep neural network (DNN) model capable of detecting the presence of an active Eav by exploiting the particular properties and features of massive MIMO.

We describe a machine learning model and DNN applied to a realistic scenario where the channel state information (CSI) of the channels (i.e., legitimate user and Eves) is unknown.

A set of different algorithms showing varying performance are compared. Moreover, the prediction complexity of the different algorithms is discussed. The chapter will explain the design issues for DNN and new machine learning-based secure transmission schemes in massive MIMO-based communication systems. Simulations proof the robustness of machine learning-based algorithms and DNN without need of feedback overhead and when the CSI of all channels is unknown. In this chapter, it will be shown that higher security of communication systems can already be achieved on the physical level.

Keywords: DNN; SVM; NB; Physical layer security; Massive MIMO

7.1 Introduction

In future sixth-generation (6G) wireless networks, mobile radios will be more than just communication devices, providing also computation, security, energy services, etc.,

[1]Faculty of Electrical Engineering and Information Technology, Chemnitz University of Technology, Germany
[2]Faculty of Electrical Engineering and Information Technology, Technische Universität Chemnitz, Germany

when appropriate. One feature of 6G [1,2] is that it will be empowered by Artificial Intelligence (AI) that allows each radio to make decisions that optimize its quality of experience over time and constructively impact the network. Massive MIMO is seen as a possible physical layer security technique to meet the 6G security requirements [3].

Eavesdropping involves two types of attack: active and passive attacks. The advantage of massive MIMO in physical layer security is to increase robustness against passive eavesdropping, which listen on the coded transmitted message without exposing their existence details, for its abilities to direct transmit power toward allowed receivers. Active eavesdropping users in massive MIMO are sensitive to pilot contamination attacks, due to the broadcast nature of wireless transmission. When data is transmitted from base station (BS) to the legitimate receiver, data will be maliciously accessed by eavesdroppers (Eves) resulting in information leakage and interference with the actual data transmitted. When the BS transmitter is aware of an Eves presence, then it can perform various solutions to degrade the Eve's channel, e.g. by sending artificial noise, etc.

Physical layer security problems occur in any type of multi-user system. Nowadays, many studies apply machine learning for wireless communications to improve the system performance such as belief propagation for channel decoding [4], resource management using deep learning algorithm for LTE [5], blind detection for MIMO systems [6], power allocation [7], end-to-end learning of communications systems [8], and apply deep learning in mobile and wireless networking (see [9] and references therein).

The first attempt deals with the physical layer secrecy [10–13]. In [10], SVM and naive-Bayes-based transmit antenna selection were proposed to select the optimal antenna that enhances physical layer security. The authors in [11] proposed a learning-based wireless-powered secure transmission to improve the secrecy throughput. In [12], a machine learning was applied to determine the activation of cooperative relays to maximize the secrecy rate in multi-hop network as in [14]. Whereas the papers [10,11] exploits multiple antennas to improve the physical layer security. In [13], an SVM-based scheme and deep learning model to improve the physical layer security in cooperative networks were developed.

Some papers in a literature have dealt with active eavesdropping [15–20]. Pei *et al.* [15] propose a linear fisher discriminant analysis (LFDA)-based scheme and SVM to provide authentication where three channel features, including the time-of-arrival (TOA), received signal strengths (RSSs), and cyclic features of the channels are used. Weinand *et al.* [16] consider a mission critical machine-type communication (MC-MTC) and Gaussian mixture model (GMM) where the normalized magnitudes of the frequently estimated channels are used as the features (input data) for the estimator. Wang *et al.* [17] introduce a feedforward neural network with a two-dimensional measurement space using the formulation of both the Pearson correlation coefficient and the Euclidean distance between two samples. In [18], the local oscillator offset and *I–Q* imbalance detected are used to create features within a neural network. Hoang *et al.* [19] introduced the artificial training data ATD and employed twin-class and a single-class SVM (TC-SVM/SC-SVM). It used

TC-SVM in the case of perfect CSI of all channels and the SC-SVM is used with the CSI of legitimate users. Ismayil *et al.* [20] used an autoencoder to learn the channel matrices and identify the best transformation of CSI where the beamforming based on this transformed channel coefficients can be decoded using this model in the receiver. Moreover, available research has shown models for detection of Eves using many channel features at the same time, i.e. TOA, RSS, and cyclic features of the channels [15]. But none of the related studies have already discussed the challenges in massive MIMO.

Our contributions can be summarized as follows:

- This approach focuses on the potential of machine learning techniques to guarantee secure transmission schemes in massive MIMO. Massive MIMO communication is very sensitive to pilot contamination attacks, so we discuss the fundamental challenges related to pilot contamination.
- This paper attempts to use integrate machine learning techniques with physical layer security to detect active Eves in massive MIMO. First, we discuss the problem of detecting active Eves in massive MIMO based on machine learning. Then, we build a machine learning model based on a realistic scenario where we do not have the CSI of any channel, i.e., of legitimate users nor Eves. This is in contrast to previous papers, traditional techniques that did not use machine learning and that assume a perfect CSI of all channels that are known at BS. Also in the previous papers, in systems without massive MIMO, many channel features were considered as input data to detect the Eve. Our model only uses the RSS as relevant feature and we exploit the features of massive MIMO to improve the performance of our detection model.
- We have developed an SVM-based scheme and NB-based scheme to classify the training received signals into one of the classes that represent the presence of the Eves.
- We propose a deep neural network (DNN)-based model to detect the active Eves.

The remainder of the paper is organized as follows: in Section 7.2, the system model is introduced. In Section 7.3, we present the proposed machine learning algorithm for detecting the presence of an active Eve. Simulation results are given in Section 7.4. Finally, Section 7.5 concludes the paper.

7.2 System model

We consider a massive MIMO model as shown in Figure 7.1. The transmitter BS (Alice), equipped with an array of M-antennas, communicates with the legitimate receiver (Bob) in the presence of an Eve. The Eve equipped with an array of N-antennas that attempts to overhear the source information. Small-scale and large-scale fading are considered for the wireless channel of BS – legitimate receiver and BS – Eve, where the channel between BS and legitimate receiver is denoted by $\sqrt{\beta_{BL}}\mathbf{h}_{BL}$, where $\mathbf{h}_{BL}$ is an $1 \times M$ vector which is a zero mean circularly symmetric

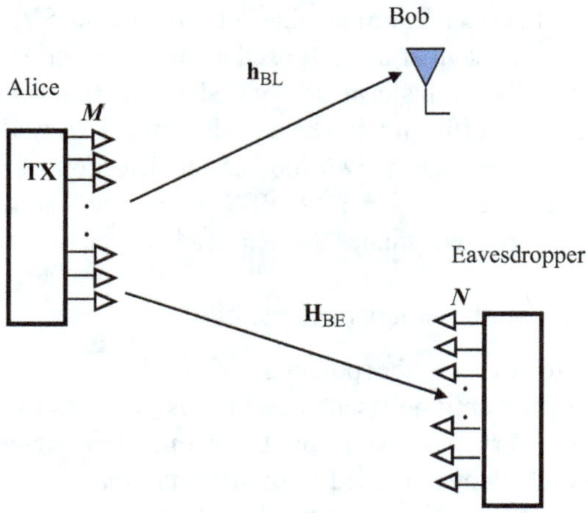

Figure 7.1 System model

complex Gaussian vector with covariance matrix **I**, β_{BL} represents the large-scale fading between BS and legitimate receiver. The channel between BS and Eve is denoted by $\sqrt{\beta_{BE}}\mathbf{H}_{BE}$, where $\mathbf{H}_{BE}$ is an $N \times M$ matrix that is zero mean circularly symmetric complex Gaussian vectors with a covariance matrix **I**.

A massive MIMO system operating in TDD mode is considered in our paper. First, the legitimate receiver sends pilots to BS where the BS assigns transmission slots for pilots for training and estimates the channel between BS and legitimate receiver. Moreover, the pilots are sent from the Eve where an active Eve is assumed, which is the weak point of the massive MIMO system. The Eve uses the antenna selection technique that can offer a good tradeoff between expense and performance [21]. The received signal at BS is denoted by:

$$r_{LB} = \sqrt{P_L\beta_{LB}}\mathbf{h}_{LB}x_P + \sqrt{P_E\beta_{EB}}\mathbf{h}_{E,B}x_P + \mathbf{n}_P \tag{7.1}$$

where x_P is the pilot signal; P_L and P_E denote the transmitted power at legitimate receiver and Eve, respectively. Furthermore, $\mathbf{n}_P$ denotes $M \times 1$ vector comprising zero mean white Gaussian noise with the covariance matrix $N_{0,BS}\mathbf{I}$.

Detection of an active eavesdropper is crucial and affects the security of the communication in a massive MIMO system. If the BS is aware of the security risks, then it can generate artificial noise to interfere with the signal reception at the Eve to drown out the Eves and it can do a separation of the signal via precoding.

7.3 Proposed machine learning algorithm for detecting the presence of an active Eve

In this section, machine learning is investigated for detecting the presence of an active Eve. First, the features are extracted from the *m* training data containing the RSS. Then, the DNN-based scheme, SVM-based model, and NB-based scheme are applied to construct the classification model and predict the class label that represents the presence of an Eve. We used a sufficiently large training data set to build our classification model.

7.3.1 DNN-based scheme

In this subsection, the proposed DNN-based scheme is presented. As shown in Figure 7.2, a network is designed with four layers. Our proposed DNN-based scheme structure consists of an input layer, two hidden layers, and one output layer. The inputs of the proposed model are the RSS where the length of the input layer is the same length of each training sequence. The hidden layer is designed to extract the different features from input dataset based on the output of the input layer. The output represents the label for the input vector **r** that identifies the presence of an eavesdropper.

Each layer consists of multiple neural nodes, the neural nodes in the input layer provide information from the training dataset to the network. Each neuron in the hidden layer receives input from the neuron of the preceding layer and each neuron has an associated weight (w) that determines the relationship between the neurons. The neural network applies the following weighting sum of its inputs:

$$q_j^{(i+1)} = \rho^{(i)}\left(\sum_{k=1}^{N_i} w_{j,k}^{(i)} u_k^{(i)} + b_k^{(i)}\right) \tag{7.2}$$

where, $w_{j,k}^{(i)}$ is the weight between the kth neuron in ith layer and jth neuron in the $(i+1)$th layer, $b_k^{(i)}$ denotes the bias of the neuron of the kth neuron in the $(i+1)$th layer, N_i gives the number of neurons, $u_k^{(i)}$ represents the input of the kth neuron in

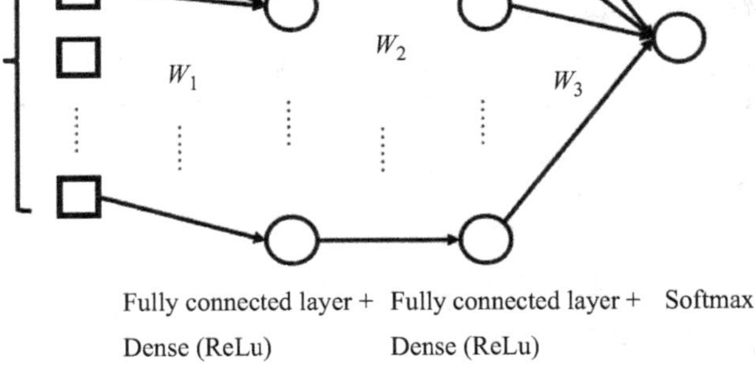

Figure 7.2 DNN model

the (i)th layer, and $\rho^{(i)}$ denotes the activation function, each neuron uses nonlinear activation function. Note that the rectified linear unit (ReLU) and the sigmoid function [22] are most commonly used for the nonlinear activation function, they are denoted as, respectively [23]:

$$f_{ReLU}(a) = \max(0, a) \tag{7.3}$$

$$f_{sigmoid}(a) = \frac{1}{1 + e^{-a}} \tag{7.4}$$

For the output layer, the softmax activation function is a well-suited activation function for classification problems, it aims to derive probabilities for different classes in the output, this is denoted as:

$$\sigma(\mathbf{z})_q = \frac{e^{z_q}}{\sum_{d=1}^{Q} e^{z_d}} \quad \text{for } q = 1, 2, \ldots\ldots Q \tag{7.5}$$

where Q is the number of classes, $\mathbf{z}$ represents the vector of the inputs to the output layer, z_q denotes the qth element of the input vector, and $\sum_{d=1}^{Q} e^{z_d}$ is a normalization term.

The categorical cross-entropy is the most common training criterion to compute the categorical cross-entropy loss for multi-class classification problems at the classification layer.

The training neural network process at the classification layer takes the values from the softmax function and assigns each input to one of a set of output classes using the cross entropy function:

$$\text{loss} = \sum_{i=1}^{V} \sum_{j=1}^{Q} t_{ij} \ln y_{ij} \tag{7.6}$$

where Q is the class number, V denotes the sample number, and t_{ij} represents the indicator that the ith sample belongs to the jth class. Furthermore, y_{ij} denotes the output for sample i for class j which is the value from the softmax function.

7.3.2 SVM-based scheme

For m samples: $S = \{(x_1, y_1), (x_2, y_2), \ldots, (x_m, y_m)\}$, $\mathbf{x}_i \in \mathcal{R}^n$, $y \in (-1, +1)$ $(i = 1, 2, \ldots, l)$. SVM transforms the classification problem as follows [24]:

$$\min_{\mathbf{w}, b} \frac{1}{2} \|\mathbf{w}\|^2 \tag{7.7}$$

s.t.

$$y_i(\phi^T(\mathbf{x}_i)\mathbf{w} + b) \geq 1,$$

where $y_i \in (-1, +1)$, $\phi(.) : \mathcal{R} \rightarrow \mathcal{H}$ is an nonlinear mapping that maps the input data from the input space χ to a higher-dimensional space $\mathcal{H}$ (the feature space). To solve problems that are not linearly separable in the higher dimensional space, SVM transforms the problem into the following optimization problem:

$$\min_{\mathbf{w},b,\xi} \; \frac{1}{2}\|\mathbf{w}\|^2 + C\sum_{m=1}^{l} \xi_i \tag{7.8}$$

s.t.

$$y_i(\phi^T(\mathbf{x}_i)\mathbf{w} + b) \geq 1 - \xi_i,$$

$$\xi_i \geq 0.$$

where C denotes a regularization parameter, penalty parameter, which controls the degree of penalty for the mis-classification samples, the error term, ξ_i represents the slack variables.

A kernel function is considered as a measure of similarity between two points that allow to construct algorithms in dot product spaces, i.e., the inner products between data vectors appears only in expressions such as $\langle\phi(\mathbf{x}), \phi(\mathbf{x}')\rangle = \phi^T(\mathbf{x})\phi^T(\mathbf{x}')$, $\mathbf{x} \in \mathcal{R}$, $\mathbf{x}' \in \mathcal{R}$. Using the kernel method, the inner product can be computed directly from $\mathbf{x}$ and $\mathbf{x}'$ without explicitly computing $\phi(\mathbf{x})$ and $\phi(\mathbf{x}')$.

There are many different types of kernel functions, like linear, nonlinear, radial basis function, sigmoid, and polynomial. The Gaussian radial basis function (RBF) is a general-purpose kernel and it is used when there is no prior knowledge about the data. The form of the Gaussian RBF function is,

$$k(\mathbf{x}, \mathbf{x}') = \exp(-\gamma\|\mathbf{x} - \mathbf{x}'\|), \tag{7.9}$$

where $\gamma > 0$ is a kernel parameter.

7.3.3 NB-based scheme

The proposed NB-based scheme is presented in this subsection. The NB-based scheme solves the problem by utilizing the conditional probability, whereas the SVM-based scheme try to maximize the margins between different classes. The NB-based scheme estimates the parameters of a probability distribution for each element of the normalized feature vector for all label classes using the training datasets in the training step.

For m samples with class variable y and feature vector $\mathbf{x}$, Bayes' theorem states the following equation:

$$\Pr(y|\mathbf{x}) = \frac{Pr(\mathbf{x}|y)Pr(y)}{Pr(\mathbf{x})} \tag{7.10}$$

Assuming the naive conditional independence. With this assumption, the NB-based scheme computes the posterior probability of that sample belonging to each class as the following equation:

$$\Pr(y|\mathbf{x}) = \frac{Pr(y)\prod_{i=1}^{m} Pr(x_i|y)}{Pr(\mathbf{x})} \tag{7.11}$$

where $\Pr(\mathbf{x})$ is constant given the probability of the occurrence of the feature of nth element of vector $\mathbf{x}$.

The NB-based scheme applies the maximum a posteriori decision rule to predict the class label that represents the presence of an Eve, which simply picks the class label that has the largest probability given the data point's features.

The NB classifier selects the class label that achieves the maximum posterior probability and it can be formulated mathematically as

$$y^* = \underset{y}{\text{argmax}}\ Pr(y)\prod_{n=1}^{N} Pr(x_n|y) \tag{7.12}$$

7.4 Simulation results and discussion

In this section, the numerical results are evaluated to validate the performance of our proposed scheme. The simulation parameters are set as shown in Table 7.1. The adaptive moment estimation (Adam) optimizer is also used in DNN.

Figure 7.3 shows the detection and false-alarm probabilities of the massive MIMO against M using the SVM based scheme. As can be seen from the figure, the SVM-based scheme significantly improves the detection probability and false alarm probability. Interestingly, SVM gives a good performance to detect the Eve without the knowledge of the CSI of all channels and with a small number of feature parameters for the SVM model.

Table 7.1 Simulation parameter

Symbols	Values
Training epoch	200
Learning rate (LR)	10^{-2}
Batch size	128
The size of training dataset	90% of datasets
The size of validation dataset	10% of datasets
The optimizer	Adam algorithm
The number of neurons in the 1*st* hidden layer	150
The number of neurons in the 2*nd* hidden layer	50

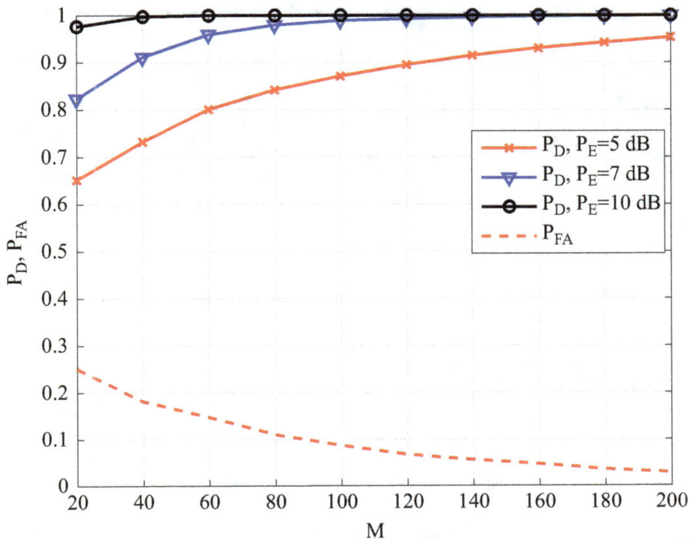

Figure 7.3 The detection probability and probability of false alarm with SVM with different values of Eve's power

Moreover, we evaluate our proposed model as M grows and use 10 values for M, between 20 and 200, with different values of Eve's power. Also, it can be seen that we need to increase M to get a higher detection probability and to approach the optimum value with small values of the Eve power.

Figure 7.4 shows the detection and false-alarm probabilities of the massive MIMO against M using the DNN-based scheme with learning rate (LR) = 0.01. From Figure 7.4, it can be seen that the system with the DNN-based scheme gives the same performance as the SVM-based scheme. They both, SVM and DNN, classify with comparable accuracy. This means that there is no reason that derives from the characteristics of the classification problem for preferring one over the other.

Figures 7.5 and 7.6 introduce both the accuracy and loss for LR = 0.01 with max. epochs = 30 and 20, respectively. Loss can be defined as the difference between the actual values and the values predicted by the model and it depends on how you predict classes for your classification problem. The accuracy is the ratio of the number of true labels in the test data matching the classifications from predict model to the number of datasets in the test data. It can be seen that the accuracy and loss are 87.5% and 0.24 for $M = 60$, LR = 0.01 and epochs = 30 and the accuracy and loss are 86.5% and 0.25 for $M = 60$, LR = 0.01 and epochs = 20. The Adam optimizer is used to update network weights iterative based in training data in Figures 7.5 and 7.6. In Figures 7.7 and 7.8, it can be seen that the accuracy and loss for LR = 0.01 with max. epochs = 30 and 20 and

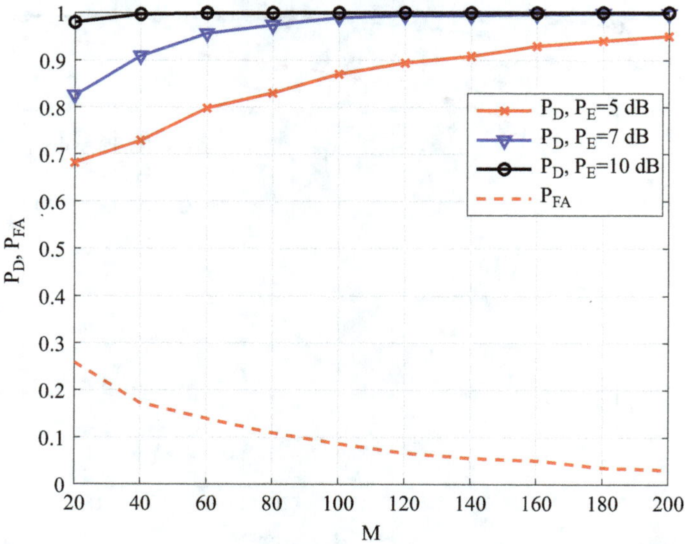

Figure 7.4 The detection probability and probability of false alarm with DNN with different values of Eve's power

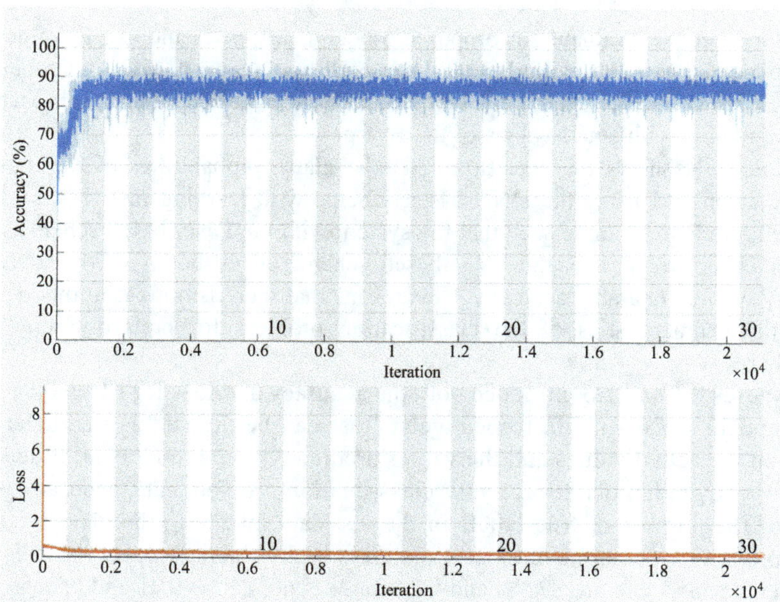

Figure 7.5 Accuracy vs. iterations and loss vs. iterations of the proposed DNN using Adam optimization algorithm, where max. epoch = 30, and LR = 0.01

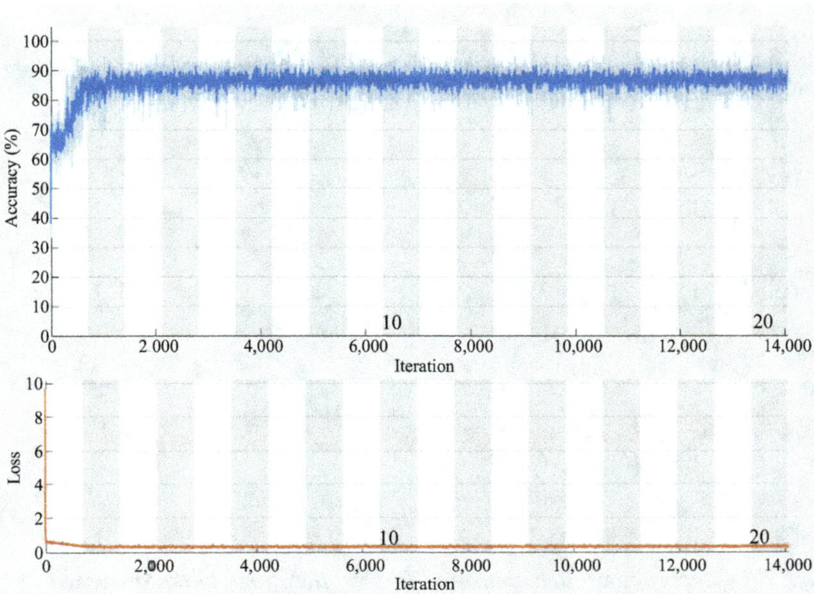

Figure 7.6 Accuracy vs. iterations and loss vs. iterations of the proposed DNN using Adam optimization algorithm, where max. epoch = 20, and LR = 0.01

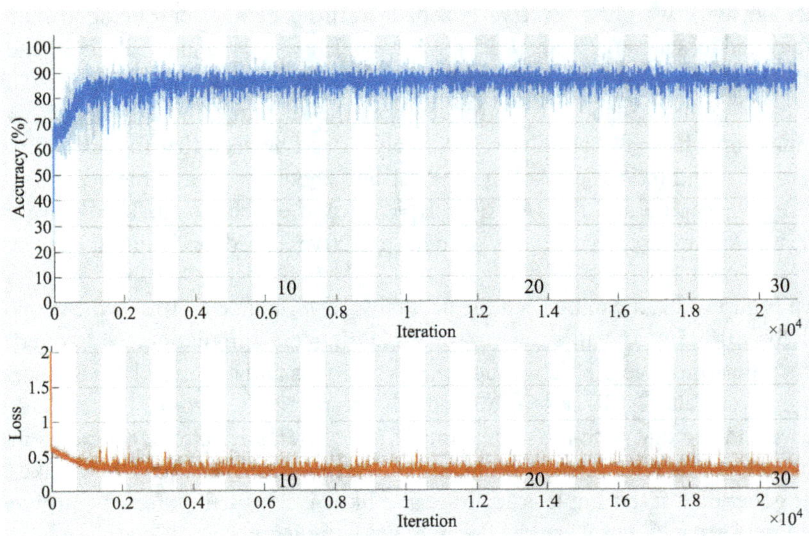

Figure 7.7 Accuracy vs. iterations and loss vs. iterations of the proposed DNN using SGDM optimization algorithm, where max. epoch = 30, and LR = 0.01

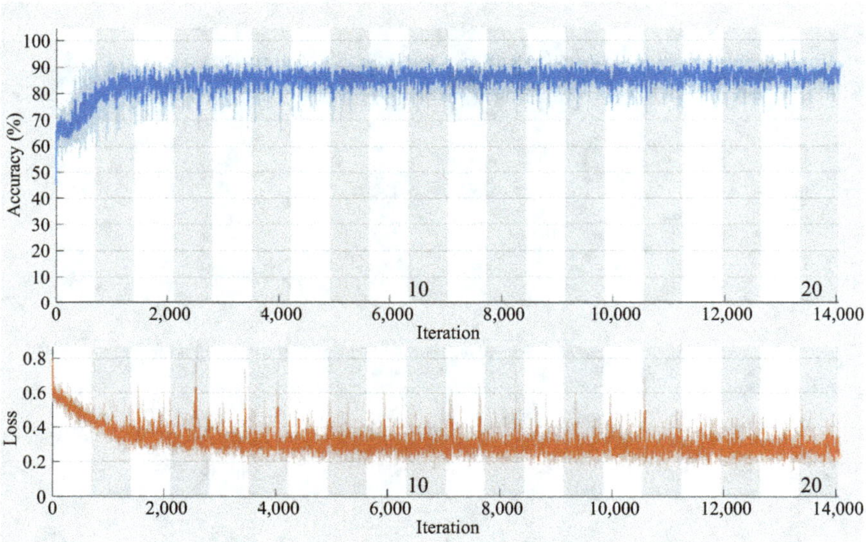

Figure 7.8 Accuracy vs. iterations and loss vs. iterations of the proposed DNN using SGDM optimization algorithm, where max. epoch = 20, and LR = 0.01

the stochastic gradient descent with momentum (SGDM) optimizer is used to update network weights iterative based in training data. It can be seen that the accuracy and loss are 86.7% and 0.25 for $M = 60$, LR = 0.01, and epochs = 30 and the accuracy and loss are 86.4% and 0.25 for $M = 60$, LR = 0.01, and epochs = 20.

Also, Figures 7.9 and 7.10 introduce both the accuracy and loss for LR = 0.1 with max. epochs = 30, and 20, respectively. It can be seen that the accuracy and loss are 67% and 0.6 for $M = 60$, LR = 0.1 and epochs = 30 and the accuracy and loss are 66.9% 0.63 and for $M = 60$, LR = 0.1 and epochs = 20. The Adam optimizer is used to update network weights iterative based on training data as in Figures 7.5 and 7.6.

In Figure 7.11, it can be seen the accuracy performance metric of the proposed DNN for different M values, as observed during our simulation. Moreover, the accuracy performance metric through the two epochs and different LR is illustrated in Figure 7.12 and 7.13 for $M = 60$ and 140, respectively. From Figures 7.12 and 7.13, we can see that with the increasing of LR, the classification accuracies of the model increase as well. Also, it is clear that a large value of LR for DNN makes the model generalize faster but less accurate and likewise, a small value of LR gives us a good performance but it makes the model a slow learner.

Figure 7.14 shows the detection and false-alarm probabilities of the massive MIMO against M using NB-based scheme. As can be seen from the figure, the NB-based scheme gives good performance at high values of Eve's power.

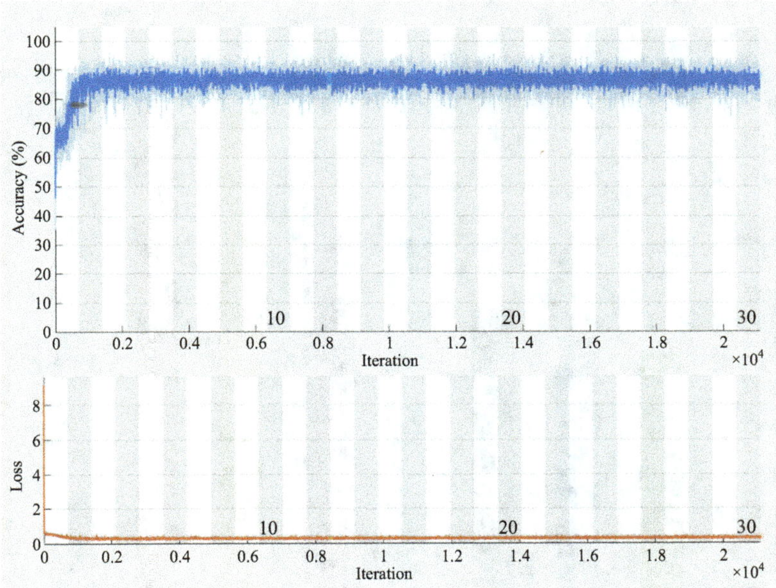

Figure 7.9 Accuracy vs. iterations and loss vs. iterations of the proposed DNN using Adam optimization algorithm, where max. epoch = 30, and LR = 0.1

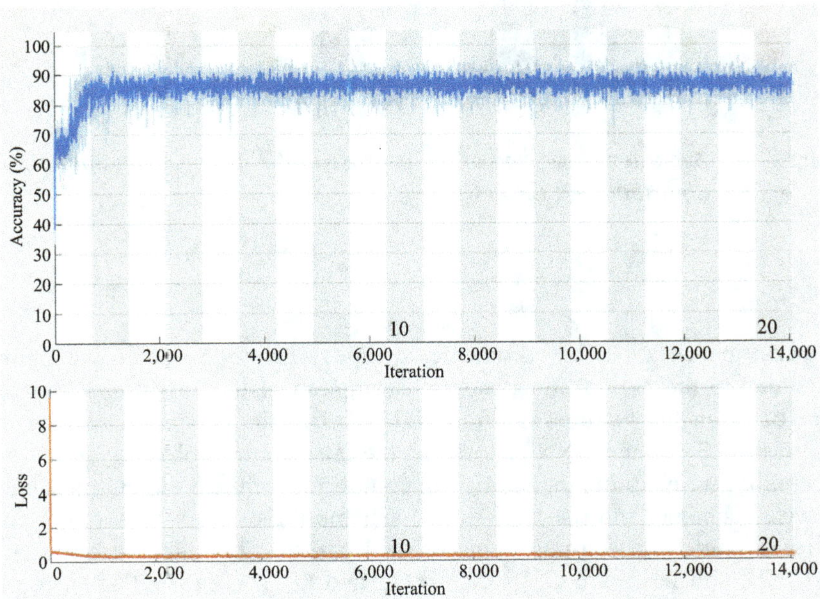

Figure 7.10 Accuracy vs. iterations and loss vs. iterations of the proposed DNN using Adam optimization algorithm, where max. epoch = 20, and LR = 0.1

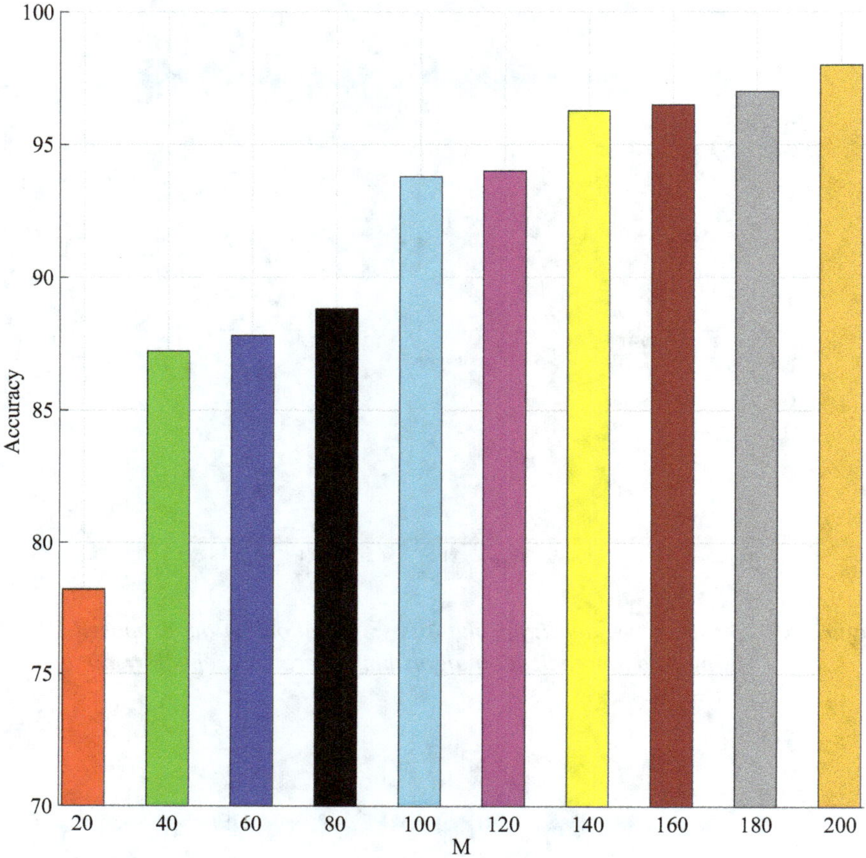

Figure 7.11 Accuracy vs. M *and of the proposed DNN using Adam optimization algorithm and LR = 0.01*

The performance of NB-based scheme degrades when the value of Eve's power is low due to the 'naive' independent assumption of the conditional independence between the features is often violated. But the NB-based scheme outperforms the SVM and DNN concerning the false alarm probability.

Finally, we examine the prediction complexity of the SVM-based scheme, NB-based scheme, and the DNN-based scheme rather than the training complexity because the training is performed offline [25]. The complexity of SVM, NB, and DNN are $\mathcal{O}(M^2)$, $\mathcal{O}(2M)$, and $\mathcal{O}(ML_1 + L_1L_2 + 2L_2)$, respectively. Where L_1 and L_2 denotes the neurons number of two hidden layers for DNN.

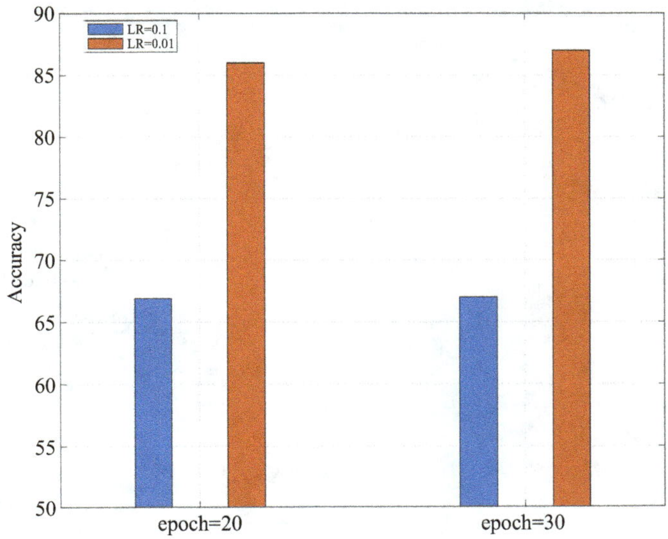

Figure 7.12 Accuracy comparison for different numbers of epochs = 20 and 30, M = *60*

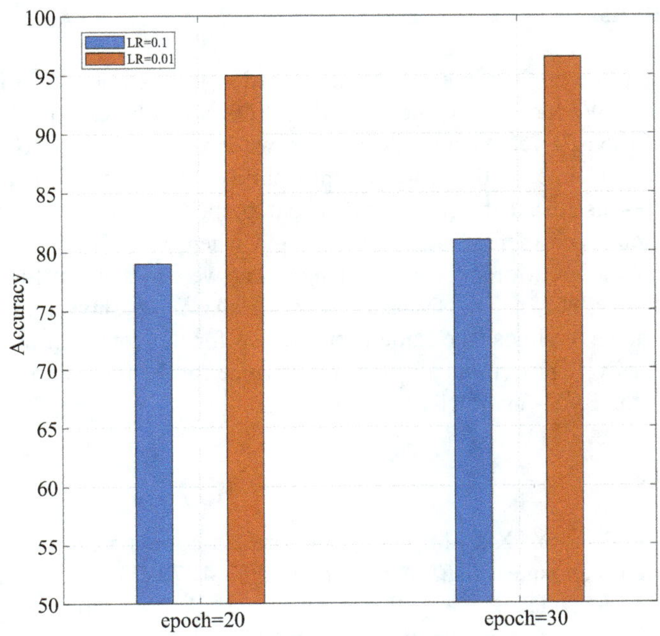

Figure 7.13 Accuracy comparison for different numbers of epochs = 20 and 30, M = *140*

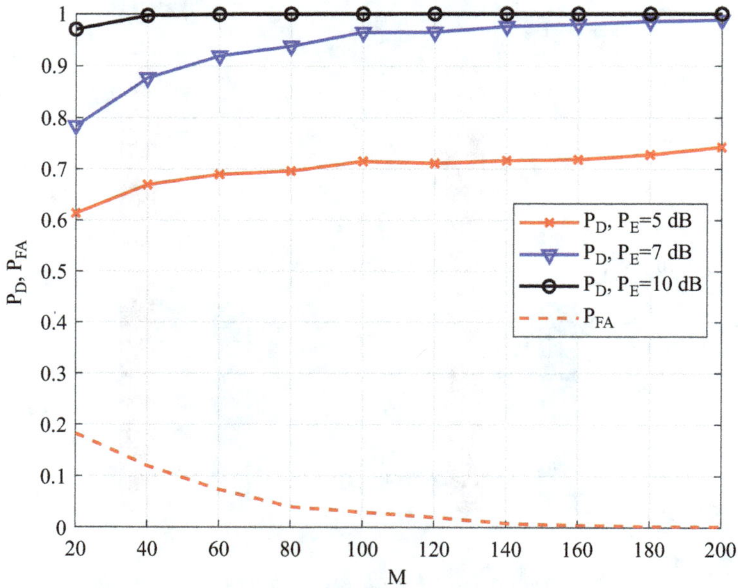

Figure 7.14 The detection probability and probability of false alarm with NB with different values of Eve's power

7.5 Conclusion

In this chapter, we apply a multiclass classification to Eves detection in massive MIMO system. We have shown how SVM and DNN can be exploited to provide efficient and robust Eve detection. Moreover, we have shown the performance of the NB-based scheme in the detection probability and false alarm probability. Simulation results reveal the robustness of our proposed machine learning-based scheme without need of feedback overhead and when the CSI of all channels is considered unknown. Moreover, we conclude that we can achieve the same performance using SVM- and DNN-based schemes. Also, we show how NB-based scheme can achieve almost the same performance for the detection probability at high values of Eve's power and a better performance for the false alarm probability compared to the SVM- and DNN-based schemes.

References

[1] Yang P, Xiao Y, Xiao M, *et al.* 6G wireless communications: vision and potential techniques. *IEEE Network.* 2019;33(4):70–75.

[2] Tariq F, Khandaker MRA, Wong KK, *et al.* A speculative study on 6G. *IEEE Wireless Communications.* 2020;27(4):118–125.

[3] Mucchi L, Jayousi S, Caputo S, *et al.* Physical-layer security in 6G networks. *IEEE Open Journal of the Communications Society.* 2021;2:1901–1914.

[4] Nachmani E, Be'ery Y, and Burshtein D. Learning to decode linear codes using deep learning. *CoRR*. 2016; abs/1607.04793. http://arxiv.org/abs/1607.04793.

[5] Challita U, Dong L, and Saad W. Proactive resource management in LTE-U systems: a deep learning perspective. *CoRR*. 2017. abs/1702.07031. http://arxiv.org/abs/1702.07031.

[6] Jeon Y, Hong S, and Lee N. Blind detection for MIMO systems with low-resolution ADCs using supervised learning. *CoRR*. 2016. abs/1610.07693. http://arxiv.org/abs/1610.07693.

[7] Amiri R, Mehrpouyan H, Fridman L, *et al.* A machine learning approach for power allocation in HetNets considering QoS. In: *2018 IEEE International Conference on Communications (ICC)*, 2018. p. 1–7.

[8] Drner S, Cammerer S, Hoydis J, *et al.* Deep learning based communication over the air. *IEEE Journal of Selected Topics in Signal Processing*. 2018;12(1):132–143.

[9] Zhang C, Patras P, and Haddadi H. Deep learning in mobile and wireless networking: a survey. *IEEE Communications Surveys Tutorials*. 2019;21(3):2224–2287.

[10] He D, Liu C, Quek TQS, *et al.* Transmit antenna selection in MIMO wiretap channels: a machine learning approach. *IEEE Wireless Communications Letters*. 2018;7(4):634–637.

[11] He D, Liu C, Wang H, *et al.* Learning-based wireless powered secure transmission. *IEEE Wireless Communications Letters*. 2019;8(2):600–603.

[12] Nguyen TT, Lee JH, Nguyen MT, *et al.* Machine learning-based relay selection for secure transmission in multi-hop DF relay networks. *Electronics*. 2019;8(9):1–14.

[13] Sakran H. Joint relay and jammer selection based on deep learning for improving the physical layer secrecy in cooperative networks. In: *2020 International Wireless Communications and Mobile Computing (IWCMC)*, 2020. p. 1124–1129.

[14] Lee J. Optimal power allocation for physical layer security in multi-hop DF relay networks. *IEEE Transactions on Wireless Communications*. 2016;15(1):28–38.

[15] Pei C, Zhang N, Shen XS, *et al.* Channel-based physical layer authentication. In: *2014 IEEE Global Communications Conference*, 2014. p. 4114–4119.

[16] Weinand A, Karrenbauer M, Sattiraju R, *et al.* Application of machine learning for channel based message authentication in mission critical machine type communication. In: *European Wireless 2017; 23rd European Wireless Conference*, 2017. p. 1–5.

[17] Wang N, Jiang T, Lv S, *et al.* Physical-layer authentication based on extreme learning machine. *IEEE Communications Letters*. 2017;21(7):1557–1560.

[18] Chatterjee B, Das D, Maity S, *et al.* RF-PUF: enhancing IoT security through authentication of wireless nodes using machine learning. *IEEE Internet of Things Journal*. 2019;6(1):388–398.

[19] Hoang TM, Duong TQ, Tuan HD, *et al.* Physical layer security: detection of active eavesdropping attacks by support vector machines. *IEEE Access.* 2021;9:31595–31607.

[20] Siyad CI and Tamilselvan S. Deep learning enabled physical layer security to combat eavesdropping in massive MIMO networks. In: Smys S, Senjyu T, and Lafata P (eds.). *Second International Conference on Computer Networks and Communication Technologies*. Cham: Springer International Publishing; 2020. p. 643–650.

[21] Uthansakul P, Promsuwanna N, and Uthansakul M. Performance of antenna selection in MIMO system using channel reciprocity with measured data. *International Journal of Antennas and Propagation*. 2011;2011:Article ID 854350.

[22] Pomerat J, Segev A, and Datta R. On neural network activation functions and optimizers in relation to polynomial regression. In: *2019 IEEE International Conference on Big Data (Big Data)*, 2019. p. 6183–6185.

[23] LeCun Y, Bengio Y, and Hinton G. Deep learning. *Nature.* 2015;521 (7553):436–444.

[24] Perez-Cruz F and Bousquet O. Kernel methods and their potential use in signal processing. *IEEE Signal Processing Magazine.* 2004;21(3):57–65.

[25] Joung J. Machine learning-based antenna selection in wireless communications. *IEEE Communications Letters.* 2016;20(11):2241–2244.

Chapter 8

Vehicular ad hoc networks employing intelligent reflective surfaces for physical layer security

Vinoth Babu Kumaravelu[1], Arthi Murugadass[2], C. Suganthi Evangeline[3], X. Anitha Mary[4], Agbotiname Lucky Imoize[5,6], R. Nandakumar[7], Stephen Ojo[8] and Joseph Isabona[9]

Abstract

A significant amount of personal data is shared by smart vehicles that are a part of vehicular ad hoc networks (VANET). As a result, security needs to be improved to stop eavesdropping and intruder attacks. The traditional encryption protocols are more complicated computationally and are intended for upper layers. These are not appropriate for applications requiring lightweight infrastructure, such as the Internet of things (IoT). Physical layer security (PLS), a popular research area, is the ideal solution for the aforementioned problems. Intelligent reflecting surfaces (IRS), one of several PLS solutions, attracted the research community because of their alluring advantages. In this chapter, we presented two different IRS configurations: the smart reflector (SR) and the access point (AP). Analytical expressions for secrecy outage probability (SOP) and secrecy rate are developed for these

[1]Department of Communication Engineering, School of Electronics Engineering, Vellore Institute of Technology, India
[2]Department of Computer Science and Engineering, Sreenivasa Institute of Technology and Management Studies, India
[3]Department of Electronics and Communication, Karunya Institute of Technology and Sciences Coimbatore, India
[4]Department of Robotics Engineering, Karunya Institute of Technology and Sciences, India
[5]Department of Electrical and Electronics Engineering, Faculty of Engineering, University of Lagos, Nigeria
[6]Department of Electrical Engineering and Information Technology, Institute of Digital Communication, Ruhr University, Germany
[7]Department of Electronics and Communication Engineering, K.S.R. Institute for Engineering and Technology, India
[8]Department of Electrical and Computer Engineering, College of Engineering, Anderson University, USA
[9]Department of Physics, Federal University Lokoja, Nigeria

arrangements. Simulations show that the IRS-assisted system outperforms the system without IRS. Additionally, it has been found that adding more IRS components increases the secrecy rate. Although relaying provides comparable benefits, the hardware and signal processing complexity associated with it makes IRS a better choice for PLS. Therefore, one of the viable components to preserve security in vehicular applications could be IRS-assisted transmission.

Keywords: Sixth-generation (6G); Intelligent reflective surfaces (IRS); Physical layer security (PLS); Secrecy capacity; Secrecy outage probability (SOP); Vehicular ad hoc network (VANET)

8.1 Introduction

The number of fifth-generation (5G) mobile subscriptions worldwide will surpass 1 billion in 2022, according to the Ericsson mobility report published in June 2022 [1]. By the end of 2022, there will be 100 million fixed wireless access (FWA) connections. The massive Internet of Things (IoT) technologies will rule the wireless era in the future. Ericsson anticipates 5.9 billion wide-area, 5.5 billion cellulars, and 24.3 billion short-range IoT connections during 2027. The compound annual growth rate (CAGR) from 2022 to 2027 will be 13%. IoT requires strong security because, without it, the devices are readily hackable [2]. When the hacker takes over, he or she can access the user's digital data and change how the gadget functions. According to reports, cyberattacks affect two out of every three US families. 3,00,000 Windows based systems suffered damage from ransomware in 2017. Medical gadgets were also hacked. As a result, IoT security is an important issue.

Many people lose their lives because of road accidents, resulting in huge medical and insurance expenses [3–6]. This has been a motivating factor for providing pleasant rides and additional services to the people in transit. Intelligent transportation system (ITS) provides a safe and dependable driving experience by enabling vehicles to communicate with other vehicles and also with the infrastructure units. The various components of the vehicular ad hoc network (VANET) are illustrated in Figure 8.1. By allowing vehicles to communicate among themselves and infrastructure, VANET offers a secure and trustworthy ride.

Current security architectures prioritize the security of upper layers such as application, transport, network, and link. Designing various encryption techniques for the upper protocol stack enables secure transmission. The complexity of the underlying mathematical issue determines how secure these encryption schemes are. The security ought to be compromised once the mathematical issue has been resolved [7]. Standard cryptosystems were developed based on conjectures that had not been verified. They are not appropriate for IoT applications since they require expensive computing. The IoT has a lightweight architecture and uses low-complexity computing devices and sensors. The traditional cryptosystems rely on encryption and key exchange, which is difficult to do when there are a lot of devices.

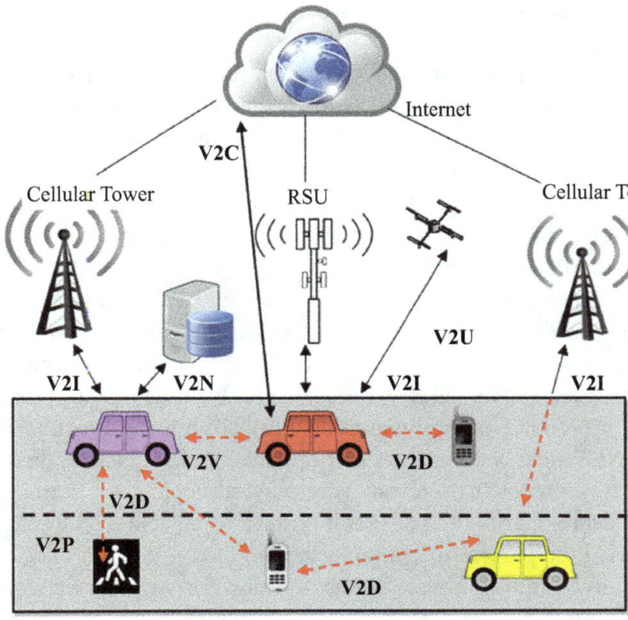

V2C – Vehicle-to-Cloud
V2D – Vehicle-to-Device
V2I – Vehicle-to-Infrastructure
V2N – Vehicle-to-Network
V2P – Vehicle-to-Pedestrian
V2U – Vehicle-to-Unmanned Aerial Vehicle
V2V – Vehicle-to-Vehicle

Figure 8.1 Various components of VANET

Physical layer security (PLS) addresses both the challenges of secrecy and reliability. The inherent randomness of a noisy wireless channel is employed as the key idea of PLS, which is used to provide authentication, privacy, message secrecy, and integrity. Here, the quality of the channel determines the feasible secrecy rate. The radio channel's physical characteristics listed below offer several security related opportunities. The first one is fading, which is the gradual deterioration of the received signal's quality. The second is the channel's randomness, which is the source of the random key generation. The third is interference, which, when employed appropriately, can alter the eavesdropper's signal reception. The fourth one creates degrees of freedom in secrecy through multiple input multiple output (MIMO), large MIMO, relays, etc. [8]. The related works on PLS assumed an irrational eavesdropper channel. Under asymptotic situations, they have proven to be effective. The PLS testing is still in its early stages.

The layout of this chapter is as follows: Section 8.2 discusses the related study on PLS. Section 8.3 elaborates on the proposed system model. Section 8.4 discusses the simulations, and Section 8.5 concludes the chapter.

8.2 Related works

With the advancement in communication technology, all VANET requirements are fulfilled by the 5G network. However, with an increase in vehicle density, the 5G network is less capable of providing ultra-high latency services, security, and reliability to the VANET network. As a result, intelligent and autonomous driving has been identified as state-of-the-art technological paradigms for sixth-generation (6G) wireless networks.

VANET supports both safety and non-safety applications like traffic management, route optimization, e-advertisements, and so on [9,10]. The safety applications in VANET are sensitive to delay as real-time decisions deal with human lives. Furthermore, non-safety applications can be delay tolerant, requiring bandwidth intensive and more computational resources. Due to dynamic topology and mobility, the connectivity in VANET is highly unstable. Furthermore, the smart vehicle, which is a part of VANET, shares a huge amount of personal data. Hence, there is a need for security advancement to prevent any eavesdropping or intruder attacks. Confidentiality, data integrity, and availability have to be achieved through security mechanisms during the process of communication.

PLS has evolved into a more efficient solution for wireless network security [11]. As seen in Figure 8.4, a source vehicle communicates a secret message to the legitimate vehicle. The eavesdropper vehicle also receives the secret message and makes an effort to decrypt it. The major goal of PLS is to investigate how the main channel and eavesdropper channel differ in terms of randomness and reciprocity while taking into account the foundations of information theory. When transmit signals are designed properly, the mutual information between the source and the legitimate vehicles is enhanced, while it is minimized between the source and the eavesdropper vehicles.

To achieve security, PLS relies on signal processing techniques like beamforming and power allocation; no encryption or decryption processes are necessary [12]. The authors of [13] demonstrated secure transmission over Gaussian channels. They showed that the low probability of detection and interception for eavesdroppers was made possible using PLS. Keyless secure transmission over fading channels is proven in [14]. By utilizing multiple antennas and relays, the authors of [15,16] were able to increase the secrecy degrees of freedom. To hinder eavesdroppers and increase security, Goel and Negi [17] generate random, artificial noise (AN) at the transmitter.

Four distinct strategies, including covert communication, directional modulation (DM), spatial modulation (SM), and intelligent reflective surfaces (IRS) or reconfigurable intelligent surfaces (RIS), are suggested in the literature to attain high degrees of PLS [11]. Covert communication masks the transmitter's transmission behavior [18,19]. Covert communication can be enabled with the aid of full duplex communication [20] and unmanned aerial vehicles [21,22]. A phased array is used in DM to transfer data toward a legitimate user. To ensure safe transmission, the received constellation diagram of the intended user is distorted, whilst the constellation diagram of the legitimate user is identical to that of the baseband signal [23–25].

SM successfully balances spectrum efficiency and hardware overhead [26–30]. Here, in addition to the modulation mapping, the chosen antenna index provides additional information, increasing the spectral efficiency. Since both spatial mapping and modulation mapping include private information, their interception could cause security to be compromised. The security of SM systems can be increased by carefully choosing the transmit antennas among the available antennas [31,32]. With more complexity, the Euclidean distance-based antenna selection (EDAS) yields a better secrecy rate [33,34]. With less computing complexity, the maximum signal-to-leakage-plus-noise ratio (SLNR) method increases the secrecy rate [35]. It is established in [17] that producing Gaussian AN at the SM transmitter can improve security performance.

IRS is anticipated to play a significant role in the design of smart environments, which is popular in 6G [36]. IRS can boost the feasible rate of legitimate users while lowering the feasible rate of eavesdroppers. IRS is effective, particularly when the eavesdropper channel is stronger than legitimate users. Figure 8.2 illustrates Alice's willingness to safely transmit information to Bob, a single antenna authorized user, in the presence of a single-antenna eavesdropper (Eve). It is considered that one IRS has N components placed between the source (Alice) and the legitimate user (Bob). A legitimate user's (Bob's) channel is presumed to be known by the IRS. Therefore, IRS can inflict the necessary phase compensation on reflected signals to maximize the received signal-to-noise ratio (SNR) at the legit user (Bob). Optimizing the beamforming at the base station (BS) (Alice) and the phase shifts at the IRS simultaneously maximizes the secrecy rate at the legitimate user. In Figure 8.3, an Alice with multiple antennas (N_T) wants to transmit secure information to a single user (Bob) in the presence of a single Eve with only one antenna. A constructive interference is introduced at Bob, and a destructive interference is introduced at Eve by controlling the phase shifts at the IRS.

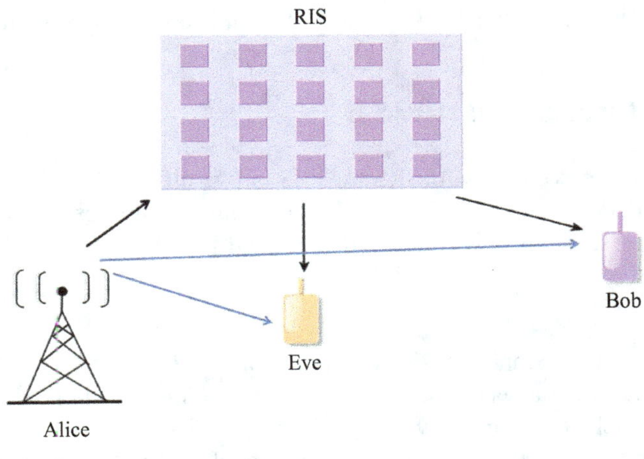

Figure 8.2 Single input-single output (SISO) system in the presence of single Eve

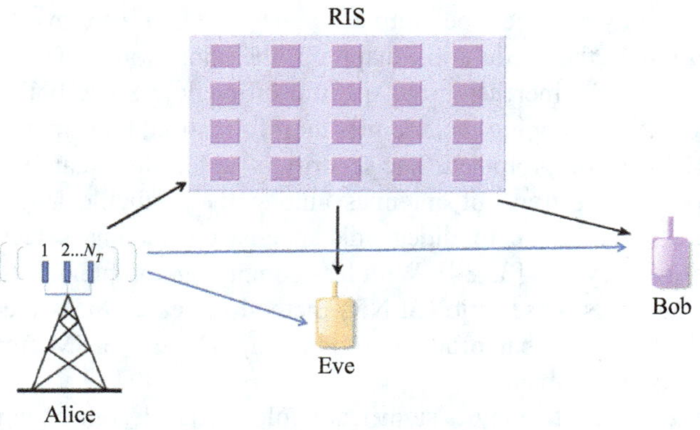

Figure 8.3 Multiple input-single output (MISO) system in the presence of single Eve

The authors examined the performance of an IRS-assisted system with a MISO configuration in [37–39]. The secrecy rate is maximized in [38] by concurrently optimizing transmit beamforming and IRS phase shifts. It is shown in [40,41] that an IRS-assisted system increases the MIMO system's secrecy rate. Here, both the user equipment (UE) and the BS are equipped with multiple antennas. For secure communication, Hong *et al.* [42] propose AN-aided MIMO with IRS.

The important contributions of this chapter, which were inspired by the aforementioned research, are listed below:

- For PLS, two distinct IRS configurations, smart reflector (SR) and access point (AP), are proposed.
- For both setups, the mathematical analysis of SOP and secrecy capacity is conducted.
- Monte-Carlo simulations are employed to validate the proposed models.

8.3 PLS through smart IRS

This section discusses the use of IRS for PLS in two different configurations: IRS as an SR and IRS as an AP. These IRS configurations are described in [43,44]. These are represented in this work by IRS-SR and IRS-AP.

8.3.1 IRS-SR for PLS

Figure 8.4 depicts a sample IRS-SR configuration for PLS in a VANET. A source vehicle (S) transmits a secret signal (x) to a destination vehicle (D). There is an eavesdropper vehicle (E) in the vicinity of the source vehicle, which also receives the same signal x. An IRS with N elements is deployed in the environment. The source and destination vehicles are assumed to have single antennas.

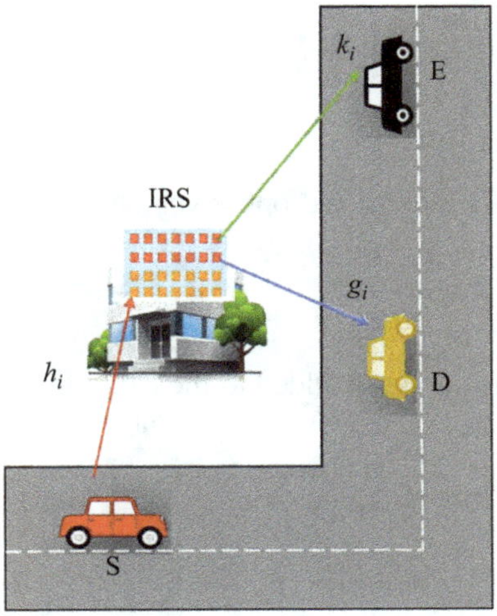

Figure 8.4 A sample IRS-SR configuration for PLS in a VANET

The channel between the source vehicle and ith IRS element is given by [45,46]

$$h_i = \alpha_i e^{j\theta_i}, i = 1 \ldots N, \tag{8.1}$$

where α_i and θ_i are the magnitude and phase of h_i, respectively. The channel associated with ith IRS element and destination vehicle is given by

$$g_i = \beta_i e^{j\varphi_i}, i = 1 \ldots N, \tag{8.2}$$

where β_i and φ_i are the magnitude and phase of g_i, respectively. The channel associated with ith IRS element and eavesdropper vehicle is given by

$$k_i = \vartheta_i e^{j\xi_i}, i = 1 \ldots N, \tag{8.3}$$

where ϑ_i and ξ_i are the magnitude and the phase of k_i, respectively. In this work, smart IRS transmission is taken into consideration, which assumes knowledge of the two-hop channel phases in advance. Each IRS element introduces a phase shift of

$$\phi_i = \theta_i + \varphi_i \tag{8.4}$$

This phase shift corrects the phase distortion of the two-hop channel for the destination vehicle. The signal received at the destination vehicle is

$$y_D = \left[\sum_{i=1}^{N} h_i e^{-j\phi_i} g_i \right] x + n_D \tag{8.5}$$

where $n_D \in \mathbb{C}\mathrm{N}(0, \sigma^2)$ is the noise added at the destination vehicle. The secret signal x is transmitted with average energy E_s. Substituting (8.1), (8.2), and (8.4) in

(8.5), the expression reduces to

$$y_D = \left[\sum_{i=1}^{N} \alpha_i \beta_i\right] x + n_D \tag{8.6}$$

The signal received at the eavesdropper vehicle is

$$y_E = \left[\sum_{i=1}^{N} h_i e^{-j\phi_i} k_i\right] x + n_E \tag{8.7}$$

where $n_E \in \mathbb{CN}(0, \sigma^2)$ is the noise added to the eavesdropper vehicle. Substituting (8.1), (8.3) and (8.4) in (8.7), the expression reduces to

$$y_E = \left[\sum_{i=1}^{N} \quad \alpha_i e^{j\theta_i} e^{-j(\theta_i+\varphi_i)} \vartheta_i e^{j\xi_i}\right] x + n_E$$

$$y_E = \left[\sum_{i=1}^{N} \quad \alpha_i \vartheta_i e^{-j(\varphi_i-\xi_i)}\right] x + n_E \tag{8.8}$$

The residual phase has an impact on the eavesdropper vehicle's received signal as a result of incorrect phase compensation. Let the dual-hop channel of the destination vehicle be

$$A = \sum_{i=1}^{N} \alpha_i \beta_i \tag{8.9}$$

Let the dual-hop channel of the eavesdropper vehicle be

$$B = \sum_{i=1}^{N} \alpha_i \vartheta_i e^{-j(\varphi_i-\xi_i)} \tag{8.10}$$

The instantaneous SNR at the destination vehicle is

$$\gamma_D = \frac{|A|^2 E_s}{\sigma^2} \tag{8.11}$$

The instantaneous SNR at the eavesdropper vehicle is

$$\gamma_E = \frac{|B|^2 E_s}{\sigma^2} \tag{8.12}$$

The instantaneous capacities of destination and eavesdropper vehicles are given by [4]

$$C_D = \log_2(1+\gamma_D) = \log_2\left(1 + \frac{|A|^2 E_s}{\sigma^2}\right) \tag{8.13}$$

$$C_E = \log_2(1+\gamma_E) = \log_2\left(1 + \frac{|B|^2 E_s}{\sigma^2}\right) \tag{8.14}$$

The secrecy capacity is given by [4]

$$C_s = \max\{C_D - C_E, 0\} \tag{8.15}$$

Substituting (8.13) and (8.14) in (8.15) results in

$$C_s = \begin{cases} \log_2(1+\gamma_D) - \log_2(1+\gamma_E), \gamma_D > \gamma_E \\ 0, \gamma_D < \gamma_E \end{cases} \tag{8.16}$$

The above expression can be further simplified to [4]

$$C_s = \begin{cases} \log_2\left(\dfrac{1+\gamma_D}{1+\gamma_E}\right), \gamma_D > \gamma_E \\ 0, \gamma_D < \gamma_E \end{cases} \tag{8.17}$$

The SOP is given by

$$P_O = \Pr[C_s < C_{Th}] \tag{8.18}$$

where C_{Th} is the secrecy capacity threshold. Substituting (8.17) in (8.18) gives [4]

$$P_O = \Pr\left[\log_2\left(\frac{1+\gamma_D}{1+\gamma_E}\right) < C_{Th}\right] \tag{8.19}$$

8.3.2 IRS-AP for PLS

Figure 8.5 depicts a sample IRS-AP configuration for PLS in a VANET. A novel IRS-AP arrangement is suggested in [45]. In this arrangement, IRS is placed closer to/on the source. As a result, there are barely any channel effects between the source and IRS. This results in communication with a single hop. Each IRS element introduces a phase shift of $\phi_i = \varphi_i$. The signal received at the destination vehicle is [45,46]

$$y_D = \left[\sum_{i=1}^{N} e^{-j\phi_i} g_i\right] x + n_D \tag{8.20}$$

Substituting (8.2) in (8.20), the expression reduces to

$$y_D = \left[\sum_{i=1}^{N} \beta_i\right] x + n_D \tag{8.21}$$

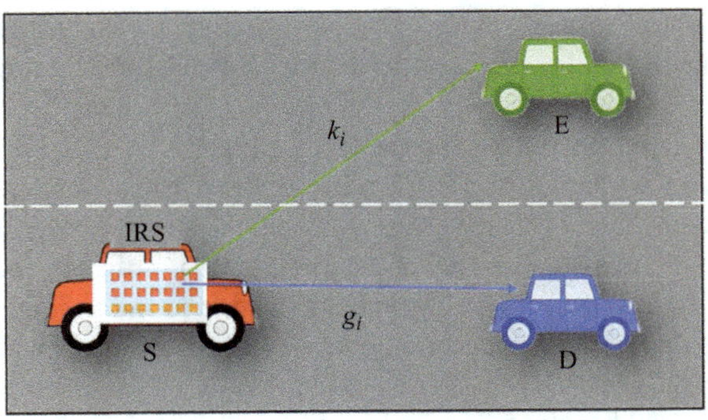

Figure 8.5 A sample IRS-AP configuration for PLS in a VANET

The signal received at the eavesdropper vehicle is

$$y_E = \left[\sum_{i=1}^{N} e^{-j\phi_i} k_i\right] x + n_E \tag{8.22}$$

Substituting (8.3) in (8.22), the expression reduces to

$$y_E = \left[\sum_{i=1}^{N} e^{-j\varphi_i} \vartheta_i e^{j\xi_i}\right] x + n_E = \left[\sum_{i=1}^{N} \vartheta_i e^{-j(\varphi_i - \xi_i)}\right] x + n_E \tag{8.23}$$

Let the one-hop channel of the destination vehicle be

$$G = \sum_{i=1}^{N} \beta_i \tag{8.24}$$

Let the one-hop channel of the eavesdropper vehicle be

$$H = \sum_{i=1}^{N} \vartheta_i e^{-j(\varphi_i - \xi_i)} \tag{8.25}$$

The instantaneous SNR at the destination vehicle is

$$\gamma_D = \frac{|G|^2 E_s}{\sigma^2} \tag{8.26}$$

The instantaneous SNR at the eavesdropper vehicle is

$$\gamma_E = \frac{|H|^2 E_s}{\sigma^2} \tag{8.27}$$

Secrecy capacity and SOP expressions for IRS-AP configuration can be derived using (8.26) and (8.27), and they are identical to (8.17) and (8.19).

8.4 Discussions on simulations

With average secrecy capacity and average SOP, the performance of the proposed IRS-assisted PLS is discussed here. The Monte Carlo simulations are done for 10^5 iterations. The channel gains of the destination vehicle and the eavesdropper vehicle are assumed to be similar. The performance of suggested IRS-assisted systems is compared with relaying-assisted systems. Table 8.1 displays the parameters used in the simulations.

Figure 8.6 compares the average secrecy capacity of IRS-SR configurations with relaying and without IRS (NIRS). It has been found that average secrecy capacity rises with SNR. The eavesdropper vehicle has a considerable impact at low SNR levels. Evidently, the IRS-assisted system significantly outperforms the conventional system NIRS. The PLS is guaranteed, and the average secrecy capacity is increased due to the increased number of IRS elements. The average secrecy capacity of $N = 8$, $N = 16$, $N = 32$, $N = 64$, and $N = 128$ element IRS is 2.03 b/s/Hz, 3.13 b/s/Hz, 4.20 b/s/Hz, 5.25 b/s/Hz, and 6.27 b/s/Hz, respectively, for −5 dB SNR. The IRS-assisted system is also compared with relaying systems having 1 and

Table 8.1 Parameters considered for simulations

Parameters	**Values**
IRS elements	8, 16, 32, 64, 128
Number of relays	1, 8
Desired rates (b/s/Hz)	1.5, 2
Channel model	Uncorrelated Rayleigh flat fading
Average channel gain of a legitimate vehicle	1
Average channel gain of eavesdropper vehicle	1
Number of eavesdropper vehicles	1
Block size	10^5
Metrics	Secrecy rate, SOP

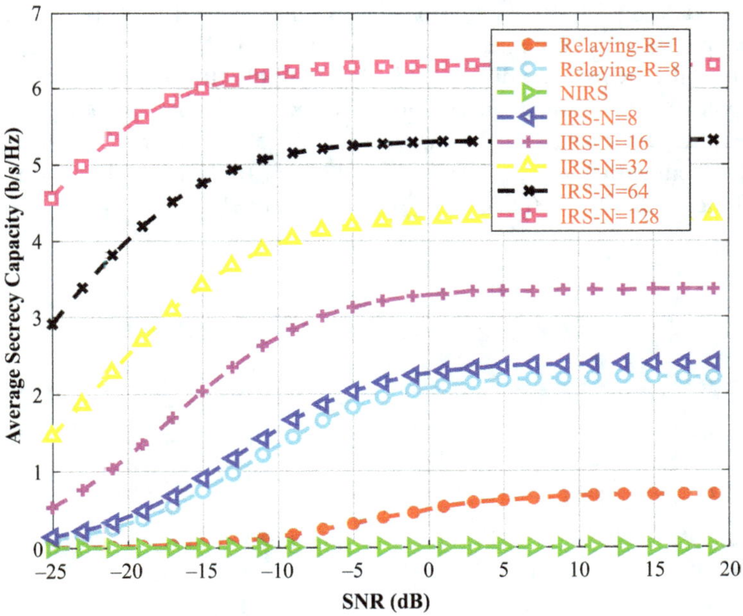

Figure 8.6 Average secrecy capacity comparison of IRS-SR with relaying and NIRS

8 relays. It is observed that relaying-assisted systems outperform conventional NIRS. This is due to the cooperative diversity achieved by relaying-assisted systems. For $R = 8$ relays, the average secrecy capacity of the relaying-assisted system is almost closer to the IRS-assisted system with $N = 8$ elements.

A non-line-of-sight (NLOS) path can be converted into multiple line-of-sight (LOS) paths using cooperative relaying [47,48]. The cooperative diversity achieved by relays grows in proportion to the number of relays deployed. Each relay requires its own power source. Additional circuitry is also required for reception, signal processing, and retransmission. As a result, network power consumption and capital expenditure for deployment rise. The duplexing protocol primarily determines the spectral efficiency of relaying-assisted schemes. Because the source and relays are not permitted to use the same physical resources simultaneously, half-duplex relays have a lower spectral efficiency. Full-duplex relaying can solve this problem. However, self-interference, co-channel interference, signal processing complexity, and power consumption are all issues with full-duplex relaying. The full-duplex relaying system achieves SOP and secrecy capacity performances comparable to the IRS-assisted system. Due to the above drawbacks of full-duplex relaying and the nearly passive IRS elements, IRS-assisted systems are preferable over relaying for PLS.

The average SOP of the IRS-SR system is compared with relaying and NIRS for a target secrecy rate of 1.5 b/s/Hz in Figure 8.7. It is evident that, regardless of SNR, the traditional NIRS system has an average SOP of 1. This is a result of eavesdropper vehicles' dominance. The recommended IRS-SR system outperforms the traditional NIRS system due to the inclusion of the IRS in the environment and proper phase compensation. In addition, the increased number of IRS elements enhances the performance of the SOP. Zero outage probability is reached by the IRS-SR system with $N = 128$, $N = 64$, $N = 32$, $N = 16$, and $N = 8$ at SNRs of −34 dB, −28 dB, −22 dB, −16 dB, and −6 dB, respectively.

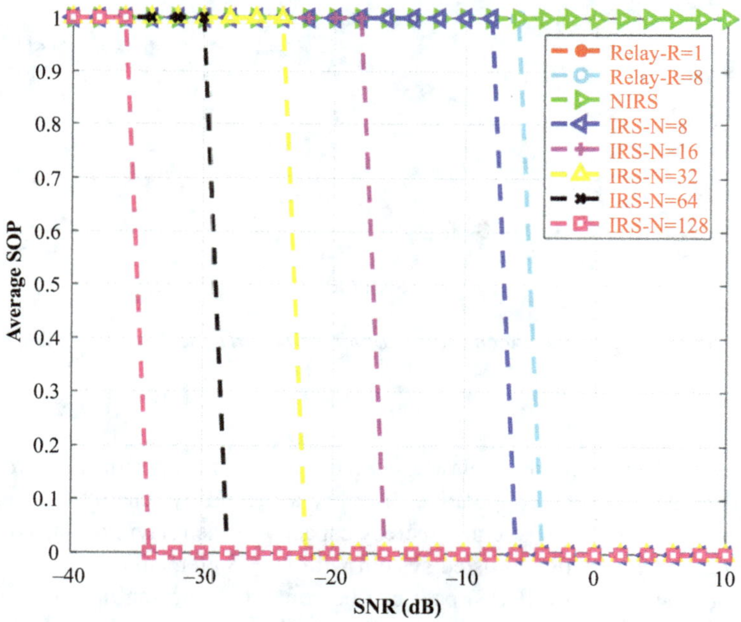

Figure 8.7 Average SOP comparison of IRS-SR with relaying and NIRS for a target secrecy rate of 1.5 b/s/Hz

Figure 8.8 compares the average SOP of IRS-SR configuration with relaying and NIRS for a target secrecy rate of 2 b/s/Hz. The conventional NIRS system has an average SOP of 1, irrespective of SNR. The IRS-SR system with $N = 8$ elements attains an SOP of 1 even at 10 dB of SNR. This is brought on by the rise in target secrecy rate demand from 1.5 b/s/Hz to 2 b/s/Hz. However, increasing the IRS elements enhances the performance of the outage probability. The IRS-SR system with $N = 128$, $N = 64$, $N = 32$, and $N = 16$ reaches zero outage probability at SNR of −30 dB, −24 dB, −16 dB, and −6 dB, respectively. When comparing Figures 8.7 and 8.8, it is clear that increasing the target secrecy rate increases the SNR required to achieve zero outage probability.

The average secrecy capacity of the IRS-AP configuration is compared with relaying and NIRS in Figure 8.9. The IRS-assisted system clearly outperforms the conventional NIRS system. Due to the increased number of IRS components, the PLS is ensured, and the average secrecy capacity is raised. $N = 8$, $N = 16$, $N = 32$, $N = 64$, $N = 128$ element IRS have average secrecy capacities of 2.32 b/s/Hz, 3.45 b/s/Hz, 4.51 b/s/Hz, 5.59 b/s/Hz, and 6.62 b/s/Hz, respectively, given a −5 dB SNR. Table 8.2 compares the average secrecy capacity of IRS-SR and IRS-AP configurations for the various number of reflective elements and −5 dB SNR. The average secrecy capacity of the IRS-AP system is ≈0.3 b/s/Hz more than the IRS-SR system due to the single-hop channel.

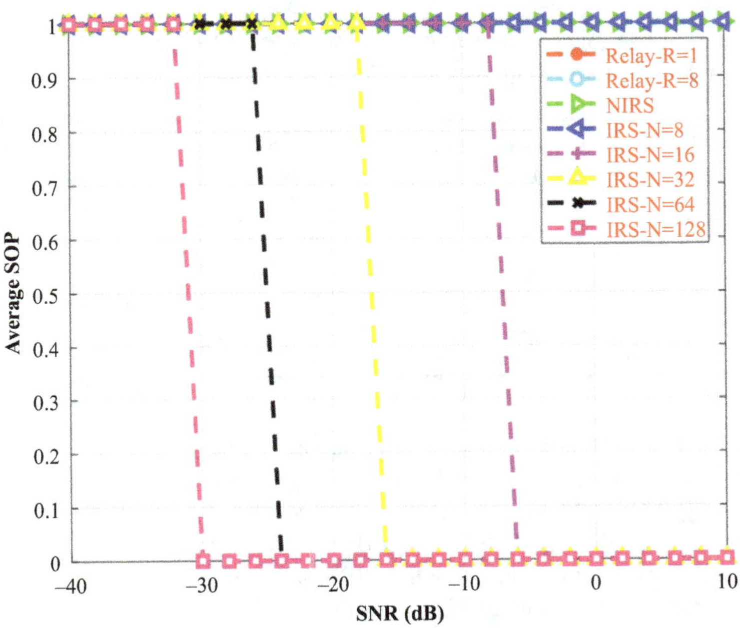

Figure 8.8 Average SOP comparison of IRS-SR with relaying and NIRS for a target secrecy rate of 2 b/s/Hz

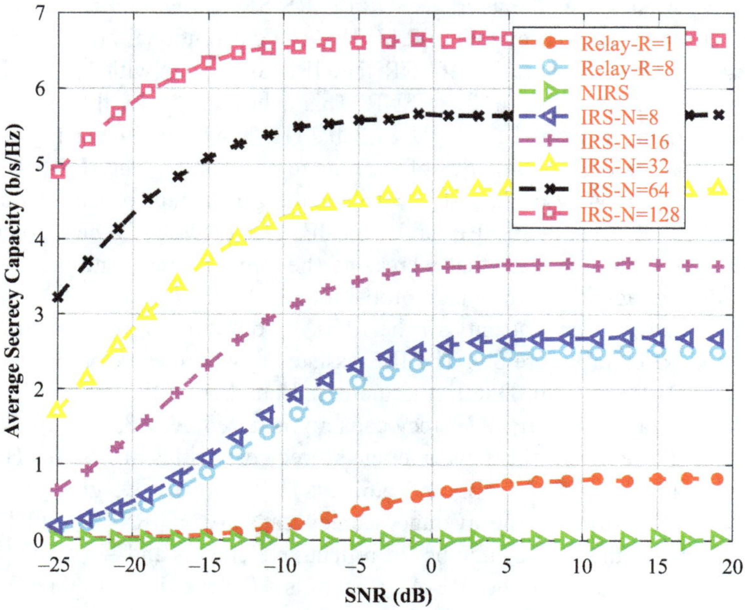

Figure 8.9 Average secrecy capacity comparison of IRS-AP with relaying and NIRS

Table 8.2 Average secrecy capacity comparison of IRS-SR and IRS-AP configurations for various IRS elements and −5 dB SNR

Number of IRS elements	Secrecy capacity (b/s/Hz) of IRS-SR	Secrecy capacity (b/s/Hz) of IRS-AP	Improvement in secrecy capacity (b/s/Hz)
$N=8$	2.03	2.32	0.29
$N=16$	3.13	3.45	0.32
$N=32$	4.20	4.51	0.31
$N=64$	5.25	5.59	0.34
$N=128$	6.27	6.62	0.35

The average SOP of IRS-AP configuration is compared with relaying and NIRS for a target secrecy rate of 1.5 b/s/Hz in Figure 8.10. Zero outage probability is reached by the IRS-AP system with $N=128$, $N=64$, $N=32$, $N=16$, and $N=8$ at SNRs of −36 dB, −30 dB, −24 dB, −18 dB, and −8 dB, respectively. Figure 8.11 compares the average SOP of IRS-AP configuration with relaying and NIRS for a target secrecy rate of 2 b/s/Hz. The IRS-AP system with $N=8$ elements attains an SOP of 1, even at 10 dB of SNR. This is due to an increase in the target secrecy rate needed from 1.5 to 2 b/s/Hz. However, increasing the IRS elements enhances the outage probability performance much more. The IRS-SR system with

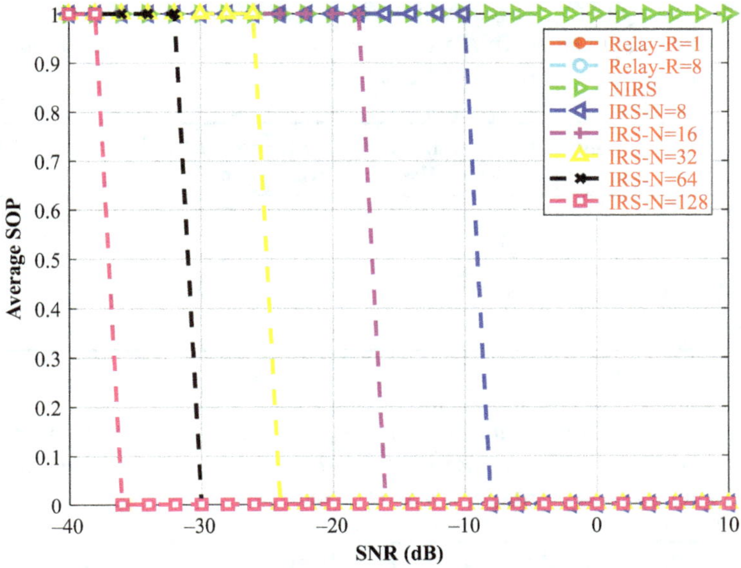

Figure 8.10 *Average SOP comparison of IRS-AP with relaying and NIRS for a target secrecy rate of 1.5 b/s/Hz*

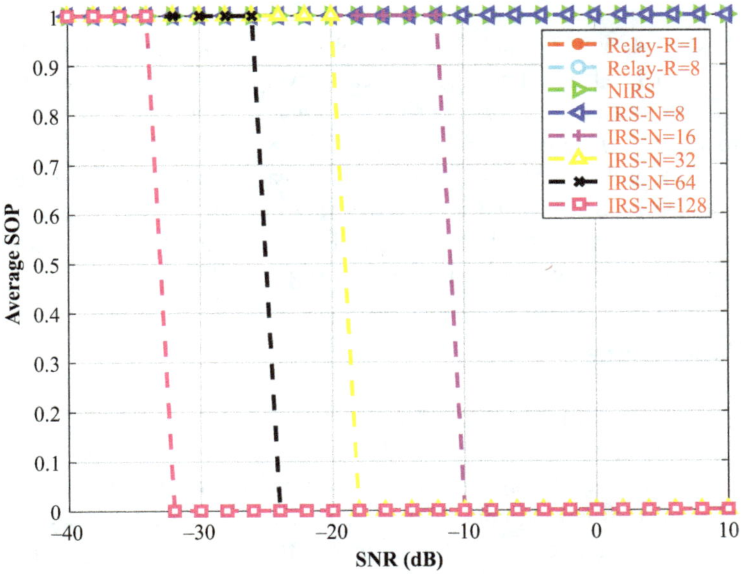

Figure 8.11 *Average SOP comparison of IRS-AP with relaying and NIRS for a target secrecy rate of 2 b/s/Hz*

Table 8.3 Average SOP comparison of IRS-SR and IRS-AP configurations with various numbers of IRS elements for zero outage probability and a target secrecy rate of 1.5 b/s/Hz

Number of IRS elements	SNR required by IRS-SR (dB)	SNR required by IRS-AP (dB)	Gain in SNR (dB)
$N = 8$	−6	−8	2
$N = 16$	−16	−18	2
$N = 32$	−22	−24	2
$N = 64$	−28	−30	2
$N = 128$	−34	−36	2

Table 8.4 Average SOP comparison of IRS-SR and IRS-AP configurations with various numbers of IRS elements for zero outage probability and a target secrecy rate of 2 b/s/Hz

Number of IRS elements	SNR required by IRS-SR (dB)	SNR required by IRS-AP (dB)	Gain in SNR (dB)
$N = 16$	−6	−10	4
$N = 32$	−16	−18	2
$N = 64$	−24	−26	2
$N = 128$	−30	−32	2

$N = 128$, $N = 64$, $N = 32$, and $N = 16$ reaches zero outage probability at SNR of −32 dB, −26 dB, −18 dB, and −10 dB, respectively. Figures 8.10 and 8.11 can be compared, and it can be seen that raising the target secrecy rate increases the SNR needed to achieve zero outage probability. Tables 8.3 and 8.4 compare the average SOP of IRS-SR and IRS-AP configurations with varying numbers of IRS elements for zero outage probability and a target secrecy rate of 1.5 b/s/Hz and 2 b/s/Hz, respectively. Due to the single hop channel, IRS-AP configuration outperforms IRS-SR configuration by a minimum of ≈2 dB gain in SNR.

8.5 Conclusions

The use of IRS on the PLS of VANET is covered in this chapter. IRS is used in two different configurations, referred to as SR and AP, respectively. When IRS is used effectively, it improves the signal quality of legitimate vehicles while deteriorating the signal quality of eavesdropper vehicles. In comparison to the system without the IRS, effective phase compensation improves the secrecy capacity of legitimate vehicles. The outage of desired vehicles decreased as more IRS components were added. However, it grew for eavesdropper vehicles. Relaying has advantages similar to those of IRS, but IRS is a better option for PLS due to its hardware and signal processing complexity. It is observed that the IRS-AP configuration offers ≈0.3 b/s/Hz improvements in secrecy rate and ≈2 dB gain in SNR over the IRS-SR configuration. The phase compensation of IRS elements is subject to a certain implementation constraint. Low-resolution IRS is accessible in real-time. Therefore, a test of the suggested system's performance for low-resolution IRS can be done in the future. IRS and non-orthogonal multiple access (NOMA) can be integrated to support massive vehicle connectivity. The received signal quality of eavesdropper vehicles can be reduced by assigning various power fractions. It is possible to test the suggested system model for generalized fading channels. Furthermore, realistic vehicular channel models could be considered in the future for testing the proposed system.

Acknowledgment

The work of Agbotiname Lucky Imoize is supported in part by the Nigerian Petroleum Technology Development Fund (PTDF) and, in part, by the German Academic Exchange Service (DAAD) through the Nigerian–German Postgraduate Program under Grant 57473408.

References

[1] https://www.ericsson.com/en/reports-and-papers/mobility-report (accessed: 28 November 2022).

[2] https://www.nextgalliance.org/white_papers/roadmap-to-6g/ (accessed: 6 October 2022).

[3] Makarfi, A. U., Rabie, K. M., Kaiwartya, O., Li, X., and Kharel, R. (2020, May). Physical layer security in vehicular networks with reconfigurable intelligent surfaces. In *2020 IEEE 91st Vehicular Technology Conference (VTC2020-Spring)* (pp. 1–6). IEEE.

[4] Makarfi, A. U., Rabie, K. M., Kaiwartya, O., *et al.* (2020). Reconfigurable intelligent surfaces-enabled vehicular networks: a physical layer security perspective. arXiv preprint arXiv:2004.11288.

[5] Mensi, N., Rawat, D. B., and Balti, E. (2021, December). Physical layer security for V2I communications: Reflecting surfaces vs. relaying. In *2021 IEEE Global Communications Conference (GLOBECOM)* (pp. 01–06). IEEE.

[6] Zheng, T. X., Wen, Y., Liu, H. W., *et al.* (2022). Physical-layer security of uplink mmwave transmissions in cellular V2X networks. *IEEE Transactions on Wireless Communications*, 21, 9818–9833.

[7] Wu, Y., Khisti, A., Xiao, C., Caire, G., Wong, K. K., and Gao, X. (2018). A survey of physical layer security techniques for 5G wireless networks and challenges ahead. *IEEE Journal on Selected Areas in Communications*, **36**(4), 679–695.

[8] Chen, X., Ng, D. W. K., Gerstacker, W. H., and Chen, H. H. (2016). A survey on multiple-antenna techniques for physical layer security. *IEEE Communications Surveys & Tutorials*, **19**(2), 1027–1053.

[9] Evangeline, C. S. and Kumaravelu, V. B. (2021). Two-phase access network selection scheme based on weighted sum and game theoretical approaches for vehicular ad hoc networks. *Journal of Circuits, Systems and Computers*, **30**(11), 2150206.

[10] Evangeline, C. S. and Kumaravelu, V. B. (2022). A two-phase fuzzy based access network selection scheme for vehicular ad hoc networks. *Peer-to-Peer Networking and Applications*, **15**(1), 107–133.

[11] Shi, W., Jiang, X., Hu, J., *et al.* (2021). Physical layer security techniques for future wireless networks. arXiv preprint arXiv:2112.14469.

[12] Wang, D., Bai, B., Zhao, W., and Han, Z. (2018). A survey of optimization approaches for wireless physical layer security. *IEEE Communications Surveys & Tutorials*, **21**(2), 1878–1911.

[13] Leung-Yan-Cheong, S. and Hellman, M. (1978). The Gaussian wire-tap channel. *IEEE Transactions on Information Theory*, **24**(4), 451–456.

[14] Wang, X., Tao, M., Mo, J., and Xu, Y. (2011). Power and subcarrier allocation for physical-layer security in OFDMA-based broadband wireless networks. *IEEE Transactions on Information Forensics and Security*, **6**(3), 693–702.

[15] Ng, D. W. K., Lo, E. S., and Schober, R. (2012). Energy-efficient resource allocation for secure OFDMA systems. *IEEE Transactions on Vehicular Technology*, **61**(6), 2572–2585.

[16] Jeong, C. and Kim, I. M. (2011). Optimal power allocation for secure multicarrier relay systems. *IEEE Transactions on Signal Processing*, **59**(11), 5428–5442.

[17] Goel, S. and Negi, R. (2008). Guaranteeing secrecy using artificial noise. *IEEE Transactions on Wireless Communications*, **7**(6), 2180–2189.

[18] Bash, B. A., Goeckel, D., Towsley, D., and Guha, S. (2015). Hiding information in noise: fundamental limits of covert wireless communication. *IEEE Communications Magazine*, **53**(12), 26–31.

[19] Yan, S., Zhou, X., Hu, J., and Hanly, S. V. (2019). Low probability of detection communication: opportunities and challenges. *IEEE Wireless Communications*, **26**(5), 19–25.

[20] Shahzad, K., Zhou, X., Yan, S., Hu, J., Shu, F., and Li, J. (2018). Achieving covert wireless communications using a full-duplex receiver. *IEEE Transactions on Wireless Communications*, **17**(12), 8517–8530.

[21] Jiang, X., Chen, X., Tang, J., *et al.* (2021). Covert communication in UAV-assisted air-ground networks. *IEEE Wireless Communications*, **28**(4), 190–197.

[22] Wang, H. M., Zhang, Y., Zhang, X., and Li, Z. (2019). Secrecy and covert communications against UAV surveillance via multi-hop networks. *IEEE Transactions on Communications*, **68**(1), 389–401.

[23] Hong, T., Song, M. Z., and Liu, Y. (2011). Dual-beam directional modulation technique for physical-layer secure communication. *IEEE Antennas and Wireless Propagation Letters*, **10**, 1417–1420.

[24] Ding, Y. and Fusco, V. F. (2013). A vector approach for the analysis and synthesis of directional modulation transmitters. *IEEE Transactions on Antennas and Propagation*, **62**(1), 361–370.

[25] Hu, J., Shu, F., and Li, J. (2016). Robust synthesis method for secure directional modulation with imperfect direction angle. *IEEE Communications Letters*, **20** (6), 1084–1087.

[26] Jaiswal, G., Gudla, V. V., Kumaravelu, V. B., Reddy, G. R., and Murugadass, A. (2020). Modified spatial modulation and low complexity signal vector based minimum mean square error detection for MIMO systems under spatially correlated channels. *Wireless Personal Communications*, **110**(2), 999–1020.

[27] Kumaravelu, V. B., Jaiswal, G., Gudla, V. V., Ramachandra Reddy, G., and Murugadass, A. (2019). Modified spatial modulation: an alternate to spatial multiplexing for 5G-based compact wireless devices. *Arabian Journal for Science and Engineering*, **44**(8), 6693–6709.

[28] Gudla, V. V. and Kumaravelu, V. B. (2019). Dynamic spatial modulation for next generation networks. *Physical Communication*, **34**, 90–104.

[29] Gudla, V. V. and Kumaravelu, V. B. (2019). Permutation index-quadrature spatial modulation: a spectral efficient spatial modulation for next generation networks. *AEU-International Journal of Electronics and Communications*, **111**, 152917.

[30] Gudla, V. V. and Kumaravelu, V. B. (2020). Enhanced redesigned spatial modulation: design and performance evaluation under correlated fading channels. *International Journal of Communication Systems*, **33**(6), e4294.

[31] Jadhav, H. K. and Kumaravelu, V. B. (2022). Transmit antenna selection for spatial modulation based on machine learning. *Physical Communication*, **55**, 101904.

[32] Gudla, V. V., Kumaravelu, V. B., and Murugadass, A. (2022). Transmit antenna selection strategies for spectrally efficient spatial modulation techniques. *International Journal of Communication Systems*, **35**(7), e5099.

[33] Rajashekar, R., Hari, K. V. S., and Hanzo, L. (2013). Antenna selection in spatial modulation systems. *IEEE Communications Letters*, **17**(3), 521–524.

[34] Pillay, N. and Xu, H. (2015). Low-complexity transmit antenna selection schemes for spatial modulation. *IET Communications*, **9**(2), 239–248.

[35] Shu, F., Wang, Z., Chen, R., Wu, Y., and Wang, J. (2018). Two high performance schemes of transmit antenna selection for secure spatial modulation. *IEEE Transactions on Vehicular Technology*, **67**(9), 8969–8973.

[36] Kumaravelu, V. B., Jadhav, H. K., BS, A., Gudla, V. V., Murugadass, A., and Imoize, A. L. (2023). Unmanned aerial vehicle-assisted reconfigurable intelligent surface for energy efficient and reliable communication. In *Unmanned Aerial Vehicle Cellular Communications* (pp. 173–201). Springer, Cham.

[37] Cui, M., Zhang, G., and Zhang, R. (2019). Secure wireless communication via intelligent reflecting surface. *IEEE Wireless Communications Letters*, **8** (5), 1410–1414.

[38] Shen, H., Xu, W., Gong, S., He, Z., and Zhao, C. (2019). Secrecy rate maximization for intelligent reflecting surface assisted multi-antenna communications. *IEEE Communications Letters*, **23**(9), 1488–1492.

[39] Yu, X., Xu, D., and Schober, R. (2019, December). Enabling secure wireless communications via intelligent reflecting surfaces. In *2019 IEEE Global Communications Conference (GLOBECOM)* (pp. 1–6). IEEE.

[40] Dong, L. and Wang, H. M. (2020). Secure MIMO transmission via intelligent reflecting surface. *IEEE Wireless Communications Letters*, **9**(6), 787–790.

[41] Jiang, W., Zhang, Y., Wu, J., Feng, W., and Jin, Y. (2020). Intelligent reflecting surface assisted secure wireless communications with multiple-transmit and multiple-receive antennas. *IEEE Access*, **8**, 86659–86673.

[42] Hong, S., Pan, C., Ren, H., Wang, K., and Nallanathan, A. (2020). Artificial noise-aided secure MIMO wireless communications via intelligent reflecting surface. *IEEE Transactions on Communications*, **68**(12), 7851–7866.

[43] Jadhav, H. K. and Kumaravelu, V. B. (2022). Blind RIS aided ordered NOMA: design, probability of outage analysis and transmit power optimization. *Symmetry*, **14**(11), 2266.

[44] Kumaravelu, V. B., Imoize, A. L., Soria, F. R. C., *et al.* (2022). Outage probability analysis and transmit power optimization for blind reconfigurable intelligent surface-assisted non-orthogonal multiple access uplink. *Sustainability*, **14**(20), 13188.

[45] Basar, E., Di Renzo, M., De Rosny, J., Debbah, M., Alouini, M. S., and Zhang, R. (2019). Wireless communications through reconfigurable intelligent surfaces. *IEEE Access*, **7**, 116753–116773.

[46] Jalaja, R. R. U., Thirumavalavan, V. C., Velmurugan, P. G. S., and Thiruvengadam, S. J. (2021). Spatially correlated dual hop ris aided next generation wireless systems: an outage perspective. *IEEE Access*, **9**, 56127–56139.

[47] Di Renzo, M., Ntontin, K., Song, J., *et al.* (2020). Reconfigurable intelligent surfaces vs. relaying: differences, similarities, and performance comparison. *IEEE Open Journal of the Communications Society*, **1**, 798–807.

[48] Lu, W. and Di Renzo, M. (2015). Stochastic geometry modeling and system level analysis & optimization of relay-aided downlink cellular networks. *IEEE Transactions on Communications*, **63**(11), 4063–4085.

Chapter 9

Physical layer security solutions and technologies

Gustavo Anjos[1], *Daniel Castanheira*[1], *Adão Silva*[1], *Suneel Yadav*[2] *and Atílio Gameiro*[1]

Abstract

To ensure confidentiality, today's communications are encrypted under the notion of computational security, using hard mathematical problems to build ciphers that apparently cannot be cracked in a useful time. However, as these constructions are not agnostic to technological advances, some of these problems may be solved efficiently with future technologies, e.g. quantum computing. In wireless networks, physical layer security emerges as a post-quantum security solution that promises to mitigate these threats. This concept of secrecy exploits the physical properties of the wireless channel to encode information so that a certain degree of statistical independence between the message and the cyphertext is observed by the eavesdropper. This notion of secrecy allows building cryptosystems that are agnostic to the computing capabilities of the attacker, being widely accepted as one of the strongest notions of secrecy created so far. The objective of this chapter is to overview the concept of physical layer security and understand how its integration could be done in future wireless networks. For that purpose, the information-theoretical framework grounding the concept is introduced, and some basic design approaches are presented. These include PHY key generation methods, secure beamforming techniques, and cooperative jamming constructions. The mechanisms to enable the integration of these technologies in future 5G and beyond networks are discussed last.

Keywords: Wireless networks; Cryptography; Physical layer security; Cooperative jamming; Information theory; Wiretap channel

9.1 Introduction

Wireless networks are a fundamental part of today's communication systems, providing not only ubiquitous connectivity to the end user but also a practical

[1]Instituto de Telecomunicações and DETI, University of Aveiro, Portugal
[2]Department of Electronics and Communication Engineering, Indian Institute of Information Technology Allahabad, India

option for setting up an infrastructure-less network. The open nature of the wireless channel is the underlying condition that enables such merits. At the same time, this broadcast nature also makes wireless networks extremely vulnerable in terms of security. With the emerging concept of the Internet of Things, billions of regular objects will exchange information using low-power wireless transceivers [1]. This massive proliferation of wireless terminals increases the risk of undetectable eavesdropping attacks, bringing new security challenges that must be handled by current wireless standards. To ensure confidentiality, current commercial systems rely on public key cryptography and on the computational hardness of solving certain mathematical problems [2]. This form of security is defined in the literature as computational security. In recent years, advances in the fields of number theory and quantum computing have begun to threaten the security strength of public key protocols, forcing the use of larger key sizes, and leading to increased implementation complexity [3]. In order to deal with these new threats, the development of more resilient security solutions has established itself as the top priority in the network security domain. One promising solution is known as physical layer security. The former concept exploits the physical properties of the wireless channel to ensure secrecy between multiple communication parties. Unlike computational security, the notion of secrecy applied in physical layer security is information-theoretic security. This security concept is grounded on the mathematical framework of information theory and is widely accepted as one of the strongest notions of secrecy created to date.

9.1.1 Shannon cryptosystem

The first attempt to formalize the concept of information-theoretic security was carried out by Claude Shannon in 1949 [4]. In his work, Shannon defined the notion of perfect secrecy by modeling the problem of secret communication using the cipher system illustrated in Figure 9.1. In this model, the legitimate parties, represented by Alice and Bob, want to exchange a message W while keeping it secret from Eve, the eavesdropper. To protect the message, Alice encodes W in a codeword X using a pre-shared secret key K; while at the receiver side, Bob retrieves the message $\widehat{W}$ by decoding X with the same secret key. In this configuration, Shannon stated that perfect secrecy is achieved if X is statistically independent of W, which can be mathematically defined as follows.

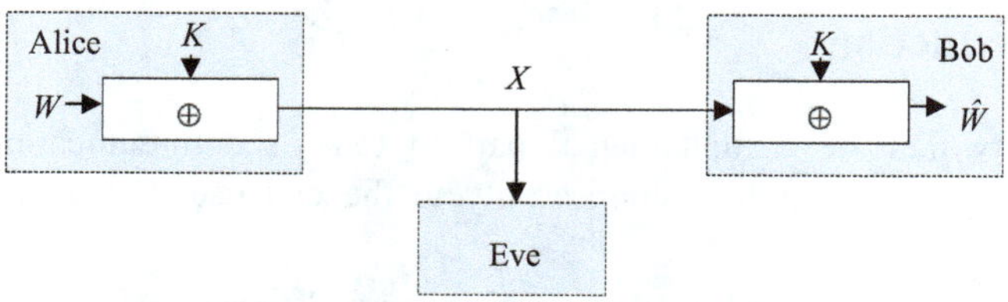

Figure 9.1 Symmetric cryptosystem

Definition 9.1. For W defined as the information source, and X the output of the eavesdropper channel, a state of *perfect secrecy* is verified under the constraint

$$I[W;X] = 0. \tag{9.1}$$

To achieve statistical independence, the entropy of the secret key $H(K)$ must be at least equal to the entropy of the message $H(W)$, which from a practical perspective is a condition difficult to satisfy [5]. For instance, if W is a binary source composed by n bits distributed uniformly, this implies that the size of K must to be at least equal to n bits, which is impractical for large values of n. Nevertheless, if the condition $H(K) \geq H(W)$ is verified, perfect secrecy can be achieved by encoding the message as $X = W \oplus K$, i.e. modulo 2 addition of each bit of W and K. The recovery of W at Bob can be done by applying the decoding operation $\widehat{W} = X \oplus K$; while in the case of Eve, the optimal decoding strategy is to discard the received codeword and try guessing the transmitted message correctly, because X is independent of W, and the secret key K is not available. The probability of successful decoding is equal to 2^{-n}. Therefore, under these circumstances, the observation or processing of X at the eavesdropper is useless. This concept of perfect secrecy, despite providing a very powerful result, is difficult to reproduce in practical systems. This stems from the difficulty of implementing the key distribution procedure, which in addition to having to be done in secrecy, it can be a quite inefficient process given the high rate at which the secret keys have to be renewed.

9.1.2 Computational security and its limitations

The difficulty of replicating in practical systems the necessary conditions to achieve perfect secrecy led to the advent of the notion of computational security, which in turn gave rise to the birth of modern cryptography. In this new concept of security, the computational effort required to calculate the solution of some hard mathematical problems is used to protect information. Therefore, an encryption scheme is classified as computationally secure if the cost or the time required to break it, exceeds the value or the lifetime of the information. This notion of security is materialized using trap-door one-way functions, which are mathematical functions that are "easy" to compute for every input, whereas the calculation of the respective inverse is an "infeasible" task [6]. In the field of computational complexity, the term "easy" is used to categorize a problem that can be computed at most in polynomial time; while the word "infeasible" is applied to classify a problem whose execution time exceeds polynomial time. The classification of a scheme as being computationally secure is grounded on the unproven assumption that such type of "infeasible" mathematical problems does exist. Nevertheless, computational security has found widespread acceptance, being, from a practical perspective, the most successful notion of security created so far [7]. In all its forms of commercial use, the field of modern cryptography establishes a clear distinction between symmetric and asymmetric encryption. While in the symmetric case, the

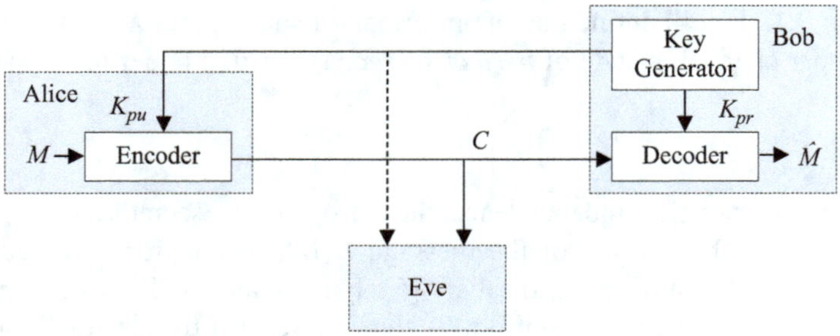

Figure 9.2 Asymmetric encryption

same key is used to encode and decode information; in asymmetric encryption, also known as public-key cryptography, the encoding and decoding operations are done using a public and a private key. As shown in Figure 9.2, the public-key protocol starts at Bob with the generation of a private and a public key, designated by K_{pr} and K_{pu}, respectively. The public key is sent to Alice via a public channel, being used to encode the message; while the private key remains secured at Bob, being applied in the decoding process. The generation of the keys is done by harnessing the structure of certain mathematical problems to produce one-way functions. By properly dimensioning the size of the keys, retrieving both the private key and the message becomes an "infeasible" problem for Eve, if only the codeword and the public key were available at the intruder. Only with the private key it is possible to retrieve the message from the codeword; therefore, as long as the private key remains protected at the legitimate receiver, the system is computationally secure. In fact, public-key encryption is the heart of modern cryptography, being widely adopted in the provision of confidentiality and authentication services. In confidentiality applications, as the execution of public key protocols is a relatively slow process, they are usually combined with symmetric encryption, not being directly applied to encode the plaintext. Instead, they take part in the key management phase, protecting the secret key used by the symmetric protocol to encrypt the messages. In this form of hybrid encryption, the size of the symmetric key is usually shorter than the size of the message. Although this approach does not allow to achieve a state of perfect secrecy, it simplifies the practical implementation of the protocol, as it does not require the permanent renewal of the keys. The implementation of public-key cryptography requires the maintenance of a public-key infrastructure (PKI), responsible for the distribution and certification of both the private and public keys. In addition to being a costly infrastructure, its operation model relies on the belief that some elements of the infrastructure are trustworthy, which implies that the respective reliability can only be measured qualitatively. Over the past few years, public-key protocols have remained relatively secure against known threats. However, as these constructions are not agnostic to technological advances, it is not possible to ensure security in the future. One of the major threats that public-key cryptography faces today is quantum computing. This new form of information processing will complement traditional computing solving some problems that classical processors are not able to handle. While such capabilities promise to bring positive changes to the world, they raise some new concerns in the field of computational security. For instance, the integer factorization of large numbers is one of the tasks that quantum computers can solve efficiently using less time complexity than the usual processing methods. As a consequence, the factoring problem will be feasible [8], and public-key protocols like the "RSA" will become obsolete. In the medium to long term, the practical implementation of the quantum technology is a major inconvenient to the notion of computational security, as some of the complexities considered "infeasible" in classical computing, can be treated as "easy" problems in the quantum world.

9.1.3 The physical layer security concept

As computer science continues to evolve, it becomes increasingly urgent to find cryptographic techniques that can operate agnostically in relation to the attacker's processing capacities [9]. The goal is to find a solution capable of exchanging information reliably between Alice and Bob, while ensuring a certain level of statistical independence between the source and the signal on Eve. In wireless communications, the imperfections of the physical channel can be used for that purpose. Through a careful manipulation of interference, noise and fading, information-theoretic security can be achieved by creating a channel advantage at the legitimate receiver. As illustrated in Figure 9.3, this advantage can be obtained in two different ways: using a keyless approach, by imposing a physical degradation of the channel at the unwanted receiver; or with a key based solution, where the random condition of the channel is used as a source for secret keys [10]. The security schemes designed according to the principles referred above belong to the field of physical layer security. Unlike public-key cryptography, in physical layer security, the secrecy performance is quantified with precision at the bit level using the mathematical framework of information theory, relying only on the statistical properties of the observed signals. In this concept of secrecy, confidentiality is achieved if the information source is statistically independent of the signals at the unwanted receiver. If such condition is satisfied, the unwanted receiver is no better informed after acquiring the channel output; therefore, observing the channel output, or observing nothing at

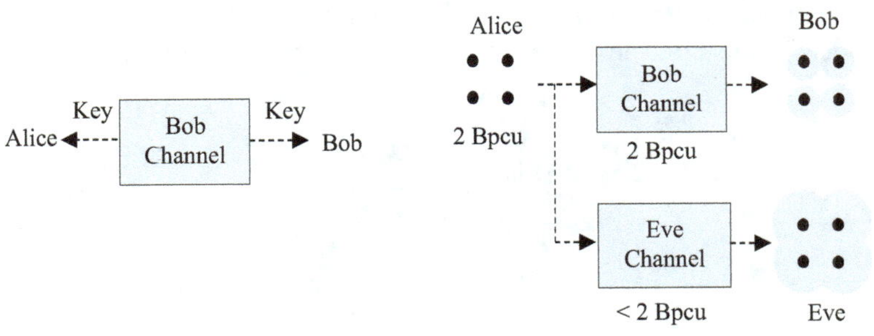

Figure 9.3 Key-based solution (left) and keyless approach (right)

all is indifferent for the intruder. This design approach allows to build security solutions that are agnostic to the technological capabilities of the opponent, providing in this way more resilience against attacks carried out by future technologies. As the name suggests, the physical layer security algorithms are executed at the physical layer of the OSI model [11], and its implementation is done imposing secrecy restrictions on the design of the constituent modules, such as the channel encoder, the source encoder, and other signal processing blocks.

9.1.4 Chapter organization

The remainder of this chapter is organized as follows: Section 9.2 presents a basic understanding of the theoretical fundamentals grounding the concept of physical layer security. Some of the most important physical layer security techniques are briefly described in Section 9.3, including PHY key generation methods, secure beamforming techniques, and cooperative jamming schemes. The integration of these techniques in 5G and beyond networks is discussed in Section 9.4. The main conclusions are presented in Section 9.5.

9.2 Fundamentals of physical layer security

This section presents some of the fundamentals grounding physical layer security. The problem of establishing secret communications over a wireless channel is formulated first by introducing the wiretap channel model. Then, the concept of secrecy capacity is defined for this channel. The last point of this section explains how this secrecy capacity can be achieved through wiretap coding.

9.2.1 The wiretap channel

As seen before, in Shannon's notion of perfect secrecy, an ideal state of confidentiality between two terminals can only be achieved if a secret key with the same entropy of the information source is available on both. However, in his work, Shannon did not account with any specific channel impairment, such as noise or fading, assuming that the secret key was the only unshared information with the eavesdropper. A few years later, Wyner defined a more realistic model, the wiretap channel, and showed that coding can be used to reach positive secrecy if the eavesdropper channel is a degraded version of the legitimate channel [12]. This model is illustrated in Figure 9.4 and incorporates a legitimate transmitter,

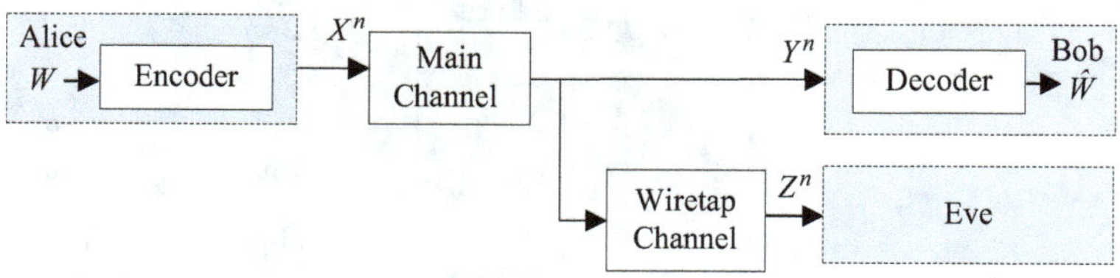

Figure 9.4 Wiretap channel model

Alice, who intends to transmit at rate R a message W to Bob, while keeping it secret from Eve, the eavesdropper. To achieve that goal, Alice maps the message $W \in \{1, 2, \ldots, 2^{nR}\}$ into X^n, a codeword of length n, and transmits it through the main channel in n channel uses. After sampling the channel output n times, Bob decodes Y^n and obtains an estimate of W, denoted as $\widehat{W}$. Eve observes the transmitted signal through an additional channel, the wiretap channel, and tries to get W from the output Z^n. If Eve's channel is degraded in relation to Bob's channel, there is a n length-code such that for every $\epsilon > 0$, the conditions

$$P[W \neq \widehat{W}] \leq \epsilon \text{ and } \frac{I[W; Z^n]}{n} \leq \epsilon \tag{9.2}$$

are mutually achieved. The first condition ensures that the message is transmitted reliably to Bob, and the second one is a secrecy constraint that ensures a vanishing of the information at Eve. The maximum rate R for which it is possible to fulfill the conditions in (9.2) is referred to as the secrecy capacity of the channel. If R does not exceed this value, it is possible to build a n length-code with ϵ as close to zero as desirable. However, setting the value of ϵ arbitrarily close to zero requires an asymptotically large codeword length, that is, $n \to \infty$. This stems from the reliability condition in (9.2), which calls for the introduction of redundancy in the system. Before advancing, it is of worth to compare the concepts of secrecy formulated by Shannon and Wyner. While the notion of perfect secrecy stated by Shannon requires exact statistical independence between the message and the signal observed at Eve, in Wyner's case, as defined in (9.2), the system is considered secure if the rate of information at Eve vanishes in the limit of large block lengths. The notion of secrecy introduced by Wyner is referred in the literature as *weak secrecy*. The latter is defined as follows.

Definition 9.2. For W defined as the information source, and Z^n a vector formed by n samples of the eavesdropper channel output, a *weak secrecy* state is verified under the constraint

$$\lim_{n \to \infty} \frac{I[W; Z^n]}{n} = 0. \tag{9.3}$$

As the secrecy constraint in (9.3) can be fulfilled even for values of $I[W; Z^n] > 0$, this notion of secrecy is referred in the literature as *weak secrecy*. Despite achieving a weaker secrecy result, this less demanding constraint avoids the requirement of exact statistical independence between W and Z^n, thus allowing to widening the scope of possible solutions to the problem. In order to strengthen the previous notion, the literature also defines the concept of *strong secrecy*, which is formulated setting $n = 1$ in expression (9.3). In the latter case, exact statistical independence between W and Z^n must be verified.

9.2.2 Secrecy capacity

The concept of secrecy capacity is for physical layer security, as well as the notion of capacity is for non-secrecy constrained channels. In its genesis, the proof in [13] guarantees that it is possible to design a code ensuring reliable communication with Bob, while keeping information secret on Eve. However, just like in the notion of capacity, that code only exists if the transmission rate is not greater than a given value, which in this case is referred as the secrecy capacity of the channel. In the form of a converse proof, the authors in [13] showed that for any rate beyond the secrecy capacity, the conditions in (9.2) cannot be achieved by any possible encoding procedure. The secrecy capacity of the wiretap channel is defined as follows.

Theorem 9.1. For V and X defined as elements of the Markov chain $V \rightarrow X \rightarrow Y, Z$, the secrecy capacity of the wiretap channel is given by

$$C_s = \max_{p(V,X)} \{I[V;Y] - I[V;Z]\}^+. \tag{9.4}$$

As stated in Theorem 9.1, the secrecy capacity is reached by selecting V and X such that the joint distribution $p(V, X)$ maximizes the achievable secrecy rate of the channel. By the definition above, it is implicit that the maximization in (9.4) can be done making the transmitted signal X a stochastic function of V, the message carrying signal. In another words, this means that the secrecy rate of the system can be improved by adding some randomness to the message carrying signal. In any case, to reach a non-zero secrecy capacity, the condition $I[V; Y] > I[V; Z]$ must be always verified, which means that the eavesdropper channel must be degraded in relation to the legitimate receiver.

9.2.3 Wiretap codes

As referred before, to put in operation the secrecy rate in (9.4), the messages must be coded so that the conditions in (9.2) are both met. If the transmission rate is not greater than the secrecy capacity of the channel, in theory, there exists a code that achieves reliability at Bob, and secrecy at Eve. In physical layer security, this type of code is designated as a wiretap code. A code for the wiretap channel can be defined as follows.

Definition 9.3. A code $(2^{nR}, n)$ for the wiretap channel consists of the following:

- One message set $\mathcal{W} = \{1, 2, ..., 2^{nR}\}$;
- A stochastic function $f : \mathcal{W} \rightarrow \mathcal{X}^n$ that encodes $w \in \mathcal{W}$ in a codeword x^n;
- A decoding function $g : \mathcal{Y}^n \rightarrow \mathcal{W}$ which maps the observation

$\mathbf{y}_n$ at Bob to the estimated message $\widehat{w}$.

Due to the antagonic nature of the constraints in (9.2), the construction of a wiretap code calls for two divergent design approaches. While the reliability condition requires a code capable of adding redundancy at Bob, the secrecy constraint calls for a code with no redundancy, and which has to create equivocation at Eve. A code that merges both approaches is feasible using a random codebook structure, formed by codewords with a large number of elements. In the case of a degraded

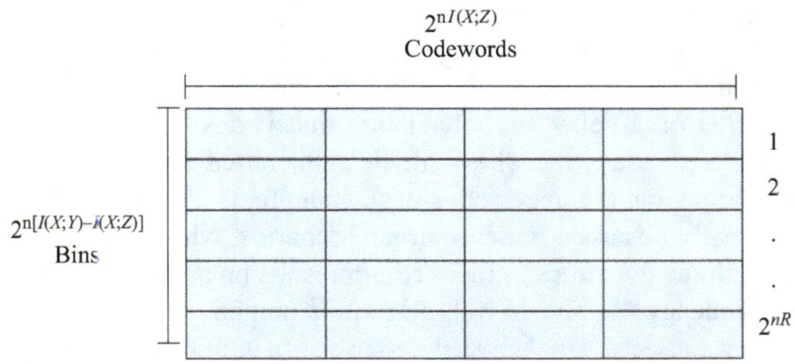

Figure 9.5 Wiretap code structure

wiretap channel, e.g. Gaussian wiretap channel, a positive secrecy rate can be achieved in theory by building a codebook [14] with the structure depicted in Figure 9.5. The construction of the codebook starts with the random generation of $2^{nI(X;Y)}$ codewords, each one with a length of n elements. Then, these codewords are grouped in $2^{n[I(X;Y)\,-\,I(X;Z)]}$ bins, so that each bin is formed by $2^{nI(X;Z)}$ codewords. Each message is mapped in a unique bin; therefore, to send a message, Bob selects the bin corresponding to the message, and chooses randomly a codeword from that bin, which is posteriorly transmitted in n channel uses. While Bob's channel allows him to distinguish all the $2^{nI(X;Y)}$ codewords, the less favorable channel conditions at Eve limits her observation to a maximum of $2^{nI(X;Z)}$ codewords; therefore, in some way, blocks of $2^{nI(X;Z)}$ codewords will have to overlap at Eve's channel output, creating equivocation at Eve. In the related literature, this encoding technique is referred as random binning. Before introducing the concept of cooperative jamming, some techniques commonly applied in multiple antenna systems are analyzed next.

9.3 Physical layer security approaches

This section introduces some basic design approaches applied in physical layer security. The idea of using the wireless channel as a source of secret keys is explained first, followed by the presentation of two important transmission techniques applied in multiple antenna systems. The use of cooperative jamming for secrecy is the last point addressed in this section. The review presented in the following is not extensive enough to be classified as a comprehensive survey; however, we should say that most of the works in the literature results from natural adaptations of these methods. In particular, these techniques find a wide range of applications in different scenarios, ranging from multi-user channel settings (e.g. broadcast channel, interference channel, multiple access channel) to multi-hop wireless networks.

9.3.1 Extracting secret keys at the physical layer

Due to the presence of physical objects in the radio channel, the transmitted signal is scattered, reflected, and diffracted. The latter phenomenon creates a multipath communication scenario, where multiple copies of the transmitted signal are observed at the receiver with different delays and attenuations. When the position of the objects and users changes over time, the channel effect on the transmitted signal is reflected on a delay and attenuation that varies randomly over multiple channel uses. As a result, the phase of the transmitted signal is rotated, and a random fluctuation on the received signal strength is observed. This dynamic condition is usually observed in dense urban scenarios, where the high user mobility together with the dynamics of the medium results on a channel state that varies in time and frequency. As shown in [15,16], this random variation can be used to establish a common secret key between two users. In a time division duplex (TDD) system, if both users exchange a reference signal within the channel coherence time, the reciprocal channel condition will allow them to observe correlated signals. These signals can be used to extract a secret key. For instance, when the users perform a reciprocal channel estimation, they can use the phase or the magnitude of the estimated channel to distill a secret key. The received signal strength (RSS) or the channel state information (CSI) can be used for that purpose. If an eavesdropper tries to tap the key observing the reference signals, the spatial decorrelation between the links will not allow him to get the correct key. This phenomenon occurs if the eavesdropper is more than half wavelength away from the users; since, in this case, the channels will be independent. The channel randomness and the link reciprocity are the underlying properties that allow the legitimate users to agree on a random key, while spatial decorrelation ensures that the latter remains confidential. In this case, the greater the entropy of the channel, higher secret key rates can be obtained from the wireless medium. To better understand this topic, next we consider a simple example where a phase-based PHY key generation procedure is applied to a complex Gaussian wiretap channel.

Example 9.1: Let us consider the wiretap channel of Figure 9.4, where Eve is more than half wavelength away from Alice and Bob. In this scenario, let us suppose that Alice and Bob wish to exchange a key, keeping it secret from Eve. To extract the secret key from the channel, Alice transmits to Bob the reference signal r_A in the coherence time i of the channel; then, in the same coherency block, Bob transmits the reference signal r_B to Alice. The signals received by Bob, and then by Alice, are defined as

$$y_{B,i} = h_{BA,i} r_A + n_{B,i} \text{ and } y_{A,i} = h^*_{BA,i} r_B + n_{A,i}, \tag{9.5}$$

respectively, where $h_{BA,i}$ denotes the channel between Alice and Bob, while $n_{B,i}$ and $n_{A,i}$ defines noise at the latter nodes. We assume that the channels and the noise components follow a complex Gaussian distribution. Then, after acquiring the signals, Alice and Bob estimate the channel phases

$$\widehat{\theta}_{BA,i} = g(y_{A,i}, r_B) \text{ and } \widetilde{\theta}_{BA,i} = f(y_{B,i}, r_A), \tag{9.6}$$

as a function of the signals observed during the channel-training phase. If the channel is in a dynamic condition, it is possible to extract a random secret key repeating the previous procedure over multiple coherency blocks. Figure 9.6 illustrates the case where a secret key of 6 bits is extracted over three coherency blocks, resulting in a secret key rate of 2 bits per coherency block. After the estimation procedure, the phases are mapped in a gray code of two bits to generate the secret key. At the eavesdropper side, the independent variation of the channel will introduce errors on the secret key.

In the previous example, it was assumed that the channel estimation is flawless, resulting in a key agreement at the legitimate users. However, because the estimation is performed at different times, in a real scenario, the channel reciprocity is not perfect, leading to errors in the secret key. Furthermore, under low signal-to-noise ratio (SNR) conditions, the noise may also cause a key mismatch. Therefore, after extracting the secret keys, an error correction phase has to take place to reconcile the keys. In practical embodiments, as illustrated in Figure 9.7, a PHY key generation protocol is implemented in multiple stages. These are described next:

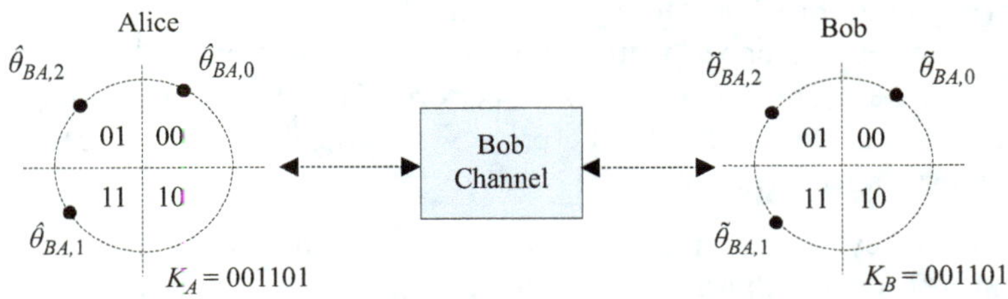

Figure 9.6 Key-based solution (left)

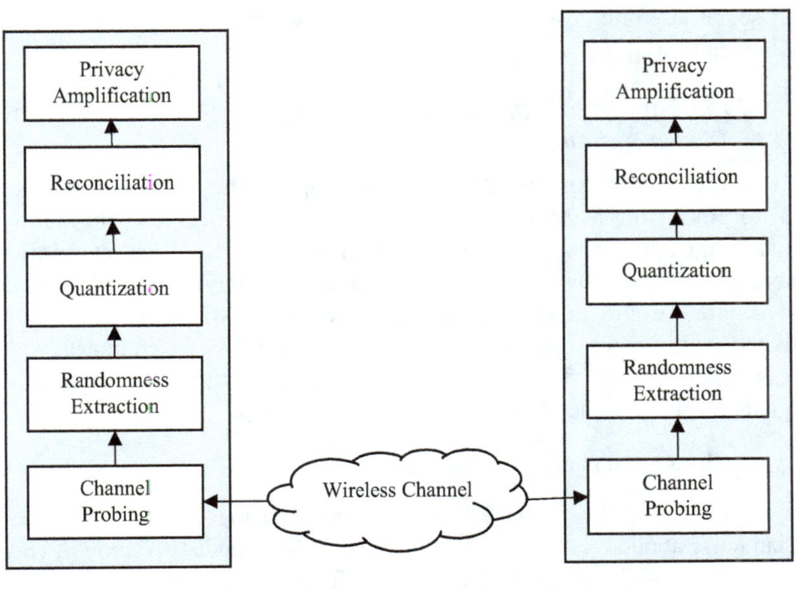

Figure 9.7 PHY key generation protocol

- **Channel probing:** The protocol starts with Alice and Bob exchanging probing signals over multiple coherency blocks. The goal is to measure the channel multiple times, extracting, for example, the RSS or phase information. The latter parameters are used then as the sources of randomness.
- **Randomness extraction:** After measuring the channel, the samples are processed in order to remove the deterministic components of the signal, increasing in this way the entropy of the key. To avoid long sequences of 0s or 1s, and to increase the security of the key, the large-scale fading effect must be removed. In this case, the randomness should be extracted from the small-scale component.
- **Quantization:** To generate the keys, the processed samples are mapped into a bit sequence. At this stage, the keys may not be identical; therefore, an error correction procedure must be carried next.
- **Reconciliation:** The errors on the keys are corrected at this stage, and key reconciliation is achieved through public discussion. This discussion is done running a reconciliation protocol [17–19].
- **Privacy amplification:** During the public discussion, some information may leak to the eavesdropper; therefore, it is necessary to remove this information from the key. The previous task is performed in the privacy amplification phase. As shown in [20,21], the privacy amplification can be implemented using hashing algorithms.

The PHY key generation protocol must be designed taking into account three important metrics, which are: key randomness, secret key rate, and key disagreement rate. The goal is to provide a uniformly distributed key while ensuring an arbitrarily low disagreement probability. Furthermore, the key should also be independent of the information exchanged during the public discussion. The highest secret key rate at which the previous conditions are met is defined as the secret key capacity [15].

9.3.2 Jamming and beamforming in multiple antenna systems

In the following, we discuss two different design approaches that were suggested in [22,23] to protect the MIMO wiretap channel, the latter illustrated in Figure 9.8. The first design approach assumes that the instantaneous CSI of the eavesdropper is available at the transmitter; while in the second case, only statistical information can be acquired. Before describing these approaches, the MIMO wiretap channel is formulated next. In the channel of Figure 9.8, node "A" is equipped with N_A antennas, while "B" and "E" have N_B and N_E antennas, respectively. In this model, the signals received by node "B" and "E" are defined as

$$\mathbf{y}_B = \mathbf{H}_{BA}\mathbf{x}_A + \mathbf{n}_B \text{ and } \mathbf{y}_E = \mathbf{H}_{EA}\mathbf{x}_A + \mathbf{n}_E, \tag{9.7}$$

where $\mathbf{x}_A \in \mathbb{C}^{N_A\times 1}$ is the message bearing signal, and $\mathbf{n}_R \in \mathbb{C}^{N_R\times 1}$ is the additive Gaussian noise at node $R \in \{B,E\}$. The channel between node "A" and $R \in \{B,E\}$ is denoted by $\mathbf{H}_{RA} \in \mathbb{C}^{N_R\times N_A}$ and is in a static fading condition. Unless otherwise stated, in this section, we consider complete CSI at the transmitter.

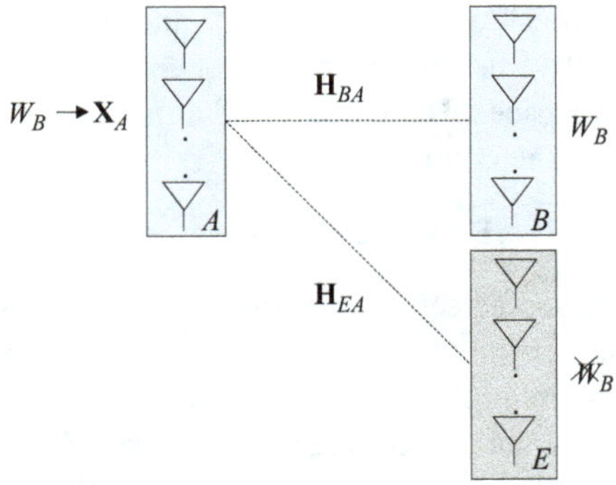

Figure 9.8 MIMO wiretap channel

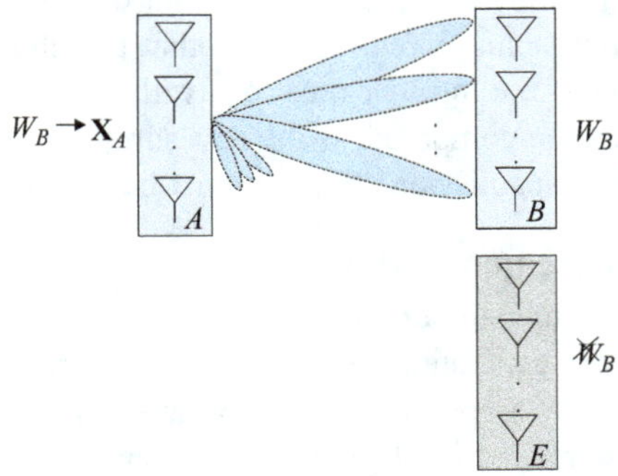

Figure 9.9 GSVD beamforming

9.3.2.1 GSVD beamforming with full CSI

In the high SNR regime, the secrecy capacity of the MIMO wiretap channel is well defined and is achieved through beamforming [22], by decomposing the system in Figure 9.9 in a set of parallel sub-channels. The acquisition of these channels and respective directions is performed at the transmitter, applying to the legitimate and eavesdropper channels a generalized singular value decomposition (GSVD) operation. This procedure allows the legitimate transmitter to obtain the direction of the sub-channels where the gain of the intended receiver exceeds the gain of the eavesdropper. As depicted in Figure 9.9, the information is then transmitted only in those directions, applying the precoding operation

$$\mathbf{x}_A = \mathbf{W}\mathbf{v}_B. \tag{9.8}$$

where $\mathbf{W}$ is the precoding matrix containing the beamforming vectors, and $\mathbf{v}_B$ is the information bearing signal. The precoding vectors are obtained by computing the GSVD of the pair ($\mathbf{H}_{BA}$,$\mathbf{H}_{EA}$), which is done in the following way.

Definition 9.4. The GSVD of $\mathbf{H}_{BA} \in \mathbb{C}^{N_B \times N_A}$ and $\mathbf{H}_{EA} \in \mathbb{C}^{N_E \times N_A}$, denoted as gsvd$(\mathbf{H}_{BA}, \mathbf{H}_{EA})$, returns two unitary matrices, $\psi_B \in \mathbb{C}^{N_B \times N_B}$ and $\psi_E \in \mathbb{C}^{N_E \times N_E}$, two non-negative diagonal matrices, $\mathbf{D}_B$ and $\mathbf{D}_E$, and a matrix $\mathbf{B} \in \mathbb{C}^{N_A \times q}$ with $q = \min(N_A, N_B + N_E)$, such that the conditions

$$\mathbf{H}_{BA}\mathbf{B} = \psi_A \mathbf{D}_A \text{ and } \mathbf{H}_{EA}\mathbf{B} = \psi_E \mathbf{D}_E \tag{9.9}$$

are verified. **W** is computed selecting the columns of **B** for which the element on the diagonal of $\mathbf{D}_A$ is larger than the one in the diagonal of $\mathbf{D}_E$.

As shown in [22], in the high SNR regime, the optimal secrecy rate is achieved with independent Gaussian wiretap coding across the sub-channels, transmitting the message-bearing codewords only in the favorable directions. The analysis of the asymptotic secrecy capacity defined in [22, Theory 2] allow us to conclude the following. If the eavesdropper is equipped with at least the same number of antennas of the transmitter, the secure degrees of freedom (s.d.o.f.) of the channel reduces to zero. A final note can be made regarding the role of the eavesdropper CSI on the capacity approaching solution. Note that to achieve an optimal result, the channel information of the intruder must be available at the transmitter, so that the GSVD precoders can be computed. If the eavesdropper is a passive element of the network, a different transmission strategy must be followed.

9.3.2.2 Artificial noise generation

In a scenario where the transmitter only knows the statistical parameters of the eavesdropper channel, the generation of synthetic noise orthogonal to the intended receiver [23] can be used to improve the secrecy rate of the system in Figure 9.10. The idea is to use the extra spatial degrees of freedom (d.o.f.) at the transmitter to degrade eavesdropper channel without harming the intended receiver. To this end, as depicted in Figure 9.10, the message bearing signal is transmitted together with artificial noise, the latter precoded in the null-space direction of the legitimate receiver. We proceed with a detailed description of this method. In this case, the transmitted

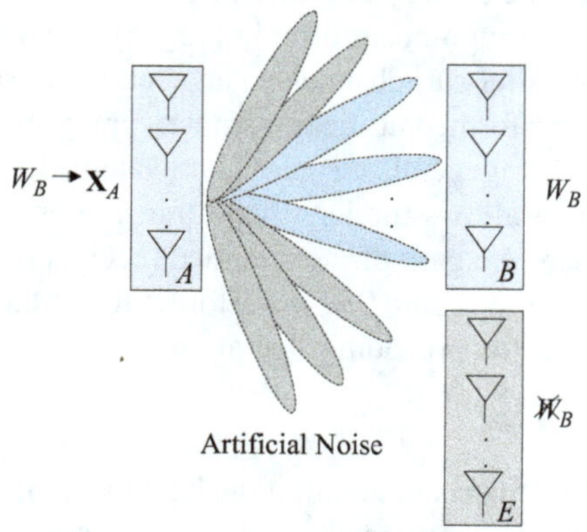

Figure 9.10 Artificial noise generation MIMO wiretap channel

signal is formed by a message and a noise component, following the structure

$$\mathbf{x}_A = \mathbf{W}\mathbf{v}_B + \mathbf{T}\mathbf{u}_J \tag{9.10}$$

where $\mathbf{v}_B$ and $\mathbf{u}_J$ denote the information and artificial noise vectors, respectively, with $\mathbf{W}$ and $\mathbf{T}$ being the precoding matrices of the latter signals. The computation of the precoders is done based on the singular value decomposition (SVD) of the legitimate channel, which is performed as follows.

Definition 9.5. The SVD of $\mathbf{H}_{BA} \in \mathbb{C}^{N_B \times N_A}$, denoted hereinafter as svd$(\mathbf{H}_{BA})$, returns a non-negative diagonal matrix $\mathbf{D} \in \mathbb{R}^{N_B \times N_A}$, and two unitary matrices $\mathbf{U} \in \mathbb{C}^{N_B \times N_B}$ and $\mathbf{V} \in \mathbb{C}^{N_A \times N_A}$, such that

$$\mathbf{H}_{BA} = \mathbf{U}\mathbf{D}\mathbf{V}^H, \tag{9.11}$$

where the columns of $\mathbf{U} \in \mathbb{C}^{N_B \times N_B}$ and $\mathbf{V} \in \mathbb{C}^{N_A \times N_A}$ contain the left and right singular vectors of $\mathbf{H}_{BA}$, respectively, and $\mathbf{D} \in \mathbb{R}^{N_B \times N_A}$ is formed by the singular values of the latter channel.

The precoding matrices $\mathbf{W} \in \mathbb{C}^{N_A \times d_M}$ and $\mathbf{T} \in \mathbb{C}^{N_A \times d_{AN}}$ are computed using the right singular vectors of the legitimate channel, with the former composed by the right singular vectors corresponding to the d_M highest singular values, and the latter is defined by the d_{AN} right singular vectors that have a zero gain to the intended receiver. In this setting, the condition

$$\mathbf{H}_{BA}\mathbf{T}\mathbf{u}_J = 0 \tag{9.12}$$

is verified, which means that the effect of the artificial noise is nulled out only at the legitimate receiver. In order to fill the entire signal space of the intruder with a jamming component, the number of dimensions d_{AN} used to transmit artificial noise should not be less than the number of antennas at the eavesdropper, i.e. the transmitter must be equipped with more antennas than the latter. The information is sent in the remaining available spatial dimensions. As shown in [23], the solution described above achieves a positive secrecy rate using Gaussian signals for the information and the jamming components.

9.3.3 Cooperative jamming

As stated above, when the transmitter is equipped with sufficient spatial degrees of freedom, it is possible to jam the eavesdropper without harming the intended receiver. However, if the transmitter has only one antenna, the latter is not feasible. In this case, the use of a cooperative jammer may be a valid solution. Due to the open nature of the radio channel, wireless networks are, in essence, an interference-limited system. In a shared communication medium, interference generated by other users affects the reliable detection of information at the intended receiver, limiting the capacity of the respective channel. For reliability-constrained channels, inter-user interference is always harmful; however, when confidentiality between users is a system requirement, this interference can be used to improve the system secrecy. In some situations, instead of scheduling all nodes to transmit information simultaneously, the overall secrecy performance can be improved by putting some of the terminals sending jamming signals. By controlling the interference patterns produced by these cooperative jammers, it is possible to achieve information theoretic security even when the unwanted receiver has better channel conditions than the legitimate receiver.

9.3.3.1 The wiretap channel with one helper

The basic model of a cooperative jamming system for the wiretap channel is illustrated in Figure 9.11. In this figure, node "A" intends to transmit a confidential message to node "B," who is a legitimate receiver. The model includes a cooperative jammer "J" that helps node "A" to conceal the message from node "E," who is an eavesdropper. It is assumed that the jammer is authenticated and is willing to cooperate with node "A" to increase the security of the system; therefore, all coalition protocols that led to the association of "J" are assumed to have been executed a priori. In the presented model, node "A" intends to transmit a message w_B to node "B," while keeping it secret from node "E," the eavesdropper. After mapping the message in a codeword, node "A" transmits the message-bearing signal x_A, while node "J" sends x_J, which defines a cooperative jamming signal. The received signals at "B" and "E" are formulated as

$$y_B = h_{BA}x_A + h_{BJ}x_J + n_B \text{ and } y_E = h_{EA}x_A + h_{EJ}x_J + n_E, \tag{9.13}$$

where h_{RT} is a constant that defines the channel coefficient between $R \in \{B,E\}$ and $T \in \{A,J\}$, and n_R is a random variable with variance σ_R^2 that represents additive Gaussian noise at the receivers. In the following, P_A denotes the power constraint at node "A," and P_J is the power constraint on the jammer.

9.3.3.2 Jamming with Gaussian noise

In the system of Figure 9.11, let us assume that node "A" established a coalition with node "J." One way of degrading the eavesdropper channel is by putting the latter terminal transmitting random noise. For instance, if node "J" starts to generate Gaussian noise, the secrecy rate in Corollary 9.1 can be achieved [24] with Gaussian codes and stochastic encoding.

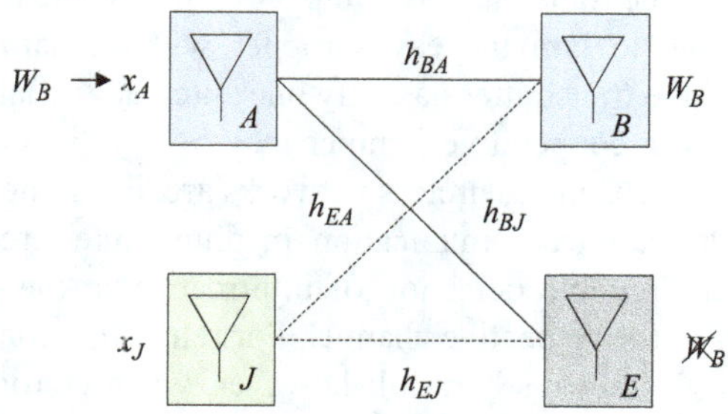

Figure 9.11 Wiretap channel with one helper

Corollary 9.1. An achievable secrecy rate for the Gaussian wiretap channel in the presence of a jammer is given by

$$R_s = \left\{ \frac{1}{2}\log_2\left(1 + \frac{|h_{BA}|^2 P_A}{\sigma_B^2 + |h_{BJ}|^2 P_J}\right) - \frac{1}{2}\log_2\left(1 + \frac{|h_{EA}|^2 P_A}{\sigma_E^2 + |h_{EJ}|^2 P_J}\right) \right\}^+ \tag{9.14}$$

The secrecy gain associated to the use of "J" can be readily understood from the analysis of (9.14). In the presence of a cooperative jammer, it is possible to achieve secrecy even when the channel between "E" and "A" is stronger than the channel between "B" and "A," as long as the condition

$$|h_{BJ}| < |h_{EJ}| \tag{9.15}$$

holds. To validate such a conclusion, the secrecy rate in (9.14) is evaluated in Figure 9.12 considering the previous scenario. As illustrated in Figure 9.12, the proper selection of a cooperative jammer can create the right conditions to produce a channel advantage at the legitimate receiver. In order to achieve (9.14), the Gaussian codewords and the jamming signal are transmitted directly without any precoding operation. It is clear that for finite power conditions, a positive secrecy rate can be reached with this scheme; however, if we analyze the performance in the high SNR regime, we realize that this transmission method does not bring any secrecy gain. For example, in the channel illustrated in Figure 9.11, the use of Gaussian signaling drives the s.d.o.f. to

$$\lim_{P_A \to \infty} \frac{2R_S}{\log_2(P_A)} = 0. \tag{9.16}$$

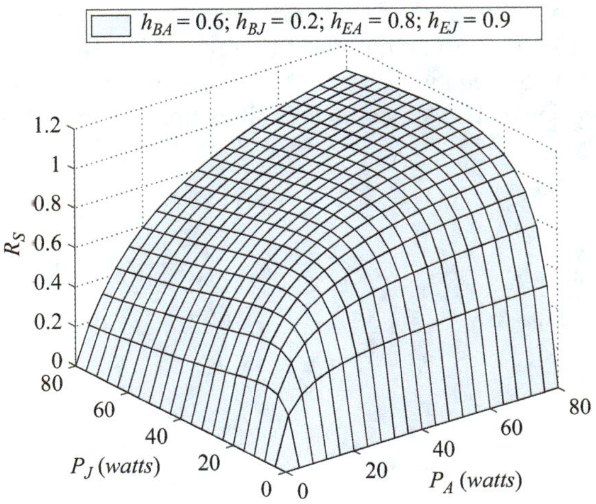

Figure 9.12 Secrecy rate of the wiretap channel with the help of jamming for $\sigma_{N_R}^2 = 1$

The result in (9.16) is valid even if the jamming power is asymptotically infinite; therefore, in this evaluation regime, the generation of Gaussian noise has no effect on the system secrecy. To overcome this problem, the authors of [25,26] proposed a different signaling method, suggesting the use of codes based on integer lattice structures. The target is to introduce some structure on the jamming procedure by combining these codes with real interference alignment (IA) techniques. As proved in [26], this type of jamming construction allows to achieve positive s.d.o.f. in several network structures. In the following, we will analyze in more detail this jamming mode.

9.3.3.3 Interference alignment

The concept of real IA was first proposed in [27] as an effective tool to manage interference in multiple user networks. In its original application, the idea was to exploit the many fractional dimensions available in single antenna systems to separate the inter-user interference from the message at a given user. Later, the work in [26] applied the concept also to physical layer security, showing that real IA could be used to improve the security level of wireless networks in the high SNR regime. As shown in Figure 9.13, in this case, the goal is to force the alignment between the message and the jamming signal at the unwanted receiver, using cooperative linear precoding techniques. If an integer lattice structure is used to signal information, it is possible to create equivocation at the unwanted receiver. The authors of [26] studied the system in Figure 9.13 and showed that the use of real IA allows to reach the maximum of 1/2 s.d.o.f. for this channel. This result was achieved by transmitting the following signals:

$$x_A = h_{EA}^{-1} v_B \text{ and } x_J = h_{EJ}^{-1} u_J, \tag{9.17}$$

where v_B is the message-bearing codeword, and u_J is the jamming component. In this case, instead of using a continuous Gaussian distribution for signal information, the codeword and the jamming signal are sampled from a pulse amplitude modulation (PAM) constellation; that is, from a set of equidistant points obtained from a one-dimensional (1D) lattice. In the follow up, "B" and "E" try to obtain the

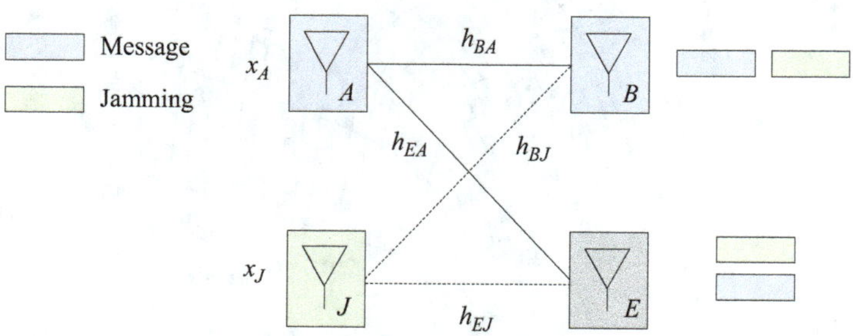

Figure 9.13 Alignment conditions

desired message through the observation of

$$y_B = \alpha v_B + \beta u_J + n_B \text{ and } y_E = v_B + u_J + n_E, \tag{9.18}$$

with α and β denoting equivalent channels coefficients. As demonstrated in [26], the superposition between v_B and u_J at the eavesdropper does not allow the rate of information at this node to scale with the logarithm of the power; therefore, node "E" experiences zero d.o.f. The legitimate receiver, on the other hand, is able to use the fractional dimensions inherent to α and β to separate the codeword from the jamming component, which is also decoded. In this way, the available d.o.f. is equally divided between the intended message and the jamming signal, resulting in a penalty of half d.o.f. at the legitimate receiver. The authors of [26] also studied the channel in Figure 9.13 for a more general setting, considering the presence of M cooperative jammers. In this study, the authors were able to show that by using real IA, it is possible to achieve the maximum s.d.o.f. of the latter channel. That maximum is formulated in Theorem 9.2.

Theorem 9.2. The s.d.o.f. of the Gaussian wiretap channel with static channel coefficients and M helpers is given by

$$D_s = M/(M+1). \tag{9.19}$$

Therefore, by increasing the number of helpers, the upper bound of 1 s.d.o.f. is reached in the asymptotic regime. This means that the secrecy penalty vanishes in the latter case. Despite being a powerful framework, the application of this type of technique faces several practical challenges. For instance, the precoding operations in (9.17) require to access the complete CSI at both transmitters, which may not be available. Moreover, the CSI distribution procedure may also introduce a significant amount of overhead on the communication, reducing the spectral efficiency of the system. To fulfill the alignment conditions, the network should be also well synchronized. In large-scale networks, ensuring precise synchronization between the network nodes can be a complex task [28].

9.4 Enabling physical layer security in 5G and beyond

This section analyzes in which way physical layer security can be incorporated into future wireless standards. To enable this integration, there are some important points that should be addressed. For instance, the design of wiretap codes of finite-block length is crucial to use the technology in practical systems. The incorporation of physical layer security in a multilayer security context, working as a complement to other cryptographic constructions, is another relevant point to be discussed. This section also analyzes how PHY key generation and cooperative jamming techniques could be initially integrated into a commercial standard.

9.4.1 Multilayer security approach

Because physical layer security is grounded on a statistical framework, which does not depend on upper layer security mechanisms, its design and operation can be

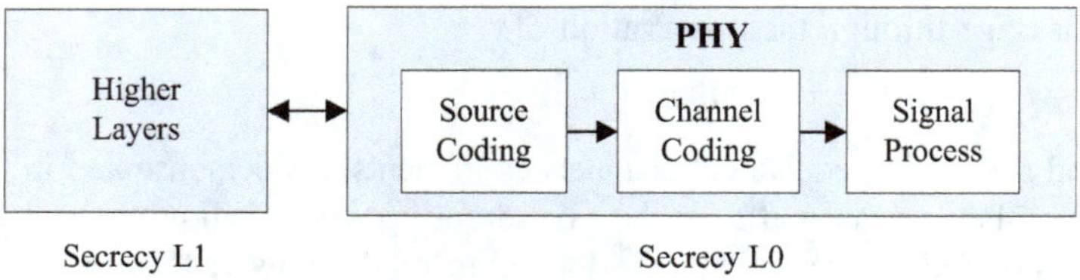

Figure 9.14 Multilayer security solution

done without affecting current protocols, thus allowing a smooth integration of commercial standards. It is also relevant to point that, although the technology is designed to run in standalone mode, its operation is only effective for specific channel states, for the other channel states, the provision of secrecy must be done with other cryptographic constructions. It is important to note that the maximum rates at which it is possible to ensure information-theoretic secrecy are lower compared to the case in which the system has only reliability constraints. The achievement of such rates is only possible when Bob's channel has higher quality than Eve's channel. In some situations, it may not be possible to ensure a channel advantage at Bob, thus making the use of the technology unfeasible. For instance, in cooperative jamming scenarios, a helper may not be always available to form a coalition. In addition, a static channel condition may also prevent secret keys from being extracted at the physical layer. Therefore, as illustrated in Figure 9.14, the embodiment of physical layer security in practical applications must be always done in the context of a multi-layer security framework. Due to the rate limitation (per user), the technology could be used as an extra layer of security designed to support low-throughput services that have stringent secrecy requirements. All type of use cases that require the exchange of small blocks of sensitive information (e.g. credit card numbers, secret keys, etc.) are potential applications for the technology. In case of adoption, physical layer security would be used as a high security service provided by the commercial standard to the application layer, allowing the service provider (e.g. telecom operator) to explore a new market segment. For instance, banking applications are a potential commercial application for the technology. In this context, a multilevel security platform would be developed.

9.4.2 Wiretap codes for 5G-NR

The theoretical study of physical layer security has been grounded on non-explicit random code constructions, which consider the asymptotic large block-length coding regime to demonstrate that a positive secrecy rate is attainable. However, to achieve in practical systems these secrecy rates, the design of explicit code constructions in a finite block-length setting is mandatory. The efforts made to address this issue have focused on adapting existing channel encoders so that they can be applied in secrecy-constrained scenarios. The low density parity check (LDPC) and the polar codes used in 5G-NR are examples of that. For instance, the combination of LDPC codes with puncturing was proposed in [29] to secure the Gaussian

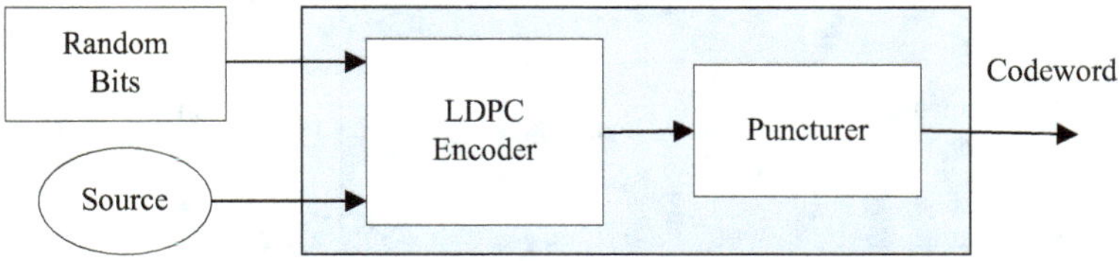

Figure 9.15 Punctured LDPC encoder [29]

wiretap channel. As illustrated in Figure 9.15, for a number of message bits lower than the code dimension, the secret bits are punctured at the transmitter, and the non-punctured independent bit locations are filled with random bits. At the legitimate receiver, the punctured bits are recovered from the non-punctured part of the codeword. Also for the Gaussian wiretap channel, Baldi *et al.* [30] suggested the use of a scrambled version of an LDPC encoder. In this case, the information is scrambled before the application of the linear block code. The authors of [31] exploited the channel polarization condition of a polar encoder to transmit information in the bit channels that are good for the legitimate receiver and bad to the eavesdropper. Meanwhile, the bit channels that are good for both are filled with random bits. Under a weak secrecy condition, the previous work achieved the secrecy capacity for a class of degraded wiretap channels. To achieve strong secrecy, Şasoglu and Vardy [32] extended Ref. [31] proposing a multi-block encoding method that also achieves the secrecy capacity. These are just a few examples of design approaches that have been considered in the literature. Note that in all of them, a source of randomness is always present in the encoding operation. In fact, this randomness is fundamental to equivocate the eavesdropper and reach the secrecy capacity of the channel.

9.4.3 Symmetric encryption with PHY key generation

In today security systems, hybrid cryptography is often used to achieve network confidentiality. In this type of systems, the secret key is exchanged using public-key encryption; then, the data is symmetrically encrypted with the latter key. As explained earlier, public-key encryption requires the maintenance of a complex infrastructure (PKI), which is responsible for distributing and certifying the keys (public and private) required to run the protocol. In some scenarios, the distribution of the symmetric key can be done using a PHY key generation protocol. In this case, the cost related to the public-key infrastructure would be avoided, and a more resilient security approach would be applied. As illustrated in Figure 9.16, in this context, the keys would be extracted from the channel and then stored in a database so that they could be used whenever necessary. By storing the keys obtained from past channel measurements, it is possible to use the technology even when the channel is in a static condition. It is also important to bear in mind that the key rates achieved with the PHY key generation protocol can be low, not allowing the immediate extraction of the key. In the literature, it is possible to find some

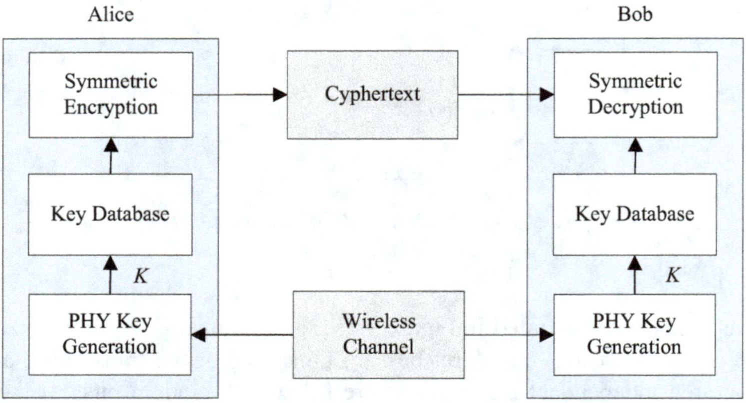

Figure 9.16 Multilayer security solution

practical implementations of the technology [33–35]. For instance, before the key reconciliation and privacy amplification stages, the authors of [33] were able to achieve key rates of 22 bits/s and 10 bits/s with key disagreement rates of 2.2% and 0.54%, respectively. In the latter case, if the Advanced Encryption Standard (AES) is used as the symmetric cypher, for a 256 bit key configuration, the key extraction would take around 25.6 s. On the other hand, if perfect secrecy is the goal, a one-time pad encryption can be applied. However, a secret key with the same size of the message must be generated for each new block of data. With such a lower key rate, it might be difficult to refresh the keys on demand. In this scenario, a database holding the keys obtained from past channel states may be a solution. It is of worth mentioning that the latter works considered single antenna systems; however, as shown in [36], it is possible to increase the key rates using multiple antennas or cooperation, i.e. increasing the spatial diversity of the system.

9.4.4 Extending CoMP to cooperative jamming

In today business models, telecom operators are forced by service level agreements (SLA) to guarantee a certain quality-of-service (QoS) in their networks. In this context, the service provider should be able to predict with some level of accuracy the expected performance of the network. In cooperative jamming systems, the helper may not be always available. Moreover, in the presence of a massive MIMO eavesdropper, or in a collusion scenario, a single base-station may not have sufficient spatial degrees-of-freedom to achieve secrecy with an artificial noise aided transmission solution. In the previous cases, an SLA violation may occur. In order to attenuate these risks, an initial deployment of the technology could be done in a non-ad hoc configuration, using as transmitting points the cluster of centrally coordinated base-stations used in 4G/5G coordinated multi-point (CoMP) systems [37]. As a solution to reduce inter-cell interference and increase cell-edge throughput, the concept of coordinated multipoint transmission and reception was introduced in long term evolution (LTE) advanced, namely in 3GPP release 11.

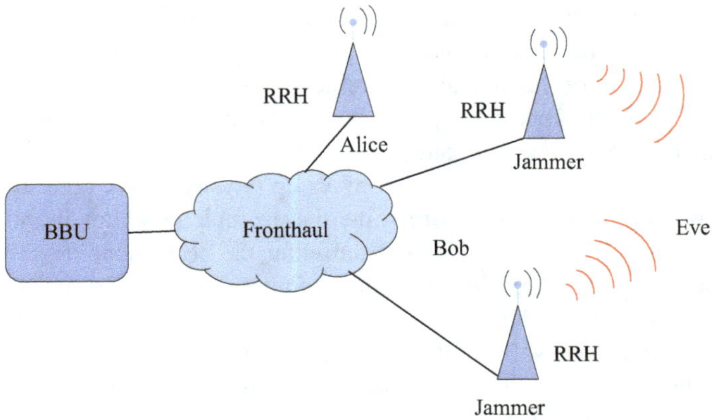

Figure 9.17 Centralized – radio access architecture

The idea is to promote network cooperation putting the constituent nodes to share data and channel state information among themselves. In this way, the signal conditions at the cell-edge can be improved with a joint transmission strategy, or interference can be mitigated with coordinated scheduling and beamforming. To efficiently implement the CoMP technology, the centralization of the baseband processing is fundamental [38]. This centralization can be achieved with the centralized – radio access network (C-RAN) infrastructure in Figure 9.17. In this case, all the baseband processing resources are centralized in a baseband unit (BBU), which connects to the remote radio headers (RRHs) through a high capacity fronthaul network. This configuration corresponds to option 8 of the functional splitting schemes proposed for the 5G radio access network. In fact, part of the CoMP infrastructure already in place in 4G and 5G could be used to enable the commercial integration of some physical layer security techniques. Note that with the distributed MIMO system in Figure 9.17, the operator could use IDLE RRHs as jammers; and it would be possible to exploit the entire range of spatial degrees of freedom of the network to improve secrecy. For instance, the multiple antenna techniques described in Section 3.2 could take advantage of the C-RAN infrastructure. The GSVD-based beamforming algorithm would have more degrees of freedom to optimize the gain difference between the legitimate and eavesdropper channels; while the artificial noise generation approach could dynamically adapt the number of transmitting points to ensure always an antenna advantage in the presence of massive MIMO eavesdroppers.

9.5 Conclusion

To motivate the topic of physical layer security, this chapter began by introducing the notion of computational security, recognizing it as the most successful security notion created to date. In this introduction, the concepts of symmetric encryption and public key encryption were explained, and their respective vulnerabilities were analyzed. As mentioned before, the main principle of computational security rests on the idea that some mathematical problems cannot be solved with existing technology. However, since the latter hypothesis is unproven, it is not possible to ensure that these complexities cannot be broken by future technologies. In this context, exploiting channel randomness by using physical layer security has been identified as a promising solution to solve these problems. After introducing the concept of physical layer security, and outlining the corresponding information-theoretic background, which included the definition of the wiretap channel and secrecy capacity, some design approaches were presented. For example, the process of extracting secret keys from the physical channel, the GSVD beamforming algorithm, the artificial noise generation technique, and the use of cooperative jamming, were given as some examples of physical layer security designs. The analysis of the previous techniques allowed us to conclude the following. The channel entropy and the amount of spatial degrees of freedom available at the transmitter are key system parameters to achieve confidentiality. In addition, it was also demonstrated how a cooperative jammer can be used to improve the secrecy in single-antenna channels. The last section of this chapter discussed how to enable the integration of the previous techniques into future wireless standards. In this regard, the multilayer security approach, the design of explicit wiretap codes, and the centralization of the baseband processing unit were pointed out as important technological enablers.

References

[1] A. Al-Fuqaha, M. Guizani, M. Mohammadi, M. Aledhari, and M. Ayyash, "Internet of Things: a survey on enabling technologies, protocols, and applications," *IEEE Commun. Surveys Tuts.* vol. 17, no. 4, pp. 2347–2376, 2015.

[2] J. Katz and Y. Lindell, "Applications of number-theoretic assumptions in cryptography," in *Introduction to Modern Cryptography*, 1st ed., CRC Press, London, 2007, pp. 273–282, Chapter 7.

[3] M. Wixey, *Quantum Computing and Encryption: What Impact Will Quantum Computing have on Encryption?*, London, UK, Fujitsu, White Paper, 2019.

[4] C. E. Shannon, "Communication theory of secrecy systems," *Bell Syst. Tech. J.*, vol. 28, pp. 656–715, 1949.

[5] R. Bassily, E. Ekrem, X. He, *et al.*, "Cooperative security at the physical layer: a summary of recent advances," *IEEE Sig. Process. Mag.*, vol. 30, no. 5, pp. 16–28, 2013.

[6] W. Stallings, "Principles of public-key cryptosystems," in *Cryptography and Network Security Principles and Practice*, 5th ed., Prentice Hall, Hoboken, NJ, 2011, pp. 269–277, Chapter 9.

[7] J. Katz and Y. Lindell, "A computational approach to cryptography," in *Introduction to Modern Cryptography*, 1st ed., CRC Press, 2007, pp. 47–59, Chapter 7.

[8] P. W. Shor, "Algorithms for quantum computation: discrete logarithms and factoring," in *Proceedings of 35th Annual Symposium on Foundations of Computer Science*, Santa Fe, USA, 1994.
[9] W. Rjaibi, S. Muppidi, and M. O'Brien, *Wielding a Double-Edged Sword Preparing Cybersecurity Now for a Quantum World*, IBM Institute for Business Value, July, 2018.
[10] A. Mukherjee, S. A. A. Fakoorian, J. Huang, and A. L. Swindlehurst, "Principles of physical layer security in multiuser wireless networks: a survey," *IEEE Commun. Surv. Tutor.*, vol. 16, no. 3, pp. 1550–1573, Third Quarter 2014.
[11] R. Deal, "OSI reference model," in *Cisco Certified Network Associate Study Guide*, Mc-Graw Hill, London, 2008, pp. 27–52, Chapter 2.
[12] A. D. Wyner, "The wire-tap channel," *Bell Syst. Tech. J.*, vol. 54, no. 8, pp. 1355–1367, 1975.
[13] I. Csiszar and J. Korner, "Broadcast channels with confidential messages," *IEEE Trans. Inf. Theory*, pp. 339–348, 1978.
[14] M. R. Bloch, "Coding mechanisms for secret communication," in *Physical Layer Security for Wireless Communications*, CRC Press, Boca Raton, FL, 2014, pp. 6–10, Chapter 4.
[15] J. Zhang, T. Q. Duong, A. Marshall, and R. Woods, "Key generation from wireless channels: a review," *IEEE Access*, vol. 4, pp. 614–626, 2016.
[16] K. Zeng, "Physical layer key generation in wireless networks: challenges and opportunities," *IEEE Commun. Mag.*, vol. 53, no. 6, pp. 33–39, 2015.
[17] Y. Liu, S. C. Draper, and A. M. Sayeed, "Exploiting channel diversity in secret key generation from multipath fading randomness," *IEEE Trans. Inf. Forensics Security*, vol. 7, no. 5, pp. 1484–1497, 2012.
[18] J. Zhang, S. K. Kasera, and N. Patwari, "Mobility assisted secret key generation using wireless link signatures," in *2010 Proceedings IEEE Infocom*, 2010, pp. 1–5.
[19] H. Liu, J. Yang, Y. Wang, Y. Chen, and C. E. Koksal, "Group secret key generation via received signal strength: protocols, achievable rates, and implementation," *IEEE Trans. Mobile Comput.*, vol. 13, no. 12, pp. 2820–2835, 2014.
[20] X. Lu, J. Lei, Y. Shi, and W. Li, "Intelligent reflecting surface assisted secret key generation," *IEEE Sig. Process. Lett.*, vol. 28, pp. 1036–1040, 2021.
[21] S. N. Premnath, S. Jana, J. Croft, *et al.*, "Secret key extraction from wireless signal strength in real environments," *IEEE Trans. Mobile Comput.*, vol. 12, no. 5, pp. 917–930, 2013.
[22] A. Khisti and G. W. Wornell, "Secure transmission with multiple antennas—Part II: the MIMOME wiretap channel," *IEEE Trans. Inf. Theory*, vol. 56, no. 11, pp. 5515–5532, 2010.
[23] S. Goel and R. Negi, "Guaranteeing secrecy using artificial noise," *IEEE Trans. Wireless Commun.*, vol. 7, no. 6, pp. 2180–2189, 2008.
[24] A. Yener and S. Ulukus, "Wireless physical-layer security: lessons learned from information theory," in *Proceedings of the IEEE*, 2015, 103, 1814–1825.

[25] X. He and A. Yener, "Providing secrecy with structured codes: two-user Gaussian channels," *IEEE Trans. Inf. Theory*, vol. 60, no. 4, pp. 2121–2138, 2014.

[26] J. Xie and S. Ulukus, "Secure degrees of freedom of one-hop wireless networks," *IEEE Trans. Inf. Theory*, vol. 60, no. 6, pp. 3359–3378, 2014.

[27] A. S. Motahari, S. Oveis-Gharan, M. Maddah-Ali, and A. K. Khandani, "Real interference alignment: exploiting the potential of single antenna systems," *IEEE Trans. Inf. Theory*, vol. 60, no. 8, pp. 4799–4810, 2014.

[28] O. El Ayach, S. W. Peters, and R. W. Heath, "The practical challenges of interference alignment," *IEEE Wireless Commun.*, vol. 20, no. 1, pp. 35–42, 2013.

[29] D. Klinc, J. Ha, S. W. McLaughlin, J. Barros, and B.-J. Kwak, "LDPC codes for the Gaussian wiretap channel," *IEEE Trans. Inf. Forensics Security*, vol. 6, no. 3, pp. 532–540, 2011.

[30] M. Baldi, M. Bianchi, and F. Chiaraluce, "Coding with scrambling, concatenation, and HARQ for the AWGN wire-tap channel: a security gap analysis," *IEEE Trans. Inf. Forensics Security*, vol. 7, no. 3, pp. 883–894, 2012.

[31] H. Mahdavifar and A. Vardy, "Achieving the secrecy capacity of wiretap channels using polar codes," *IEEE Trans. Inf. Theory*, vol. 57, no. 10, pp. 6428–6443, 2011.

[32] E. Şasoglu and A. Vardy, "A new polar coding scheme for strong security on wiretap channels," in *Proceedings of IEEE International Symposium on Information Theory*, July 2013, pp. 1117–1121.

[33] N. Patwari, J. Croft, S. Jana, and S. K. Kasera, "High-rate uncorrelated bit extraction for shared secret key generation from channel measurements," in *IEEE Trans. Mobile Comput.*, vol. 9, no. 1, pp. 17–30, 2010.

[34] Y. Wei, K. Zeng, and P. Mohapatra, "Adaptive wireless channel probing for shared key generation based on PID controller" *IEEE Trans. Mobile Comput.*, vol. 12, no. 9, pp. 1842–1852, 2013.

[35] S. Sun, Y. Wu, B. S. Lim, and H. D. Nguyen, "A high bit-rate shared key generator with time-frequency features of wireless channels," in *2017 Proceedings of the IEEE Globecom*, 2017, pp. 1–6.

[36] K. Zeng, D. Wu, A. Chan, and P. Mohapatra, "Exploiting multiple-antenna diversity for shared secret key generation in wireless networks," in *2010 Proceedings of the IEEE Infocom*, 2010, pp. 1–9.

[37] D. Lee, H. Seo, B. Clerckx, *et al.*, "Coordinated multipoint transmission and reception in LTE-advanced: deployment scenarios and operational challenges," *IEEE Commun. Mag.*, vol. 50, no. 2, pp. 148–155, 2012.

[38] L. M. P. Larsen, A. Checko, and H. L. Christiansen, "A survey of the functional splits proposed for 5G mobile crosshaul networks," *IEEE Commun. Surv. Tutor.*, vol. 21, no. 1, pp. 146–172, First Quarter 2019.

Chapter 10

Steganography-based secure communication via single carrier frequency division multiple access (SC-FDMA) transceiver

Avila Jayapalan[1], Prem Savarinathan[1] and Swetha Thennavan[1]

Abstract

In the past few decades, wireless communications have empowered communication between people. The communication has been achieved through various transmission systems, of which single-carrier frequency division multiple access (SC-FDMA) is one prominently featured among them. It is an enhanced version of the orthogonal frequency division multiple access (OFDMA) system with low PAPR and high-power capabilities. Despite SC-FDMA being one of the best means for transmission, it requires inalienable safety efforts from intruders. Information that is imparted remotely is less safe as opposed to wired communication. Thus, the security of SC-FDMA turns into vitality. Here modified-least significant bit (MLSB) algorithm has been proposed to incorporate security in the SC-FDMA system. The data is first embedded with the proposed algorithm and then is transmitted through the SC-FDMA system over an Additive White Gaussian Noise (AWGN) channel. Then using the same algorithm, the information is rooted at the receiver end of SC-FDMA system. The performance of the proposed algorithm is analyzed through various parameters of image quality assessment and bit error rate (BER). The proposed method boasts higher PSNR and SSIM values of 68.9102 and 0.9995, respectively. MSE, AD, and NAE were observed to be far lesser, with the metrics being 0.0084, 0.00021, and 0.000060, respectively.

Keywords: Wireless communication; Long-term evolution (LTE); SC-FDMA; Security; Authentication; LSB algorithm

10.1 Introduction

Wireless communication is the powerful and effective means of creating innovative territory in the communication field-transmitting data from one point to another.

[1]Department of Electrical & Electronics Engineering, Sastra Deemed University-India, India

The transmitter and the receiver can be installed in part of the world, and the distance between them may be metered to thousand kilometers. Wireless communication systems have responded through the innovation of technologies since the introduction of first-generation mobile networks. Wireless communication standards have accelerated rapidly due to their high demand.

The first wireless communication standard is first-generation (1G) technology which has consummated the fundamental voice transmission in mobile with frequency ranges from 800 MHz to 900 MHz. The channel bandwidth is only 25 kHz. It is known as advanced mobile phone service (AMPS) in North America, and Europe, it is known as The European Total Access Communication System (ETACS). At the end of 1991, US digital cellular (USDC) system was set up as the digital version of the 1G system. It offered three times higher capacity than AMPS. Then the next standard is a second generation (2G), in which the data rate is increased from 14 to 64 kbps, giving way to SMS and e-mail. The channel bandwidth for Global System Mobile (GSM) is 200 kHz, and eight users are allowed.

General Packet Radio Service (GPRS) supports data and is a packet-based network. All eight slots of GSM are allocated for GPRS if required. Enhanced data rates for GSM (EDGE) are the upgraded version of GSM. Following the 2G technology, the next standard 3G introduces the first-ever mobile broadband concept in the communication with a data rate of 390 kbps and the transmission of video and audio from one mobile to another mobile. The 3G technology is considered the giant leap in the wireless communication. It supported voice-over Internet protocol (VoIP), web sessions, live music, etc.

Following 3G the fourth-generation (4G) is introduced, with a higher data rate of about 1 Gbps and supports high-quality streaming video and audio without buffering. Due to the advantage of high mobility and low latency, it provides a way for the transmission of a large amount of data at less data rate.

The 4G long-term evolution (LTE)-advanced standard [1,2] gives rise to the fastest wireless broadband service with rich quality video and audio having high data rates. The 4G technology has the following features:

I. Higher data rate up to 1 Gbps
II. Flexible in spectral allocation
III. Improvement in spectral efficiency
IV. Economically supportive
V. Improvement in quality of service
VI. Higher throughput
VII. Reduction in latency
VIII. Infrastructure is built in a way to reuse the existing cell
IX. Optimized IP traffic and services
X. Enhanced security and mobility
XI. Cognitive radio

From Table 10.1, it is clear that as the generation changes from lower to higher the spectral efficiency is improved. LTE advanced adopts a transmission scheme for uplink known as single carrier frequency division multiple access (SC-FDMA), and for downlink, it is OFDMA. SC-FDMA, a multiple access system, has been utilized

Table 10.1 Spectral efficiency of communication systems

Year	Service	Standard	Maximum link spectral efficiency (bits/s/Hz)
1981	1G	NMT 450	0.45
1991	2G	GSM	0.52
2000	2.75G	CDMA2000	0.0078
2001	3G	WCDMA(FDD)	0.077
2002	3G	CDMA2000 1x EV-DO	2.5
2007	3.5G	HSDPA	4.22
2009	4G	LTE	16.32
2013	4G	LTE-Advanced	30.00

for data transmission, allowing different clients to mutually utilize a commonly accessible resource. Every subcarrier contains the data of every single transmitted data and has support for a wide scope of information rates. It is an improved form of the OFDMA system with a lower peak-to-average power ratio (PAPR) and higher power efficiency. For example, SC-FDMA system with quadrature phase shift keying (QPSK) has PAPR of 10 dB when compared to the OFDM system [3].

For secure data transmission, various approaches for securing the data are available. These include techniques such as cryptography, steganography, security codes, obfuscation and many more. In this chapter, steganography is applied to SC-FDMA as a means of security. Steganography is the technique of covertly placing user information within any form of advanced media such as audio and video images [4]. The premise behind steganography is to cover up the presence of information in any medium. The idea of image steganography is rather simple. The original image and the stego-image (the image that has undergone steganography) are expected to be similar and it must be challenging to detect the difference between them. The hidden message may assume the form of plaintext, cipher text, or anything that can be represented as a piece stream. Steganography algorithms hide the message covertly, leaving no hint of the original message being covered in the cover object.

10.1.1 Related works

An audio signal is transmitted through the SC-FDMA system in [5]. Chaotic encryption schemes are used to encrypt the data. The performance of this scheme is then analyzed by qualitative metrics, which include but are not limited to histogram responses, PSNR, MSE, and SNR. The expanded least significant bit (LSB) has been utilized in [6] and, the results were compared with a traditional LSB algorithm. In [7], both cryptography and steganography are used in tandem for secure data transmission. The cipher text is embedded in the cover image by employing the LSB substitution algorithm. This manuscript focused on reducing the PAPR of both SC-FDMA and OFDMA. In that regard, clipping and filtering (CAF) and selective mapping (SLM), which are the PAPR reduction techniques, have been utilized [8]. In [9], the authors analyzed the performance of SC-FDMA for different modulation schemes such as binary phase shift keying (BPSK), QPSK, 16-quadrature amplitude modulation (QAM), 64-QAM in LTE uplink. Optimal performance is observed with BPSK modulation for LTE uplink transmission in the SC-FDMA system. In article [10], confidential information is transmitted through the SC-FDMA system in the third party's presence. In [11], the least significant bit (LSB) and Blowfish algorithms are combined to provide secure data transmission. In [12], audio steganography is done using a novel LSB algorithm. The overall process is based on randomness and non-sequencing. In [13], selective embedding is used as a stego key in image steganography and the data is added to the image without affecting the quality of the image. In [14], the secret information is converted into the cipher text using advanced encryption standard (AES) algorithm. The secret information is then embedded into the cover image using the LSB algorithm. The encrypted stego-image is transmitted through OFDM system illustrated in [15], AES and data encryption standard (DES) are the encryption algorithms utilized. This article takes advantage of the distinction of two pixels last bits of the spread picture [16]. In [17] the information is hidden inside the cover image using the LSB algorithm.

Having collected the information through the related works this manuscript focusses on integrating both a wireless communication systems and steganography techniques between the source and the destination via the wireless medium. SC-FDMA transceiver utilized in 4G techniques is employed because of the advantages it offers and the stego-image is transmitted. It paves the way for secure and successful communication.

10.1.2 Security

Nowadays, the Internet is an essential thing for sharing information. The fundamental issue is passing the information in a secure manner from one point to another point. In this case, confidentiality and data integrity are the two key aspects being maintained to safeguard unauthorized access and to provide security.

Wireless networks are also prone to attack because their medium of transmission is open air. The main aim is to mitigate malicious users from hacking the network. The data transmitted must be encrypted. Even if the malicious users hack the data, they must not be able to decode the information. For secure data transmission, various security options are available, such as cryptography, steganography, security codes, and obfuscation.

Cryptography scrambles the information, generating cipher text that can be understood only by the sender. Thus, the technique of hiding and securing data using codes is termed cryptography. The person who is part of the information only can understand it and process it further. Thus, this technique will be a promising solution for unauthorized access to information. Varieties of algorithms are available to generate the key which is the main step in the cryptographic process. The success of any method depends upon the robustness of the key. The stronger the key (i.e.) more time it can withstand the malicious attack superior the algorithm. In this computer, age cryptography is a method in which ordinary plain text is changed into cipher text. This is termed an encryption process. This cipher text could be realized only by the authorized receiver because they can only decode it. The key is necessary to decode the information and the key is transmitted to the intended receiver in a secure manner prior to the transmission of this information. Further the conversion of cipher text to plain text is termed a decryption process.

Steganography is the technique of covering up secret information inside any form of advanced media. The main concept behind steganography is to cover up the presence of information in any medium like video, picture, audio, etc. The idea of image steganography is quite simple. It is the process of masking the information within the given image. The image used for this purpose is called a cover image, and, after hiding the data, it is known as a stego-image. The original image and the stego-image should be the same, and hard to detect the difference between them. The hidden message might be plaintext, cipher text, or anything representing a piece stream. The performance is analyzed by various metrics such as histogram, PSNR, MSE, and SNR. Sometimes both methods are combined with enjoying the advantages of both of them.

In image steganography, the original image is termed as the carrier because it is used to conceal the secret information into secure transmission. An image is denoted as an N*M matrix and the matrix elements denote the pixels intensity value. The secret message is concealed in the carrier by altering the values of the pixel. It is done properly by utilizing the stego key, which is the steganography algorithm, and the overall process is called the data-embedding process. The image with the secret message is called a stego-image. Then with the help of the same stego key, the secret message is decoded from the stego-image. This process is known as the data retrieving process. The data embedding process is shown in Figure 10.1(a) and the data retrieval process is shown in Figure 10.1(b).

10.1.3 Multiple access scheme

Signal transmission can be broadly classified into two categories: single-carrier transmission and multicarrier transmission. In single-carrier transmission, only one signal is carried by each carrier. The peak-to-average power ratio (PAPR) of single-carrier system is of lesser value and it does not suffer from phase and frequency offset issues. Multiple access techniques enable multiple users to access the channel simultaneously. It is classified into three types, namely frequency division multiple access (FDMA), time division multiple access (TDMA), and code division multiple access (CDMA).

In FDMA, the entire frequency band is equally split into sub-channels and allocated to many users. In order to avoid interference between multiple users,

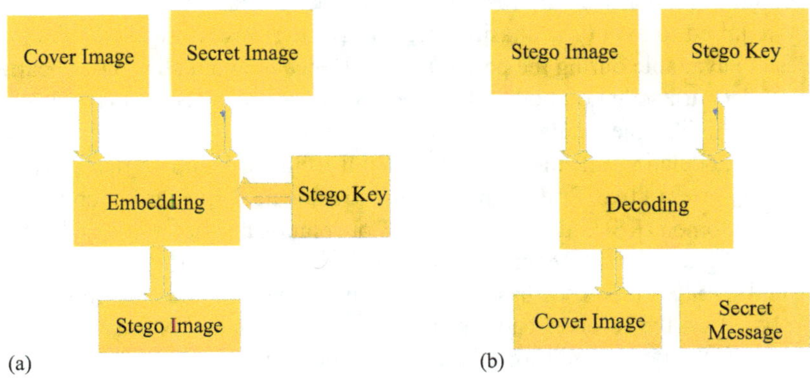

Figure 10.1 (a) Data embedding. (b) Data retrieving.

guard band is provided to the users. Also, to avoid interference with other wireless systems which coexist in the same wireless environment, a guard band is provided at the ends of the spectrum. Guard bands lead to wastage of bandwidth and reduce capacity. In FDMA, there is no storage facility. FDMA is simple because no synchronization is required, and it can be easily implemented.

In TDMA, the entire bandwidth is allocated to one user on a timely basis. Due to one user occupying the entire bandwidth, there is no interference among the users. It can support both voice and data transmission. Only one frequency is shared among many users. Storage of user information is possible. It does not require narrowband filters and there is no need for a guard band. Synchronization is required between the transmitter and the receiver. Overhead is high in TDMA.

CDMA is based on spreading codes known as Pseudonoise (PN) sequence. Many users can occupy the available bandwidth with overlapping of spreading code. Various types of codes, such as Kasami, Gold, and Walsh Hadamard, are available. The power requirement is less, and also it is immune to hackers. The receiver is complex when compared to other multiple-access methods. With the increase in the number of users, the quality of the system decreases.

The orthogonal frequency division multiplexing (OFDM) is one of the multiple access techniques which finds application in the latest telecommunication standards such as WIMAX and 5G cellular technology. Multi-carrier CDMA (MC-CDMA) is a technique that incorporates both OFDM and CDMA. It takes advantage of both CDMA and OFDM. It is suitable for places with high noise content due to multipath links. At the same time, it adapts the disadvantages of both techniques.

10.1.4 OFDM

OFDM is the modified version of frequency division multiplexing (FDM). In FDM, the entire bandwidth is divided into sub-channels. In order to avoid interference between the subchannels guard band is provided between them. Due to this guard band, bandwidth is wasted. This problem has been overruled in OFDM. In OFDM, the subcarriers are orthogonal to each other. The spacing between the carriers is set as $1/T$, which is the inverse of symbol duration. The peak of the carrier overlaps with the other carrier at zero crossings. In the case of baseband OFDM, the frequency required to avoid the aliasing effect is $2N/T$. However, with IFFT, only N samples are available during the period time T. Hence to avoid aliasing, subcarriers with null data are added on both sides. These carriers are called virtual carriers.

OFDM divides the available data into N parallel bit stream, modulates it with any digital modulation schemes, and transmits it using each sub-carrier. N value is chosen as a power of two. Prior to modulation, to mitigate fading, error control codes like Reed Solomon (RS) code, Turbo codes, Convolutional code, etc. are added. It is followed by interleaving. Modulated schemes like Phase Shift Keying (BPSK), QAM, and QPSK can be used based on the requirement. The modulation scheme is chosen based on the receiver end's signal-to-noise ratio level. Normally the same modulation technique is used for all subcarriers but it is possible to have different modulation formats for different subcarriers. This method works well in places where the noise is too high. The modulated information is fed as input to an inverse fast Fourier transform (IFFT) circuit, which gets converted into an OFDM symbol. In order to eliminate the intersymbol interference (ISI) from the previous symbols the cyclic prefix is included. Cyclic prefix means adding a symbol to the first portion of the OFDM symbol taken from the last portion of the OFDM symbol. By doing so, the time duration of the OFDM symbol is extended. Hence, the name cyclic prefix. Cyclic prefix, in general occupies 10% of the total symbol duration. Cyclic prefix is selected so that it is greater than the root mean square (RMS) delay spread of the channel. These features convert the frequency selective fading into flat fading. Thus, OFDM is a very efficient technique for broadband data transmission over radio frequency and it can be implemented at a simple way and in a low cost.

In general, in OFDM, all the subcarriers are assigned to single user. It is also possible to assign the subcarriers to multiple users. Also, each user can be assigned multiple subcarriers. This is known as OFDM-based multiple access (OFDMA). It is complex when compared to other multiple-access schemes.

Some of the merits of OFDM are

I. High diversity
II. High efficiency
III. Low interference
IV. More flexibility
V. Better coverage

Some of its demerits are

I. Complex
II. Requires extra power
III. High sensitivity

10.1.5 SC-FDMA

The 3GPP has declared single carrier frequency division multiple access (SC-FDMA) as the suitable technique for LTE which supports broadband. PAPR is defined as the maximum of a given sample in the OFDM symbol to the average power of the OFDM symbol, and is expressed in dB. Due to the large number of subcarriers in OFDM system, the peak power is normally higher than the average power. To support this high peak power highly sophisticated power amplifiers are required. If the amplifiers are not chosen properly, it leads to adjacent channel interference. In addition, the hardware complexity and cost of the components also increase. Hence the PAPR must be minimized. SC-FDMA system does this and this feature makes it popular.

SC-FDMA is analogous to the OFDMA system [18]. The only difference between them is the presence of additional FFT and IFFT. Hence, SC-FDMA can be called as OFDM with DFT mapper. It is also known as DFT-precoded OFDM. The input is modulated by using QPSK, as it yields better performance in terms of error rates among other modulation schemes. In QPSK, two bits make one symbol. The modulated information makes its way to a serial to parallel converted from where the parallelized data is fed to the fast Fourier transform (FFT) block. The subsequent FFT block result is further fed into the subcarrier mapper, which maps the symbols over the whole bandwidth across different sub-carriers. Then the output from the mapping is applied to IFFT block. Then the output of it is given to parallel to serial converter. The cyclic prefix is then included to reduce ISI. Then the output from the cyclic prefix is transmitted through the AWGN channel.

With the cyclic prefix, the transmitted OFDM signal is denoted by

$$S(x) = \begin{Bmatrix} \text{if}ft(do, d1, \ldots.dN-1, & x = 0, 1, \ldots..N-1 \\ S(x+K), & x = -NeP, \ldots\ldots., -1 \end{Bmatrix} \tag{10.1}$$

$$y(t) = \sum_{k=-NeP}^{N-1} s(k)w\left(t - \frac{k}{n}T\right), -\frac{NeP}{N}T \leq t \leq T$$

where $w(t)$ is the time-domain window function, such as the rectangular window or the raised cosine function.

The receiver is a reciprocal of the transmitter. On the receiver end, the cyclic prefix is first removed and the resulting output is fed into the serial to parallel converter. The converter's output is then fed as an input to the FFT block. This block provides the frequency domain representation of the signal. The transformed signal is then passed to the subcarrier de-mapping block. As the name suggests, it de-maps the mapping performed at the transmitter side. The output is given to the IFFT, which provides the time domain representation. Parallel to serial conversion is carried out, then the signal is demodulated using QPSK demodulator.

The block diagram of SC-FDMA is illustrated in Figure 10.2.

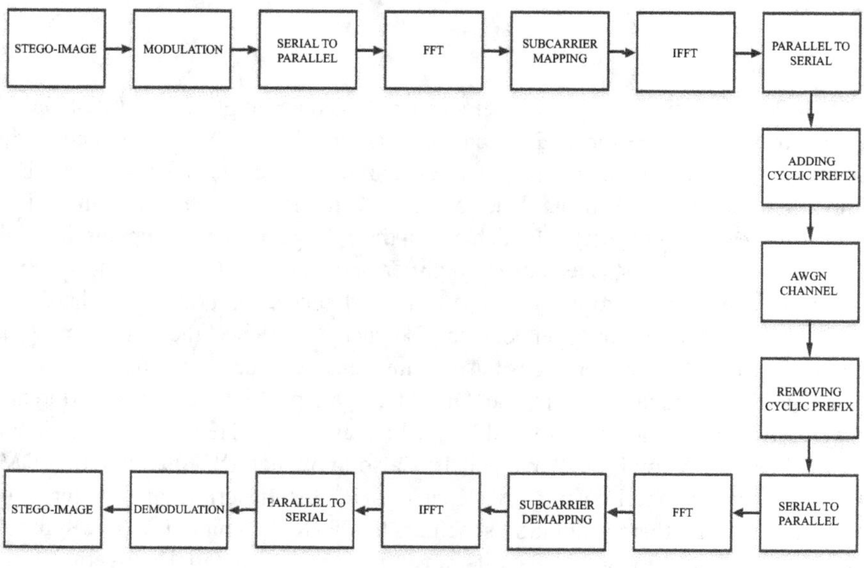

Figure 10.2 SC-FDMA

10.1.6 Least significant bit (LSB) algorithm

Images can be classified into gray-scale images and color images. In a gray-scale image, each pixel is denoted by eight bits. The last bit in the pixel is known as LSB. One technique of interest in image steganography is the LSB algorithm. The basic premise behind the LSB algorithm is to alter the least significant bit of an image pixel with the binary ASCII value of data that is required to be secured. The image may assume the form of a cluster of pixels as such, every pixel is constituted by 8 bits. In the proposed method, the last bit of pixel value is modified to assume the values 0 or 1. By doing so, it would not noticeably modify the nature of the image and it is the requirement of stego-image, but the level of safety that this algorithm yields would render the stego-image extremely vulnerable. In this regard, there arises the necessity for the alteration of the LSB algorithm to enhance the safety level further.

Figure 10.3 illustrates the flow chart of the LSB algorithm embedding and extraction process.

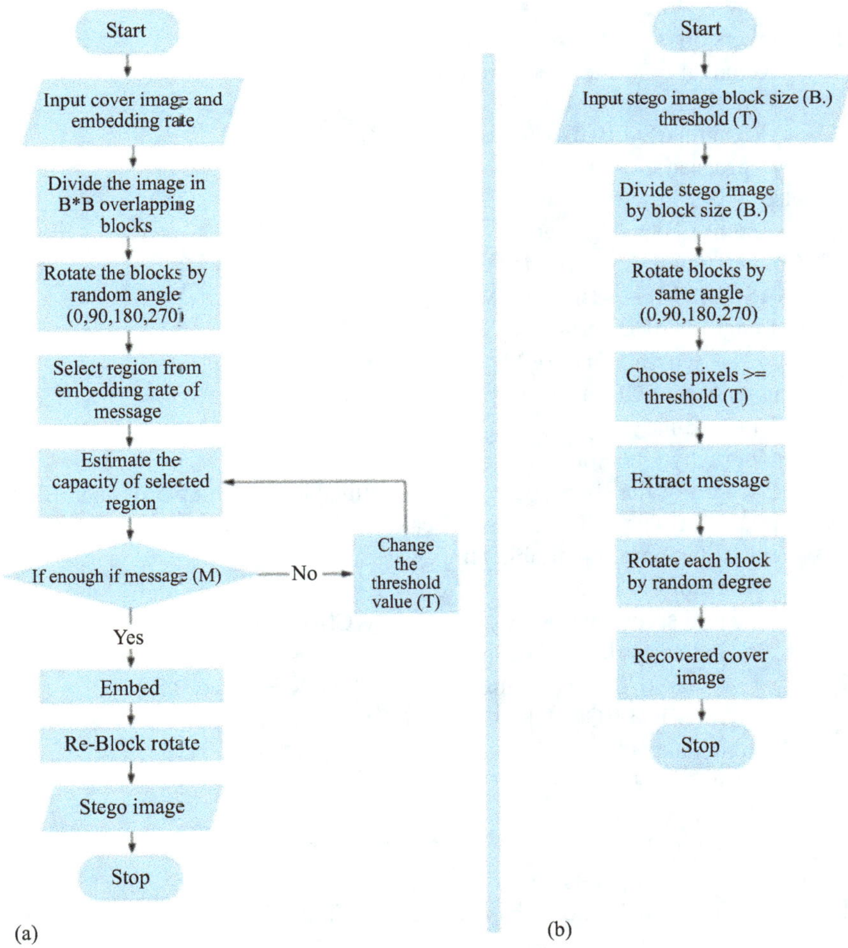

Figure 10.3 Flowchart of LSB algorithm: (a) embedding process and (b) extraction process

10.1.7 Modified LSB algorithm

To overrule the drawbacks of the LSB algorithm, significant changes are incorporated into the modified LSB algorithm to mitigate the issues of vulnerability in security.

10.1.7.1 Data embedding procedure

The modified-LSB algorithm is in certain ways similar to the LSB algorithm. The steps of the proposed embedding technique are as follows:

```
  I. Data acquisition
       Acquisition of carried image CI and confidential message CM
         CI = Carrier Image
         CM = Confidential message
       Acquisition of carrier image metrics height H and width G
         H = Height
         W = Width
 II. Conversion of CM into an ASCII
       Removal of white spaces in CM.
       CM –> ASCII(CM)
       if(length(ASCII(CM) )< length(CI)/7)
         proceed
     else
       Acquire New Message
III. Conversion of ASCII(CM) into Binary
       ASCII(CM) –> Binary(CM)
       Length N of Binary(CM)
         N = length(Binary(CM))
 IV. Conversion of CI into Binary
       CI –> Binary(CI)
  V. Stenography Operation
       Each bit of the Binary(CM) is to be embedded in the LSB of each Binary
       pixel of Binary(CI) after modification
       Initial temporary Variable temp.
         for loop (N iterations)
           left_shift(Binary(CM) [8], Binary(CM) [7])
           indicates the bit position
           Binary(CM) [8] = Binary(CM) [8] XOR Binary(CM) [7]
           if ( Binary(CM)[i] == Binary(CI)[i+8] )
             next = 0
         else
             next = 1
         Binary(CI)[i+8] = next
     Continue for all pixels
 VI. Revert into ASCII and reshape by H * W dimensions.
```

10.1.7.2 Data retrieving procedure

Acquire the Stego-Image SI

```
VII.  SI = Stego-Image
Acquire the Height and width of the SI
VIII. H = Height of SI
      W = Weight of SI
Acquire K, the number of pixels
 IX.  K = H * W
      for loop(K iteration)
         for Each pixel
            Pixel [8] = Pixel [7] XOR Pixel [8]
            right_shift( Pixel [7] , Pixel [8] )
            Read and Store pixel [8] in array temp
               temp[i] = Pixel [8]
               i = i+ 1
            Temp array has 8 bits which make up one character. Repeat for all
            characters.
```

Convert each temp array into a character that makes up the secret message.

Example:

Let us assume that the first eight pixels of the cover image may have the following gray-scale values:

01110010, 01100010, 00010111, 11101100, 11010100, 01110111, 00000010, 01110011

To hide the letter C which has the ASCII code value of 67, first the decimal value is converted into a binary value which is 01000011. The least significant bit of pixel is replaced into this ASCII value and is

01110010, 01100011, 00010110, 11101100, 11010100, 01110110, 00000011, 01110011

Then 7th and 8th bits are left shifted to have the values

01110100, 01100110, 00010100, 11101001, 11010001, 01110100, 00000110, 01110110

The 7th and 8th bits are XORed and replaced in the 8th bit

01110100, 01100111, 00010100, 11101001, 11010001, 01110100, 00000111, 01110111

It is the final stego-image pixel. The reverse process of the above method extracts the original cover image and the embedded secret message.

10.2 Proposed methodology

Figure 10.4 shows the block diagram of the proposed method. The information to be transmitted is hidden using proposed modified-LSB algorithm. It is done by embedding the data into the cover image, which gives rise to the stego-image. The stego-image is then transmitted utilizing the SC-FDMA transmitter with AWGN added with Rayleigh fading as the noise source and at the receiving end, it is received and decoded and using the reverse process, the original information is taken out from the stego-image. The difference between the cover image and the stego-image is visually imperceptible to the human eye.

10.3 Performance metrics

The overall image quality will degrade after embedding the secret information within it. In order to qualitatively evaluate the extent of degradation, image quality metrics known as performance metrics are used. By comparing the reference stego-image to the original image, the metrics can reveal the extent of degradation by analyzing the image quality of a degraded image. The performance metrics are the following.

10.3.1 Mean square error (MSE)

The MSE represents the aggregate squared difference between the pixels in the stego-image and the cover image. It sheds light on the degree to which the cover image has changed after embedding with confidential data. The MSE between the carrier image and the stego-image is determined by the equation:

$$MSE = \frac{1}{MN}\sum_{x=1}^{M}\sum_{y=1}^{N}\left(S_{xy} - C_{xy}\right)^2 \tag{10.2}$$

where M and N are the resolution values of the carrier images.

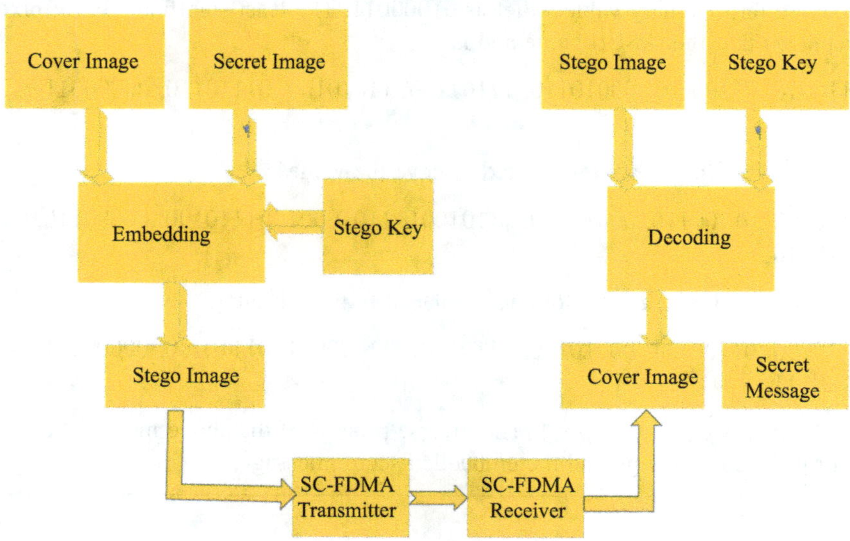

Figure 10.4 SC-FDMA

The lesser the value of the MSE, the more alike the images in comparison are assumed to be.

10.3.2 Peak signal-to-noise ratio (PSNR)

PSNR is used to analyze the image compression quality. It measures the mutilation of the stego-image with the original cover image as a reference. It is denoted by:

$$\mathrm{PSNR(dB)} = 10\log\frac{255^2}{\mathrm{MSE}} \tag{10.3}$$

PSNR is calculated in dB. PSNR values of 40 dB and higher indicate good fidelity. Values ranging from 30 to 40 dB are acceptable. Anything lesser than 30 dB is deemed unacceptable. PSNR values increase exponentially with the image quality. Ideal PSNR value is infinity [19].

10.3.3 Structural Similarity Index (SSIM)

The SSIM evaluates image quality deterioration resulting from processing techniques like data compression. It is expressed by

$$\mathrm{SSIM}(x,y) = \frac{\left(2\mu_x\mu_y + C_1\right)\left(2\sigma_{xy} + C_2\right)}{\left(\mu_x^2 + \mu_y^2 + C_1\right)\left(\sigma_x^2 + \sigma_y^2 + C_2\right)} \tag{10.4}$$

where μ_x, μ_y is the mean intensity of carrier image and stego-image pixel; σ_x, σ_y is the mean intensity of carrier image and stego-image block.

When the SSIM value approaches 1, both the cover image and the stego-image are more or less like each other. As mentioned earlier, the ideal value is one, this is challenging to achieve in practice.

10.3.4 Average difference (AD)

The AD represents the average sum of the difference between the original image and the stego-image. It is defined as:

$$\mathrm{AD} = \frac{1}{MN}\sum_{x=1}^{M}\sum_{y=1}^{N}\left(S_{xy} - C_{xy}\right) \tag{10.5}$$

The ideal value of AD is zero, indicating high similarities between stego-images and cover-images.

10.3.5 Normalized cross-correlation (NCC)

NCC is the metric used to gauge the degree of affinity and contrast between the original image and the stego-image. NCC is expressed as

$$\mathrm{NCC} = \sum_{x=1}^{M}\sum_{y=1}^{N}\frac{\left(S_{xy} * C_{xy}\right)}{S_{xy}^2} \tag{10.6}$$

The ideal value is 1, but in practice, it is challenging to achieve the ideal value. However, values approaching 1 indicate good performance.

10.3.6 Normalized absolute error (NAE)

NAE is the quality metric that shows the absolute error between the original image and the stego-image. NAE is expressed as

$$NAE = \frac{\sum_{x=1}^{M}\sum_{y=1}^{N}\left(S_{xy} - C_{xy}\right)}{\sum_{x=1}^{M}\sum_{y=1}^{N} S_{xy}} \tag{10.7}$$

Image quality is inversely proportional to NAE. The ideal value is 0.

10.3.7 Maximum difference (MD)

To measure the MD (error) between the original image and the stego-image MD is used. It is expressed as

$$\mathrm{MD} = \mathrm{Max}\left(\left|S_{xy} - C_{xy}\right|\right) \tag{10.8}$$

where x=1, 2...m; y=1,2...n.

MD is inversely proportional to the image quality. The ideal value is 0.

10.4 Results and discussion

Figure 10.5 displays the graph of the bit error rate (BER) to the signal-to-noise ratio (SNR) graph plotted between the SC-FDMA system and the conventional OFDM system for 16 QAM. For a BER of 10^{-4}, the SNR in case of SC-FDMA is 13 dB, whereas for OFDM, it is 24 dB. From this graph, it can be concluded that SC-FDMA outperforms OFDM.

Figure 10.6 shows the BER versus SNR graph plotted between QPSK, 16-QAM, and 64 QAM modulation schemes over AWGN channel in the SC-FDMA system. From the figure, it is evident the QPSK modulation scheme gives good BER performance compared to QAM. For BER of 10^{-3} QPSK modulation scheme offers SNR of

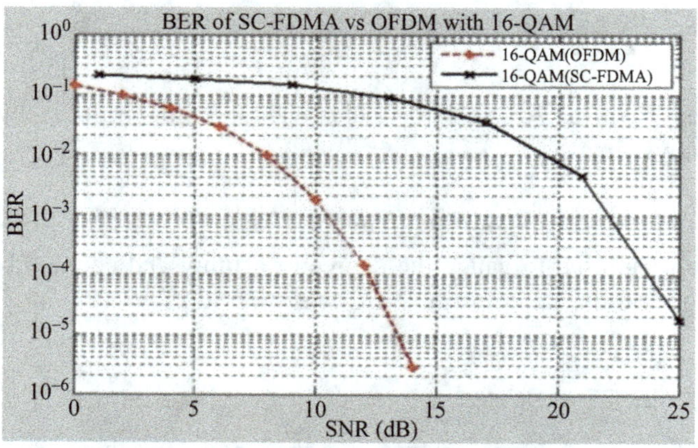

Figure 10.5 BER versus SNR

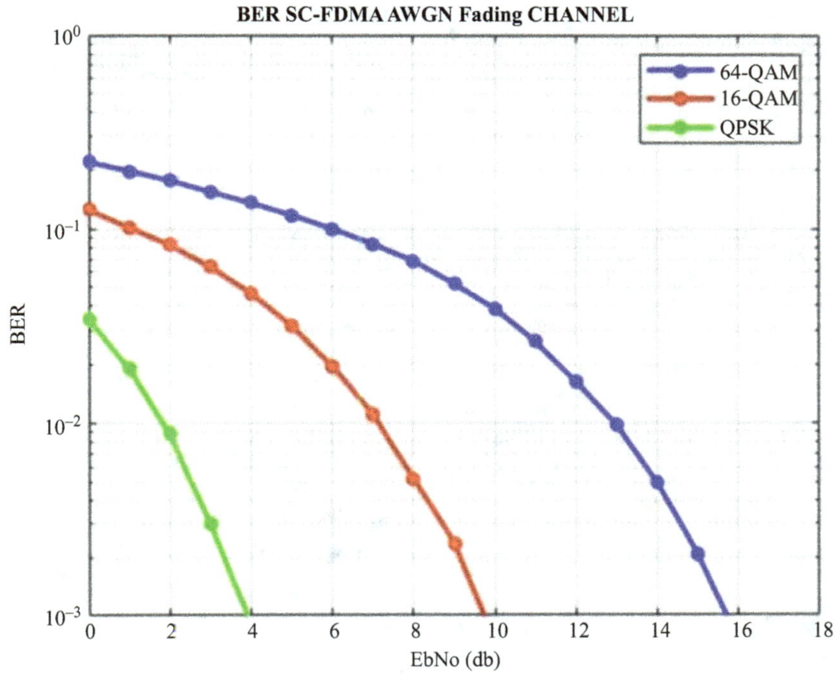

Figure 10.6 Comparison between QAM and QPSK in AWGN channel

4 dB, whereas in the case of 16-QAM, it is 9.8 dB and, for 64-QAM, it is 15.7 dB. QPSK supports two bits simultaneously and QAM supports more bits based on the *M* value. In terms of error, QPSK has less probability of error when compared to the other two methods. This is because the distance between the constellation points and the noise line is higher in the case of QPSK when compared to 64-QAM. Closer to the noise line, the constellation points become more sensitive to noise. Any small noise variations which change the binary ones to zeros and binary zeros to one thus increasing the probability of error. Hence, it is a tradeoff in choosing the appropriate modulation scheme.

Figure 10.7 denotes the BER versus SNR graph drawn for QPSK and QAM modulation schemes over a Rayleigh Fading channel. From the figure, it is evident that QPSK modulation schemes outperform QAM. For a BER of 10^{-3}, SC-FDMA system having QPSK modulation schemes offers an SNR of 23.5 dB, whereas for 16-QAM, it is 29 dB and, for the 64-QAM scheme, it is 33.7 dB.

Figure 10.8 shows the BER versus SNR graph plotted for QPSK modulation schemes over AWGN and Rayleigh fading channels. From the figure, it is evident that QPSK modulation schemes over the AWGN channel yield better performance when compared to the Rayleigh fading channel. For BER of 10^{-3}, QPSK modulation schemes over AWGN channel offer SNR of 4 dB, whereas for Rayleigh fading channel, it is 17 dB.

Table 10.2 shows various performance metrics like MSE, PSNR, SSIM, AD, NCC, NAE, and MD which aid in analyzing the robustness of the proposed

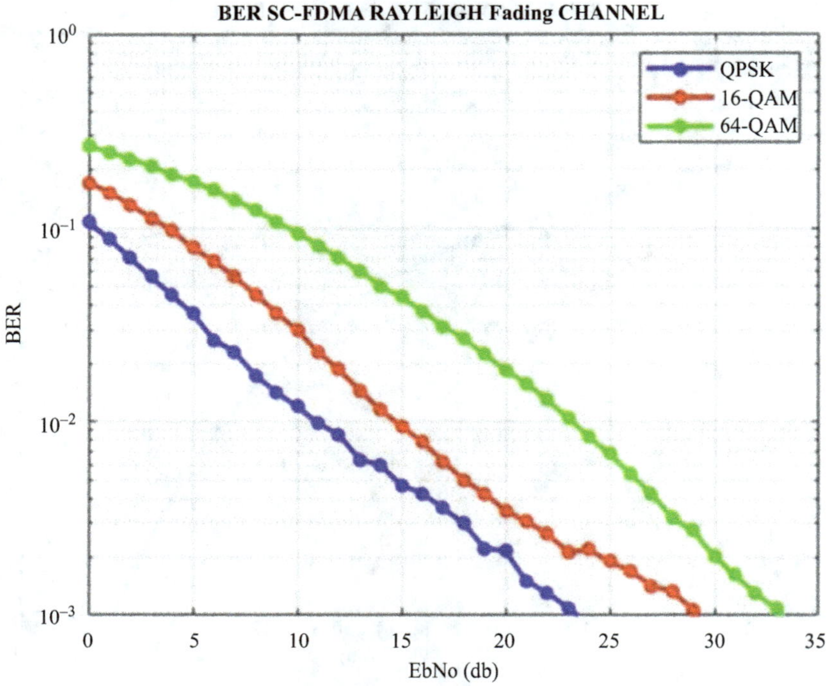

Figure 10.7 Comparison between QAM and QPSK in Rayleigh fading channel

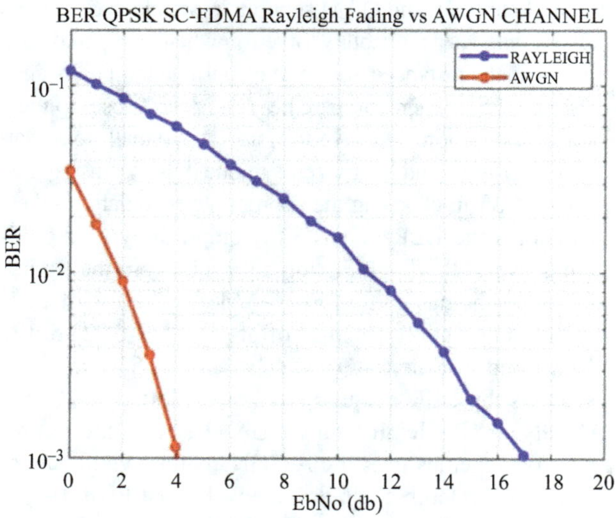

Figure 10.8 Comparison between AWGN and Rayleigh fading channel

Table 10.2 Various metrics for different gray-scale images

Data size	Cover image	Stego-image	MSE	PSNR	SSIM	AD	NCC	NAE	MD
1 kb			0.0558	60.69	0.999	0.0043	0.989	0.0001	0.01
			0.0446	61.49	0.999	0.0086	1	0.0003	0.01
			0.0188	65.42	0.998	0.0013	1.0	0.0002	0.02
			0.0504	61.13	0.997	0.0121	0.999	0.0003	0.01
	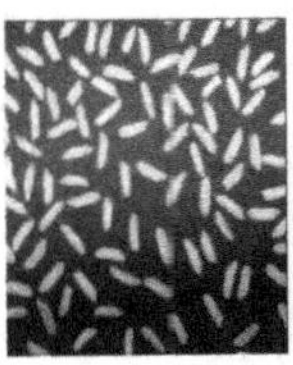	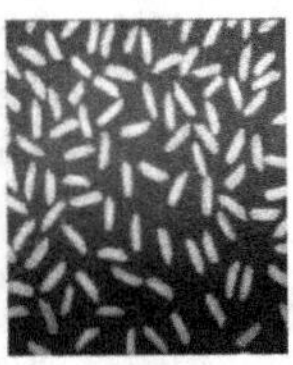	0.0546	60.79	0.997	0.0037	0.997	0.0002	0.01
			0.0020	75.05	0.995	0.0011	1	0.0001	0.01

algorithm. The common message size of 1 kb is applied across various gray-scale images. As the PSNR increases, the MSE diminishes, demonstrating the good image quality. Normalized absolute error is the quality metric which indicates the absolute error between the original image and the stego-image. Ideally, it should be equal to 0. From the table, the values are observed to approach 0. Likewise, the maximum difference and average difference also approach 0, indicating that the cover image and stego-image have slight variances between them. The SSIM evaluates image quality deterioration. Normalized cross-correlation is used to gauge the degree of affinity and contrast between the cover image and the stego-image. Both these metrics should be equal to 1 or approach 0 which is an indicator of good performance. From the table, both the values approach 1.

Table 10.3 lists the image quality degradation as message size is incremented in steps. Various performance metrics for different message sizes are measured for a gray-scale image of Saturn. From the table, it may be inferred that an increment in message size results in a reduction of PSNR values and the promotion of MSE value which in turn give rise to poorer image quality. Normalized absolute error should be ideally equal to 0 but it can be seen that as the message size increases, the NAE also increases correspondingly. Following this trend, maximum difference and average difference expected to be ideally 0 also increase as the message size increases. It can be concluded that the cover image and stego-image have a significant amount of difference between them. Similarly, SSIM and normalized cross-correlation should be equal to 1 for better performance but it can be seen from the table that both the values are increasing as the image size increases which in turn results in image quality degradation.

Table 10.4 illustrates the different histogram responses for the original and the stego-images. It also features the result of bitwise XOR operation perform between the original and the stego-images. As such, due to the highly similar natures of the

Table 10.3 Various metrics for different message sizes

Image	Data size	MSE	PSNR	SSIM	AD	NCC	NAE	MD
	1 kb	0.0020	75.0543	0.995	0.0011	1	0.00010	0.01
	2 kb	0.0041	72.0305	0.9927	0.0021	1	0.00012	0.01
	3 kb	0.0061	70.2941	0.9901	0.0029	0.99	0.00015	0.01
	4 kb	0.0081	69.0580	0.9876	0.0034	0.99	0.00020	0.015
	5 kb	0.0101	68.1006	0.9857	0.0038	0.98	0.00020	0.023
	6 kb	0.0121	67.3248	0.9833	0.0040	0.98	0.00031	0.037
	7 kb	0.0141	66.6714	0.9813	0.0040	0.98	0.00036	0.045
	8 kb	0.0161	66.1019	0.9791	0.0041	0.97	0.00040	0.055
	9 kb	0.0181	65.5972	0.9771	0.0042	0.97	0.00046	0.063
	10 kb	0.0200	65.1486	0.9751	0.0044	0.96	0.00051	0.08

Table 10.4 Histogram responses and bitwise XORs of the original and stego-images

Image name	Original image histogram	Stego-image histogram	Original-stego XORed image	Image
Baby	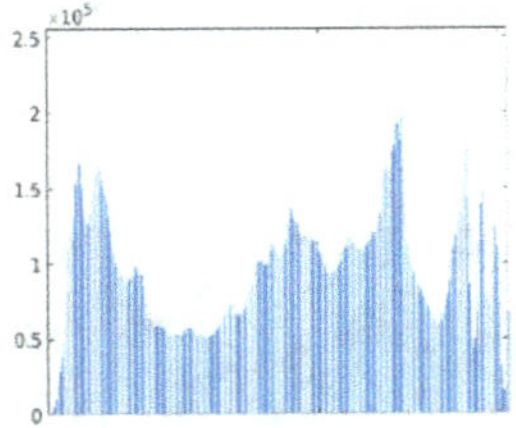	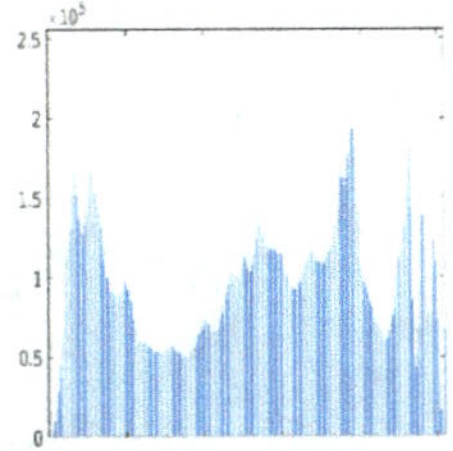		
Camera man	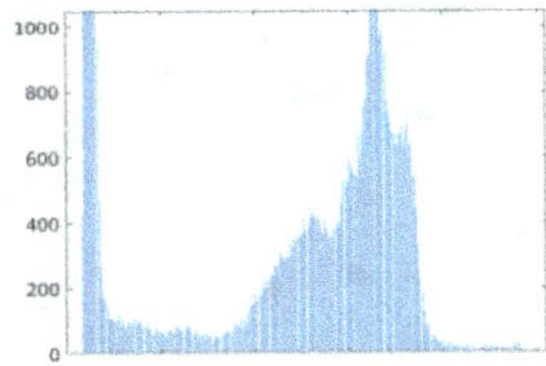	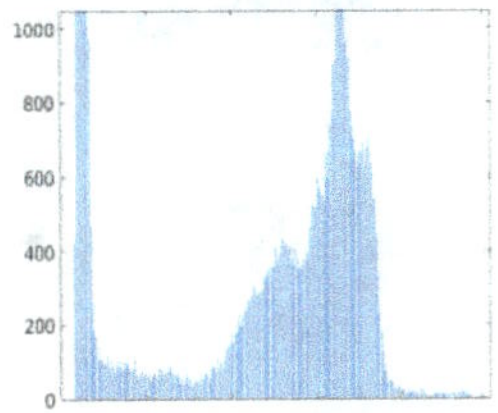		
Moon	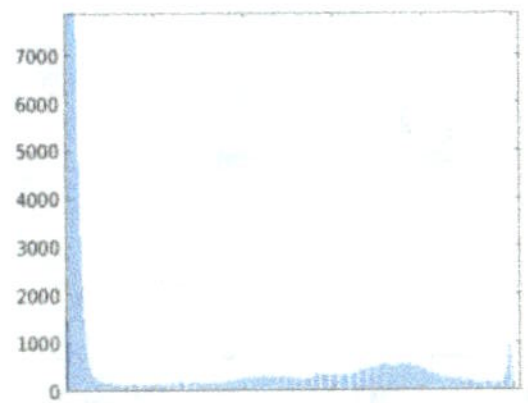	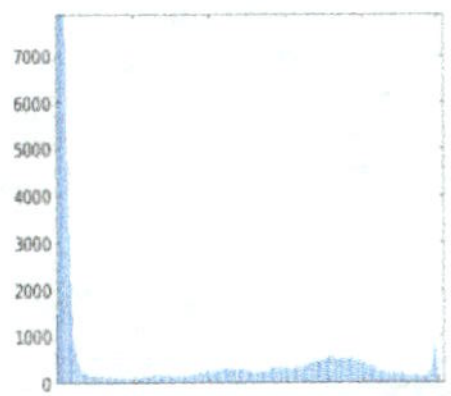		

(Continues)

Table 10.4 (Continued)

Image name	Original image histogram	Stego-image histogram	Original-stego XORed image	Image
Pout	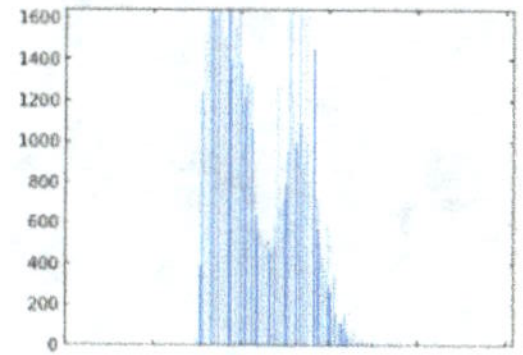	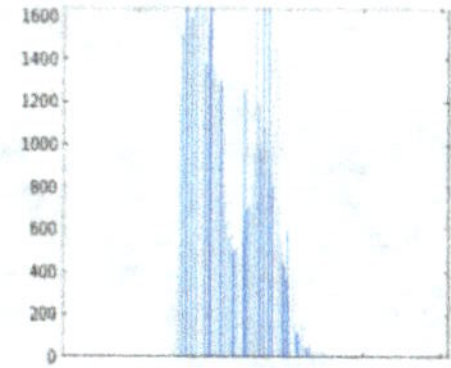		
Rice	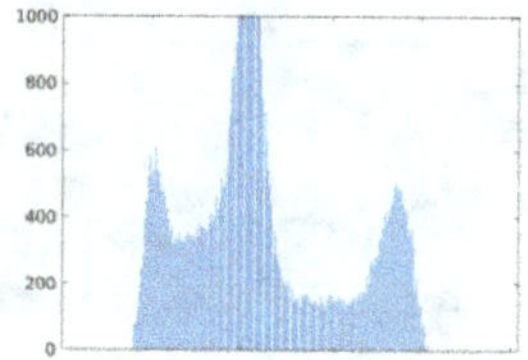	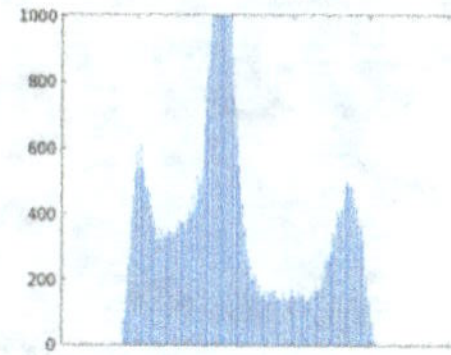		
Saturn	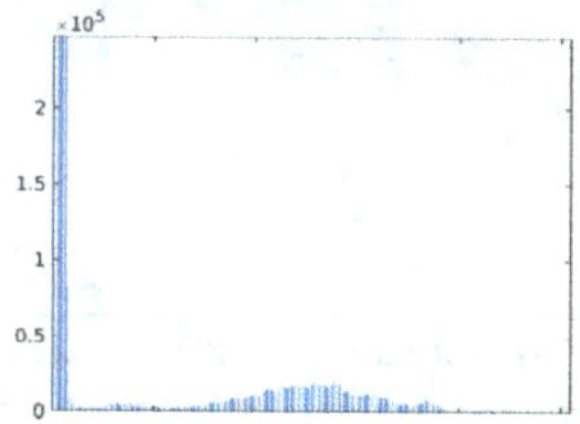	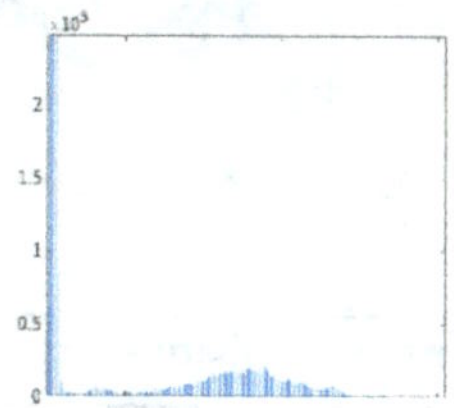		

histogram responses between the original and the stego-images, it can be inferred that the information is hidden well and is virtually imperceptible to the human eye. To the effect, the images seem indistinguishable, which indicates the robust performance of the modified LSB algorithm. The result of the bitwise XOR operation further cements this fact. As the bits in both the original and the stego-image are very minutely different, the XOR operation would yield a low or 0, which appears black as the equivalent representation of the pixel value.

Table 10.5 shows that the proposed method yields a better performance compared to the LSB method. The PSNR, MSE, and SSIM values for the LSB method are 52.67, 0.35, and 0.88, respectively, while the proposed method gives 68, 0.0084, and 0.99, respectively. Likewise, other various parameters confirm that the proposed algorithm outperforms the traditional LSB algorithm across the board.

Figure 10.9 shows PSNR versus the message length graph plotted for the values in Table 10.2. The figure shows that the PSNR value decreases as the

Table 10.5 Comparison of proposed method and LSB method for Lenna image

Parameters	Proposed method	LSB method
PSNR	68.9102	52.6709
MSE	0.0084	0.35
SSIM	0.995	0.8897
AD	0.00021	0.0043
NCC	1	0.93
NAE	0.000061	0.00030
MD	0.01	0.033

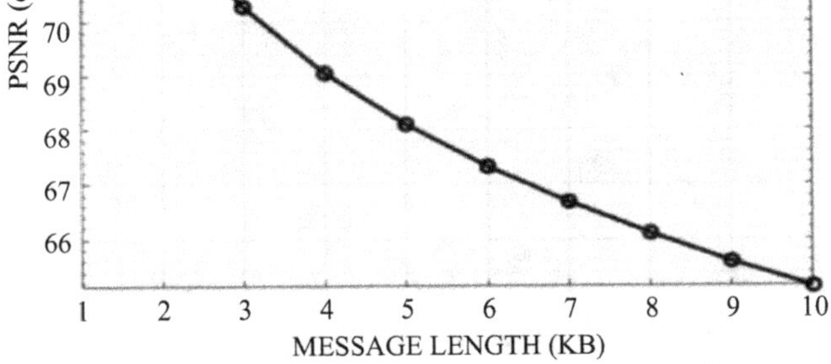

Figure 10.9 Relation between PSNR and message length

message length increases. It results in the overall degradation of image quality. For a message length of 1 kb, the PSNR is 75 dB, whereas for 2 kb, it is 72 dB, for 5 kb, it is 68 dB and, for 8 kb, it is 66 dB, and so on.

Figure 10.10 shows MSE versus the message length graph plotted for Table 10.2. From the figure, MSE value will increase when message length is increased which yet again results in the degradation of the image quality. For a message length of 1 kb, MSE is 0.0020, whereas for 2 kb, it is 0.0041 and, for 5 kb, it is 0.0101 and, for 8 kb, it is 0.016 dB and so on.

A comparative analysis was carried out between the proposed modified LSB approach and the conventional steganographic technique. The main qualitative parameter that persisted as a common denominator across all the works was the PSNR values. As such was used to objectively compare and contrast the performances across several algorithms. Another metric worth noting was the timing analysis from which the algorithms performance may be ascertained in terms of how fast it is able to run. However, this metric would be subjective as it would depend on the hardware capabilities of the particular system it was executed. Nevertheless, the timing analysis of standard system specification is carried out. The comparative table is illustrated in Table 10.6.

From Table 10.6, it can be understood that the proposed modified LSB algorithm has outperformed many other techniques in terms of PNSR values while boasting an impressive 68.91 dB, far higher than the 30 dB minimum acceptable threshold for maintaining image fidelity.

As such, the modified LSB has proven to be robust approach toward image steganography by sharply overcoming the drawbacks of the traditional LSB approach, and to the effect has resounded a good performance across several evaluation metrics. This method, coupled with the SC-FDMA modulation scheme with its higher efficiency, has proven to be a reliable, efficient and secure communication system for exchanging messages reliably.

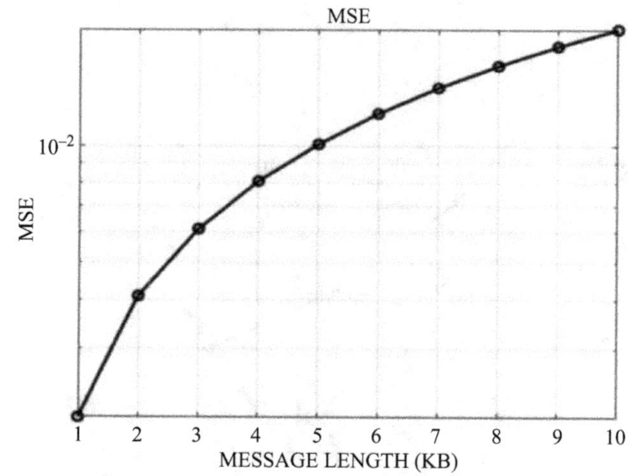

Figure 10.10 Relation between MSE and message length

Table 10.6 Comparison of the proposed method and LSB method

Image (Lenna)	PSNR
Proposed algorithm	68.91
[20]	54.19
[21]	27.09
[22]	53.76
[23]	57.00
[24]	35.19
[25]	37.32
[26]	36.59
[27]	32.68

10.5 Conclusion and future scope

SC-FDMA transceiver is considered to be a better option in terms of their low PAPR when compared to other transceivers like OFDM. BER performance of the AWGN channel and Rayleigh fading channel is compared and simulated, proving that the AWGN channel is better than Rayleigh fading channel. Also, it is concluded that the QPSK modulation scheme is better than QAM. Once the transceivers performance is enhanced, the next step is enhancing the algorithms robustness. The secret data is embedded into the cover image using the proposed algorithm and converted into stego-image which is transmitted through an SC-FDMA transmitter. At the receiving end, it is received and extracted. The performance of the proposed algorithm is analyzed through various parameters and proved that the proposed method is better than the LSB algorithm.

5G is the latest technology to support a number of applications in wireless communication like wireless gadgets, autonomous systems, smart city, auto-driving vehicle with the high data rates, shorter delay, huge capacity, and better quality of service. The non-orthogonal multiple access (NOMA) with successive interference cancellation (SIC) is well-suited promising radio access technique to satisfy the demands. The above-discussed algorithms can be implemented in NOMA for secure data transfer.

References

[1] V.M. Padmapriya, K. Thenmozhi, P. Praveenkumar, and R. Amirtharajan. "Misconstrued voice on SC-FDMA for secured comprehension–a cooperative influence of DWT and ECC". *Multimedia Tools and Applications*. 2022, 815:7201–7217.

[2] V.M. Padmapriya, K. Thenmozhi, P. Praveenkumar, *et al.* "ECC joins first time with SC-FDMA for mission security". *Multimedia Tools and Applications*. 2020, 79:17945–17967, doi.org/10.1007/s11042-020-08610-5.

[3] K. Vidhya. "Performance of MIMO-OFDM systems". *International Journal of Operational Research*. 2020, 372:293–306.

[4] A.F. Abd El-Rahman, F.E. Abd El-Samie, S.E. A. Khamis, and S.A. Napoleon. "A new method to enhance SC-FDMA communication with channel estimation errors". *Wireless Personal Communication*. 2019, 104: 527–542.

[5] S. Ebadinezhad and S. Hasan. "BER evaluation in LTE SC-FDMA under multipath channels". *International Journal of Recent Technology and Engineering*. 2019, 84:3539−3547.

[6] S. Kaur, S. Bansal, and R.K. Bansal. "Computing for sustainable global development". In *International Conference*. 2014, pp. 870−875.

[7] V.M. Padmapriya, M. Priyanka, K.S. Shruthy, S. Shanmukh, K. Thenmozhi, and R. Amirtharajan. "Chaos aided audio secure communication over SCFDMA system". In *International Conference on Vision Towards Emerging Trends in Communication and Networking*. 2019, pp. 1–5.

[8] T. Bhuiyan, A.H. Sarower, Md. Rashed Karim, and Md. Maruf Hassan. "An image steganography algorithm using LSB replacement through XOR substitution". In *International Conference on Information and Communications Technology*. 2019, pp. 1–7.

[9] B. Karthikeyan, V. Abbinaiya, and T. Sumathi. "A novel approach in Steganography combining random key and substitution cipher". In *International Conference on Intelligent Computing and Control Systems*. 2019, pp. 673–678.

[10] V. Montalvo, A.F. Reyes, and M.C. Paredes-Paredes. "Comparison and analysis of PAPR reduction techniques in OFDMA and SC-FDMA systems". In *IEEE Fourth Ecuador Technical Chapters Meeting*. 2019, pp. 1–6.

[11] J.S. Roy and S.S. Mishra. "Performance of SC-FDMA for LTE uplink under different modulation schemes". *International Conference on Mechatronics*, Robotics and Systems Engineering. 2019:202–206.

[12] M.F. Marzban, A.El. Shafie, N. Al-dhahir, and R. Hamila. "Security-enhanced SC-FDMA transmissions using temporal artificial-noise and secret key aided schemes." *IEEE Access*. 2019, 7:14807–14824.

[13] A.D.P. Ariyanto, E.H. Rachmawanto, De R.I.M. Setiadi, and C.A. Sari. "Performance analysis of LSB image steganography combined with blowfish-RC4 encryption in various file extensions," In *Fourth International Conference on Informatics and Computing*. 2019, pp. 1–6.

[14] M. Mustafa, M. Mahmoud, H. Tagelsir, and I. Elshoush. "A novel enhanced LSB algorithm for high secure audio steganography". In *10th Computer Science and Electronic Engineering*. 2018, pp. 125–130.

[15] S. Kumar, S. Kumar, N.K. Singh, A. Majumder, and S. Changder. "A novel approach to hide text data in colour image". In *7th International Conference on Reliability, Infocom Technologies and Optimization (Trends and Future Directions)*. 2018, pp. 1–5.

[16] P.P. Bandekar and G.C. Suguna. "LSB based text and image steganography using AES algorithm". In *International Conference on Communication and Electronics Systems*. 2018, pp. 782–788.

[17] B.V. Naik, N.L.K. Sai, and Ch. Manohar Kumar. "Efficient transmission of encrypted images with OFDM system". In *IEEE International Conference on Power, Control, Signals and Instrumentation Engineering*. 2017, pp. 2383–2388.

[18] A. Ul. Islam, F. Khalid, M. Shah, *et al.* "An improved image steganography technique based on MSB using bit differencing". In *The Sixth Conference on Innovative Computing Technology*. 2016, pp. 265–269.

[19] A.K. Singh, J. Singh, and Dr. H.V. Singh. "Steganography in images using LSB technique". *International Journal of Latest Trends in Engineering and Technology*. 2015, 51:426–430.

[20] K. Raghunath and A. Chockalingam. "SC-FDMA versus OFDMA: sensitivity to large carrier frequency and timing offsets on the uplink". In *Proceedings of Global Telecommunications Conference*. 2009, pp. 1–6.

[21] M. Hussain, A.W.A. Wahab, Y.I.B. Idris, A.T. Ho, and K.H. Jung. "Image steganography in spatial domain: a survey". *Journal of Signal Processing: Image Communication*. 2018, 65:46–66.

[22] M. Anusha, K.N. Bhanu, and D. Divyashree. "Secured communication of text and audio using image steganography". In *International Conference on Electronics and Sustainable Communication Systems (ICESC)*. 2020, pp. 284–288.

[23] Shrilekha and N.V.S. Reddy. "Improving security in image steganography using MSB bit differencing and cryptographic algorithm". In 2018 *International Conference on Computational Techniques, Electronics and Mechanical Systems (CTEMS)*. 2018, pp. 228–230.

[24] P. Kanojia and V. Choudhary. "LSB based image steganography with the aid of secret key and enhance its capacity via reducing bit string length". In *3rd International conference on Electronics, Communication and Aerospace Technology (ICECA)*. 2019, pp. 257–262.

[25] A. Rodrigues and A. Bhise. "Reversible image steganography using cyclic codes and dynamic cover pixel selection". In *International Conference on Wireless Communications, Signal Processing and Networking (WiSPNET)*. 2017, pp. 509–513.

[26] S. Khan, M.A. Irfan, M. Ismail, T. Khan, and N. Ahmad. "Dual lossless compression based image steganography for low data rate channel". In *2017 International Conference on Communication Technologies (ComTech)*, Rawalpindi. 2017, pp. 60–64, doi:10.1109/COMTECH.2017.8065751.

[27] G. Swain. Digital image steganography using eight-directional PVD against RS analysis and PDH analysis. *Advances in Multimedia*, 2018, 2018: 4847098, doi:10.1155/2018/4847098.

Chapter 11

A lightweight algorithm for the detection of fake incident reports in wireless communication systems

Yuichi Sei[1], *Akihiko Ohsuga*[1] *and Agbotiname Lucky Imoize*[2,3]

Abstract

Sensor devices in 6G technology are an affordable method for identifying target incidents within large wireless communication systems (WCSs). However, they face potential compromise or capture by compromised actors. For instance, fake incident reports in a compromised device can result in congested networks that hinder the passage of valid incident data. Fake incident reports can be identified, although this approach can prove tricky because of the requirement for certification tokens, which do not always offer a workable solution to congested networks. One suggested strategy would involve creating space-efficient Bloom filters. Their creation would result from correctly combining the correct devices and placing them in each device in advance. The next stage would see an incident report appear, featuring an XOR of the tokens (XT), with all devices confirming the information according to its Bloom filter. Illegally acquiring a device can prove costly because it would compromise the Bloom filter data and the XT allocated to the correct incident report. Thus, this study suggests using a secure algorithm to update the data. Unlike existing studies, detecting a fake incident report would only increase by approximately one hop; however, the amount of traffic created by a compromised device would decrease by around 60%. Thus, the suggested method would lessen the traffic resulting from attacks featuring file incident reports, which would, in turn, make the network less congested.

Keywords: Wireless communication systems; Fake incident reports; Security mechanism; Bloom filter

[1]Department of Informatics, Graduate School of Informatics and Engineering, The University of Electro-Communications, Japan
[2]Department of Electrical and Electronics Engineering, Faculty of Engineering, University of Lagos, Nigeria
[3]Department of Electrical Engineering and Information Technology, Institute of Digital Communication, Ruhr University, Germany

11.1 Introduction

The new 6G technology will facilitate the construction of wireless communication systems with low latency. Sensor devices act as a core functionality that can pinpoint and delineate incidents, such as crimes or natural disasters [1,2]. Connecting these devices in a wireless network forms a wireless communication system (WCS). Multihop wireless paths in WCSs also send reports to their final recipient, the base station. When sensor devices detect incidents of interest, they transmit the relevant data (Figure 11.1). Research has shown that sensor devices in hostile settings face the possibility of attack, which can leave them open to compromise and capture. Network congestion can also transpire. In such situations, compromised devices can produce fake incident reports if hackers acquire the secret keys within the sensor devices (Figure 11.2). Congested networks can hinder the transmission of valid reports to base stations, which camouflages hackers' illegal activities.

Numerous studies [3–6] have reviewed the notion of quickly identifying fake incident reports in networks. Such identification can ease the issue of congested networks. This approach revolves around allowing devices to possess symmetric keys. When incidents occur, T proximal sensor devices produce a report containing multiple message authentication codes (MACs). A device's alignment with the report becomes apparent in a MAC generated by a device utilizing a symmetric

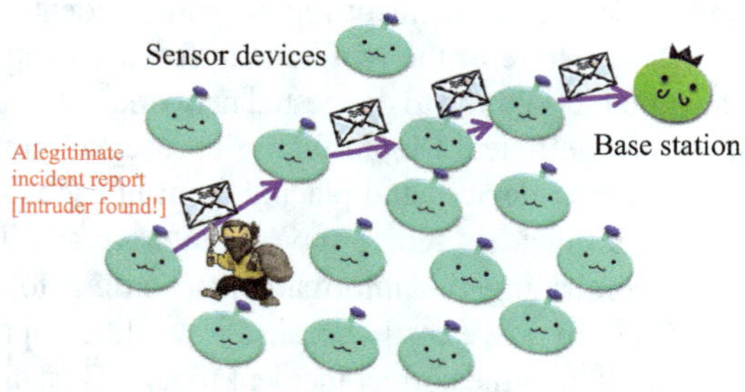

Figure 11.1 An incident is detected in a WCS

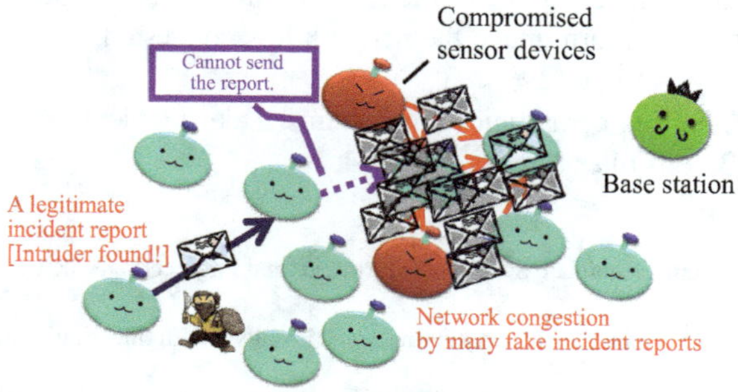

Figure 11.2 Fake incident report attacks

key. Different approaches can incorporate additional certification data into reports. Creating a report can enable the probabilistic validation of MACs by forwarding every device to the base station over several hops. All reports feature sizable portions of certification data, meaning they encompass large amounts of traffic even though they can recognize fake incident reports in-network. Thus, a compromised network produces numerous fake incident reports, which can, in turn, congest networks. However, correct reports consume resources, as seen by how sensor devices' battery power levels can diminish; restricted batteries can also impede sensor devices' functions and reduce WCSs' operational capacity. A complete security breakdown can also occur in situations involving T or more compromised devices.

This study proposes a streamlined strategy that can demonstrably detect fake incident reports in-network in situations when aspects other than T devices are compromised. However, despite such an approach's capacity to significantly reduce occurrences of congested networks, the possibility remains that the same total report hops would appear before a fake incident report is identified.

Pre-deployment, the sensor devices receive a Bloom filter containing correct XORs of tokens (XTs), which can identify fake incident reports even when more than T sensor devices are compromised. This approach, as supported by mathematical analyses, could cut traffic in the incident of compromised devices eliciting fake incident reports.

The paper's primary contributions comprise the following*.

- First, this study outlines an algorithm with a high likelihood of identifying fake incident reports in WCSs. This algorithm means all reports feature incident data and a one-off certification system. The suggested XT has a shorter bit length than comparable data in current research, meaning that the necessary amount of traffic for incident reports is also reduced. Consequently, the outlined algorithm can ease the threat of congested networks due to compromised devices. This study seeks to lessen traffic volume and congested networks in WCSs.
- Second, all forwarding devices can pinpoint, based on the Bloom filter, the legitimacy of the one-time XTs in the forwarding report. Bloom filters are data structures that identify whether or not the provided data forms part of a set, which can result in high levels of spatial efficiency and allows for a few false positives. Network managers control Bloom filters' detection capacity. It facilitates the identification of fake incident reports in two hops on an average.
- Third, this study proposes a refined mechanism involving the one-off XTs. This mechanism would impair any illegally acquired tokens and input new, more unpredictable tokens to stave off potential attacks. The revamped mechanism would only play a role in the outlined method, which has a similar capacity to detect fake incident reports using one-time XTs as existing

*An earlier version of the chapter appeared in [7].

techniques. Moreover, the mechanism, including its refinements, would not need much traffic, especially in comparison to current methods.

- Fourth, this study involves experiments and comparisons of the suggested algorithm with four current iterations. According to the findings, the revamped strategy can cut the amount of traffic generated during attacks.

This study takes the following course: Section 11.2 portrays a model of sensor networks and fake incidents; Section 11.3 assesses related strategies and related concerns; Section 11.4 showcases the method's design; Section 11.5 evaluates the efficiency of the approach adopted in the study, utilizing the results from the prior section as the foundation of research; and Section 11.6 summarizes the research. The outlined strategy would help tackle the issue of fake incident reports because it would cut the amount of traffic necessary to pinpoint such reports while retaining the capacity to identify them in the first place.

11.2 Related work

Numerous studies [8–10] have assessed secured data aggregation in WCSs. Although the methods espoused in these studies cannot detect in-network fake incident reports, they can identify fake incident reports at the base station. Thus, WCSs cannot identify trespassing or analogous incidents because compromised devices have congested networks.

Studies evaluating in-network fake incident report detection have outlined a strategy to distribute secret keys to sensors. All sensor devices that possess symmetric keys embody this strategy's theoretical basis. The sensor devices' locations generate the keys post-deployment; alternatively, the devices come preloaded with the keys. Sensor devices can also share keys, and the sensors can generate incident reports that contain key IDs, the MACs, and sensor device IDs close to an incident. The sensor devices' alignment with reports becomes apparent in MACs produced by the devices via a symmetric key. The MACs' probabilistic validation, as shown in the corresponding report, occurs because all forwarding devices in the reporting process have been sent to the base station over numerous hops. When reports contain fake MACs, delivery does not transpire. Studies have attempted to improve the chances of a forwarding device possessing a key that can create MACs in the incident report.

Fake incident report detection strategies involve the random distribution of keys to sensor devices via randomized key distribution approaches [3,11]. Incidents only become legitimized when sensor devices gather more than or equal to T discrete MACs. Such a mechanism has its uses in situations involving moving base stations. Regardless, the suggested randomized key distribution mechanism indicates threshold behavior. Moreover, should actors gather more than T keys, fake incidents cannot be detected.

Sensor devices' positions drive the allocation of keys to all devices as part of fake incident report detection strategies that employ key distribution strategies based on locations [4]. Although their methods can detect fake incident reports in-network, they also generate a significant amount of traffic, attributable to the long reports created by compromised devices in such scenarios. Numerous studies have established the default of $T = 5$. In WCSs, a MAC tends to be eight bytes long [12].

Other studies have posited that the amount of traffic of a report without security mechanisms in WCSs amounts to 40 bytes [13,14]. Compromised devices can induce sizes doubling the original packet in a fake incident report, meaning they can reach sizes of 80 bytes. Research has found that identifying fake incident reports helps control situations that involve altered base station positions and numerous compromised devices, as shown in the study [5]. However, as this method employs T MACs, attacks involving many fake incident reports can occur, as in comparable methods.

Babu *et al.* outlined the secure data aggregation based on principle component analysis (SDA-PCA) [8]. This approach involves selecting a reporting leader based on device quality and energy levels. This reporting leader collates sensing data from the neighboring sensor devices and sends such information in the form of a report to the base station. SDA-PCA considers highly mobile sensor devices as compromised, and these devices cannot become reporting leaders. Such an approach impedes the passage of fake incident reports to the base station. The transparent criteria for determining whether or not a device is compromised makes it easier to place devices so that they do not meet these criteria. However, compromised devices can become reporting leaders, which can, in turn, delineate incorrect data. The algorithm assumes the veracity of reports created by the reporting leader and cannot identify when reporting leaders generate fake incident reports.

Wang *et al.* outlined an algorithm for secure aggregation in WCSs [9], which chooses a reporting leader from neighboring devices according to link quality and residual energy. This approach keeps data confidential because of the encrypted communication between devices, reporting leaders, and base stations. All devices and clusters have unique keys, which can prevent spoofing, but also causes potential issues: only the base station can identify this spoofing, which cannot be detected during the report's transfer to the base station. Thus, if an attacker compromises even a single device, the in-network loses its capacity to identify the numerous fake incident reports that could appear.

Pedroso *et al.* [15] also proposed a CONFINIT algorithm that blocks fake incident report attacks. CONFINIT compares all devices' sensing values with one another; if a device has vastly different sensing values from the other devices, there is a high possibility that the device is compromised. Consequently, the non-compromised devices would ignore the possibly compromised device. CONFINIT seeks to stop compromised devices' creation of fake incident reports. However, the lack of a mechanism to identify fake incident reports at the forwarding device or base station means that even if a compromised device creates this type of report, it would not be identified as such.

Although the strategies introduced thus far can impede the creation of fake incident reports or detect them in the base station, they cannot identify fake incident reports in-network.

Ye *et al.* suggested a statistical en-route filtering (SEF) strategy [3] involving an algorithm that could detect fake incident reports in-network. When sensor devices identify an incident, they create MACs according to the incident information and device ID; the reporting leader gathers at least T MACs. This approach deploys keys for generating MACs randomly to each device, and data about which key is held within which device is not maintained. Therefore, a problem could transpire because only T keys have been compromised, meaning attackers can create fake incident reports anywhere.

One proposed method involved distributing "correct key ID combinations" to each sensor device beforehand [5]. This strategy forwarded an incident report to the base station, after which the forwarding devices would determine whether the report has T or more MACs with the correct key IDs attached. Although this approach has a high rate of detecting fake incident reports, it can encounter issues because of the requirement to attach T MACs and T key IDs to all reports, which lengthens them considerably.

Kumar *et al.* outlined CD-PEFS, which selects three reporting leaders in each class [6]. A reporting leader collates sensing data from the neighboring devices and creates a report with T tokens, T device IDs, and ($k' + 1$) MACs. On an average, k's value is 10 when the class number 150. An increased number of classes prompts a rise in the value of k'. Despite the high detection rate of fake incident reports in [6], sizable amounts of data are attached to the report.

Yi outlined EMAS [16] wherein the sensor devices send data featuring incident information, key ID, location ID, and the corresponding MAC to a reporting leader. The reporting leader employs hash functions to map T MACs into a Bloom filter and creates a report with the Bloom filter, T sensor IDs, T location IDs, and T key IDs. This approach means the forwarding devices can identify fake incident reports according to the Bloom filter, device IDs, key IDs, and location IDs. EMAS can compress T MACs into one filter. However, the number of reports can still prove significant because of the requirement for much information for the certification process.

Research focusing on identifying fake incident reports in-network can add data to the incident reports for verification, meaning the in-network can quickly identify any fake incident reports created by compromised devices. However, problems can arise because the significant amount of information in the reports means the compromised devices can congest the network. Section 11.5 compares algorithms [3].

A number of methods have been proposed to control congestion in wireless communication systems, although they do not target unauthorized reporting. For example, Grover *et al.* proposed a rate-aware congestion control function at the transport layer [17], and Revathi *et al.* proposed a method to determine the appropriate communication path by collecting information on report delivery delay accordingly [18]. These methods can be used in conjunction with the proposed method to further reduce network congestion.

11.3 Assumptions

11.3.1 Sensor networks

Small sensor devices are densely populated. Detection by multiple devices can adapt to equipment failures and improve detection accuracy. A device acts as the reporting leader, and all detecting devices report the signals they identify to the reporting leader. The incident report uses a multihop path to describe the report delivered by the reporting leader to the base station. The sensor devices employ a localization strategy to determine their location following their deployment [19–21].

The sensor devices' design aligned with cost-effective strategies, prompting the assumption that tamper-resistant hardware is not featured. It is presumed that the sensor devices do not move in the devices post-deployment. The existing studies mirror these assumptions [22–25].

Common key cryptosystem acts as the foundation for many WCSs due to the computational costs of public-key approaches [26]. Additionally, we assume that all sensor devices validate the base station's report and that the base station validates reports to the sensor devices using a mechanism that ensures the base station's integrity [27]. It is also presumed that correct incidents rarely occur, e.g., twice monthly. The target incidents include non-frequent occurrences, such as forest fires and break-ins.

The suggested method would help identify criminal acts, such as break-ins, arson, robbery, and other situations, in WCSs. The expectation is that transgressive activities do not often occur, meaning the suggested method is well-suited to such usage. Thus, the necessary traffic to successfully employ the method is lower than in current strategies. Should such incidents occur more frequently, the one-time XT updates would increase, as would the amount of necessary traffic. This eventuality would lessen the advantage of the suggested method because it would then employ similar amounts of traffic to current strategies.

11.3.2 Attack model

A hacker can compromise multiple sensor devices in a network. The hacker could steal all data resulting from the compromised device. A compromised device erroneously identifies nearby incidents when, in reality, no such occurrence has taken place. Such erroneous reports can influence users to make potentially damaging decisions and result in congested networks. It is assumed that WCSs feature replicated device detection methods [28]. Thus, if a hacker takes more than one device, they may acquire control of that particular device but cannot reproduce any other compromised devices.

Although hackers can create arbitrary incident reports, each forwarding device checks the incident report's legitimacy and can thus detect if an incident report created in an attack is incorrect. However, a hacker can use data from a compromised device to trick the forwarding device into believing that the incident report generated by the hacker is legitimate.

11.4 Proposed method

11.4.1 Overview

A regular geographic class separates the target region. Let n_i represent the sensor device with ID i. The base station generates token r_i for each sensor device n_i

beforehand. Sensor devices detecting an incident report cooperate to generate the XT R from T tokens in the same class. The base station recognizes all correct mergers of T tokens in advance, meaning that the base station also recognizes all correct XTs. Should an attack compromise T devices, the attacker would lack the ability to create correct XTs when the compromised devices do not exist in the same class: a correct XT would require T devices in the same class.

All devices come preloaded with Bloom filters generated from correct XTs. These Bloom filters can identify whether the XT included in an incident report is correct with high probability.

Two primary challenges appear in this scenario. First, if the devices share Bloom filters, they have the same detection capacity; the failure to identify fake incident reports at the first hop means they would not be identified until the base station. Second, the hacker could acquire the XT data of a correct incident via a leak. Such situations highlight the need for an updated mechanism.

Figures 11.3 and 11.4 show this process and the connection between the process, tokens, XTs, and Bloom filters.

Tables 11.1 and 11.2 show examples of the Tx and the Tb. The following section illustrates the proposed algorithm's process and outlines the approach to solving two issues.

$q(q \geq T)$ represents the number of tokens assigned to devices in a class, and g represents the number of classes. A device is provided to each of the prepared qg tokens—$r_1, r_2, \ldots, r_{qg}$. The variable b indicates the tokens' bit length. A leader device n_w in G_w is selected by a local leader election method such as [29]. Subordinate devices that identify an incident send a report to the leader device, n_w. The leader device creates XTs by adopting the T tokens' XOR operation.

The data sent to each forwarding sensor device determines an XT's validity. If a hacker compromises less than T devices, the fake incident report becomes

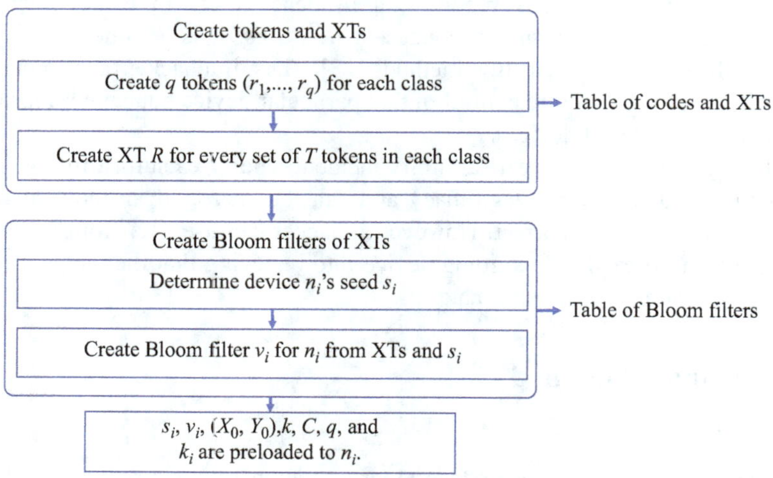

Figure 11.3 The base station's process before the deployment

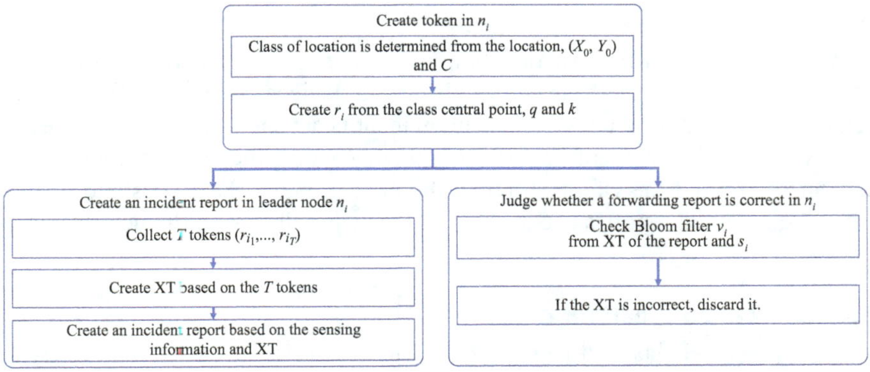

Figure 11.4 Each device's process

Table 11.1 Example of Tx

Class ID	XT ID per class	Set of T tokens	XT
1	1	{0101…., 0001…., …}	$R_1 = \ldots$
…	…	…	…
1	${}_qC_T$	{1110…., 0111…., …}	$R_qC_T = \ldots$
…	…	…	…
g	1	…	$R_{(g-1)*_qC_T+1} = \ldots$
…	…	…	…
g	${}_qC_T$	…	$R_{g*_qC_T} = \ldots$

Table 11.2 Example of Tb

Device ID	Seed	Bloom filter
1	s_1 = a34398dd	v_1 = 100110…
2	$s_2 = \ldots$	$v_2 = \ldots$
…	…	
N	$s_N = \ldots$	$v_N = \ldots$

detectable. This study's suggested method, similar to the methods expounded in other studies, cannot identify fake incident reports when an attack compromises T devices in a class.

The XOR operation of the T tokens assigned to T devices generates an XT. Tokens can only be used once because the report containing the XT might be forwarded by a compromised sensor device. Such a situation indicates the need for a suitable updated method used by the tokens to generate the XT.

11.4.2 Processes

All devices store their tokens, and Bloom filters are generated from a correct set of XTs in advance at the base station. Multiple devices collaborate to create an incident report featuring an XT when an incident is detected. A forwarding device then evaluates the incident report's legitimacy using the XT attached to the incident report and the Bloom filter in the device. Next, the base station determines the received report's accuracy. The following describes the details of this process.

11.4.2.1 Token creation

Let G_w represent a class with an ID of w. Each device n_i has a symmetric key k_i, shared with the base station. Post-deployment, device n_i generates cord r_i based on the class ID. Let g represent the number of classes, with each class having a q token; each device in the class receives one of these tokens.

The devices' class determines the tokens created by the devices according to the proposed method. C and (X_0, Y_0) denote the class's size and the central point of the target region, respectively. Class G_i is defined as:

$$\begin{aligned} X_{x_i} &= X_0 + x_i \cdot C \\ Y_{y_i} &= Y_0 + y_i \cdot C \\ (i,j &= 0, \pm 1, \cdots). \end{aligned} \tag{11.1}$$

The devices come preloaded with the central point $(X_0, >Y_0)$, a hash function, h, a shared master secret key, κ, and class size, C. Related studies have also employed a method that identifies each device's post-deployment location [30]. A non-complex calculation allows all devices to determine their class's central point (X_{x_i}, Y_{y_i}).

The equation below calculates token r_i of device n_i:

$$r_i = h(\kappa || X_{x_i} || Y_{y_i} || \Phi(i) q). \tag{11.2}$$

The variable $\Phi(i)$ conveys the rank of ID i among all device IDs in the same class in ascending order, and || represents concatenation.

Post-deployment, each device n_i broadcasts the device ID to one another in the class G_w. Device n_i, which assembles IDs in ascending order, ascertains the rank of the device n_i's ID i.

All pairs of neighboring devices share a unique pairwise key. The variable $k_{i,j}$ represents the pairwise key of devices n_j and n_i.

11.4.2.2 Bloom filter

All devices use a Bloom filter [31] to validate XTs. Let A represent a set containing n elements. $H[A]$ denotes the Bloom filter created from A. $H[A]$ can be used to test whether element α is contained in A. False positives can occur, but false negatives cannot. That is, if $\alpha \in A$, $H[A]$ surely outputs "true." On the contrary, if $\alpha \notin A$, $H[A]$ outputs "false" with a high probability.

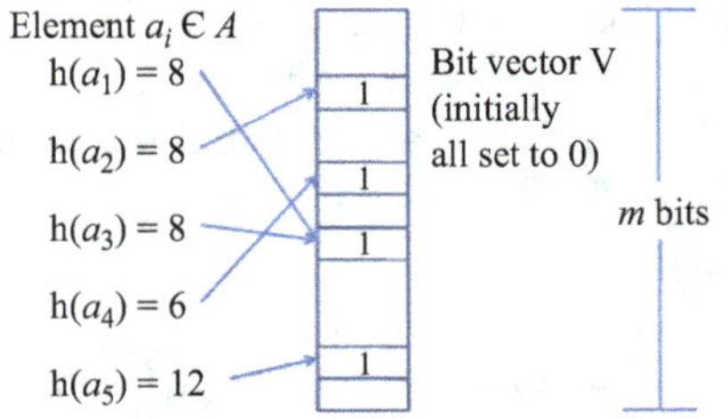

Figure 11.5 A Bloom filter's execution process where the number of hash functions is one

Figure 11.5 illustrates the execution of a Bloom filter. The filter length is m, and all bits are 0. Several functions are prepared. The range of each hash function is from 1 to m. In this research, the number of hash functions is one.

The bit at position $h(a)$ in the filter is set to 1 for all elements $a \in A$. The bit position $h(\alpha)$ undergoes checking when a query for α takes place. The findings show that α is not in set A when the value amounts to 0.

Upon hashing all n elements of $a \in A$ into a Bloom filter, the probability that a bit remains at 0 is $(1 - 1/m)^n$. According to the natural logarithm, $\lim_{m\to\infty}(1 + 1/m)^m = e$. In other words, for large m, $(1 + 1/m)^m \approx e$. Thus, $(1 - 1/m)^n = ((1 + 1/(-m))^{-m})^{-n/m} \approx e^{-n/m}$. The following equation illustrates a Bloom filter's false positive rate (FPR):

$$R(n, m) \approx e^{-n/m}. \tag{11.3}$$

The target FPR is set to F_t. Solving the equation of $F_t = e^{-n/m}$ in relation to m shows:

$$m = \left\lfloor -\frac{n}{ln(F_t)} \right\rfloor. \tag{11.4}$$

The pre-deployment stage sees Bloom filters created at the base station from all XTs potentially generated in the WCS. The following subsections describe the specific XT preparation. Before deployment, each Bloom filter is provided to each device. A unique seed, s_i, generates each Bloom filter v_i at device n_i. Moreover, using $h(s_i||\alpha)$ instead of $h(\alpha)$ as a hash value of α generates and checks the Bloom filter, v_i.

11.4.2.3 Report generation

The variable E denominates the incident's description. A device n_i creates a report as follows:

$$M_i = Enc(k_{i,u}, r_i)||E. \tag{11.5}$$

The variable $Enc(k_{i,u},r_i)$ represents encryption of the report, r_i, using key $k_{i,u}$. u and $||$, respectively, represent the leader device's ID in class G_w and concatenation. Subsequently, device n_i delivers M_i to n_u. The leader device extracts and collates all

$Enc(k_{i,u}, r_i)||E$ located in the detecting devices. Random selections of T reports occur when the leader device gathers more than T reports, represented by the term $M_{i_1}, ..., M_{i_T}$. The term $r_{i_1}, ..., r_{i_T}$ is also attained after the reports undergo decryption by device n_u, which creates the following:

$$R_w = r_{i_1} \oplus r_{i_2} ... \oplus r_{i_T}. \quad (11.6)$$

$\oplus$ denotes the XOR operation. The resulting R_w is the class's XT.

Let n_v denote the leader device's neighbor device. The report $E||Enc(K_{u,v}, R_w)$ is forwarded from n_u to n_v.

11.4.2.4 En-route filtering

When device n_j delivers the incident report $M = E||Enc(k_{i,j}, R_w)$ to device n_i, the decryption of $Enc(K_{i,j}, R_w)$ means device n_i can attain R_w. The device then employs the Bloom filter to determine the validity of the XT R_w. The report is dropped should the device deem the XT R_w to be a fake. On the contrary, when the device deems the XT R_w correct, the report goes to the next device, n_s.

11.4.2.5 Data regulated at the base station

Given that the incident report delivered to the base station collates all data about XTs, the base station can validate the XTs. Should it fail this validity assessment, a report is considered the work of a hacker.

11.4.3 Update of tokens and Bloom filters

The tokens employed to create an XT in a report require updating to deliver a correct report to the base station. This change must occur because of the possibility that the compromised device learned the XT from one of the report's forwarding devices. The base station regulates several data types to update the XTs.

11.4.3.1 The process at the base station

The base station manages the two tables shown below.

Table of tokens and XTs (Tx)

Here, *RS* represents the set of all XTs generatable in the WCS. The variable R_i constitutes an element of *RS*. The tokens $r_{i_1}, ..., r_{i_T}$ generate R_i. The Tx links the ID of the class apportioned R_i and R_i to the aforementioned T tokens.

Table of Bloom filters (Tb)

Pre-deployment, all devices n_i have a Bloom filter, v_i, assigned to them; the base station creates the latter. A random seed, s_i, is generated and used upon the base station creating the Bloom filter, v_i. The Tb connects a device n_i to each seed s_i and Bloom filter v_i.

11.4.3.2 Updating Bloom filters and token procedure

Upon delivery of a correct report to the base station with the R_w XT by class G_w, the base station creates the ID w of the class apportioned R_w and R_w as the base station ascertains the T tokens $(r_{i_1}, ..., r_{i_T})$ through Tx.

The base station creates novel T tokens and XT $R'_w = r'_{i_1} \oplus r'_{i_2} \oplus ... r'_{i_T}$ using pseudorandom number generators. The base station sends reports $M_j = Enc(k_j, r'_{i_j})$ $(j = 1, ..., T)$ to the devices of class G_w. The base station then sets R'_w and r'_{i_j} while removing R_w and r_{i_j} to update RT.

The new Bloom filter v' should replace the old Bloom filter v. However, compromised devices may decrypt the information and obtain v' if the base station sends v' as it stands, even if it has undergone encryption. Thus, the proposed algorithm only sends various parts of v and v' from the base station, which, in turn, delivers $v_i \oplus v'_i$, consisting of different aspects of v' and v, to the devices. For instance, in the incident that v = 0100100 and v' = 0100001, the data {4, 6} goes from the base station to the equivalent device. Equation (11.7) illustrates such a procedure. The subscript "(2)" outlines the value presented in a binary number, and m symbolizes a Bloom filter's bit length:

$$\left\{d \in [0, m-1] | ((v_i \oplus v'_i) \ll d) \& 10...0_{(2)} = 10...0_{(2)}\right\}. \tag{11.7}$$

The symbol & represents a bitwise AND, and $\ll$ stands for a left arithmetic shift.

11.5 Analysis

11.5.1 Hop counts are required until the devices identify fake incident reports

The required mean hop count undergoes evaluation when detecting a fake incident report. The calculation's first step involves clarifying the probability, p_1, of such a report in a device. No compromised device means $(1 - F_t)$ represents the target probability of recognizing a fake incident report in a device. Refining a Bloom filter's bit length can obtain an arbitrary value of F_t.

Compromised N_c devices mean hackers can obtain the N_c Bloom filters, meaning that RS elements become searchable. The variable m_l symbolizes how many elements are in RS. The number of potential T token combinations per class is qC_T because of the assignment of q tokens to each class and the selection of T tokens from the q tokens for generating an incident report.

The presence of g classes means the potential combinations of all classes' T tokens can be represented by

$$m_l = g \cdot {}_qC_T. \tag{11.8}$$

b represents an XT or a token's bit length. The potential representations of b bits total 2^b. Because the number of correct XTs is m_l, the compromised XTs total $2^b - m_l$. If N_c devices become compromised by a hacker, the attack can be said to have N_c Bloom filters. The likelihood of N_c Bloom filters failing to detect compromised XTs as compromised is $F_t^{N_c}$ because the FPR of each Bloom filter has a setting of F_t. Assume m_f totals the compromised XTs that a hacker may consider correct. The value is represented by

$$m_f = F_t^{N_c}(2^b - m_l). \tag{11.9}$$

Hackers can m_l correct XTs and m_f compromised XTs to be correct, meaning the likelihood of an attack resulting in a compromised XT is represented by $m_f/(m_l + m_f)$. All forwarding devices can identify compromised XTs with probability $(1 - F_t)$ if the device acquires a compromised XT. Thus, the chance that all forwarding devices can judge the received XT resulting from the attack is represented by

$$p_1 = \frac{m_f}{m_l + m_f} \times (1 - F_t). \tag{11.10}$$

In this scenario, p_h' signifies the total hops until the forwarding devices recognize a fake incident report, and H signifies the maximum hops to the base station. $(1 - p_1)^{i-1} \cdot p_1$ delineates the likelihood of the ith device identifying a fake incident report because $(i - 1)$ devices do not recognize it with probability $(1 - p_1)^{i-1}$ and the ith device recognizes it with probability p_1. Thus, the expected hop count is provided by

$$\begin{aligned} p_h'(H) &= \sum_{i=1}^{H} i \cdot (1 - p_1)^{i-1} \cdot p_1 \\ &= \frac{1 - (1 - p_1)^H}{p_1}. \end{aligned} \tag{11.11}$$

11.5.2 The amount of traffic generated per class in an attack

The variable D signifies the most times an incident can occur within the same class. The incident data's bit length is $|E|$, and XT's bit length is b. D incident reports are shown by

$$Q = (|E| + b) \cdot D. \tag{11.12}$$

11.5.3 The amount of communication generated by correct incident reports

Reports about updated tokens from the base station require delivery by the devices to the base station in conjunction with incident reports that align with the suggested method. Should a device in a class identify an incident, the leader device receives the incident data E and its token from q devices. The leader device then creates an incident report for delivery to the base station. In the incident of the hop count being h, the amount for creating and delivering the incident report to the base station is provided by

$$LT = (|E| + b)(q + h). \tag{11.13}$$

Variable Λ symbolizes the anticipated number of bits in a Bloom filter requiring alteration upon the base station changing an *RS* element. The variables m and n signify the Bloom filters' bit length and the total elements, respectively.

Suppose that S symbolizes a set with $(n-1)$ elements, S_α denotes $S \cup \{\alpha\}$, S_β shows $S \cup \{\beta\}$, and $S_{\alpha\beta}$ indicates $S \cup \{\alpha, \beta\}$. In this scenario, $\alpha \notin S$ and $\beta \notin S$. $H[S]$ represents a Bloom filter generated from set S.

Suppose that α symbolizes the deleting element, and β shows the new element. The following five cases would require consideration:

1. $H[S] = H[S_\alpha] = H[S_\beta]$
2. $H[S] = H[S_\alpha]$ and $H[S] \neq H[S_\beta]$
3. $H[S] \neq H[S_\alpha]$ and $H[S] = H[S_\beta]$
4. $H[S] \neq H[S_\alpha]$, $H[S] \neq H[S_\beta]$ and $H[S_\alpha] = H[S_\beta]$
5. $H[S] \neq H[S_\alpha]$, $H[S] \neq H[S_\beta]$ and $H[S_\alpha] \neq H[S_\beta]$

The bits in the Bloom filter requiring alteration when the base station changes one element of the RS differ in all cases. In case 1, no bits need changing because deleting α and the addition of β fail to impact the Bloom filter. Additionally, the number of bits requiring variation in cases 2, 3, 4, and 5 is 1, 1, 0, and 2, respectively.

$R(n, m)$ symbolizes the FPR of a Bloom filter with a bit length m and number of elements n. Thus, the likelihood is that $H[S] = H[S_\alpha]$ is illustrated by $R(n-1, m)$ due to the characterization of the compromised positive of $H[S]$ against $\alpha \notin S$. This development means the likelihood of each case occurring is shown as follows:

1. $R(n-1, m)R(n-1, m)$
2. $R(n-1, m)(1 - R(n-1, m))$
3. $(1 - R(n-1, m))R(n-1, m)$
4. $(1 - R(n-1, m))(1/m)$
5. $(1 - R(n-1, m))\left(1 - R(n-1, m) - \frac{1}{m}\right)$

The item $(1/m)$ in case 4 signifies that the likelihood of $h(\alpha) = h(\beta)$. Hash function h's output bit length is m_f.

Thus, this expression emerges:

$$\Lambda(n, m) = R(n-1, m)(1 - R(n-1, m)) + (1 - R(n-1, m))\left(R(n-1, m) + 2 * \left(1 - R(n-1, m) - \frac{1}{m}\right)\right)$$

$$= e^{\frac{1-n}{m}}\left(1 - e^{\frac{1-n}{m}}\right) + \frac{e^{\frac{1-2n}{m}}\left(e^{\frac{n}{m}}(-2 + m) + e^{\frac{1}{m}}m\right)}{m} = \frac{2e^{\frac{1-n}{m}}(m-1)}{m}. \tag{11.14}$$

Equation (11.4) determines a Bloom filter's bit length, m. The term m shows the necessary number of bits to express the Bloom filter's arbitrary bit location. The terms N and $\log_2 N$ show the number of sensor devices and bits to personify a device, respectively. Resultantly, the following expression shows the requisite communication traffic quantity to refine Bloom filters and tokens:

$$ET = (N \cdot \Lambda(m_l, m)m + bT) \cdot N \quad) \cdot H. \tag{11.15}$$

11.5.4 Energy consumption

The reception, transmission, and power-down phases saw the leaked current amount to 16, 18, and 0.01 mA, respectively, with mica2 Berkeley motes acting as sensor devices [32]. These sensor devices may not be the most current, but their reliable performance levels mean they still have a role to play in research [33–35]. It is assumed that a voltage of 3 V and a bit rate of 19.2 kbps represent typical figures for the two measurement units.

$LT + ET$ symbolizes the bits required to update XTs and send an incident report. The bit rate of 19.2 kbps means the necessary time to treat $(LT + ET)$ bits amounts to $(LT + ET)/(19.2 \cdot 1{,}000)$.

In addition, since sending and receiving reports necessitates $(16 + 18)$ mA and the voltage amounts to 3V, the necessary energy to send an incident report and update XTs is provided by $3(LT + ET)/(19.2 \cdot 1{,}000)$. Here, $3(LT + ET)/(19.2 \cdot 1{,}000)v$ denotes the consumed energy when v incident reports emerge.

Even if no incidents occur, sensor devices still consume energy. There are N sensor devices, with each device requiring 0.01 mA. Thus, $N \cdot 3 \cdot 0.01 \cdot 3{,}600 \cdot 24 \cdot 30$ represents the monthly consumed energy in the WCS, where $3{,}600 \cdot 24 \cdot 30$ symbolizes the total seconds in each month.

The following expression delineates the monthly energy consumption of each sensor device when v incidents take place, and there are N sensor devices:

$$E_I(v) = \frac{3(LT + ET)(18 + 16)}{19.2 \cdot 1{,}000} v + N \cdot 3 \cdot 0.01 \cdot 3{,}600 \cdot 24 \cdot 30 \,[\text{mJ}].$$

11.6 Evaluation

11.6.1 Parameter selection

These experiments involve default parameter values set as $D = 1$, $H = 50$, $T = 5$, $|E| = 64$, $g = 1{,}000$, $q = 10$, and $N = 10{,}000$. They were identified as follows.

The assumption is that the target incidents do not regularly take place. Therefore, as a default value, D was set to 1.

H represents the hop counts from a device to the base station, the value of which changes based on the WCS's organization. In many studies, this value was set to roughly 50. Therefore, referring to these pieces of research, this study set H to 50. The hop count to the base station does not significantly impact the assessment because each approach (other than SEF), such as the suggested method, identifies the fake incident report within one or two hops.

In such a situation, T or more devices should identify the same incidents. Therefore, should the T values prove too large, the capacity to identify correct incidents lessens. Conversely, too-small T values increase attacks' success rates. This balance requires consideration. Numerous studies [3] have presumed a default value of $T = 5$, a parameter followed during these experiments, wherein the T values ranged from 1 to 10.

Circumstances dictate the total bits necessary to represent an incident's contents. A WCS with no security mechanisms necessitates roughly 40 bytes to signify a detected incident [13]. This study set the $|E|$ value to 64 bytes based on such a value and the assumption that further data, such as in-depth sensing information, could bolster the research.

In this instance, N symbolizes the total devices in a WCS. In [4], the default N value stood at 10,000. This study adopted the same approach by setting $N = 10{,}000$ as the default value, with the N values ranging from 1,000 to 10,000.

11.6.2 Evaluation results

The FPR_t and the number of compromised devices, N_c, influence the detection rate. Figure 11.6 shows the results. This study featured compromised devices picked at random from the network, which is a common approach for detecting methods' performance levels [3].

The detection rate approximately matches the F_t value. However, when Nc amounts to roughly 100 and F_t is set to 0.5, the detection rate proves lower than the F_t value. However, the unlikelihood of a scenario involving more than 100 devices becoming compromised without the WCS administrator's awareness means F_t can be set to 0.5.

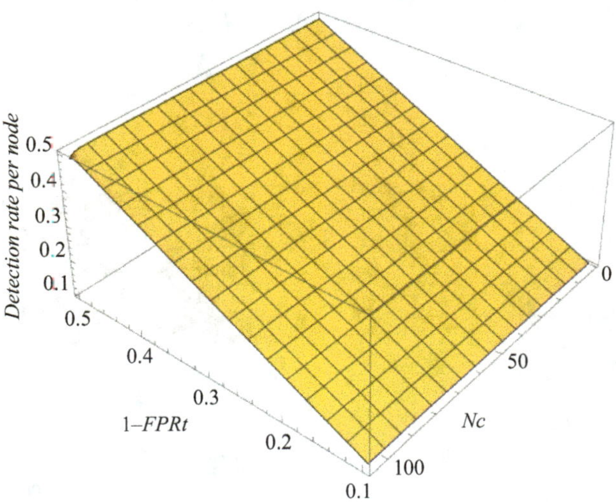

Figure 11.6 Detection rate of a device

The detection rate stands at approximately $Nc = 100$ and $F_t = 0.5$ because the hacker can calculate the number of correct XTs by compromising multiple sensor devices. This development allows the hacker to examine the acquired Bloom filters regardless of the randomly generated tokens' legitimacy. Large Nc means the hacker can examine more Bloom filters. Moreover, for a small F_t, the value increases the opportunities to judge the Bloom filters. Thus, large Nc and small F_t increase the possibility of an attacker generating correct XTs.

Figure 11.7 illustrates the detection rate per device with different T and g values. In this study, $N = 10{,}000$. The detection rate is notably low when g and T stand at roughly 10. The symbol g symbolizes the number of classes; in this instance, each class has 1,000 devices. If each class has multiple sensor devices and T is large, the number of elements in RS is significant, based on (11.8). Consequently, the detection rate lowers, as per (11.10). Thus, this parameter setting is not worth considering. This study indicates that the number of sensor devices in each class varies from 10 to 20. In other words, if N is 10,000, g should range from 500 to 1,000.

Figure 11.8 shows that the estimated hop count before a fake incident report is identified. The higher the F_t value, the higher the detection rate and the smaller the necessary total of hops pre-detection; when F_t is 0.5, the average hop count amounts to 2, suggesting a suitably high detection rate.

This study compared [3,5,6,16] with the suggested method. Figure 11.9 illustrates the findings. When $Nc > T$, the [3]s security mechanism was found to be inoperative, as the hop count totaled 50. However, given the sizable authentication data given to all reports, [6,16] required a maximum of one hop to identify fake incident reports. The findings also showed that a large hop count is required until devices can identify compromised [3] reports due to [3]s security mechanism breaking down when more than T devices are compromised. Conversely, other approaches, including the suggested method, can identify fake incident reports even when more than T devices are jeopardized.

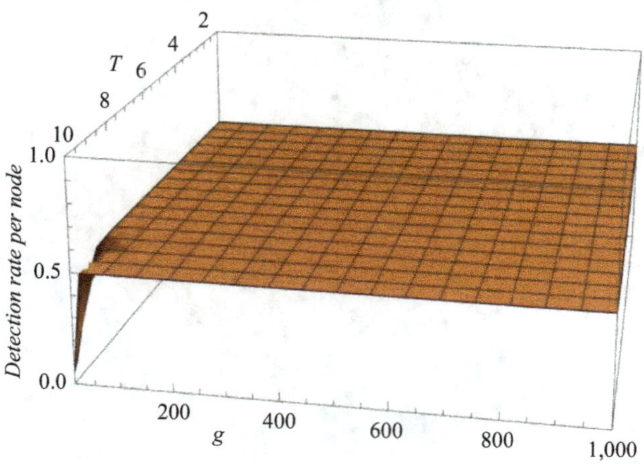

Figure 11.7 Detection rate of a device

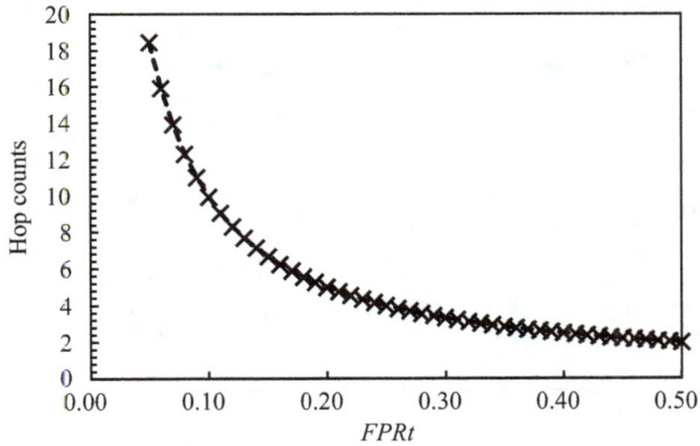

Figure 11.8 Hop count until fake incident reports are detected

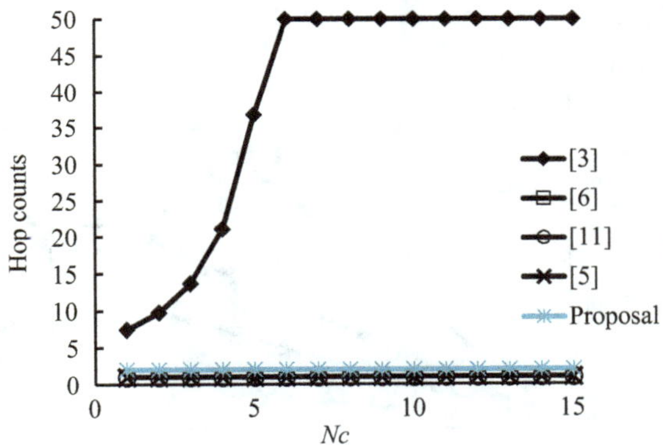

Figure 11.9 Hop counts until fake incident reports are detected

The sturdy security measures of [5,6,16] mean only one hop is required to identify a fake incident report. Moreover, because the detection rate of the suggested rate stands at 0.5 (a value that equals $1 - F_t$), the average hop count in the proposed method amounts to 2. Although the proposed method raises the necessary number to identify a fake incident report by a hop, the study's findings indicate that it lessens the traffic resulting from an attack.

If no devices were compromised, each device's mean energy consumption was calculated. Figure 11.10 shows the findings. According to related research, the suggested method necessitates increased energy expenditure. Regardless, according to studies, even if ten incidents took place each month, the rate increased only negligibly, from 77.8 to 78.3. This study presumed that no compromises would occur. The findings show the required traffic to run the WCS during typical procedures: the traffic per incident report was the smallest in the suggested method, necessitating more traffic to update XTs and Bloom filters. However, despite such

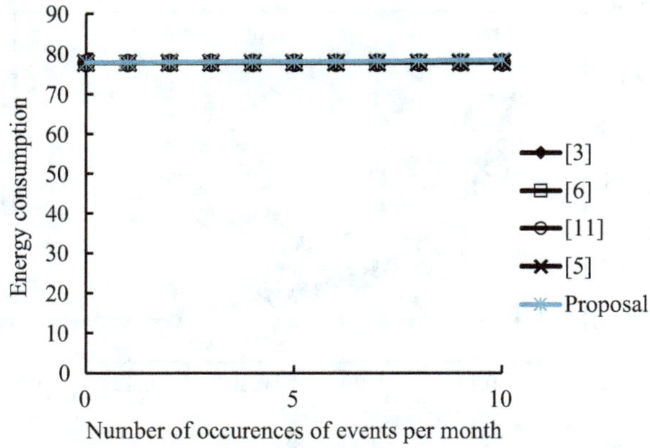

Figure 11.10 Amount of energy consumed per month

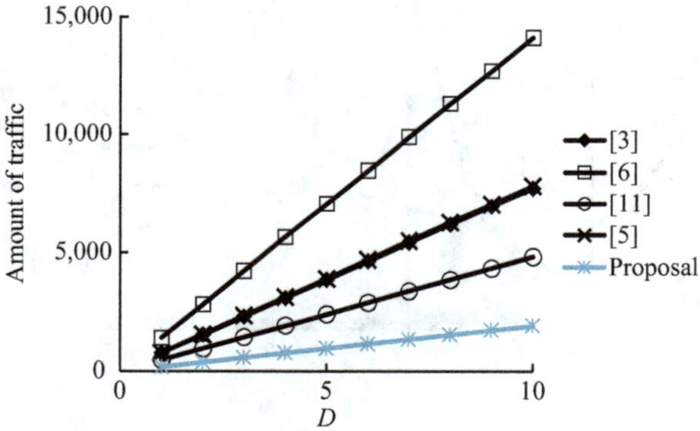

Figure 11.11 Amount of traffic a compromised device can generate per second within a class (varying D)

differences in the procedural methods, little difference exists in the amount of energy consumed by the WCS because of the comparatively large amount of energy used by the devices during routine operations.

The final step involved an analysis of the suggested method [3, 6] to determine how much traffic a hacker could generate in each class. Figure 11.11 shows the results. Since the report in [3,6] could feature T MACs, a hacker could create long reports, resulting in significantly more traffic than the suggested approach. Conversely, the suggested method involved a lightweight authentication mechanism with minimal traffic from compromised devices. Research has shown that incident reports need T MACs or T key IDs to authenticate reports, meaning a significant amount of traffic. However, the method in this study only needed one XT for verification. Such a mechanism can substantially lessen fake incident report traffic. Consequently, in comparison to the existing approaches, the suggested method provides the lowest maximum amount of traffic to a hacker.

Figure 11.9 illustrates the detection rate, indicating the superior performance of [6]. However, Figure 11.11 also reveals the amount of traffic, an area where [6] performs less well.

Other experiments have involved different *T* and *N*, as shown in Figures 11.12 and 11.13. Each method shows a reciprocal increase between the traffic and *T*. The findings also show that the value does not impact the suggested method's traffic due to *T* tokens being mapped to a sole XT. The *T* value does not affect such an XT's bit length. Equally, the *N* value does not noticeably influence the methods. Although a method may need device ID data, the device ID's bit length only tends to rise on a log scale in instances of increased device numbers.

According to these findings, the traffic potentially generated by a hacker in the suggested method is the lowest.

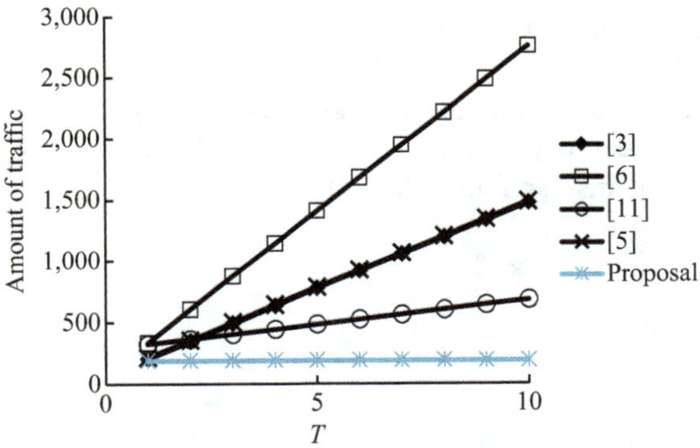

Figure 11.12 Amount of traffic a compromised device can generate per second within a class (varying T*)*

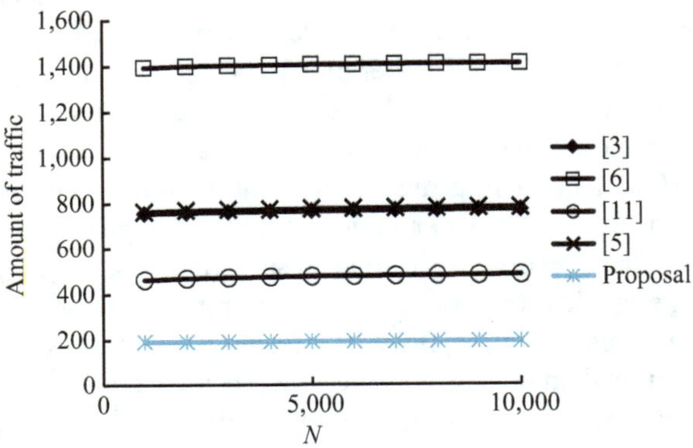

Figure 11.13 Amount of traffic a compromised device can generate per second within a class (varying N*)*

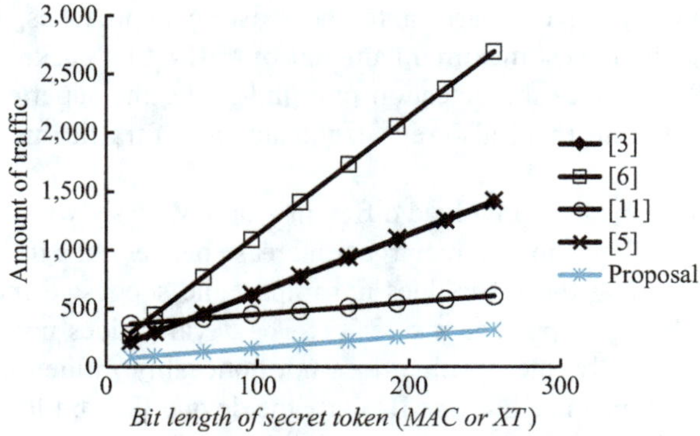

Figure 11.14 Amount of traffic a compromised device can generate per second within a class (varying the bit length of the secret token)

Finally, the impact of MAC and XT bit lengths on traffic volume was evaluated (Figure 11.14.) The longer the MAC or XT bit length, the more secure the system is. Therefore, these bit lengths should be determined after a preliminary investigation of the required security by the system administrator. Regardless of the value of these bit lengths, the figure shows that the proposed method has the lowest amount of traffic.

11.7 Discussion

Symmetric key encryption schemes offer a lower computation cost [36]. Moreover, calculating hash values can involve less computation [37]. In terms of XOR operation and assessing Bloom filters' specified bits, computation does not involve significant costs due to the simple bit operations. Thus, the sensor devices' computation costs in the suggested method prove minimal.

The sensor devices require the base station to generate seeds and tokens before employing XOR to create XTs. This process also demands the generation of Bloom filters for all sensor devices. The number of XTs is $g\ qC_T$. Thus, large g or q values require the base station to generate multiple XTs. However, since generating an XT does not need complicated XOR, the process is not a long one.

Figures 11.10 and 11.11 illustrate the amount of traffic: the former shows no compromised devices, and the latter indicates that a compromised device creates numerous fake incident reports. According to such findings, the suggested method can withstand fake incident report attacks by compromised devices, although when there are no compromised devices, the traffic matches the existing method's traffic.

The suggested method presumes that correct incidents do not occur frequently. The communication overhead becomes non-negligible if multiple correct incidents take place due to the tokens of several sensor devices requiring an update whenever the base station takes delivery of a correct report. Evidence suggests that hackers need a congested network to impede the delivery of correct incident reports to the base station. Numerous correct incidents create challenges for hackers seeking to attack such incidents. Thus, it can be assumed that a scenario involving such an attack would feature a small number of correct incident reports.

The maximum number of reports produced by a hacker mirrors the number of reports received by correct devices. The WCS application also means that the WCS manager influences how many reports devices can generate. For instance, employing the WCS in infrequent contexts, such as detecting crime, places an upper limit on the number of generatable reports, and there are no normal device operation restrictions.

11.8 Conclusion

In high-density 6G wireless communication networks, there is a need to control network congestion. In this chapter, we focus on burst attacks with a large number of infringement reports. This study outlined an algorithm to identify fake incident reports in a large WCS with a small number of hops. In contrast to other research, the incident reports had a single, one-time XT assigned to them for verification in the suggested method. This approach can substantially reduce the quantity of report traffic. Current research needs T MACs, T key IDs, or additional data for verification, and hackers can create substantial traffic with a single fake incident report. Equally, although current approaches can identify fake incident reports in a single hop, the suggested method's default setting would require two hops to achieve the same result. However, the suggested method can substantially cut traffic, which, in turn, reduces the chances of network congestion attacks. Studies have found that the suggested method can cut traffic volume by upwards of 60%. Therefore, future research should conduct thorough testing involving hundreds of sensor devices in order to obtain more comprehensive results.

Acknowledgment

The work of Agbotiname Lucky Imoize is supported in part by the Nigerian Petroleum Technology Development Fund (PTDF) and, in part, by the German Academic Exchange Service (DAAD) through the Nigerian–German Postgraduate Program under Grant 57473408.

References

[1] X. Zou, L. Li, H. Du, and L. Zhou, "Intelligent sensing and computing in wireless sensor networks for multiple target tracking," *J. Sensors*, vol. 2022, no. 2870314, pp. 1–11, 2022, doi: 10.1155/2022/2870314.

[2] Y. Wang, Y. X. Liu, S. J. Zhu, X. F. Gao, and C. Tian, "Approximation designs for energy harvesting relay deployment in wireless sensor networks," *J. Comput. Sci. Technol.*, vol. 37, no. 4, pp. 779–796, 2022, doi: 10.1007/S11390-022-1964-5.

[3] F. Ye, H. Luo, S. Lu, and L. Zhang, "Statistical en-route filtering of injected false data in sensor networks," *Proc. IEEE INFOCOM*, vol. 4, pp. 2446–2457, 2004, doi: 10.1109/INFCOM.2004.1354666.

[4] J. Wang, Z. Liu, S. Zhang, and X. Zhang, "Defending collaborative false data injection attacks in wireless sensor networks," *Inf. Sci. (NY).*, vol. 254, pp. 39–53, 2014.

[5] Y. Sei and A. Ohsuga, "False event detection for mobile sinks in wireless sensor networks," in *Proceedings of the European Intelligence and Security Informatics Conference (EISIC)*, 2013, pp. 52–59, doi: 10.1109/EISIC.2013.15.

[6] A. Kumar, N. Bansal, and A. R. Pais, "A partial key pre-distribution based en-route filtering scheme for wireless sensor networks," *J. Ambient Intell. Humaniz. Comput.*, vol. 12, no. 1, pp. 1471–1486, 2021, doi: 10.1007/S12652-020-02216-3/FIGURES/6.

[7] Y. Sei and A. Ohsuga, "False event message detection robust to burst attacks in wireless sensor networks," *IEEE Open J. Commun. Soc.*, vol. 3, pp. 1630–1642, 2022, doi: 10.1109/OJCOMS.2022.3208088.

[8] M. V. Babu, J. A. Alzubi, R. Sekaran, R. Patan, M. Ramachandran, and D. Gupta, "An improved IDAF-FIT clustering based ASLPP-RR routing with secure data aggregation in wireless sensor network," *Mob. Networks Appl.*, vol. 26, no. 3, pp. 1059–1067, 2021, doi: 10.1007/S11036-020-01664-7/FIGURES/7.

[9] Y. Wang, F. Li, P. Ren, S. Yu, and Y. Sun, "A secure aggregation routing protocol with authentication and energy conservation," *Trans. Emerg. Telecommun. Technol.*, vol. 33, no. 1, p. e4387, 2022, doi: 10.1002/ETT.4387.

[10] X. Liu, J. Yu, K. Yu, G. Wang, and X. Feng, "Trust secure data aggregation in WSN-based IIoT with single mobile sink," *Ad Hoc Networks*, vol. 136, p. 102956, 2022, doi: 10.1016/J.ADHOC.2022.102956.

[11] Z. Yu and Y. Guan, "A dynamic en-route scheme for filtering false data injection in wireless sensor networks," *Proc. IEEE INFOCOM*, pp. 1–12, 2006, doi: 10.1109/INFOCOM.2006.304.

[12] S. Ruj, A. Nayak, and I. Stojmenovic, "Pairwise and triple key distribution in wireless sensor networks with applications," *IEEE Trans. Comput.*, vol. 62, no. 11, pp. 2224–2237, 2013, doi: 10.1109/TC.2012.138.

[13] M. T. Hansen, "Asynchronous group key distribution on top of the cc2420 security mechanisms for sensor networks," in *Proceedings of ACM WiSec*, 2009, pp. 13–20.

[14] C. Gezer, M. Niccolini, and C. Buratti, "An IEEE 802.15.4/ZigBee based wireless sensor network for energy efficient buildings," in *Proceedings of IEEE WiMob*, 2010, pp. 486–491.

[15] C. Pedroso and A. Santos, "Dissemination control in dynamic data clustering for dense IIoT against false data injection attack," *Int. J. Netw. Manag.*, e2201, pp. 1–26, 2022, doi: 10.1002/NEM.2201.

[16] C. Yi, "En-route message authentication scheme for filtering false data in WSNs," *Secur. Commun. Netw.*, vol. 2021, no. 4068507, p. 1018, 2021, doi: 10.1155/2021/4068507.

[17] A. Grover, R. Mohan Kumar, M. Angurala, M. Singh, A. Sheetal, and R. Maheswar, "Rate aware congestion control mechanism for wireless sensor networks," *Alexandria Eng. J.*, vol. 61, no. 6, pp. 4765–4777, 2022, doi: 10.1016/J.AEJ.2021.10.032.

[18] A. Revathi and S. G. Santhi, "Gateway-based congestion avoidance using two-hop node in wireless sensor networks," *Lect. Notes Data Eng. Commun. Technol.*, vol. 126, pp. 17–32, 2022, doi: 10.1007/978-981-19-2069-1_2/COVER.

[19] M. Saska, T. Krajnik, and L. Pfeucil, "Cooperative μUAV-UGV autonomous indoor surveillance," in *International Multi-Conference on Systems, Signals & Devices*, March 2012, pp. 1–6, doi: 10.1109/SSD.2012.6198051.

[20] I. B. F. De Almeida, M. Chafii, A. Nimr, and G. Fettweis, "Blind transmitter localization in wireless sensor networks: a deep learning approach," in *Proceedings of IEEE International Symposium on Personal, Indoor and Mobile Radio Communications (PIMRC)*, 2021, pp. 1241–1247, doi: 10.1109/PIMRC50174.2021.9569361.

[21] N. Bacanin, M. Antonijevic, T. Bezdan, M. Zivkovic, and T. A. Rashid, "Wireless sensor networks localization by improved whale optimization algorithm," in *Proceedings of International Conference on Artificial Intelligence: Advances and Applications*, 2022, pp. 769–783, doi: 10.1007/978-981-16-6332-1_62.

[22] M. Tanveer, G. Abbas, Z. H. Abbas, M. Bilal, A. Mukherjee, and K. S. Kwak, "LAKE-6SH: lightweight user authenticated key exchange for 6LoWPAN-based smart homes," *IEEE Internet Things J.*, vol. 9, no. 4, pp. 2578–2591, 2022, doi: 10.1109/JIOT.2021.3085595.

[23] M. Alizadeh, M. H. Tadayon, and A. Jolfaei, "Secure ticket-based authentication method for IoT applications," *Digit. Commun. Networks*, 2022, doi: 10.1016/J.DCAN.2021.11.003.

[24] P. Wang, F. Xue, H. Li, Z. Cui, L. Xie, and J. Chen, "A multi-objective DV-Hop localization algorithm based on NSGA-II in Internet of Things," *Mathematics*, vol. 7, no. 2, pp. 184, 2019, doi: 10.3390/MATH7020184.

[25] Y. Jin, L. Zhou, L. Zhang, Z. Hu, and J. Han, "A novel range-free node localization method for wireless sensor networks," *IEEE Wirel. Commun. Lett.*, vol. 11, no. 4, pp. 688–692, 2022, doi: 10.1109/LWC.2021.3140063.

[26] N. Verma, A. Kaushik, and P. Nayak, "A lightweight secure authentication protocol for wireless sensor networks," in *Proceedings of International Conference on Innovative Computing and Communications*, 2021, vol. 1165, pp. 291–299, doi: 10.1007/978-981-15-5113-0_21.

[27] D. Liu and P. Ning, "Multilevel μTESLA: broadcast authentication for distributed sensor networks," *ACM Trans. Embed. Comput. Syst.*, vol. 3, no. 4, pp. 800–836, 2004, doi: 10.1145/1027794.1027800.

[28] Y. Sei and S. Honiden, "Distributed detection of node replication attacks resilient to many compromised nodes in wireless sensor networks," in *Proceedings of 4th Annual International Conference on Wireless Internet (WICON)*, November 2008, pp. 1–8.

[29] N. Elsakaan and K. Amroun, "Distributed and reliable leader election framework for wireless sensor network (DRLEF)," *Lect. Notes Networks Syst.*, vol. 378 LNNS, pp. 123–141, 2022, doi: 10.1007/978-3-030-95918-0_13.

[30] S. Wang, X. Jiang, and H. Wymeersch, "Cooperative localization in wireless sensor networks with AOA measurements," *IEEE Trans. Wirel. Commun.*, vol. 21, pp. 1–13, 2022, doi: 10.1109/TWC.2022.3152426.

[31] B. H. Bloom, "Space/time trade-offs in hash coding with allowable errors," *Commun. ACM*, vol. 13, no. 7, pp. 422–426, 1970.

[32] G. Anastasi, A. Falchi, A. Passarella, M. Conti, and E. Gregori, "Performance measurements of motes sensor networks," in *Proceedings of the 7th ACM International Symposium on Modeling, Analysis and Simulation of Wireless and Mobile Systems (MSWiM)*, October 2004, p. 174, doi: 10.1145/1023663.1023695.

[33] S. R. Jondhale, R. Maheswar, and J. Lloret, "Fundamentals of wireless sensor networks," in *Received Signal Strength Based Target Localization and Tracking Using Wireless Sensor Networks*, 2022, pp. 1–19.

[34] J. Singh, R. Kaur, and D. Singh, "Energy harvesting in wireless sensor networks: a taxonomic survey," *Int. J. Energy Res.*, vol. 45, no. 1, pp. 118–140, 2021, doi: 10.1002/ER.5816.

[35] A. W. Bhat and A. Passi, "Wireless sensor network motes: a comparative study," in *Proceedings of the 9th International Conference on Computing for Sustainable Global Development (INDIACom)*, 2022, pp. 141–144, doi: 10.23919/INDIACOM54597.2022.9763269.

[36] G. De Meulenaer, F. Gosset, F. X. Standaert, and O. Pereira, "On the energy cost of communication and cryptography in wireless sensor networks," in *Proceedings of the IEEE WiMob*, 2008, pp. 580–585, doi: 10.1109/WIMOB.2008.16.

[37] B. Qin, R. H. Deng, S. Liu, and S. Ma, "Attribute-based encryption with efficient verifiable outsourced decryption," *IEEE Trans. Inf. Forensics Secur.*, vol. 10, no. 7, pp. 1384–1393, 2015, doi: 10.1109/TIFS.2015.2410137.

Chapter 12

A real-time intrusion detection system for service availability in cloud computing environments

Kolawole Abubakar Sadiq[1], *Aderonke Favour-Bethy Thompson*[2], *Olaniyi Abiodun Ayeni*[2] *and Gabriel Junior Arome*[2]

Abstract

The spike in Internet usage and outsourcing of computing needs, such as databases, networking, storage, among others, to third parties, also known as cloud computing, poses a significant security threat to cloud users due to the cloud deployment medium, the Internet. The Internet exposes cloud users' data confidentiality, integrity, and availability to cybercriminals, who gather cloud users' personal information for illicit activities or sometimes make the cloud service unavailable for legitimate users. The intrusion detection system (IDS) is a prominent second-line approach for monitoring illicit activities like distributed denial of service attacks (DDoS) over cloud communication networks. However, it faces challenges in areas of false alarm, detection time and accuracy, primarily attributed to the enormous amount of attributes the machine learning (ML) algorithm needs to process within a short period. Feature selection (FS) using statistical and meta-heuristic algorithms is a promising method to overcome the IDS challenges. This chapter explores the binarization of the continuous data in the UNSW_NB15 network attack dataset to enhance the efficiency of the ML algorithms. Also, the work optimizes the statistical FS method, maximum relevance, and minimum redundancy (MrMr) with a nature-inspired algorithm known as Cuckoo search. The experimental evaluation of the proposed algorithms was done using Python IDLE 3.7.1. Various performance metrics, like detection time, false alarm, and accuracy, using the confusion matrix were obtained from four selected algorithms: K-nearest neighbor (KNN), logistic regression (LR), decision tree (DT), and multi-layer perceptron (MLP). Among all the four algorithms, DT produced the best result with an accuracy of 96%, a precision of 96% and 97% (training and testing), and recall

[1]Department of Computer Science, Kwara State Polytechnic, Nigeria
[2]Department of Cyber Security Science, Federal University of Technology Akure, Nigeria

scores of 96% and 97% (training and testing). A detection time of 1.60 s was obtained, making the model the most suitable among the four algorithms for real-time IDS. The model obtained an improved accuracy and detection time compared with the selected works of literature.

Keywords: Cuckoo search; Intrusion detection system; Feature selection; K-nearest neighbor; Logistic regression; Decision tree; Multi-layer perceptron

12.1 Introduction

Cloud computing has been unanimously accepted in the global IT market because of its bi-overlay benefits (low capital and operational expenditures) over conventional client/server networks [1,2]. Cloud computing distinguishes itself from the conventional client/server architecture with five distinct features: measure service, on-demand self-service, rapid elasticity, broad access network, and resource pooling [3,4]. Cloud technology integrates with existing paradigms such as virtualization, grid computing, web 2.0, distributed computing, and many more, which are accessible through the Internet to provide on-demand, scalable, reliable resources to prospective users on pay-as-you-use or free for limited features [5]. Cloud model primary drivers are cloud service providers (CSP), responsible for providing capital and operational expenditures needed for smooth cloud operations, and cloud users (CU), individuals or organizations that subscribe to cloud services. Among the prominent cloud services are software-as-a-service (SaaS), platform-as-a-service (PaaS), and infrastructure-as-a-service (IaaS). At the same time, the cloud deployment models available are private, public, community, and hybrid clouds [6,7]. Figure 12.1 indicates the different services of the cloud network, the deployment strategies, and the unique characteristics of cloud computing.

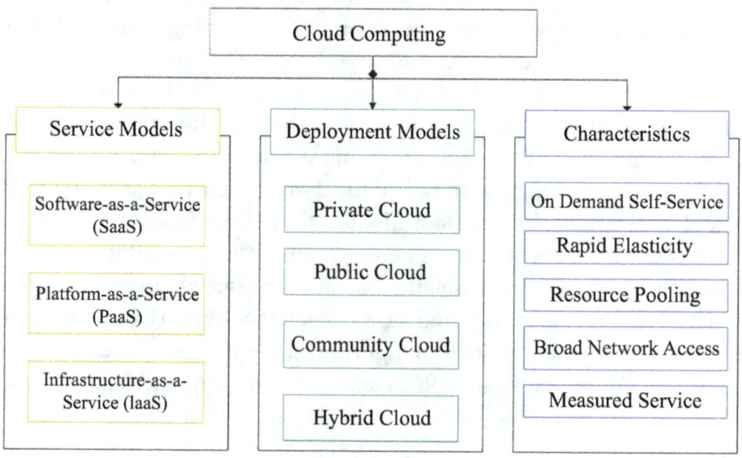

Figure 12.1 Cloud computing service, deployment models, and characteristics

Despite the benefits provided by cloud computing, some cloud features like resource pooling, broad network access, and rapid elasticity expose the CU data confidentiality, integrity, and availability (CIA) to cyber criminals. The resource pooling uses the multi-tenant model for better hardware utilization, where single hardware such as storage, processing, memory, and network bandwidth can serve millions of CU, thereby threatening the CU data confidentiality [8,9]. The broad network access allows thin client devices from any location to access the cloud networks, which makes the CU data vulnerable to unauthorized deletion, modification, and falsification. Rapid elasticity allows CU to scale upward and downward computing resources when needed and can threaten cloud availability through economic denial of sustainability (EDoS) or distributed denial of service (DDoS) attacks [1,8].

Cloud service availability remains the most prominent issue in the cyber security field among all the CIA threats. Many researchers identify DDoS as one of the most challenging attacks on cloud computing service availability. DDoS prevents legitimate CU from accessing the cloud resources or degrading the cloud services with unsolicited requests, which overwhelm the cloud network resources such as memory, processing, network bandwidth, and database pool [10,11]. The attacker uses botnets, which consist of masters, handlers and agents, to perpetrate illicit acts. The master, through the handlers, recruits vulnerable devices called agents and communication anonymously, making the location of the master challenging to detect. The handler takes advantage of the agent's vulnerabilities, such as hardware weakness, outdated patches, and protocol defects, to install a malicious program, eventually making the agent act like a zombie and obeys directives given by the master through the handlers [10]. Different incentives motivate DDoS attackers ranging from revenge, cyber espionage, finance, and intellectual contest [12]. Many works of the literature suggested different DDoS preventive and detection measures like packet filtering (IG) [10,13] that filters all ingress and egress packets, and intrusion detection system (IDS) [14–16], which checks for actions that violate the security policies. The IDS can detect known, unknown, or both attacks through its available variants known as signature-based and anomaly-based [17]. Among other methods presented across numerous literature are software-defined network (SDN) [18,19], which allows global configuring and monitoring of network traffic.

IDS remains the favorite defensive model across literature but faces time complexity and accuracy shortcomings [20–22]. Many researchers attribute the time complexity and accuracy challenges to the large dimensionality of the dataset and suggested feature selection (FS) as a solution to an effective IDS model [15,23,24]. Feature selection helps identify and remove noise or irrelevant attributes from the feature space. The classifications of FS are filter, wrapper, embedded, and hybrid [25,26], as shown in Figure 12.2.

The filter FS uses statistical techniques like chi-square, rough set, and information gain to select highly contributing attributes from the set using either forward select or backward elimination techniques [27]. Wrapper uses search algorithms like meta-heuristic or swarms intelligence algorithms for the iterative process until certain constraints are satisfied [28]. The hybrid combines the features of filter and

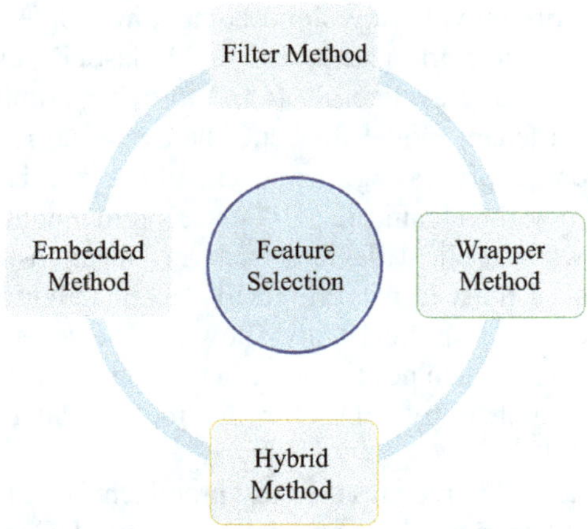

Figure 12.2 Feature selection methods

wrapper FS, while the embedded selects each subset during the training phase, though the embedded has a better predictive capability but increases the training time of the model [29].

This work develops a real-time IDS model for service availability in cloud computing using the hybrid FS to select the most informative attributes from the UNSW_NB15 dataset [30]. The proposed defensive models consist of preprocessing, model development, and model evaluation stages. The first stage performs data cleansing using normalization, discretization, and binarization. In the second stage, the cuckoo search proposed in [31] serves as the searching technique for selecting feature attributes to be evaluated randomly, and it is an iterative process. In contrast, the MrMr algorithm in [32] is the objective function for the cuckoo search algorithm evaluation process. The features with high gains are classified using four different algorithms: K-nearest neighbor (KNN), logistic regression (LR), decision tree (DT), and multi-layer perceptron (MLP) classifier model. Lastly, the IDS false alarm rate, accuracy, detection rate, and execution time were evaluated.

12.1.1 Key contributions of the chapter

The chapter contributions are as follows:

(i) The chapter uses binarization of continuous data to improve the efficiency and detection time of the model, as all continuous data were first converted to 0 and 1 before usage, making it easy for training and testing.
(ii) The work uses MrMr to eliminate redundant features rather than information gain (IG), which only ranks features but ignores redundant features evaluation.
(iii) It uses only 20 features to achieve better accuracy and false rate than existing methods in the literature.

12.1.2 Chapter organization

Section 12.2 present the related work on intrusion detection and defensive methods. Section 12.3 discuss the theoretical background of security issues. Section 12.4 presents the research methodology. Section 12.5 discusses the results. Section 12.6 presents the conclusion and feature scope.

12.2 Related work

The research [16] presents an anomaly detection system to detect novel attacks using binarization for preprocessing, information gain to select the most informative attributes from the UNSW_NB15 dataset and artificial neural network (ANN) multilayer perceptron (ANN-MLP) for the prediction. The ANN-MLP implementation was carried out in matrix laboratory 8.1 (MATLAB®) and attained an accuracy of 76.96% and an error of 23.95%. However, the work uses limited preprocessing techniques and does not consider faster classifiers suitable for real-time intrusion detection. In [33], the researchers implement anomaly-based network intrusion detection using binary-based particle swarm optimization (BPSO) to select the most informative feature attributes from the NSL-KDD dataset. Support vector machine (SVM) was adopted as the model, and the standard-based PSO (SPSO) serves as a turner for adjusting the SVM control parameters. The python environment was used to implement the model, and the results show an accuracy of 91.69% using SVM alone and 97.75% using BPSO and SVM. Lastly, 99.10% accuracy with BPSO, SPSO, and SVM. Also, the training and testing time obtained using BSPO, SPSO, and SVM is lower than SVM, BPSO, and SVM, respectively. However, the NSL-KDD data set lacks network attributes that address current network traffic sensitivity and comprehensive network traffic information. In [34], the research uses ensemble feature selection and classification techniques on three separate datasets: real-time honeypot, NSL-KDD, and Kyoto, to validate the model performance. The univariant ensemble filter feature selection (UEFFS) combines five statistical methods: information gain, chi-square, gain ratio, symmetric uncertainty, and relief in selecting informative attributes from the feature set. The selected features from the UEFFS are analyzed using ensemble classifiers: SVM, Naïve Bayes (NB), logistic regression (LR), and decision tree (DT). The majority voting serves as an ensemble combiner. The python language and Scikit-learning (ML library) were used for the experimental setup and evaluated different metrics like accuracy, recall, precision, *F*-measure, and area under the curve (AUC). Results indicate that the UEFFS-based model accuracy scores 96.62%, 99.93%, and 99.89% against 95.99%, 99.88%, and 99.90% without UEFFS using NSL_KDD, Kyoto, and real-time honeypot datasets, respectively. Also, the attack detection time using UEFFS was reduced with the NSL-KDD and Kycto. However, the UEFFS model did not positively impact the real-time honeypot dataset but degraded its performance. The UEFFS did not use any searching algorithm for the feature selection iterative process and lacked interaction with the ensemble classifier, as selected features are handpicked before being passed to the ensemble classifier.

In [35], the work uses IDS as the second line of defence using three filter feature selection methods: Pearson's correlation coefficient (PCC), Spearman's correlation coefficient (SCC), and Kendall's tau coefficient (KTC). Also, different datasets, KDD Cup 99, NSL-KDD, and UNSW_NB15, were used to validate the dynamism of the model. The selected feature set was tested on different classifiers: SVM, NB, DT, KNN, and random forest (RF). The result compared the accuracy of each classifier with and without feature selection. The KNN and DT perform best among other classifiers in accuracy, but the KNN takes longer to classify than the DT. However, the report did not state the experimental setup that produced the results. The authors of [36] deploy filter and wrapper approaches to select relevant features from the feature set using the firefly algorithm at the wrapper and mutual information (MI) at the filter stage (MIFA). Two classifiers, C4.5 and Bayesian Network (BN), were used to validate the selected features on KDD CUP 99 dataset using three strategies; the first stage uses MI only, the second uses MIFA with C4.5 as an evaluator and lastly, MIFA with BN as an evaluator. A voting method selects the most frequent feature from the three strategies. The method improved the accuracy by 99.98% using MIFA C4.5 against 99.95% when using all features. However, the KDD CUP 99 is obsolete. In [37], the researchers hybridized filter and wrapper feature selections with correlation feature selection (CFS) to solve the problem of redundancy in IG and gain ratio (GR) while deploying three search algorithms at the wrapper: best-first, greedy stepwise, and genetic algorithm (GA). KDD CUP 99 and DARPA 1999 were used to validate the developed models. The first stage evaluates the FS with CFS, the second stage evaluates FS using filter and wrapper, and lastly, classify the selected features using RF. Results show that among the three search algorithms, GA performs best with an accuracy of 0.42% compared with other search algorithms. The filter FS also showed significant performance improvement but did exceptionally well when combined with the wrapper method. The research did not perform outlier detection operations.

The researchers in [38] identify FS as a method of enhancing the IDS performance by selecting optimal attributes from the feature set to reduce the IDS detection time and improve its accuracy. Two heuristic algorithms, the fruity algorithm fly (FFA) and ant lion optimizer (ALO), were hybridized and tested with KDD CUP 99, NSL-KDD, and UNSW-NB15 on MATLAB 2017a. The experimental setup consists of three states, FFA, ALO, and FFA-ALO. It uses four classifiers, SVM, KNN, NB, and DT, to check for matrices like accuracy, elapsed time, specificity, and sensitivity. The FFA-ALO reduced the number of features from 41 to 12, 16, and 15 in the three datasets, therefore reducing the central processing unit (CPU) time and memory usage of the IDS. However, the research did not consider other critical performance metrics, such as the false alarm rate. The authors of [39] use IDS as a monitoring tool to check malicious activities in communication networks. It suggests wrapper FS using tabu search and RF (TS-RF) to enhance the IDS performance. Tabu search performs the searching and weighting of each feature, while the RF is the classifier. The first phase compared TS-RF with other feature selection techniques, such as GR, chi-square (CS), and Pearson's correlation (PA), using the UNSW-NB15 dataset. In contrast, the second phase compared results from the TS-RF model with other results from relevant literature. Three performance matrices, accuracy, false positive rate, and the number of features, were used for the result analysis. TS-FS has the lowest false positive rate with fewer features but with slightly lower accuracy compared to other works of literature using Rule-based, GA with logistic regression (LR), hybrid [PS, ant colony optimization (ACO), GA]. The work did not handle the imbalance challenges of the UNSW-NB15 dataset, which significantly impacts the model classification accuracy.

In [40], the pigeon algorithm is proposed and tested across different datasets, KDD CUP 99, NSL KDD, and UNSW-NB 15, using DT as the classifier. The research suggests a new binarizing technic known as cosine pigeon-inspired optimizer (PIO) and compares it with the conventional sigmoid function used in many swarm intelligence algorithms. The PIO reduces the features from 41 to 7, 41 to 5, and 49 to 5, respectively. Results from the confusion matrix indicate that cosine PIO uses fewer features with better accuracy and detection time. Swarm intelligence algorithms mostly get trapped during the local search, and different literature suggested methods like levy flight, random walk, and uniform distribution to address this issue. However, the research did not mention controlling the PIO model's exploration and exploitation challenges. The authors of [41] compares two different FS types: filter and wrapper. The filter FS evaluates the NSL KDD dataset using IG, principal component analysis (PCA), and CFS, while the wrapper method uses GA, artificial bee colony (ABC), and PSO. The wrapper FS uses Python language for the evaluation, while the filter FS uses WEKA and test matrices like accuracy, training and testing time, and recall, among others. The wrapper FS shows an accuracy of 97.75%, 97.87%, and 98.04% against the filter FS with 95.76%, 92.64%, and 96.20%, respectively. Also, the wrapper FS uses less training and testing time than the filter FS. However, the work did not state the fitness function used for the selected wrapper algorithms.

The work in [42] selected relevant features using IG and two ensemble techniques, bagging and boosting, using a tree-based classifier, was used to evaluate FS selected from the two FS approaches mentioned earlier using WEKA. Results show that the IG FS using bagging ensemble learning with a J48 base classifier performs better than other methods, with an accuracy and a false alarm rate of 84.25% and 2.79%, respectively. However, the model's dynamism cannot be ascertained as the model was tested on just one dataset. The authors of [43] improve the performance of the IDS by introducing a feature ranking approach using IG filter FS and ensemble classifiers (KNN, random tree, J48 graft, and RF) on the NSL KDD dataset. The WEKA tool was used to compare the traditional classifiers with the ensemble model. The results show an accuracy of 99.72% compared with 98.07% from KNN using traditional classifiers. The approach uses filter FS known to be independent of classifiers, making it unsuitable for real-time attack detection.

In [44], the research implements CFS on the UNSW-NB15 dataset. Four classification algorithms, NB, RF, J48, and ZeroR, were used to evaluate the model's accuracy, false positive rate, recall, precision, and F-measure. The model proposes two clustering techniques, K-means and expectation maximization (EM), to cluster the dataset into normal and malicious traffic. Results indicate that RF and J48 perform more excellently than other classifiers, with an accuracy of 9.60% and 93.78%, respectively. Although the CFS enhanced the model's performance by 7.8%, the research did not address the imbalance challenges of the UNSW-NB15 dataset. Khorram and Baykan's [45] work uses PSO, ACO, and ABC to select the relevant features for detecting network attacks. KNN and SVM algorithms are used as classifiers to evaluate the performance of these feature selection algorithms. This study uses different metrics on the standard NSL-KDD dataset for training and testing in the WEKA tool. The result shows that PSO, ACO, and ABC algorithms perform best. The ABC feature selection provides a 98.9% accuracy rate and 0.78% false alarm with the KNN algorithm as the classifier, which is the best result among the examined algorithms. However, the NSL-KDD dataset lacks network attributes that address current network traffic sensitivity.

12.3 Theoretical background of security issues in cloud computing

Cloud services are only deployable through the network, typically the Internet, which makes the cloud viable for cybercrimes. Many security challenges that violate the cloud users' confidentiality, integrity, and availability have been recorded globally. This section covers relevant security issues and mitigation approaches related to cloud computing.

12.3.1 Cyber attacks

Cyber-attacks are unlawful acts by individuals or groups that violate or endanger a network infrastructure's hardware or software security policies [17]. Cyber-attack motives range from financial, fun, cyber-espionage, and intellectual contests [27,46]. The cyber-attacks can be insider attacks carried out by someone with authorized access within the network and motivated mainly by revenge or greed. In contrast, external attacks hire an insider or external criminals on a mission to cost organizations financial and reputational losses. The DDoS remains prominent among cyber-attacks and costs organizations between $20,000 and $40,000 per hour.

12.3.2 DDoS in cloud computing

A DDoS uses thin client device software or hardware vulnerabilities to install malware. The malware takes the device captive and acts like "Zombies" to perform illicit acts on the attacker's instructions [12]. The captive devices continuously send unrequested messages to the victim, eventually making the victim's services unavailable for legitimate users. Figure 12.3 shows a DDoS attack scenario.

DDoS attackers either target the application or network/transport layer. The network/transport layer attacks exploit protocol weakness, obsolete hardware, and software vulnerabilities to lunch attackers on victim machines [10,12]. The application-level flooding attacks target the hypertext transfer protocol (HTTP) and session initiation protocol SIP [12]. Detecting DDoS attacks is difficult because

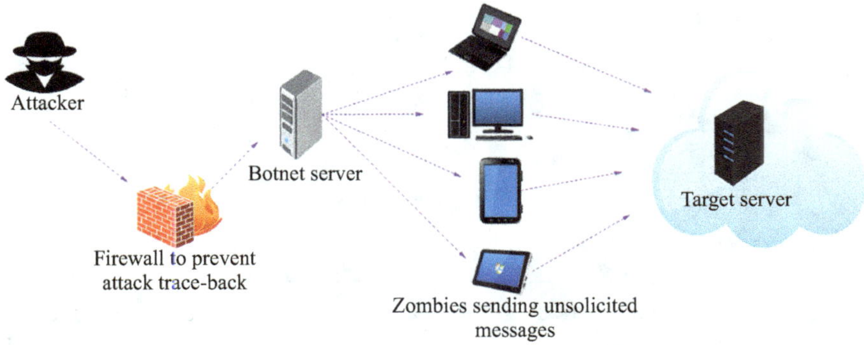

Figure 12.3 DDoS attack scenario

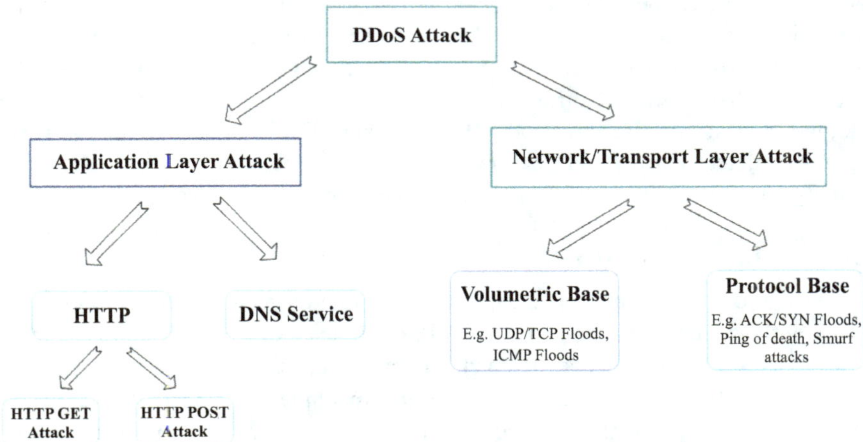

Figure 12.4 DDoS attack taxonomy

attackers mostly spoof legitimate nodes' Internet protocol (IP) addresses to send packets, making it a rigorous task to detect or trace back [10]. Figure 12.4 gives the taxonomy of DDoS attacks.

12.3.3 IDS

Intrusion detection is the act of analyzing and detecting actions like threats, attacks, and security bridges that violate or are harmful to the well-being of a network communication system [47]. The IDS aims to safeguard the confidentiality, integrity, and availability (CIA) of the network users, and IDS are either host-based or network-based. The host-based installs the IDS software on the user's device and reports any actions violating security policies. In contrast, the network-based software or hardware resides on the network server [46]. The detection mechanisms of the IDS are signature-based or misuse detection, anomaly-based detection, and

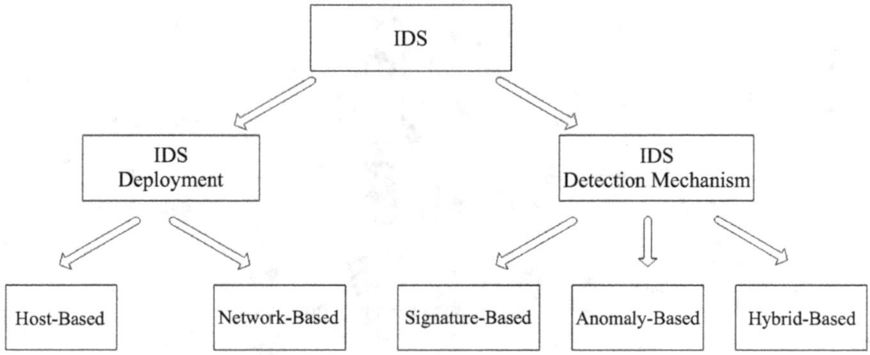

Figure 12.5 IDS deployment taxonomy

hybrid-based detection [48]. In signature-based detection, the IDS maintains a knowledge-based (rules) set to detect actions that violate security policies but can only detect known attack types. The anomaly-based IDS studies the hypothesis of the attackers' behavior which differs from normal users' behavior and helps detect unknown attacks. Lastly, the hybrid IDS uses signature-based and anomaly-based features to detect known and unknown attacks. This research proposal focuses on anomaly-based IDS. Figure 12.5 shows IDS deployments and detection mechanisms.

12.3.4 Anomaly-based IDS

Anomaly or outlier detection is a variant of IDS that checks network activities that significantly deviates from normal network activities. Network anomalous events, in most cases, indicate events such as technical glitches, DDoS attacks, and noise, among others [49]. Many methods, including machine learning (ML), statistical techniques, information theory, and spectral techniques, have been used to build anomaly model [50]. This research explores the machine-learning approach to build IDS that detect unknown attacks.

12.3.5 ML in security

ML allows the computer to learn based on experience rather than being explicitly programmed. Unlike the conventional computer system that draws a conclusion from input and process, ML maps the input data with the relevant output, and predictive results are the output using hypothesis. ML is mostly deployed in fraud detection, medical diagnosis, pattern recognition, and computer security. The ML types are supervised ML, unsupervised ML, semi-supervised ML, and reinforcement ML.

1. **Supervised ML**: The supervised method uses a mathematical model to investigate the relationship between data input, producing an output iteratively. At each iteration, the mathematical model learns a useful pattern used for decision at the next iteration. The dataset in supervised learning is divided into training and testing, usually in a ratio of 70–30. Among the various algorithms used for hypothesis detection in ML are SVM, KNN, LR, and MLP.

2. **Unsupervised ML**: Unlike supervised ML, unsupervised ML uses only the input to form a hypothesis. The unlabeled test data are after that supplied for the classification or categorical task, and conclusions are drawn from the similarities learnt from the input set. The unsupervised ML also deploy a clustering task by dividing the input into subsets, and all test data are clustered in their appropriate subsets. The K-means algorithm is a common example in this category.
3. **Semi-supervised ML:** The semi-supervised ML combines the attributes of supervised and unsupervised ML. Missing or incomplete data are injurious to the accuracy of supervised ML, especially when working with fewer data. Since the semi-supervised ML works in both situations (with or without input labels), it can fill this gap by giving an appropriate hypothesis and is mostly deployed in areas like natural language programming (NLP).

12.3.6 Ensemble learning

Like ML, ensemble learning uses numerous algorithms to obtain a better predictive model. The ensemble model combines the output of each classification algorithm using a combiner and outputs the aggregated results of all the classification algorithms, as shown in Figure 12.6. The ensemble model has proven to have better

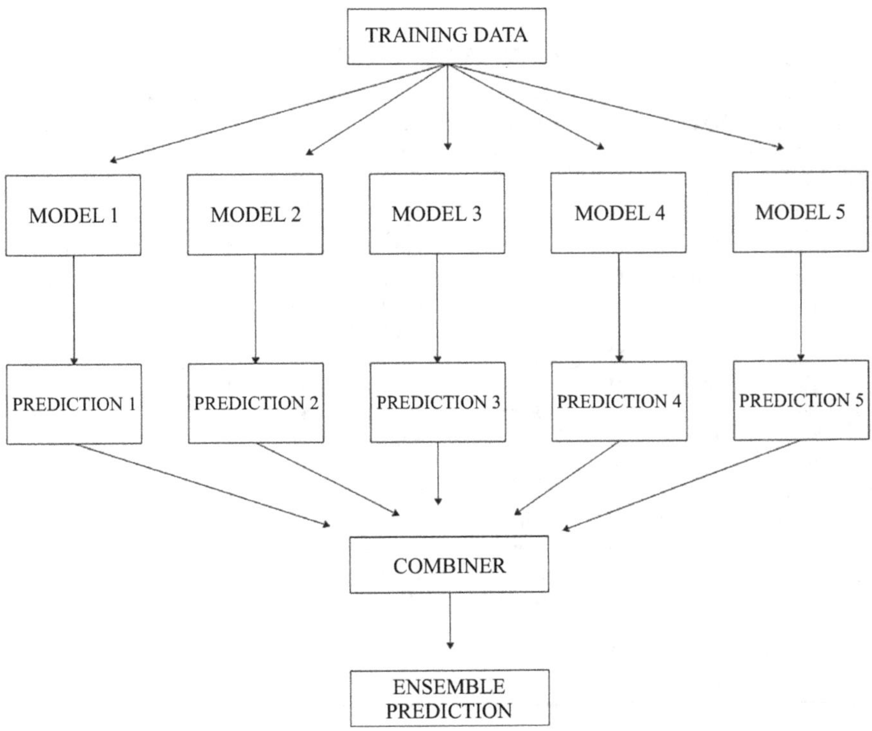

Figure 12.6 Ensemble learning

accuracy than using a single learner, while the types of ensemble combiners are bagging, boosting, stacking, and voting [50–54].

12.3.7 Dataset description

This research that uses the UNSW_NB15 network intrusion dataset was taken at the University of New South Wales (UNSW) cyber security laboratory in 2015 [55]. The dataset consists of 45 feature attributes as presented in Table 12.1 and nine different types of attacks: Fuzzers, Analysis, DoS, Exploits, Generic, Reconnaissance, Shellcode, backdoors, and worms. The feature attributes consist of data types: nominal, integer, float and binary, as shown in Table 12.1 and divided into 82,332 training sets. The UNSW_NB15 dataset addressed the drawbacks of KDD-98, KDD-CUP99, and NSL-KDD, which lack network attributes that address current network traffic sensitivity and lack comprehensive network traffic information [16]. The UNSW_NB15 is also not wholly immune from flaws and characterized by two challenges: class imbalance and class overlap [55]. This research addresses the class imbalance and class overlap issues using various preprocessing tasks like normalization, binarization, discretization, and feature selection. Figure 12.7 shows a sample of the UNSW_NB15 set.

Table 12.1 UNSW_NB15 feature attributes and data types

S. no.	Feature attribute	Data type	S. no.	Feature attribute	Data type	S. no.	Feature attribute	Data type
1	srcip	Nominal	16	Dload	Float	35	ackdat	Float
2	sport	Integer	17	Spkts	Integer	36	is_sm_ips_ports	Binary
3	dstip	Nominal	18	Dpkts	Integer	37	ct_state_ttl	Integer
4	dsport	Integer	19	swin	Integer	38	ct_flw_http_mthd	Integer
5	proto	Nominal	20	dwin	Integer	39	is_ftp_login	Binary
6	state	Nominal	21	stcpb	Integer	40	ct_ftp_cmd	Integer
7	dur	Float	22	dtcpb	Integer	41	ct_srv_src	Integer
8	sbytes	Integer	23	smeansz	Integer	42	ct_srv_dst	Integer
9	dbytes	Integer	24	dmeansz	Integer	43	ct_dst_ltm	Integer
10	sttl	Integer	25	trans_depth	Integer	44	ct_src_ ltm	Integer
11	dttl	Integer	26	res_bdy_len	Integer	45	ct_src_dport_ltm	Integer
12	sloss	Integer	27	Sjit	Float	46	ct_dst_sport_ltm	Integer
13	dloss	Integer	28	Djit	Float	47	ct_dst_src_ltm	Integer
14	service	Nominal	29	Stime	Timestamp	48	attack_cat	Nominal
15	Sload	Float	30	Ltime	Timestamp	49	Label	Binary
12	sloss	Integer	31	Sintpkt	Float			
13	dloss	Integer	32	Dintpkt	Float			
14	service	Nominal	33	tcprtt	Float			
15	Sload	Float	34	synack	Float			

id	dur	proto	service	state	spkts	dpkts	sbytes	dbytes	rate	sttl	dttl	sload	dload	sloss	dloss	sinpkt
1	0.000011	udp	-	INT	2	0	496	0	90909.1	254	0	180363632	0	0	0	0.011
2	0.000008	udp	-	INT	2	0	1762	0	125000	254	0	881000000	0	0	0	0.008
3	0.000005	udp	-	INT	2	0	1068	0	200000	254	0	854400000	0	0	0	0.005
4	0.000006	udp	-	INT	2	0	900	0	166667	254	0	600000000	0	0	0	0.006
5	0.00001	udp	-	INT	2	0	2126	0	100000	254	0	850400000	0	0	0	0.01
6	0.000003	udp	-	INT	2	0	784	0	333333	254	0	1045333312	0	0	0	0.003
7	0.000006	udp	-	INT	2	0	1960	0	166667	254	0	1306666624	0	0	0	0.006
8	0.000028	udp	-	INT	2	0	1384	0	35714.3	254	0	197714288	0	0	0	0.028
9	0	arp	-	INT	1	0	46	0	0	0	0	0	0	0	0	60000.7
10	0	arp	-	INT	1	0	46	0	0	0	0	0	0	0	0	60000.7
11	0	arp	-	INT	1	0	46	0	0	0	0	0	0	0	0	60000.7
12	0	arp	-	INT	1	0	46	0	0	0	0	0	0	0	0	60000.7
13	0.000004	udp	-	INT	2	0	1454	0	250000	254	0	1454000000	0	0	0	0.004
14	0.000007	udp	-	INT	2	0	2062	0	142857	254	0	1178285696	0	0	0	0.007
15	0.000011	udp	-	INT	2	0	2040	0	90909.1	254	0	741818176	0	0	0	0.011
16	0.000004	udp	-	INT	2	0	1052	0	250000	254	0	1052000000	0	0	0	0.004
17	0.000003	udp	-	INT	2	0	314	0	333333	254	0	418666656	0	0	0	0.003
18	0.00001	udp	-	INT	2	0	1774	0	100000	254	0	709600000	0	0	0	0.01
19	0.000002	udp	-	INT	2	0	1568	0	500000	254	0	3136000000	0	0	0	0.002
20	0.000004	udp	-	INT	2	0	2054	0	250000	254	0	2054000000	0	0	0	0.004
21	0.00001	udp	-	INT	2	0	2170	0	100000	254	0	868000000	0	0	0	0.01
22	0.000009	udp	-	INT	2	0	202	0	111111	254	0	89777776	0	0	0	0.009
23	0.00001	udp	-	INT	2	0	1334	0	100000	254	0	533600000	0	0	0	0.01
24	0.000005	udp	-	INT	2	0	2058	0	200000	254	0	1646400000	0	0	0	0.005
25	0.000003	udp	-	INT	2	0	286	0	333333	254	0	381333312	0	0	0	0.003
26	0.000007	udp	-	INT	2	0	1500	0	142857	254	0	857142848	0	0	0	0.007
27	0.000006	udp	-	INT	2	0	902	0	166667	254	0	601333312	0	0	0	0.006
28	0.000002	udp	-	INT	2	0	1626	0	500000	254	0	3252000000	0	0	0	0.002
29	0.000007	udp	-	INT	2	0	364	0	142857	254	0	208000000	0	0	0	0.007
30	1.434166	tcp	-	FIN	10	6	530	268	10.459	254	252	2660.780029	1249.51	2	1	151.243

Figure 12.7 Screenshot of UNSW_NB15 feature attributes

12.4 Research methodology

The methodology proposed in this work covers three stages: preprocessing, model development, and model evaluation.

12.4.1 Preprocessing

The preprocessing performs dataset cleansing by removing outliers using normalization techniques in (12.1)

$$X_n = \left(\frac{X - x_{\min}}{x_{\max} - x_{\min}} \right) \tag{12.1}$$

The normalization helps to scale down the data attributes within a specified range, such as 0 to 1, and the normalization process excludes the four nominal feature attributes: id, proto, service, states, and attack_cat. The normalization result using the Python IDLE 3.7 edition is shown in Figure 12.8. The binarization output is given in Figure 12.9.

The feature attributes in Figure 12.8 are converted into binary value vectors to make the classifier more efficient [56] using (12.2) and (12.3):

$$t = \frac{\sum_{x=1}^{n} x}{n} x = \begin{cases} 1, \text{ if } z > t \\ 0, \text{ otherwise} \end{cases} \tag{12.2}$$

$$s = \sqrt{\frac{\sum (x - t)^2}{n}} \tag{12.3}$$

where n, x, and s are the numbers of items, the total sum of all numbers and the standard deviation, respectively. Algorithm 12.1 denotes the binarization steps.

```
Normalization :
        dur  spkts  dpkts  ...  ct_src_ltm  ct_srv_dst  is_sm_ips_ports
0      1.00    1.0    1.0  ...        1.00        0.98              1.0
1      1.00    1.0    1.0  ...        1.00        0.98              1.0
2      1.00    1.0    1.0  ...        1.00        0.97              1.0
3      1.00    1.0    1.0  ...        0.98        0.97              1.0
4      1.00    1.0    1.0  ...        0.98        0.97              1.0
...     ...    ...    ...  ...         ...         ...              ...
82327  1.00    1.0    1.0  ...        0.98        1.00              1.0
82328  0.98    1.0    1.0  ...        0.97        0.98              1.0
82329  1.00    1.0    1.0  ...        1.00        1.00              0.0
82330  1.00    1.0    1.0  ...        1.00        1.00              0.0
82331  1.00    1.0    1.0  ...        1.00        1.00              1.0

[82332 rows x 39 columns]
```

Figure 12.8 Normalization output of 40 UNSW_NB15 feature sets

```
Binarization :
        dur  spkts  dpkts  ...  ct_src_ltm  ct_srv_dst  is_sm_ips_ports
0         1      1      1  ...           1           1                1
1         1      1      1  ...           1           1                1
2         1      1      1  ...           1           1                1
3         1      1      1  ...           1           1                1
4         1      1      1  ...           1           1                1
...     ...    ...    ...  ...         ...         ...              ...
82327     1      1      1  ...           1           1                1
82328     1      1      1  ...           1           1                1
82329     1      1      1  ...           1           1                0
82330     1      1      1  ...           1           1                0
82331     1      1      1  ...           1           1                1

[82332 rows x 39 columns]
```

Figure 12.9 Binarization output

Algorithm 12.1 Binarization algorithm

Input: Select the attribute z from dataset s
Output: $x_i = 0$ or 1

Procedure:
Select the attribute z from dataset s
Compute the threshold t using (12.2)
For each subset x_i of attribute z
If $x_i > t$, the value is set to 1
Else the value is 0
End if
End for
End

The continuous features of the dataset are discretized such that values 0, 1, 2, 3, 4, and 5 can be assigned to five different values within the same feature set. The binarization task computes the threshold t and standard deviation s. The features are assigned value 0 if z is below the value of t and value 1 when above t, as shown in Figure 12.10. This process increases the efficiency of the classifier algorithm, thereby improving the speed and accuracy of the model [41,51].

The proposed MrMr measures the impurity or uncertainty subsets of the dataset using the entropy proposed in [57]. Let i be discrete random values. The entropy of $E(i)$ measures the information impurity or uncertainty using (12.4)

$$E(i) = -\sum_{i \in I} p(i) \log_2 p(i) \tag{12.4}$$

$\boldsymbol{p}(i)$ is i probability density function and computed using (12.5)

$$p(i) = \frac{\sum_i^n x}{n} \tag{12.5}$$

```
      proto service state attack_cat
0       udp       -   INT     Normal
1       udp       -   INT     Normal
2       udp       -   INT     Normal
3       udp       -   INT     Normal
4       udp       -   INT     Normal
...     ...     ...   ...        ...
82327   udp       -   INT     Normal
82328   tcp       -   FIN     Normal
82329   arp       -   INT     Normal
82330   arp       -   INT     Normal
82331   udp       -   INT     Normal

[82332 rows x 4 columns]
      proto service state attack_cat
0         1       1     1          0
1         1       1     1          0
2         1       1     1          0
3         1       1     1          0
4         1       1     1          0
...     ...     ...   ...        ...
82327     1       1     1          0
82328     3       1     2          0
82329     2       1     1          0
82330     2       1     1          0
82331     1       1     1          0

[82332 rows x 4 columns]
```

Figure 12.10 Discretization output

where x represents elements of the same class and n represents the total number of elements.

i and j are expressed by (12.6) as joint entropy.

$$E(i;j) = -\sum\nolimits_{i\in I, j\in J} p(i,j)\log_2 p(i,j) \tag{12.6}$$

where $p(i,j)$ is the joint probability density function of $E(i;j)$ and expressed individually using (12.5).

The conditional entropy expressed in (12.7) measures the uncertainty when one variable is known, and the other is unknown:

$$E(i|j) = -\sum\nolimits_{i\in I, j\in J} p(i,j)\log_2 p(i|j) \tag{12.7}$$

E(I|j) is the order probability of *i* and *j*.

The entropy of a group where all elements belong to the same class is zero (0), while elements in either class are denoted as one.

The information gain *G* of two variables, *i* and *j* is expressed in (12.8)

$$G(i;j) = E(i) - E(i|j) \tag{12.8}$$

E(*i*) and *E*(*i*|*j*) are expressed in (12.4) and (12.7), respectively.

The *E*(*i*|*j*) value from (12.7) is high when *i* and *j* are closely related. Otherwise, 0 if unrelated. Unfortunately, the mutual information expressed in (12.8) can only measure each dataset's attribute. However, it cannot determine each attribute's relevance *D* and redundancy *R*. This research minimizes redundancy and maximizes the relevance proposed by [26], as expressed in (12.9) and (12.10):

$$\min\ R(Q), R = \frac{1}{|q|^2}\sum\nolimits_{xi,xj\in Q} G(xi;xj) \tag{12.9}$$

where $|q|$ is the number of features in the feature set *Q* and *G*(*i*;*j*) is the mutual information between *i* and *j*

$$\max\ D(Q,u), D = \frac{1}{|q|}\sum\nolimits_{xi\in Q} G(xi;u) \tag{12.10}$$

where *u* is the target activity, *G*(*i*;*u*) is the mutual information between *i* and *u*, and (12.8) and (12.9) can be optimized simultaneously using (12.11):

$$\max\ \Phi(D,R), \Phi = D - R \tag{12.11}$$

Suppose *F* is the original feature set and Q_{m-1} is the selected feature set. The task is to select the *m*th feature from the set $\{F - Q_{m-1}\}$, which maximizes Φ. The incremental search procedure for the task expressed in (12.12) computes the cross-validation classification error e_k and finds the smallest error size denoted as Ω:

$$\max_{xi\in F-Qm-1}\left[G(xi;u) - \frac{1}{m-1}\sum\nolimits_{xj\in Qm} G(xi;xj)\right] \tag{12.12}$$

The MrMr procedure is denoted in Algorithm 12.2.

Algorithm 12.2 MrMr algorithm

Input: A training dataset T = D(F, C), the number of features to be selected as n
Output: Selected feature Q_n

Procedure:
Initialize relevant parameters; D ← are the feature set instances such as labelled and unlabeled, F ← is the original feature set, C ← is the target activity or class, Q = null, e_k ← error, and Ω ← is the smallest error ;
For each $f \in \boldsymbol{F}$ do;
Calculate $\max_{xi \in F - Qm-1} \left[G(xi; u) - \frac{1}{m-1} \sum_{xj \in Qm} G(xi; xj) \right]$;
Get sequential feature sets $Q_1 \subset Q_2 \subset Q_3 \ldots\ldots\ldots\ldots \subset Q_{n-1} \subset Q_n$;
For each $k \in Q_1 \ldots\ldots\ldots Q_k, \ldots\ldots\ldots\ldots\ldots\ldots Q_n$ do
Select the smallest error Ω from the large set of e_k
n_k is selected as the smallest k that corresponds to e_k
Return selected feature Q_n

The cuckoo search algorithm is an optimization algorithm inspired by the lifestyle of a bird called the cuckoo bird. The cuckoo bird uses other birds' nests for egg-laying and brooding by mimicking the host bird's eggs. If the host bird discovers the alien egg, such a solution is considered weak because the host bird either removes the alien egg or abandons the nest. The undiscovered eggs are reproduced and carried to the next generation. Such a solution is considered the best solution. The aim is to use a new and better solution (Cuckoo) to replace a not-so-good solution in the nest.

The probability of discovering the cuckoo egg is denoted in (12.13)

$$P a \in (0, 1) \tag{12.13}$$

where P_a is the probability of discovering the alien egg.

The cuckoo optimization uses three constraints for its implementation:

1. Each cuckoo bird lays only one egg at a time and places it in a random nest. i.e. one feature attribute from the feature set is selected for evaluation at a time.
2. Eggs with high quality will carry over to the next generation, i.e. only feature attributes that score high are selected.
3. The available next is fixed, and the host bird discovers an alien egg with probability in (12.13). The host bird either throws away the alien egg or leaves the nest, i.e. the feature attributes from the dataset are fixed.

The cuckoo search process is as shown in Algorithm 12.3. The selected features using the cuckoo search and MrMr feature selection is given in Figure 12.11.

Algorithm 12.3 Cuckoo search algorithm with MrMr

Input: Initialize value of host nest n, probability P_a and maximum number of iteration Max_{itr}

Set $t = 0$ {*counter initialization*}
For (i = 1: $i \leq n$)

Output: Produce the best solution

1. Generate initial n host, $x_i^{(t)}$
2. Evaluate $fx_i^{(t)}$ using the objective function in (12.12)
   Repeat
3. Randomly select a cuckoo by levy flight using (12.14)

$$x_i^{(t+1)} = x_i^t + \alpha \oplus levy(\lambda) \tag{12.14}$$

4. Evaluate $fx_i^{(t+1)}$ using the objective function in (12.12)
5. If $fx_i^{(t)} < fx_i^{(t+1)}$ then
6. Replace $x_i^{(t)}$ with $x_i^{(t+1)}$
7. Search a new nest with levy's flight (λ) using (12.14)
8. Keep the best solution
9. Iterate steps 1 to 8 while (t < MaxGeneration or stop Criterion)

```
'dinpkt', 'dbytes', 'spkts', 'dloss', 'ackdat', 'synack', 'tcprt
n', 'ct_ftp_cmd', 'is_sm_ips_ports', 'ct_flw_http_mthd']
explained_variance_ratio_ is 0.9995382828841535
shape after: (257673, 32)
combined_data.dur is scaled up by 10,000
before: (257673, 32)
Feature ranking using cuckoo search and MrMr:
1. feature 7 (0.105157)
2. feature 2 (0.072822)
3. feature 8 (0.055564)
4. feature 27 (0.052154)
5. feature 17 (0.050276)
6. feature 29 (0.046487)
7. feature 25 (0.045896)
8. feature 21 (0.044636)
9. feature 24 (0.043946)
10. feature 20 (0.043338)
11. feature 5 (0.035413)
12. feature 23 (0.034343)
13. feature 28 (0.031006)
14. feature 13 (0.030514)
15. feature 30 (0.030006)
16. feature 18 (0.026528)
17. feature 22 (0.026246)
18. feature 6 (0.024412)
19. feature 26 (0.024183)
20. feature 9 (0.023638)
>>> |
```

Figure 12.11 Selected features using the cuckoo search and MrMr feature selection

12.4.2 Model development

This research proposes four widespread learning algorithms: KNN, LR, DT, and MLP Figure 12.12 shows the model architecture.

These learning algorithms were chosen based on their accuracy and frequent usage in several works of literature for building an ML model.

The proposed ML method uses the following steps:

1. The selected features from the cuckoo search MrMr algorithm are classified using an ML model with four classifiers.
2. The UNSW_NB15 dataset is split into 70% training and 30% testing using the Python IDLE tool.
3. The result of each classification algorithm is ranked in accuracy, detection time, true positive (TP), true negative (TN), false positive (FP), and false negative (FN).

12.4.3 KNN

The KNN is an ML algorithm that can solve classification and regression problems by assuming every similar event has close relationships. It calculates the distance between similar events using mathematical methods like Euclidean distance, Hamming distance, Jaccard distance, and Minkowski distance in making predictions. Among the algorithm's advantages is simplicity and versatility, but not a suitable algorithm for prediction when the volume of data is enormous. The steps below are used to implement the KNN algorithm.

Step 1: Load the dataset (UNSW_NB15).
Step 2: Select the population of the neighbor by initializing K.

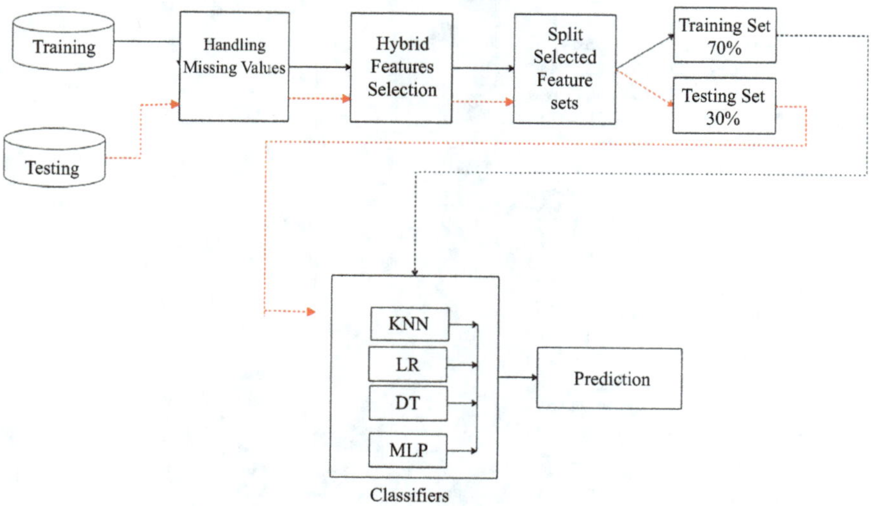

Figure 12.12 Model architecture

Step 3: Index the data samples collected.
Step 4: arrange distance and indices based on arrival (ascending).
Step 5: Select the K and use its label to return the mean of K or mode when predicting regression or classification problems, respectively.

12.4.4 Logistic regression

LR belongs to the supervised ML family and uses power probability to forecast the outputs of a binary event. The algorithm usually assigns discrete variables to an event outcome using the sigmoid function (0 and 1) to predict if the event is true (1) or false (0). After the data points are given to each variable using the sigmoid function, a hyperplane is drawn to separate the variables. Depending on the events to be analyzed, the LR has different variants, such as binary LR, ordinal LR, and multinomial LR.

12.4.5 Decision tree

Unlike other supervised ML such as LR and SVM, the DT works well with categorical and continuous data samples using various powerful algorithms like C4.5, ID3, classification and regression tree (CART) to split data samples into nodes and sub-nodes. The hierarchical structure of the DT consists of the root node, which gave birth to the branch node, and base evaluation criteria by the root and branch nodes form a homogenous node called the leaf node. DT employs a greedy strategy to divide data into branch and leaf nodes. After dividing the data, DT uses a top-down approach recursively until all or most events are classified into a specific branch and leaf nodes.

12.4.6 Multi-layer perceptron

The MLP belongs to the ANN family that generates an event output from an event input. It uses backpropagation to train multiple input layers connected as a direct graph to the multiple outputs. Like other variants of ANN, the MLP consists of input, hidden, and output nodes.

12.5 Results and discussions

This work uses the hybridized feature selection technique (cuckoo search and MrMr) to select the most informative features from the UNSW NB15 dataset. The developed feature selection model selected 20 out of the 41 feature attributes of the dataset, and the 20 features were divided into ratios 70 to 30 for training and testing the model. Four base classifier algorithms, KNN, LR, DT, and MLP, were used for the experimental evaluation in classifying the UNSW_NB15 packets as normal or attack using Python IDLE 3.7. tool. The motivation for selecting UNSW_NB15 is stated in Section 12.3.7, and different performance metrics like accuracy, detection time, TP, TN, FP, and FN were carried out.

Accuracy: The accuracy specifies the ratio between the corrected classified labels and the entire dataset, using (12.15)

$$Accuracy = \frac{TP + TN}{TP + TN + FP + FN} \tag{12.15}$$

True positive rate (TPR): The true positive or sensitivity is the ratio between the correctly classified labels and the entire labels in the dataset and obtained using (12.16)

$$Sensitivity = \frac{TP}{TN + FN} \tag{12.16}$$

True negative rate (TNR): This, also known as specificity, measures the ratio between the negative labels and the entire negative labels (attacks) in the entire dataset:

$$Specificity = \frac{TN}{TN + FP} \tag{12.17}$$

False negative rate (FNR): FNR measures the proportion between the misclassified labels and the total number of negatives (attacks) in the dataset:

$$False\ Negative = \frac{FN}{FN + TP} \tag{12.18}$$

False alarm: This represents the total number of attack warnings given compared to the actual number of attacks:

$$False\ Alarm = \frac{FP + FN}{TP + TN + FP + FN} \tag{12.19}$$

Table 12.2 indicates individual base classification algorithm performance, and Figure 12.13 shows the comparison of the model's TP, TN, FP, and FN across the four base classification algorithms.

The result shows that DT has the highest number of TP and TN and recorded the lowest FP and FN out of the base classification algorithms using 20 features from the UNSW_NB15 dataset.

Table 12.2 Performance metrics measurement

Base classifiers	Accuracy	TP	TN	FP	FN	Detection time (s)
KNN	81	8,917	11,097	2,207	2,479	147.8
LR	70	9,497	8,955	1,627	4,621	0.8
DT	**96**	**10,675**	**13,140**	**449**	**436**	**1.60**
MLP	74	7,255	11,095	3,869	2,481	8.1

Note: The bold values indicate the DT performance is higher than the performances of other algorithms.

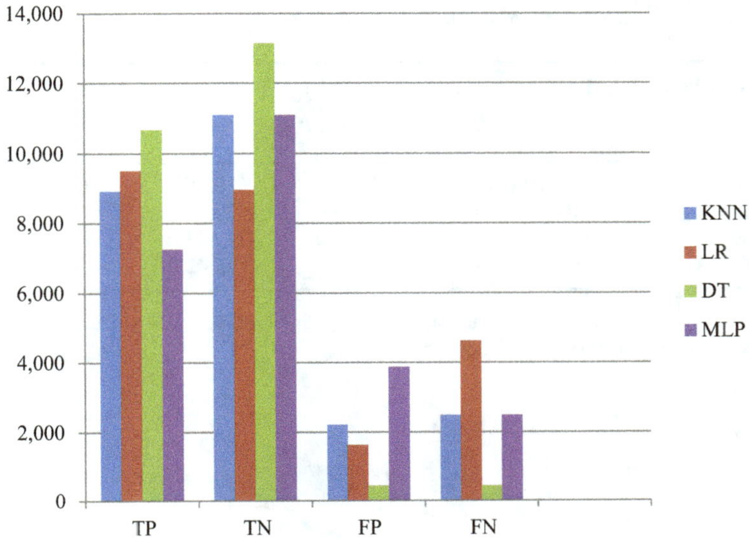

Figure 12.13 Performance comparison of the models' TP, TN, FP, and FN

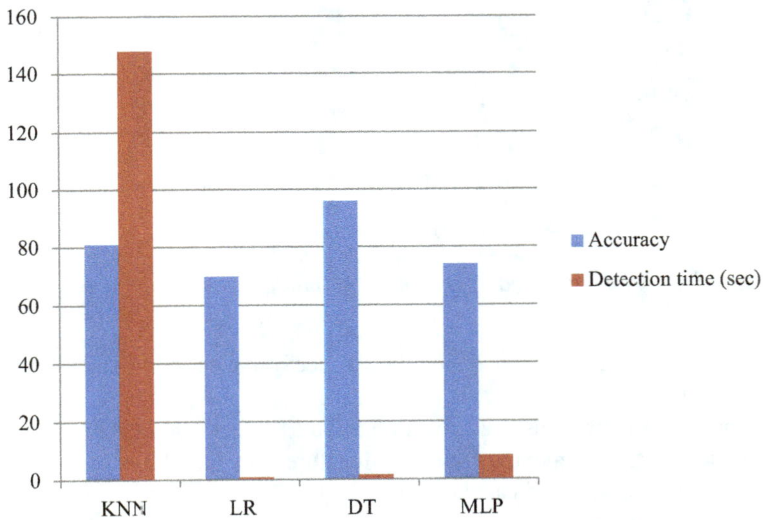

Figure 12.14 Performance comparison of the models' accuracy and detection time

In Figure 12.14 above, DT ranks the highest at 96.42% using 20 features from 45 feature attributes and a second detection time of 1.60 s. The LR has the lowest classification time of 0.8 s but is ranked the lowest in terms of an accuracy at 70%.

Finally, the work compared the results of the cuckoo search with MrMr with some literature that used only statistical methods like IG. Table 12.3 and

Table 12.3 Performance metrics measurement

Reference	Method	No. of features use	Accuracy	Detection time
[56]	IG and MLP	30	76.96%	**0.25**
[58]	XGBoost and DT	19	90.85%	–
[59]	RF	41	95.43%	–
DT	Proposed method	20	**96.42%**	1.60

Note: The bold under accuracy indicates our method perform better than others in accuracy, but the detection time of [56] is better.

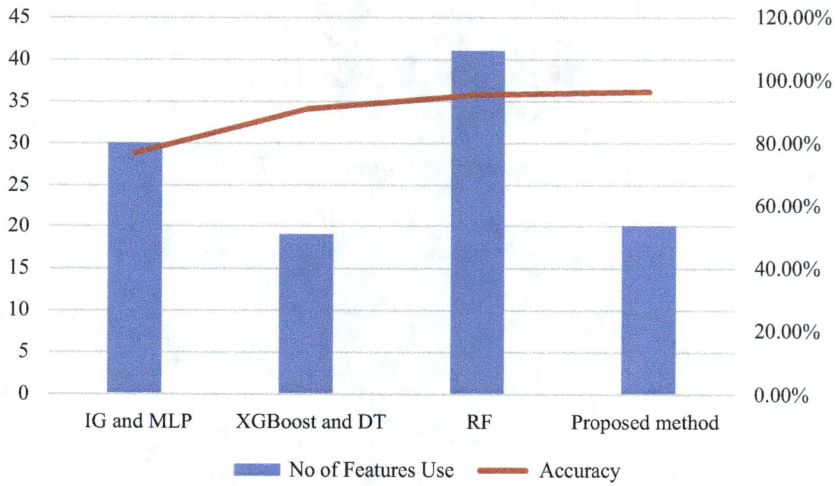

Figure 12.15 Performance comparison of the model accuracy and detection time

Figure 12.14 show our proposed method's performance compared with some existing research.

The proposed method uses 20 feature attributes to achieve 96.42% accuracy in 1.60 s, while the closest method from the literature uses all 41 feature attributes to attain 95.43% accuracy, as indicated in Figure 12.15. The result shows the efficiency of the proposed method in improving the detection time and accuracy of IDS with relevant feature attributes, thereby making it suitable for real-time IDS detection [60–62].

12.6 Conclusions and future scope

This chapter proposes a real-time anomaly-based IDS to detect attacks within a short period in a cloud-based environment. Binarization and discretization are used to preprocess the continuous and categorical features of the UNSW_NB15 dataset.

After the preprocessing processes, the cuckoo search algorithm with MrMr as the objective function selects the most informative attributes from the dataset. Twenty feature attributes with high scores were trained and tested using four classification algorithms: KNN, LR, DT, and MLP. The experimental setup was carried out using the Python IDLE 3.7.1 and performance matrices like accuracy, TP, TN, FP, FN and detection time. The result indicates that DT has the highest accuracy of 96.42% and a good detection time of 1.60 s. However, the research did not test other performance matrices like area under the curve (AUC) and receiver operating characteristics (ROC). Future studies will test more matrices and use recent datasets and ensemble learners to ascertain the dynamicity of the model.

References

[1] K. A. Sadiq, F. S. Oyedepo, and J. K. Ayeni, "A lightweight economic denial of sustainability (EDOS) DEFENCE," *Eur. J. Comput. Sci. Inf. Technol.*, vol. 8, no. 3, pp. 57–64, 2020.

[2] T. Salman, D. Bhamare, A. Erbad, R. Jain, and M. Samaka, "Machine learning for anomaly detection and categorization in multi-cloud environments," in *Proceedings of the 4th IEEE International Conference on Cyber Security and Cloud Computing, CSCloud 2017 and 3rd IEEE International Conference of Scalable and Smart Cloud, SSC 2017*, vol. June 26–28, no. EEE CSCloud 2017, pp. 97–103, doi: 10.1109/CSCloud.2017.15.

[3] S. Parikh, D. Dave, R. Patel, *et al.*, "Security and privacy issues in cloud, fog and edge computing security and privacy issues in cloud, fog and edge computing," *Procedia Comput. Sci.*, vol. 160, pp. 734–739, 2019, doi: 10.1016/j.procs.2019.11.018.

[4] Y. Wang, "Cloud-dew architecture cloud-dew architecture," *Int. J. Comput.*, vol. 4, no. September, pp. 199–210, 2015, doi: 10.1504/IJCC.2015.071717.

[5] K. Skala, "Cloud, fog and dew computing: a distributed hierarchy grid computing (GC) cloud computing (CC)," *Open J. Cloud Comput.*, vol. 2, no. 1, pp. 16–24, 2015.

[6] N. Sutradhar, M. K. Sharma, and G. Sai Krishna, "Cloud computing: security issues and challenges," *Lect. Notes Electr. Eng.*, vol. 692, no. 3, pp. 25–32, 2021, doi: 10.1007/978-981-15-7486-3_4.

[7] V. Taran, O. Alienin, S. Stirenko, Y. Gordienko, and A. Rojbi, "Performance evaluation of distributed computing environments with Hadoop and Spark frameworks," in *2017 IEEE International Young Scientists Forum on Applied Physics and Engineering. YSF 2017*, January, vol. 2017, pp. 80–83, 2017, doi:10.1109/YSF.2017.8126655.

[8] O. Osanaiye, K. K. R. Choo, and M. Dlodlo, "Distributed denial of service (DDoS) resilience in cloud: review and conceptual cloud DDoS mitigation framework," *J. Netw. Comput. Appl.*, vol. 67, pp. 147–165, 2016, doi: 10.1016/j.jnca.2016.01.001.

[9] J. Jang-Jaccard and S. Nepal, "A survey of emerging threats in cybersecurity," *J. Comput. Syst. Sci.*, vol. 80, no. 5, pp. 973–993, 2014, doi: 10.1016/j.jcss.2014.02.005.

[10] A. Sahi, D. Lai, Y. Li, and M. Diykh, "An efficient DDoS TCP flood attack detection and prevention system in a cloud environment," vol. 5, pp. 6036–6048, 2017, doi: 10.1109/ACCESS.2017.2688460.

[11] J. Mirkovic and P. Reiher, "A taxonomy of DDoS attack and DDoS defense mechanisms," *ACM SIGCOMM Comput. Commun. Rev.*, vol. 34, no. 2, pp. 39, 2004, https://doi.org/10.1145/997150.997156. Available: http://portal.acm.org/citation.cfm?doid=997150.997156.

[12] K. A. Sadiq, A. F. Thompson, and O. A. Ayeni, "Mitigating DDoS attacks in cloud network using fog and SDN: a conceptual security framework," *Int. J. Appl. Inf. Syst.*, vol. 12, no. 32, pp. 11–16, 2020.

[13] M. Turčaník, "Packet filtering by artificial neural network," in *ICMT 2015 – International Conference on Military Technologies (ICMT 2015)*, July 2015, 2015, doi: 10.1109/MILTECHS.2015.7153739.

[14] A. Kaur, S. K. Pal, and A. P. Singh, "Hybridization of K-means and firefly algorithm for intrusion detection system," *Int. J. Syst. Assur. Eng. Manag.*, vol. 9, no. 4, pp. 901–910, 2018, doi: 10.1007/s13198-017-0683-8.

[15] W. L. Al-Yaseen, "Improving intrusion detection system by developing feature selection model based on firefly algorithm and support vector machine," *IAENG Int. J. Comput. Sci.*, vol. 46, no. 4, pp. 1–7, 2019.

[16] J. O. Mebawondu, O. D. Alowolodu, J. O. Mebawondu, and A. O. Adetunmbi, "Network intrusion detection system using supervised learning paradigm," *Sci. African*, vol. 9, e00497, 2020, doi: 10.1016/j.sciaf.2020.e00497.

[17] O. O. Olasehinde, O. C. Olayemi, and B. K. Alese, "Multiple model tree meta algorithms improvement of network intrusion detection predictions accuracy," *Int. J. Inf. Secur. Res.*, vol. 9, no. 3, pp. 891–897, 2019, doi: 10.20533/ijisr.2042.4639.2019.0102.

[18] F. S. Dantas Silva, E. Silva, E. P. Neto, M. Lemos, A. J. Venancio Neto, and F. Esposito, "A taxonomy of DDoS attack mitigation approaches featured by SDN technologies in IoT scenarios," *Sensors (Switzerland)*, vol. 20, no. 11, pp. 1–28, 2020, doi: 10.3390/s20113078.

[19] M. Latah and L. Toker, "A novel intelligent approach for detecting DoS flooding attacks in software-defined networks," *Int. J. Adv. Intell. Informatics*, vol. 4, no. 1, pp. 11–20, 2018, doi: 10.26555/ijain.v4i1.138.

[20] H. Gharaee and H. Hosseinvand, "A new feature selection IDS based on genetic algorithm and SVM," in *2016 8th International Symposium on Telecommunications IST 2016*, 2017, pp. 139–144, doi: 10.1109/ISTEL.2016.7881798.

[21] M. Alkasassbeh, G. Al-Naymat, A. B.A, and M. Almseidin, "Detecting distributed denial of service attacks using data mining techniques," *Int. J. Adv. Comput. Sci. Appl.*, vol. 7, no. 1, pp. 436–445, 2016, doi: 10.14569/ijacsa.2016.070159.

[22] T. Gu, H. Chen, L. Chang, and L. Li, "Intrusion detection system based on improved abc algorithm with tabu search," *IEEJ Trans. Electr. Electron. Eng.*, vol. 14, no. 11, pp. 1652–1660, 2019, doi: 10.1002/tee.22987.

[23] K. M. Alwan, A. H. AbuEl-Atta, and H. H. Zayed, "Feature selection models based on hybrid firefly algorithm with mutation operator for network intrusion detection," *Int. J. Intell. Eng. Syst.*, vol. 14, no. 1, pp. 192–202, 2021, doi: 10.22266/IJIES2021.0228.19.

[24] H. Xu, S. Yu, J. Chen, and X. Zuo, "An improved firefly algorithm for feature selection in classification," *Wirel. Pers. Commun.*, vol. 102, no. 4, pp. 2823–2834, 2018, doi: 10.1007/s11277-018-5309-1.

[25] S. Jeyasingh and M. Veluchamy, "Modified bat algorithm for feature selection with the Wisconsin Diagnosis Breast Cancer (WDBC) dataset," *Asian Pacific J. Cancer Prev.*, vol. 18, no. 5, pp. 1257–1264, 2017, doi: 10.22034/APJCP.2017.18.5.1257.

[26] F. Kamalov, S. Moussa, R. Zgheib, and O. Mashaal, "Feature selection for intrusion detection systems," in *Proceedings of the 2020 13th International Symposium on Computational Intelligence and Design Isc. 2020*, 2020, pp. 265–269, doi: 10.1109/ISCID51228.2020.00065.

[27] O. Osanaiye, O. Ogundile, F. Aina, and A. Periola, "Feature selection for intrusion detection system in a cluster-based heterogeneous wireless sensor network," *Facta Univ. – Ser. Electron. Energ.*, vol. 32, no. 2, pp. 315–330, 2019, doi: 10.2298/fuee1902315o.

[28] A. C. Enache, V. Sgarciu, and A. Petrescu-Nita, "Intelligent feature selection method rooted in Binary Bat Algorithm for intrusion detection," in *SACI 2015 – 10th Jubilee IEEE International Symposium on Applied Computational Intelligence and Informatics, Proceedings*, 2015, pp. 517–521, doi: 10.1109/SACI.2015.7208259.

[29] D. Sheela Jeyarani and A. Pethalakshmi, "Optimized feature selection algorithm for high dimensional data," *Indian J. Sci. Technol.*, vol. 9, no. 31, pp. 1–8, 2016, doi: 10.17485/ijst/2016/v9i31/79656.

[30] "unsw-nb15-dataset." https://research.unsw.edu.au/projects/unsw-nb15-dataset (accessed Oct. 18, 2022).

[31] X. S. Yang and S. Deb, "Cuckoo search via Lévy flights," in *The World Congress on Nature and Biologically Inspired Computing NABIC 2009 – Proc.*, 2009, pp. 210–214, doi: 10.1109/NABIC.2009.5393690.

[32] G. De Tre, A. Hallez, and A. Bronselaer, "Fast-mRMR: fast minimum redundancy maximum relevance algorithm for high-dimensional big data," *Int. J. Intell. Syst.*, vol. 29, no. 2, pp. 495–524, 2014, doi: 10.1002/int.

[33] M. M. Sakr, M. A. Tawfeeq, and A. B. El-Sisi, "Network intrusion detection system based PSO-SVM for cloud computing," *Int. J. Comput. Netw. Inf. Secur.*, vol. 11, no. 3, pp. 22–29, 2019, doi: 10.5815/ijcnis.2019.03.04.

[34] S. Krishnaveni, S. Sivamohan, S. S. Sridhar, and S. Prabakaran, "Efficient feature selection and classification through ensemble method for network intrusion detection on cloud computing," *Cluster Comput.*, vol. 24, no. 3, pp. 1761–1779, 2021, doi: 10.1007/s10586-020-03222-y.

[35] S. Fenanir, F. Semchedine, and A. Baadache, “A machine learning-based lightweight intrusion detection system for the Internet of Things,” *Revue d' Intell. Artif.*, vol. 33, no. 3, pp. 203–211, 2019.

[36] B. Selvakumar and K. Muneeswaran, “Firefly algorithm based feature selection for network intrusion detection,” *Comput. Secur.*, vol. 81, pp. 148–155, 2019, doi: 10.1016/j.cose.2018.11.005.

[37] S. R. Hasani, Z. A. Othman, and S. M. M. Kahaki, “Hybrid feature selection algorithm for intrusion detection system,” *J. Comput. Sci.*, vol. 13, no. 2, pp. 232–240, 2019, doi: 10.3844/jcssp.2014.1015.1025.

[38] M. Samadi Bonab, A. Ghaffari, F. Soleimanian Gharehchopogh, and P. Alemi, “A wrapper-based feature selection for improving performance of intrusion detection systems,” *Int. J. Commun. Syst.*, vol. 33, no. 12, pp. 1–26, 2020, doi: 10.1002/dac.4434.

[39] A. Nazir and R. A. Khan, “A novel combinatorial optimization based feature selection method for network intrusion detection,” *Comput. Secur.*, vol. 102, pp. 102164, 2021, doi: 10.1016/j.cose.2020.102164.

[40] H. Alazzam, A. Sharieh, and K. E. Sabri, “A feature selection algorithm for intrusion detection system based on Pigeon Inspired Optimizer,” *Expert Syst. Appl.*, vol. 148, 2020, doi: 10.1016/j.eswa.2020.113249.

[41] M. M. Sakr, M. A. Tawfeeq, and A. B. El-Sisi, “Filter versus wrapper feature selection for network intrusion detection system,” in *Proceedings – 2019 IEEE 9th International Conference on Intelligence and Computational Information Systems ICICIS 2019*, 2019, pp. 209–214, doi: 10.1109/ICICIS46948.2019.9014797.

[42] Q. R. S. Fitni and K. Ramli, “Implementation of ensemble learning and feature selection for performance improvements in anomaly-based intrusion detection systems,” in *Proceedings – 2020 IEEE International Conference on Industry 4.0, Artificial Intelligence, and Communications Technology IAICT 2020*, 2020, pp. 118–124, doi: 10.1109/IAICT50021.2020.9172014.

[43] Kunal and M. Dua, “Attribute selection and ensemble classifier based novel approach to intrusion detection system,” *Procedia Comput. Sci.*, vol. 167, no. 2019, pp. 2191–2199, 2020, doi: 10.1016/j.procs.2020.03.271.

[44] F. Hussain, M. Hammad, W. El-medany, and R. Ksantini, “Cardiovascular diseases classification via machine learning systems,” 2021 International Conference on Innovation and Intelligence for Informatics, Computing, and Technologies (3ICT), Zallaq, Bahrain, 2021, pp. 51–56, doi: 10.1109/3ICT53449.2021.9581384.

[45] T. Khorram and N. A. Baykan, “Feature selection in network intrusion detection using metaheuristic algorithms,” *Int. J. Adv. Res. Ideas Innov. Technol.*, vol. 4, no. 4, pp. 704–710, 2018.

[46] P. Mishra, V. Varadharajan, U. Tupakula, and E. S. Pilli, “A detailed investigation and analysis of using machine learning techniques for intrusion detection,” *IEEE Commun. Surv. Tutor.*, vol. 21, no. 1, pp. 686–728, 2019, doi: 10.1109/COMST.2018.2847722.

[47] O. O. Oladimeji, A. B. Kayode, A. A. Olusola, and A. O. Isaiah, "Evaluation of selected stacked ensemble models for the optimal multi-class cyber-attacks detection," *Int. J. Cyber Situational Aware.*, vol. 5, no. 1, pp. 26–48, 2021, doi: 10.22619/ijcsa.2020.100132.

[48] M. C. Belavagi and B. Muniyal, "Performance evaluation of supervised machine learning algorithms for intrusion detection," *Procedia Comput. Sci.*, vol. 89, pp. 117–123, 2016, doi: 10.1016/j.procs.2016.06.016.

[49] Aiza Chaudry, "Know Your Network Anomalies," Published in Nvisible, 2019. https://medium.com/nvisible/know-your-network-anomalies-5dd4c7cd49b5. (accessed Oct. 24, 2022).

[50] O. Osanaiye, H. Cai, K. K. R. Choo, A. Dehghantanha, Z. Xu, and M. Dlodlo, "Ensemble-based multi-filter feature selection method for DDoS detection in cloud computing," *Eurasip J. Wirel. Commun. Netw.*, vol. 2016, no. 1, 2016, doi: 10.1186/s13638-016-0623-3.

[51] Y. Shen, K. Zheng, C. Wu, M. Zhang, X. Niu, and Y. Yang, "An ensemble method based on selection using bat algorithm for intrusion detection," *Comput. J.*, vol. 61, no. 4, pp. 526–538, 2018, doi: 10.1093/comjnl/bxx101.

[52] N. M. Baba, M. Makhtar, S. A. Fadzli, and M. K. Awang, "Current issues in ensemble methods and its applications," *J. Theor. Appl. Inf. Technol.*, vol. 81, no. 2, pp. 266–276, 2015.

[53] G. Mageswary and M. Karthikeyan, "Modified firefly algorithm based optimum feature selection and ensemble tree based model for network intrusion detection using data mining technique," vol. 6, pp. 604–610, 2020, doi: 10.35940/ijitee.E3215.049620.

[54] A. Gupta and A. R. Thakkar, "Optimization of stacking ensemble configuration based on various metahueristic algorithms," in *Souvenir of the 2014 IEEE International Advance Computing Conference IACC 2014*, 2014, pp. 444–451, doi: 10.1109/IAdCC.2014.6779365.

[55] Z. Zoghi and G. Serpen, "UNSW-NB15 Computer Security Dataset: Analysis through Visualization," 2021. Available: http://arxiv.org/abs/2101.05067.

[56] J. O. Mebawondu, O. D. Alowolodu, J. O. Mebawondu, and A. O. Adetunmbi, "Network intrusion detection system using supervised learning paradigm," *Sci. African*, vol. 9, p. e00551, 2020, doi: 10.1016/j.sciaf.2020.e00497.

[57] C. E. Shannon and W. Weaver, "The theory of mathematical communication," *Int. Bus.*, p. 131, 1949. Available: https://pure.mpg.de/rest/items/item_2383164_3/component/file_2383163/content.

[58] S. M. Kasongo and Y. Sun, "Performance analysis of intrusion detection systems using a feature selection method on the UNSW-NB15 dataset," *J. Big Data*, vol. 7, no. 1, 2020, doi: 10.1186/s40537-020-00379-6.

[59] G. Kocher and G. Kumar, "Performance analysis of machine learning classifiers for intrusion detection using UNSW-NB15 dataset," in *6th International Conference on Signal and Image Processing (SIGI 2020)*, pp. 31–40, 2020, doi: 10.5121/csit.2020.102004.

[60] A. Wani and S. Revathi, “DDoS detection and alleviation in IoT using SDN (SDIoT-DDoS-DA),” *J. Inst. Eng. Ser. B*, vol. 101, no. 2, pp. 117–128, 2020, doi: 10.1007/s40031-020-00442-z.

[61] A. Ferriyan, A. H. Thamrin, K. Takeda, and J. Murai, “Feature selection using genetic algorithm to improve classification in network intrusion detection system,” *Proceedings – 2017 International Electronics Symposium on Knowledge Creation and Intelligent Computing (IES-KCIC)*, January, vol. 2017, pp. 46–49, 2017, doi: 10.1109/KCIC.2017.8228458.

[62] K. A. Sadiq, A. F. Thompson, and O. A. Ayeni, “Toward healthcare data availability and security using fog-to-cloud networks,” in *Intelligent Interactive Multimedia Systems for e-Healthcare Applications*, A. K. Tyagi, A. Abraham, and A. Kaklauskas (eds.). Singapore: Springer Singapore, 2022, pp. 81–103.

Chapter 13

Addressing the security challenges of IoT-enabled networks using artificial intelligence, machine learning, and blockchain technology

Garima Verma[1] *and Shiva Prakash*[1]

Abstract

The Internet of Things (IoT) usage in 6G networks is one of the most trending and rapidly growing domains, amalgamating with many new technologies like machine learning (ML), deep learning, and blockchain. Due to this, smart devices are improving in terms of various parameters like efficiency, complexity, reliability, and so on. IoT helps track and monitor the 6G networks in communication and processing and tackles various security challenges. The implementation system of these technologies comes with many heterogeneous challenges that require specific protocols to overcome these issues. Furthermore, the advent of such a huge amount of data being generated with these systems has resulted in different types of security issues in such 6G frameworks. Therefore, this survey addresses the major challenges in implementing and deploying these IoT-enabled 6G frameworks. In this survey, an exhaustive literature review has been done. Various attacks that can take place in any IoT-enabled 6G platform have also been covered in detail. Also, the challenges related to the security aspects of 6G networks have been discussed, and their corresponding solutions have been proposed. Further, critical analysis of these issues being addressed with the trending technologies has also been discussed.

Keywords: IoT-enabled 6G network security; IoT protocols; IoT applications; IoT attacks in 6G; Artificial intelligence; Blockchain technology; Machine learning

[1]Department of Information Technology and Computer Application, Madan Mohan Malaviya University of Technology, India

13.1 Introduction

There have been many advantages and disadvantages in the domain of 5G networks which led to the invention of 6G networks. Until now, there have been no standardized specifications for the 6G network, but many issues could be addressed in 6G networks, which are already there in 5G networks. Security issues may be related to infrastructure, models, architectures, or platforms. These issues of 6G networks could be easily resolved using the latest artificial intelligence (AI), machine learning (ML), and blockchain technologies. Execution of Internet of Things (IoT) accompanies parts of difficulties. The normalization, interoperability, information capacity, handling, trust of the executives, character, classification, honesty, accessibility, security, and security are a portion of the open difficulties in different IoT applications [1]. IoT comprises many arising innovations which are regularly developing, and its purposes are increasing in various application regions. There are many difficulties related to IoT-enabled frameworks' security and privacy aspects. Many projects are working in this domain where IoT is being integrated with all recent technologies like ML, blockchain, etc. The 6G network would be considered as the amalgamation of existing 5G functionalities, tools and equipment's of latest technologies, and other domains of virtualization functions. In all scenarios, if any new innovation comes into the market, it has so many security challenges. In fact, the 6G networks would be considered a network that will use major functionalities of AI. Moreover, 6G networks could be treated as an AI-driven network that is supposed to use the tools and techniques of AI. Therefore, this work clearly describes the three ongoing innovations that help address all types of security issues. The workflow of the chapter is given in Figure 13.1.

There are so many challenging issues of 6G networks that large organizations capture on a regular basis where the requirement of suggesting new IoT security approaches has been increased. Some numerous companies and organizations have taken keen interest in deployment of 6G networks like Ericsson, LG, Samsung, and so on. The important aspect of 6G networks is that they will

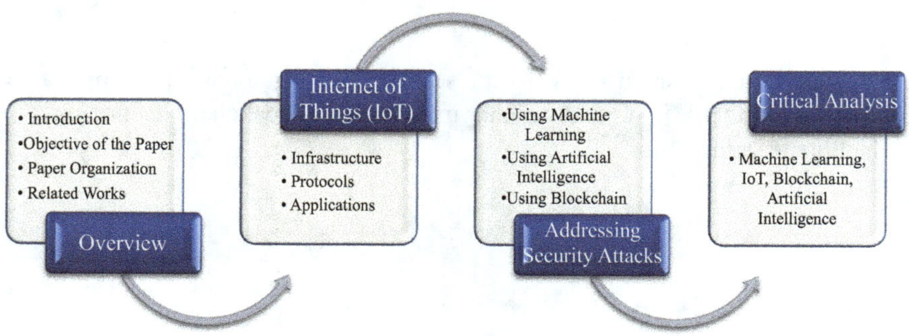

Figure 13.1 Workflow of the chapter

support more diverse applications than the previous ones and enhance the productivity and parameters of new enabling technologies like augmented reality (AR) and virtual reality (VR). Further, these technologies have a broader scope in AI, ML, IoT, and blockchain fields. Also, the mobile operators will adapt to a flexible network developing a decentralized business process models for 6G network with features like compatibility, automation, better communication features, decentralized architectures, etc. Since there is no standard till now which should be accepted for 6G networks, the standards and features are still in the developing phase. Therefore, IoT incorporated various methodologies, approaches, standards, and functionalities included in 6G networks giving rise to various weak parameters to these running organizations. Due to the functionalities and vast expansion of IoT gadgets, it has become very important to understand and suggest some new approaches regarding security concern of 6G networks because the IoT data is very sensitive and crucial. High level cryptographic algorithms, specifically designed APIs, proper verification, and authentication channels are required to tackle these arising cyber threats and crimes in weak IoT-enabled 6G networks and components. IoT infrastructure is mainly used to achieve more human to machine interaction and efficient sensor-based operations for better utility. It enables the user's better data management and data analytics for 6G networks. Since these areas are divided into small regions, it becomes a challenging task to first ensure data and security breaches in smaller regions. Then, these regions are amalgamated and security measures are applied over large geographical domains to ensure data transmission security and privacy. IoT-based infrastructure for 6G networks helps in better decision making and benefits many industries and organizations in various ways. There are many research gaps in IoT domain for 6G networks and the most crucial one is the security aspect. As the online data is increasing day-by-day, more secure methods are demanded by the researchers to improve data privacy and data authentication in 6G networks services could not be delivered to the users.

There is a great problem of interoperability and design-related issues in 6G networks, which should be resolved early. There are other aspects where IoT parameters get influenced like security, reliability, robustness, compatibility, heterogeneity, homogeneity and so on in 6G networks. In multimedia and telecommunications, 6G technology is the latest for mobile cellular networks, which uses wireless sensors for data transmission. The data transmission rate will be much faster than the 5G and others. Basically, the 6G networks will be using the broadband networks where the service areas are divided into small geographical domains for data transfer called as cells. Another area of research gap is the monitoring and analysis of real-time data and then protecting it from unintended users simultaneously is a tedious task in 6G networks. Also, the quality of services provided should be up to the mark, otherwise quality will be degraded. Figure 13.2 discusses the overview of different generations of cellular wireless networks. Their privacy and security concerns of each network have also been shown.

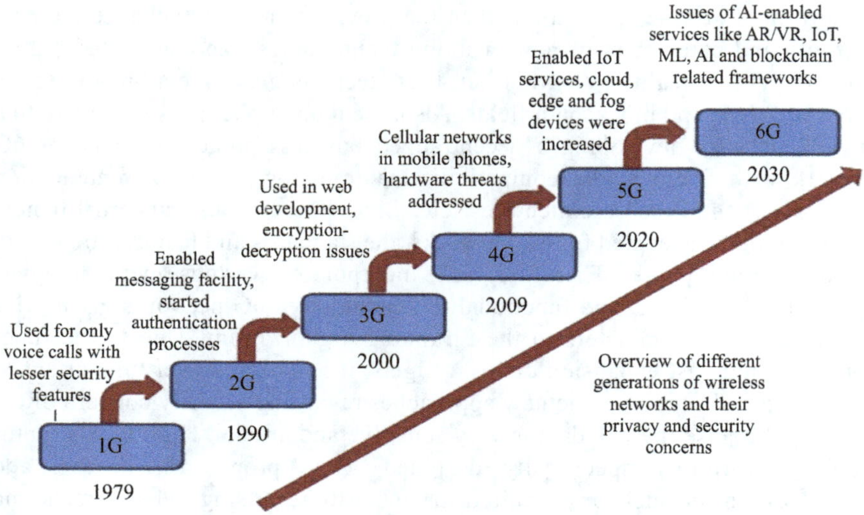

Figure 13.2 Overview of security and privacy issues in wireless networks

13.1.1 Objective

The major reason for selecting this domain is to examine different types of security challenges and suggesting some important innovations of 6G networks which can help to improve the major parameters of IoT-enabled systems. The authors have identified various domains where security challenges are causing major problems in implementing IoT frameworks in 6G networks. Coming up next are the commitments of the chapter:

- This chapter comprises of brief introduction and exhaustive study of 6G networks for IoT domain and its significance.
- Further, the security challenges on each layer of the IoT framework for 6G networks have been discussed.
- A brief overview of the 6G networks has been discussed along with the introduction. With the emerging trend of latest technologies, the privacy and security concerns of the 6G networks have also been discussed.
- A broad study on comparative advances like AI, artificial knowledge, and blockchain innovation in context with the 6G networks that have been amalgamated with IoT-enabled systems.
- The major issues and relating arrangement approach in 6G networks involving the emerging technologies (blockchain, AI, and ML) are likewise made sense of.

13.1.2 Chapter organization

The underneath section of the survey is starting with the related work and the literature review in Section 13.2. Further, Section 13.3 comprises the major components comprising the IoT infrastructure, various types of protocols and applications used in it. In Section 13.4, different kinds of IoT attacks and their corresponding solutions have been discussed. Section 13.5 discusses the major IoT issues with the trending technologies like ML, AI, and blockchain in context with 6G networks. In Section 13.6, summary and critical analysis of the review has been done. Also, some key areas have been elaborated further. In Section 13.7, the chapter winds up with the conclusion part.

13.2 Related work

Various literature works have already been done on privacy and security-related aspects of 1G, 2G, 3G ,4G, and 5G networks. With much better flexibility, these networks covered global geographical locations enabling better communication processes using cellular networks. The creators make sense of the fundamental framework design and security issues in paper [2]. Beforehand a few works connected with a security issue in IoT applications frameworks are as of now done. Table 13.1 references a synopsis of a portion of the overview works. Albeit a few works as of now exist in such a manner according to alternate points of view, for execution purposes, there is no such review done. One of the major drawbacks of these networks is one way authentication which only ensures that the network can easily authenticate the number of users but it becomes difficult to authenticate the network and people by the user. So, in this chapter, creators have recognized the new arising innovation (ML, AI, and blockchain), which can be tended to security issues in IoT. Some of the work incorporating late innovation and IoT have been finished. In [4], the authors have analyzed different security issues of perception, application, and transportation layers. Also, some relevant solutions to the security problems of these layers like integration of cross-layers have been proposed. Ngu *et al.* [3] proposed an IoT-enabled framework for detecting blood alcohol content based on real time scenario using sensor data through smartwatch. The authors have also worked on challenges and capabilities of developing any IoT middleware framework exhibiting qualities like adaptability, reliability, security, etc. Mosenia *et al.* [5] have surveyed different vulnerabilities and security threats on edge level layers especially on edge nodes, communication layers, and edge computing paradigms. Then, they have gone through possible scenarios of potential attackers and applications along with crucial attacks and their threats. Also, they have explained the two most emerging threats which were not noticed before in previous literature. In [8], the authors have first explained the relationship among IoT and cyber physical systems which are important parameters in constructing cyber-physical frameworks. Then, the state-of-the-art of IoT domain includes architectures, technologies, security aspects, and privacy issues. Further, the authors have investigated the relationship among fog/edge computing paradigms and IoT and explained some real-life scenarios like smart grid system, smart cities, smart transportation, etc. Alaba *et al.* [7] have discussed the taxonomy of existing IoT security threats, applications, architectural layers, communication layers, etc. Also, they have examined major security frameworks

Table 13.1 Related works on IoT security for 6G networks

Reference papers	Years	Contribution of the authors
Jing *et al.* [4]	2014	Various security concerns based on 3-tier architecture of IoT and their corresponding solutions have been discussed.
Ngu [3]	2016	The authors have proposed a middleware-based architecture of IoT and further discussed each layer. Further, they have proposed security and reliability features in detail.
Mosenia [5]	2016	In this paper, a survey based on a reference model and security threats has been discussed. Also, the paper reviewed various solutions so that these threats can be addressed properly and further appropriate solutions can be suggested
Lin *et al.* [8]	2017	This paper first discusses the relationship between IoT and cyber physical systems (CPS) integrated systems. Further, security and privacy issues of edge and fog computing has been discussed.
Alaba *et al.* [7]	2017	This paper comprises the security issues in IoT domain. Various threats and countermeasures of architecture, issues and challenges have been reviewed. Further, the paper concludes with different solutions for these security challenges.
Yang [6]	2017	In this paper, the authors have proposed a detailed survey on different issues of IoT security model. Further, a detail study about various security countermeasures and solutions have been discussed.
Das *et al.* [11]	2018	Various models and architectures based on IoT security and their applications have been carefully examined in this survey paper. The paper has carefully observed the issues in IoT-based systems related to access control, trust management, verification, etc.
Di Martino *et al.* [12]	2018	This survey paper has examined different types of architectures of IoT systems and also suggested various solutions to the problems that arise due to security and compatibility issues.
Hassija *et al.* [10]	2019	This paper presents various security related threats in IoT domain. Different solutions have been proposed based on ML, deep learning, edge computing, AI, blockchain, etc.
Mohanta [13]	2020	In this survey paper, first, various algorithms, infrastructures, protocols, and applications of IoT domain have been discussed. Then, major concerns related to IoT security have been discussed in detail. Now for resolving these issues, some advance techniques that can be used in order to resolve them has been discussed. After doing exhaustive literature review, the authors have proposed some latest techniques like deep learning, blockchain, and so on.
Vijaya Kumari [14]	2021	In this survey, many healthcare startups of India have been mentioned. New frameworks have been discussed which plays a major role in many important startups in healthcare industry.
Pranav Ratta [15]	2021	The researchers have identified various healthcare-related issues and done a deep survey on problems in healthcare domain. Also, there are various problems that people are suffering now a days in healthcare domain, such areas have been explored in this survey paper.
This chapter	2023	In this survey, first, a brief introduction of 6G networks and its related security concerns have been discussed thoroughly. Then, how these security breaches could easily be handled using latest technologies like IoT, ML, AI, and blockchain has been deeply discussed. Also, various infrastructure, protocols have been discussed. Further, the chapter covers various applications of IoT domains of 6G networks.

implementation challenges in IoT heterogeneous environment, security threats, and security-related architectures. In [6], the review mainly consists of four segments. The first segment consists of IoT limitations and their specific solutions. Second segment comprises of IoT attacks and third one explains architectures of access control mechanism and authentication. The layer examines the major security issues at various layers. Das *et al.* [11] have discussed threat model and major attacks of IoT environment including taxonomy, security services like user authentication, verification, access control, identity management, etc. Also, the authors have discussed a comparative study of various protocols and functionalities of IoT security protocols. Martino *et al.* [12] have discussed about security and interoperability parameters and analyzed a detailed comparative study on various architectures of IoT-enabled environment and also proposed solutions for major security issues. Hassija *et al.* [10] have discussed various end-to-end secure IoT architectures along with a detailed review on security-related challenges and issues. Also, the role of blockchain, fog computing, ML, etc. has been discussed in order to improve the security of IoT-enabled frameworks. Mohanta [13] has discussed the overview of IoT technology and major security issues comprising of Confidentiality, Integrity and Availability (CIA) triad.

Also, the authors have examined the domains of ML, AI, blockchain, etc. Vijaya Kumari [14] has proposed a novel communicational framework for analyzing the home environment using advanced features and switching functionalities. Also, they have demonstrated the effectiveness and functionalities of different sensors used in lights, switches, temperature sensors, motion sensors, etc. Pranav Ratta [15] has discussed various functionalities of healthcare systems using recent frameworks of blockchain and IoT. The authors have explained the applicability of these technologies in major healthcare domains like patient monitoring, drug traceability, etc. Basically, the healthcare scenarios of deploying major IoT frameworks have been discussed. The related works on IoT security and their key contribution are summarized in Table 13.1.

13.3 IoT architecture, protocol, applications for 6G networks

Internet of Things (IoT) has different possibilities to implement in various constant areas. It coordinates actuators, sensors, shrewd gadgets distinguishing proof, and the Internet to fabricate a canny framework. According to a report, Goldman Sachs assessed that around 29 billion savvy things would be associated with an alternate organization by 2020. The developments of 6G networks somewhat recently have helped to gather the data for distributed computing middle of the road with haze/edge figuring. The IoT has various kinds of an organization like disseminated, lattice, and vehicular networks [7]. The uses of IoT had a tremendous effect on everyday life like sensors convey in the patient body to observing in basic condition, observing gas spillage in shrewd kitchen, farming field, brilliant vehicle leaving, savvy transportation, following products subtleties in inventory network framework.

13.3.1 IoT infrastructure

IoT domain comprises of various advanced things that gather, process, store, compute and speak with other shrewd processes. IoT consists of three main layers: physical, organization, and application. As of late, businesses are created numerous things which are inserted with savvy things. As displayed in Figure 13.2, the IoT architecture comprises of sensors, however it likewise coordinates with some arising innovation. The security issues like information privacy [16], end-to-end communication [17], ongoing tracking [18], and IoT-based testbed [19] should be tended to for efficient IoT applications. IoT application handling and registering in real time is quite possibly the most difficult issue. Distributed computing gives more capacity and guarantees security to the information. The different shrewd gadgets are associated with an application utilizing a few standard conventions. The information interoperability and reliability [20] in the IoT framework work utilizing a shrewd calculation.

13.3.2 Standard protocols

For a 6G network that needs to be deployed, numerous protocols are used for data transfer and information processing. Many protocols are required to control and monitor the overall processing of the IoT frameworks. Some of the important protocols that are widely used for all such mechanisms are listed in Figure 13.3.

13.3.2.1 Message queuing telemetry transport (MQTT)

MQTT represents transportation of MQ telemetry. It is a clear what's more, lightweight informing convention for distributing/buy in, intended for restricted gadgets and low data transfer capacity, high inertness, or untrustworthy organizations. The plan standards are to limit the prerequisites for network data transfer capacity and

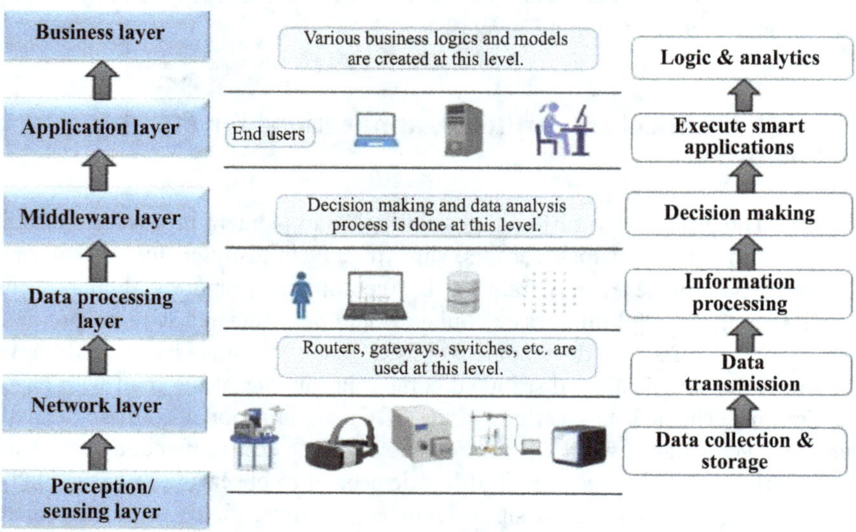

Figure 13.3 Layer-wise components of IoT infrastructure for 6G networks

gadget assets while likewise attempting to guarantee dependability and some level of delivery confirmation. These standards additionally bring about making the convention ideal for the arising universe of low end associated gadgets "machine-to-machine" (M2M) or "Web of Things."

13.3.2.2 Constrained application protocol (COAP)

CoAP protocol is one of the very important protocols which is a part of Internet Application protocol. This protocol allows only the authorized users to enter into a network and perform various tasks. It has been characterized in RFC 7252. Further, it limits the number of gadgets called as hubs to enter into a network for better performance. It is used by many gadgets for equal sharing of the network.

13.3.2.3 REpresentational State Transfer (REST)

This protocol is used to represent the State Transfer Member. It uses HTTP convention and follows all the norms of Internet. It treats every component present in a network as an asset and then follows up a certain HTTP protocol technique for better network interaction. This protocol came in 2000 where REST servers are used for data storage. Here, REST clients are adjusted according to the assets that are utilized in a particular network. Specific URIs or IDs are used to characterize every component in a network for depicting files like XML, HTML, JSON, etc.

13.3.2.4 Advanced message queuing protocol (AMQP)

This protocol is an open-source protocol for middleware layer and used for message transmission from source to destination. It helps in improving various parameters like security, reliability, orientation, and so on. It uses the concept of publishers and consumers for message transmission. Role of publisher is to create the message and consumer helps in processing the message.

13.3.2.5 Transmission control protocol (TCP)

One of the important protocols belongs to the Internet protocol suite and is for reliable data transmission. In this, client–server architecture is being followed, ensuring the secure and reliable data transmission.

13.3.2.6 User datagram protocol (UDP)

This protocol is basically used for better data communication in any network. It helps in achieving low-latency communications. Also, it helps in preventing data loss between various connections available on different applications through www. This protocol tends to speed up the data transfer rate by signing the receiver's agreement before the process starts.

13.3.2.7 Datagram congestion control protocol (DCCP)

DCCP allows a blockage control of different components without taking them to the application layer. It works on the transport layer and ensures reliable connections for set up in a network. It is published in RFC 4340. It greatly checks the congestion control and connection tear down activities. Also, it manages the traffic present in the network.

13.3.2.8 RSVP

It is a resource reservation protocol used by all the routers and hosts in a network. Basically, the hosts tend to use this protocol to request some specific type of service, also known as CoS for better message transmission. Further, the routers use this protocol for transmitting the data through all the routers in the same path. This protocol has been established in RFC 2205.

13.3.2.9 Quick UDP Internet connections (QUIC)

It is a general-purpose protocol used at the transport layer. This protocol quickly resolves all the issues related to the advanced Internet applications occurring at application layer without any prior consent. It was developed by Google in 2013 and works on Google's server. Further, Chrome has implemented this protocol for general purpose usage for public.

13.3.3 Applications of IoT-enabled 6G networks

Nowadays, the whole world is somehow equipped with the latest technologies build with IoT and ML. A lot of open-source free platforms have been developed like Amazon Web Services (AWS), Microsoft Azure, Oracle IoT, and many more for building industry compatible devices which can be further used to develop large-scale applications. Many open-source platforms have been developed with AI and ML-driven technologies for rapid processing and managing a large amount of data in 6G networks. IoT is emerging as one of the most promising domains nowadays. The development of smart devices like smart sensors, actuators, etc. has helped researchers read, store, and process the data more efficiently for any 6G-enabled network. A lot of applications of IoT in any 6G network can be easily seen in Figure 13.4, which positively impacts the efficiency and productivity of the system.

13.3.4 Key areas of 6G networks

There are several components of 5G network that has already been deployed. Among these components, major components have been deployed using AI. This has practically made many things possible like detection of threats and estimation of better option at the MAC layer and looking for data breaches in various networks. For developing more advanced architectures, AI-enabled tools have already been used in 5G networks but a number of constraints are there in traditional 5G

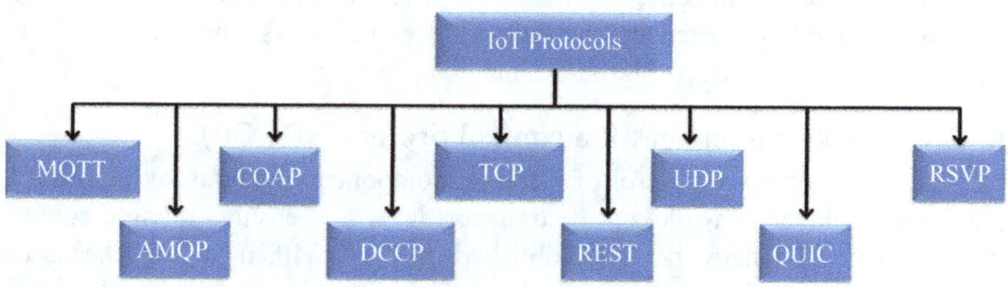

Figure 13.4 Standard IoT protocols used for 6G networks

network like latency, Quality of Service (QoS), throughput, etc. It will be challenging to accomplish such methods that could be easily deployed in 6G networks using the intelligent AI and ML features.

Moreover, the framework of real-time systems has already been deployed in 5G networks like in vehicular ad hoc networks but in this, the urgent situations could not be tackled early or in real time due to many latency-related issues.

13.3.4.1 Smart homes

The IoT is one of the most demanding fields, converting a normal traditional home into an intelligent one. People are using LEDs, refrigerators, cameras based on sensors, smart watches, smart alarms, etc., which collaborate all together and connect the entire environment to sense the surroundings and further connect and communicate with the other devices through wired or wireless mediums in 6G networks. Due to the usage of these smart sensors and devices in 6G networks, people's lives have become more luxurious and comfortable. In [21], the authors have discussed about how to design a smart home with the help of IoT technology and 6G networks. These latest technologies like the IoT, fog, and edge computing have helped people develop such applications where an intelligent home can be monitored easily and the data can be processed at any time. But, for the better processing and working of these applications of 6G networks, the framework needs to be protected from the unauthorized users into the IoT network. Hence, in papers [22,23], the researchers have discussed various authentication schemes for running and processing the data over a large IoT network. Further in paper [24], many security issues have been discussed in IoT-enabled smart homes.

13.3.4.2 Smart hospitals

With the help of advancements in IoT domain in 6G networks, now it is possible to track the status of any patient remotely with the use of sensors, actuators, and other tracking devices. Like, in paper [25], the authors have demonstrated a framework based on IoT which is using cloud-based features for data collection and processing in the healthcare domain. Further, in paper [26], the authors have worked on authentication and authorization of incorporating smart devices in the healthcare frameworks. In these healthcare systems, privacy is one of the major concerns where security and privacy of the healthcare system should be ensured.

13.3.4.3 Smart city

With the ever-growing cities, a lot of problems of 6G networks have been arising like traffic management, waste management, and so on. For developing a smart city, there needs to be a solution for monitoring and controlling traffic management problems [27]. Further, in [28], the authors have discussed various challenges and issues which are arising for creating and developing a smart city. Also, a proper survey has been done to find solutions for these existing problems in implementing such a smart city. By using these smart IoT-enabled devices of 6G networks, smart city can be implemented to make people's lives more efficient, secure, and comfortable.

13.3.4.4 Smart transportation

Nowadays, traffic management is one of the major concerns in a city. So, the need of the hour is to develop and work upon such an intelligent traffic management system which can collect the data of vehicles on prior basis from the roadside unit and further process that data to retrieve the information about vehicles, distance, and all the traffic details in any 6G network. The authors have discussed some smart transportation issues based on IoT in papers [29,30]. Further, in [31], the authors addressed the IoT-enabled ITS system for the transportation system. In paper [32], authors have discussed a framework known as "Magtrack" where one can find the road conditions by using sensors and IoT-enabled ML techniques.

13.3.4.5 Smart grid

A smart grid is an IoT-enabled 6G network application where a grid-like system can be made using automation features and sensors. By these systems, generating the electric power and distributing it among different users in real time becomes much easier. The status can be monitored and tracked in real time for determining various factors of 6G networks. One of the cyber security approaches for grid management has been discussed in paper [33]. Further, another architecture has been proposed in [34] based on IoT cloud system. Various factors like efficiency, cost, speed, accuracy of IoT-enabled 6G network systems can be improved further for developing any IoT-based grid system.

13.3.4.6 Supply chain system

IoT gadgets can monitor and trace the transport system with the latest devices like sensors, actuators, and so on in 6G networks. This has replaced the traditional way of tracking the activities of the transport system with the advanced approach. With this, now the manufacturer can know about the exact details of the product's location and also can easily monitor and track the current status of the product like where is the product packed or where it has reached till now through supply chain management using 6G network. It has increased the efficiency and storage [35] parameters and also this paper [35] also discussed the real-time product tracking and monitoring using the supply chain management. Further, in [36,37], the authors have discussed various architectures and their associated risks in detail.

13.3.4.7 Smart retails

Smart retailing is one of the major applications of IoT where manufacturers, wholesalers, and buyers will get a platform to adapt real-time operations, which in turn will increase the productivity and operations of the inventory system in any 6G network. In [38], different retail sectors have been discussed which are developed using the IoT-enabled devices and AI.

13.3.4.8 Smart agriculture

Nowadays, smart agriculture is a very trending domain of IoT-enabled 6G systems. Many farmers are availing of the latest facilities provided by these gadgets like crop monitoring, alert systems in fields, soil irrigation, and so on. Also, the smart IoT devices help the farmers to check the soil quality in an efficient way, providing proper water to the crop fields; check the real-time status of the crops growing in a specific season which is somehow reducing the burden of the farmers in terms of time, cost, and efficiency of 6G networks. In [39,40], the authors have developed a smart irrigation system using ML and IoT to enhance the features of smart farming. Further, in papers [41,42], the authors have discussed another type of smart irrigation system with smart features. With a lot of advents of smart agricultural devices, a lot of data is required to be managed regularly, and many security issues associated with these 6G networks and devices is a major challenge for the researchers.

13.4 Attacks in IoT-enabled 6G systems

Various IoT attacks that occur in any IoT-enabled 6G system and cause several problems by capturing the data are discussed. Different security attacks in IoT platforms are presented in Table 13.2.

Table 13.2 Possible security attacks in IoT platforms in 6G networks

References	Attacks	Description
[43]	DoS attacks	The DoS attack generally occurs whenever simultaneous systems try to block the resources of any targeted system.
[44]	Malicious nodes	These are generally faulty nodes that occur in a system due to some specific reason which results in abnormal behavior of a system.
[45]	Tempering attack	This attack modifies the major parameters between a client and a server for manipulating the data being circulated in a network.
[46]	Jamming attacks	This attack tries to transmit various radio signals which disrupt the communication of a network by decreasing the SINR value.
[47]	Intrusion detection attack	This is an attack which helps in identifying any specific attack long before a successful attack is likely to take place.
[48]	Wormhole attack	In this attack, majorly two attackers try to listen to a network and then locate themselves strategically and further record the information circulated in the network.
[49]	Distributed DoS	It is a large DoS attack where the intruder tries to used more than one IP address and many host machines become malware infected.
[50]	Man-in-the-middle attack	It is a cyberattack when different attackers try to intercept a network and manipulate the conversation by pretending or by eavesdropping.
[51]	Side channel attack	Instead of targeting the software, this attack majorly exploits the program execution by

(Continues)

Table 13.2 (*Continued*)

References	Attacks	Description
		indirect effects on a system's hardware.
[52]	Access control attacks	Such attack happens when a missing access control check or wrong access control check occurs.
[53]	Impersonation attacks	In this attack, an attacker pretends as a trusted person to steal a network's information by carrying out fraudulent activities.
[54]	Spoofing attack	In this, a spoofer pretends as a trusted person to get the details about the target, then steals the shared data and performs certain actions based on that.
[55]	Sybil attacks	In this attack, the attacker tries to subvert the communication network by building various pseudonymous identities and further uses its data to perform attack.
[56]	Active attacks	The attacker tries to change or modify the content being shared in a network and the victim gets to know about the attack on a prior basis.
[57]	Deceptive attacks	In this, the attacker tries to obtain the confidential data from the users and then uses his data to implement other attacks as well like fake e-mail.
[58]	Botnets attacks	It consists of large-scale attacks to disrupt a network's normal functioning and tries to grab unauthorized access of critical systems for remotely operating the devices.
[59]	Identity and data theft attacks	This attack includes usage of another person's details or identities for personal profits for example, malicious software, grabbing bank account details, etc.
[60]	Ransomware attacks	It is a malware attack where the intruder locks and encrypts the user's data using some important keys and then demands money to decrypt the data. For example, malicious links available on different sites.
[61]	Advanced persistent threats	A group of people is involved in such a type of attack where they target any specific audience or organization for any unique objective.
[62]	Brute force attack	This attack happens when an attacker tries to guess victim's login detail by trying all possible combinations of the passwords. For example, login details of any account.
[63]	Privilege escalation attack	For obtaining the elevated access to of different resources of any operating system, the attacker tries to first diagnose the bugs or any flaw and then enters into the application domain of a system.

13.5 Analysis of security challenges and issues in 6G networks

A 6G network will enable numerous sophisticated tasks to be done easily like tracking the precise location of a person in a room in order to find a consecutive pattern of a patient's daily routine. This crucial data needs to be secured from unwanted people or hackers. Fraudulent attempts should be minimized with the latest techniques introduced by ML, IoT, etc. Nowadays, business data plays a leading role in predicting future benefits and losses for an organization.

The 6G network will help to access and locate the business data more frequently using smart control devices. So, the confidentiality and privacy of these attacks is a major concern for customer-related processes and business data. In such scenarios, security threats, ransomware attacks, and other deep fakes problems will be challenging to manage and handle. Other domain of security is the unparalleled processing of data and information in the era of quantum computing where attackers and unauthorized hackers can easily penetrate into a 6G network and hack the cryptographic algorithms and security barriers.

13.5.1 Using ML techniques for 6G-enabled IoT security issues

ML is the latest technology that is being used in all sorts of developments and applications of 6G networks. It helps to perform the tasks intelligently and flexibly. A lot of computations are there in the development process of any IoT framework in 6G networks using the traditional method. So, ML plays a major role in solving these computations of 6G networks. Whenever a model is being created, it needs to be designed, trained and tested using different ML approaches. Figure 13.5 explains the basic working of ML process and further, it also describes the integration of ML with 6G technologies using IoT devices. Since, many applications of IoT domains in 6G networks are already discussed above. Such applications also need prior decision making for better aspects before the original event takes place in 6G networks. For example, predicting the risks involved in the stock market will help the investors better understand the stocks at an initial phase or predict fire in a kitchen and using alarm sound to prevent the fire. Now, all these things are possible by using the integration of ML concepts with IoT in any 6G networks. Further, security is again one of the major issues that need to be addressed for creating and developing a tamper-proof system of 6G networks. In paper [64], an efficient framework needs to be designed to using 6G networks to collect, store, and process huge data using IoT and ML techniques. In [65], the researchers have properly reviewed the major security issues which mainly occur on deploying the ML and IoT devices. In paper [66,67], the researchers have discussed about a type of intrusion detection system using IoT. Figure 13.5 shows different ML techniques integrated with 6G networks and IoT devices. In Figure 13.5, the first step that has been reflected is to collect and aggregate the data at a specific location using sensors, actuators and other wireless devices. Once, the data has been gathered, it is filtered through algorithms. The relevant data is taken forward for processing and the

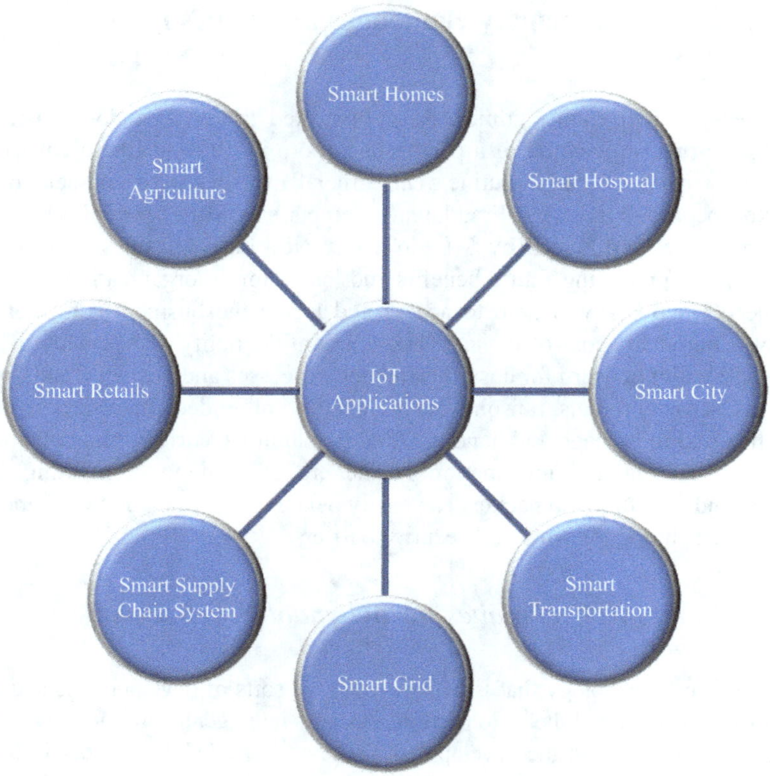

Figure 13.5 IoT-enabled 6G-related applications

unwanted data is left. Further, the processed data is taken and some categories of ML algorithms are applied. These algorithms are categorized into supervised learning, unsupervised learning, and reinforcement learning.

In Figure 13.5, various components of ML techniques have been discussed. With the help of IoT devices in 6G networks, the data is being collected from all over to further process and enhance the clarity of data. The data is being accumulated at a specific server or location which is then filtered and checked for important data. With the use of specific sensors and devices, the data is being categorized and then specific algorithm is being used which helps in data processing and data classification for 6G networks.

13.5.2 Using AI techniques for 6G-enabled IoT security issues

The capability of smart IoT devices like sensing and acting makes the 6G network frameworks more usable worldwide. Since a large number of devices are being connected all together to store and process the data further in 6G networks, a huge amount of data is also required. Therefore, it includes a major challenging task of

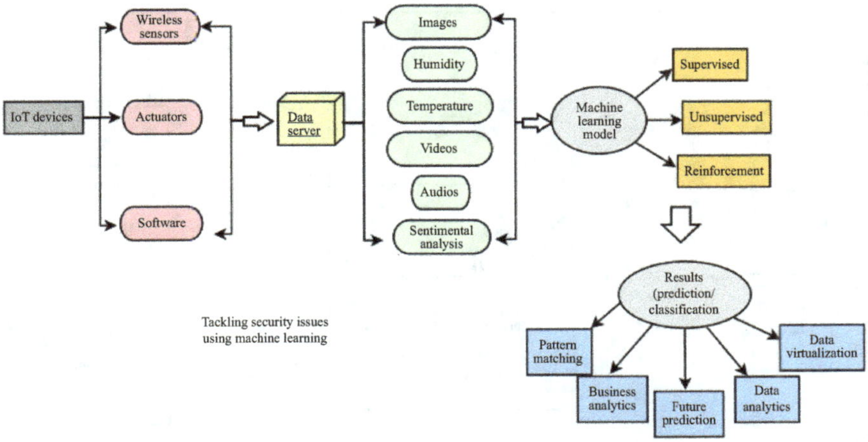

Figure 13.6 ML techniques for IoT-related issues for 6G networks

processing and performing calculation in a 6G network for IoT environment. Therefore, for addressing these issues, AI comes into picture. Since the integration of 6G networks with IoT and AI helps the system to improve its capability, efficiency, accuracy, and also improves the overall analysis rate of the system. In [68], the authors have explained how AI may help large IoT data, heterogeneous data and unstructured data to improve system accuracy in a real-time scenario. Further, in [69], researchers have discussed Large Margin Cosine Estimation Technique (LMCE) to detect different IoT security-related concerns in IoT-based environments. Further, in [70], an integration of IoT devices with blockchain technology has been discussed for making an IoT-enabled framework tamperproof. In Figure 13.6, some common integration of IoT devices with AI has been shown. Major examples of these integrations of 6G networks include robotics, driverless cars, smart city applications, and so on. In Figure 13.6, once the data is being collected, then the data is being processed using any major algorithm of AI for better data analytics. The IoT-enabled infrastructure of 6G networks helps improve the system's decision making and performance so that the IoT-enabled infrastructure could work properly.

AI helps incorporate the IoT integration components in 6G networks, leading to better data collection and decision-making process. Also, this domain explores many aspects of IoT healthcare like hospitals, clinics, and healthcare organizations enabling faster data collection and data processing of 6G networks. These fields are providing tremendous applications and software tools to many healthcare fields using 6G networks.

13.5.3 Using blockchain technology for 6G-enabled IoT security issues

Blockchain is a kind of distributed structure that stores the data/transactions in form of records, also called blocks in different public ledgers through a peer-to-peer connected network. Further, the data is broadcasted in this blockchain network.

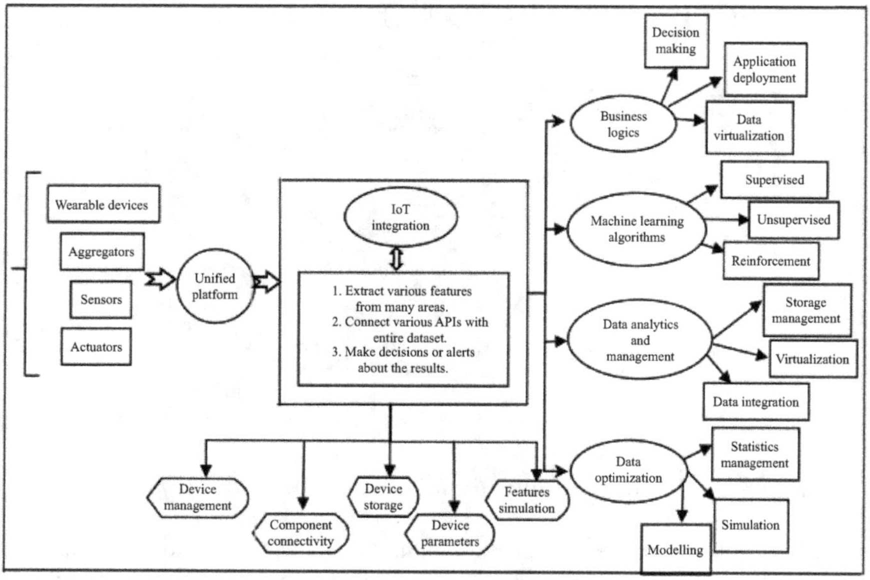

Figure 13.7 AI techniques for IoT-related issues for 6G networks

A single block contains many valid transactions and different sets of attributes. A very famous application of blockchain is smart contract which can be used for 6G networks for better communication. These contracts are executable programs which help in deploying any business logic into a communication network. Further, these blockchain networks use the consensus algorithm to gain mutual consensus among various nodes in any 6G networks. In paper [71], various blockchain architectures and applications have been discussed in detail. In paper [72,73], the work related to IoT security by integrating blockchain as a solution has been discussed. Many securities-related challenges in 6G networks for IoT can be further solved using blockchain technology. For example, healthcare records can be maintained easily with the help of blockchain technology in any 6G networks.

In Figure 13.7, smart execution of IoT healthcare data has been shown. Blockchain helps to preserve important data's confidentiality and privacy for better data management in any 6G network. Smart contracts reduce the security risks of threats and cybercrimes of healthcare data in 6G networks. Blockchain helps to eliminate the security risks by removing the central authority and using a decentralized framework. It works on forming a common consensus by executing different transactions using pre-defined operations in 6G networks.

13.6 Summary of the review

In this chapter, the authors have mainly tried to focus on all the security aspects of 6G networks for IoT domain and further suggest solutions for these problems of 6G

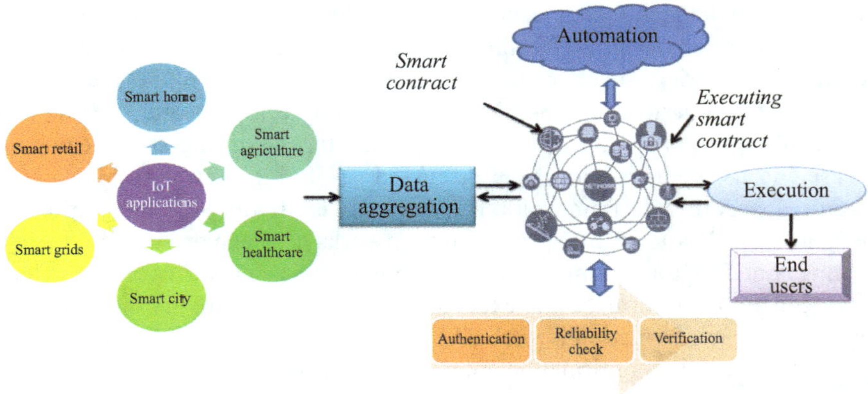

Figure 13.8 Blockchain technique for IoT-related issues for 6G networks

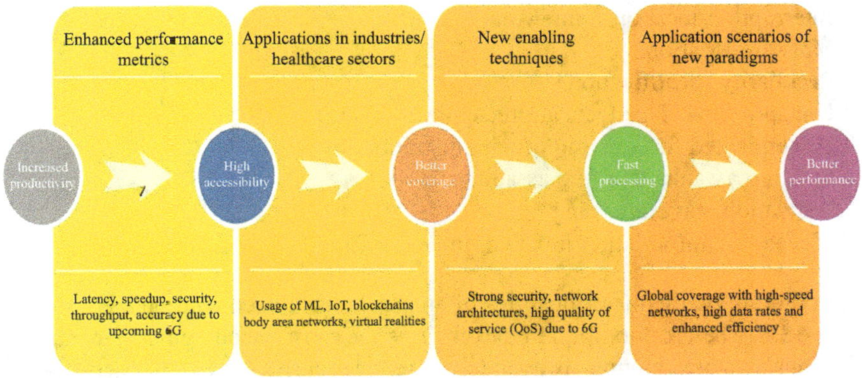

Figure 13.9 Key points of the survey for 6G networks

networks using ML, AI, and blockchain technology. The major goal of this survey is to identify and analyze the solutions for many securities related problems of 6G networks. Since, security is one of the major challenging problems of 6G networks and needs to be addressed as soon as possible in order to develop a successful framework for 6G networks. Figure 13.8 discusses various features and advantages of the 6G network for fast data processing over a communication channel. Although there are many features of 6G networks over the 5G networks, a lot of privacy and security challenges are there that need to be addressed (Figure 13.9).

13.6.1 Critical analysis of ML, AI and blockchain technology

The ML techniques majorly consist of two types: supervised learning and unsupervised learning. In paper [74,75], the authors have discussed various issues and challenges which occur while implementing these IoT-integrated networks. Further, in [76–78], the researchers have discussed major architectures and security aspects which appear in IoT domains. In [79], the authors have suggested various green-IoT methods for developing and creating an ecofriendly environment. In [80], an approach to classifying the pneumonia disease and its prediction has been discussed. In [81], a survey on blockchain towards a secure ML framework has been covered. In [82], integration of blockchain with various software-defined networks has been discussed. The IoT systems contain a ton of information which is being handled further for calculation. Since, this information is exceptionally urgent and private thus, holding security of such information is vital. The information, first and foremost, should be separated by the confirmation interaction and afterward, the overt repetitiveness of the information ought to be checked. Many articles [83] have been investigated to get into the profundity of the serious issues emerging in IoT security. In this way, AI methods can distinguish and give answers for different issues referenced beneath:

- Interruption location framework.
- Malware location.
- Peculiarity identification.
- Unapproved IoT gadgets are recognizable proof.
- Disseminated forswearing of administration.
- Sticking assault, spoofing assault.
- Validation, eavesdropping.
- Misleading information infusion, impersonation.

In this study, the major qualities of an article have been recorded as shown in Figure 13.8. The creators found blockchain is the most encouraging innovation as of late analysts are chipping away at to tackle the security issue of 6G networks for IoT applications:

- Identity confirmation.
- Firmware recognition and self-recuperating in 6G networks.
- Data uprightness and secure correspondence.
- Verification and approval for 6G networks.
- Information sharing and authorization control.
- Secure capacity and calculation.
- Trust issues and complications of 6G networks.

According to the overview done in this chapter, creators considered around 83 papers for analyzing and concluding the future scope and research challenges of 6G networks for IoT domain (Figure 13.10). The expectation is fundamental in an application like shrewd transportation and savvy weather conditions estimating. The AI gives some of the security issues of 6G networks like malware location, protection and conservation of IoT devices, and approval.

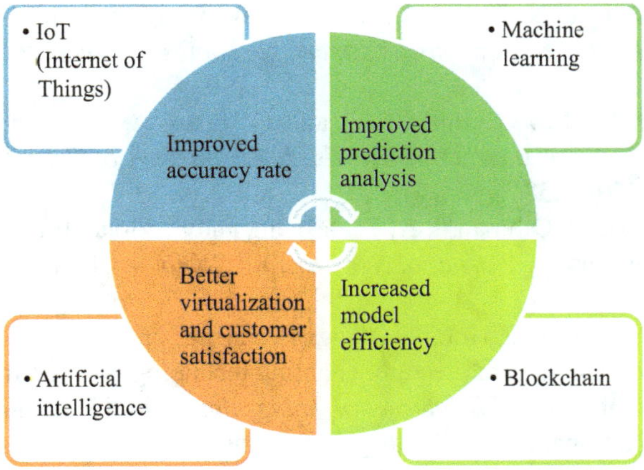

Figure 13.10 Key points of the survey for 6G networks

13.7 Conclusion and future scope

In this chapter, the authors first and foremost concentrate on the different security challenges in 6G networks concerned with IoT applications. Also, the creators have overviewed to address existing security challenges of 6G networks. The study found that some examination has previously been finished in different innovations like ML, artificial knowledge, and blockchain innovation for 6G networks, which are equipped for tending to the current security issue of 6G networks. Since the growth of IoT is exponentially increasing, the need is to examine different parameters of 6G networks like connectivity, reliability, storage, efficiency, complexity, security, etc. So exhaustively, the concentration has been made in three innovations. A portion of the exam challenges notice eventually. Apart from this, there are many more applications and domains of 6G networks for which future directions are still required for better productivity and innovations based on IoT-enabled infrastructure.

References

[1] A. Čolaković and M. Hadžialić, Internet of things (IoT): a review of enabling technologies, challenges, and open research issues, *Computer Networks* 144 (2018) 17–39.

[2] D. Mocrii, Y. Chen, and P. Musilek, IoT-based smart homes: a review of system architecture, software, communications, privacy and security, *Internet of Things* 1 (2018) 81–98.

[3] A. H. Ngu, M. Gutierrez, V. Metsis, S. Nepal, and Q. Z. Sheng, IoT middleware: a survey on issues and enabling technologies, *IEEE Internet of Things Journal* 4 (1) (2016) 1–20.

[4] Q. Jing, A. V. Vasilakos, J. Wan, J. Lu, and D. Qiu, Security of the Internet of Things: perspectives and challenges, *Wireless Networks* 20 (8) (2014) 2481–2501.

[5] A. Mosenia and N. K. Jha, A comprehensive study of security of Internet-of-Things, *IEEE Transactions on Emerging Topics in Computing* 5 (4) (2016) 586–602.

[6] Y. Yang, L. Wu, G. Yin, L. Li, and H. Zhao, A survey on security and privacy issues in Internet-of-Things, *IEEE Internet of Things Journal* 4 (5) (2017) 1250–1258.

[7] F. A. Alaba, M. Othman, I. A. T. Hashem, and F. Alotaibi, Internet of Things security: a survey, *Journal of Network and Computer Applications* 88 (2017) 10–28.

[8] J. Lin, W. Yu, N. Zhang, X. Yang, H. Zhang, and W. Zhao, A survey on Internet of Things: architecture, enabling technologies, security and privacy, and applications, *IEEE Internet of Things Journal* 4 (5) (2017) 1125–1142.

[9] P. I. R. Grammatikis, P. G. Sarigiannidis, and I. D. Moscholios, Securing the Internet of Things: challenges, threats and solutions, *Internet of Things.* 5 (2019) 41–70.

[10] V. Hassija, V. Chamola, V. Saxena, D. Jain, P. Goyal, and B. Sikdar, A survey on iot security: application areas, security threats, and solution architectures, *IEEE Access* 7 (2019) 82721–82743.

[11] A. K. Das, S. Zeadally, and D. He, Taxonomy and analysis of security protocols for Internet of Things, *Future Generation Computer Systems* 89 (2018) 110–125.

[12] B. Di Martino, M. Rak, M. Ficco, A. Esposito, S. Maisto, and S. Nacchia, Internet of Things reference architectures, security and interoperability: a survey, *Internet of Things* 1 (2018) 99–112.

[13] B. Mohanta, D. Jena, and S. Sobhanayak, Multi-party computation review for secure data processing in IoT-fog computing environment, *International Journal of Security and Networks* 15 (2020) 164–174. 10.1504/IJSN.2020.109697.

[14] J. Vijaya Kumari and P. Neelam, IoT based smart home automation system, *International Journal of Emerging Technologies and Innovative Research* 8 (1) (2021) 668–672 (www.jetir.org), ISSN:2349-5162, Vol. 8, Issue 1, page no.668-672, January-2021, Available: http://www.jetir.org/papers/JETIR2101091.pdf.

[15] A. K. Pranav Ratta, M. S. Sparsh Sharma, and G. Dhiman, Application of blockchain and Internet of Things in healthcare and medical sector: applications, challenges, and future perspectives, *Journal of Food Quality*, 2021 (2021) 20. ArticleID 7608296. https://doi.org/10.1155/2021/7608296

[16] A. Al-Hasnawi, S. M. Carr, and A. Gupta, Fog-based local and remote policy enforcement for preserving data privacy in the Internet of Things, *Internet of Things* 7 (2019) 100069.

[17] K.-C. Chen and S.-Y. Lien, Machine-to-machine communications: technologies and challenges, *Ad Hoc Networks* 18 (2014) 3–23.

[18] V. Casola, A. De Benedictis, A. Riccio, D. Rivera, W. Mallouli, and E. M. de Oca, A security monitoring system for Internet of Things, *Internet of Things* 7 (2019) 100080.

[19] S. Siboni, V. Sachidananda, Y. Meidan, *et al.*, Security testbed for Internet-of-Things devices, *IEEE Transactions on Reliability* 68 (1) (2018) 23–44.

[20] R. Nawaratne, D. Alahakoon, D. De Silva, P. Chhetri, and N. Chilamkurti, Self-evolving intelligent algorithms for facilitating data interoperability in IoT environments, *Future Generation Computer Systems* 86 (2018) 421–432.

[21] K. Bing, L. Fu, Y. Zhuo, and L. Yanlei, Design of an internet of things-based smart home system, in: *2011 2nd International Conference on Intelligent Control and Information Processing, Vol. 2,* IEEE, 2011, pp. 921–924.

[22] U. Satapathy, B. K. Mohanta, D. Jena, and S. Sobhanayak, An ECC based lightweight authentication protocol for mobile phone in smart home, in: *2018 IEEE 13th International Conference on Industrial and Information Systems (ICIIS)*, IEEE, 2018, pp. 303–308.

[23] S. S. Panda, D. Jena, and B. K. Mohanta, A remote device authentication scheme for secure communication in cloud based IoT, in: *2019 2nd International Conference on Innovations in Electronics, Signal Processing and Communication (IESC)*, IEEE, 2019, pp. 165–171.

[24] R. K. Kodali, V. Jain, S. Bose, and L. Boppana, Iot based smart security and home automation system, in: *2016 International Conference on Computing, Communication and Automation (ICCCA)*, IEEE, 2016, pp. 1286–1289.

[25] K. Jaiswal, S. Sobhanayak, B. K. Mohanta, and D. Jena, Iot-cloud based framework for patient's data collection in smart healthcare system using Raspberry-pi, in: *2017 International Conference on Electrical and Computing Technologies and Applications (ICECTA)*, IEEE, 2017, pp. 1–4.

[26] S. R. Moosavi, T. N. Gia, A.-M. Rahmani, *et al.*, Sea: a secure and efficient authentication and authorization architecture for IoT-based healthcare using smart gateways, *Procedia Computer Science* 52 (2015) 452–459.

[27] Y. Mehmood, F. Ahmad, I. Yaqoob, A. Adnane, M. Imran, and S. Guizani, Internet-of-things-based smart cities: recent advances and challenges, *IEEE Communications Magazine* 55 (9) (2017) 16–24.

[28] H. Arasteh, V. Hosseinnezhad, V. Loia, A. *et al.*, IoT-based smart cities: a survey, in: *2016 IEEE 16th International Conference on Environment and Electrical Engineering (EEEIC)*, IEEE, 2016, pp. 1–6.

[29] A. J. Neto, Z. Zhao, J. J. Rodrigues, H. B. Camboim, and T. Braun, Fog-based crime-assistance in smart IoT transportation system, *IEEE Access* 6 (2018) 11101–11111.

[30] L. F. Herrera-Quintero, J. C. Vega-Alfonso, K. B. A. Banse, and E. C. Zambrano, Smart ITS sensor for the transportation planning based on IoT approaches using serverless and microservices architecture, *IEEE Intelligent Transportation Systems Magazine* 10 (2) (2018) 17–27.

[31] S. Muthuramalingam, A. Bharathi, N. Gayathri, *et al.*, IoT based intelligent transportation system IoT-ITS for global perspective: a case study, in: *Internet of Things and Big Data Analytics for Smart Generation*, Springer, 2019, pp. 279–300.

[32] M. R. Dey, U. Satapathy, P. Bhanse, B. K. Mohanta, and D. Jena, Magtrack: detecting road surface condition using smartphone sensors and machine

learning, in: *TENCON 2019–2019 IEEE Region 10 Conference (TEN-CON)*, IEEE, 2019, pp. 2485–2489.

[33] X. C. Yin, Z. G. Liu, L. Nkenyereye, and B. Ndibanje, Toward an applied cyber security solution in IoT-based smart grids: an intrusion detection system approach, *Sensors* 19 (22) (2019) 4952.

[34] A. Meloni, P. A. Pegoraro, L. Atzori, A. Benigni, and S. Sulis, Cloud-based IoT solution for state estimation in smart grids: exploiting virtualization and edge-intelligence technologies, *Computer Networks* 130 (2018) 156–165.

[35] A. O. Akmandor, Y. Hongxu, and N. K. Jha, Smart, secure, yet energy efficient, Internet-of-Things sensors, *IEEE Transactions on Multi-Scale Computing Systems* 4 (4) (2018) 914–930.

[36] Z. Li, G. Liu, L. Liu, X. Lai, and G. Xu, IoT-based tracking and tracing platform for prepackaged food supply chain, *Industrial Management & Data Systems* 117 (9) (2017) 1906–1916.

[37] Y. P. Tsang, K. L. Choy, C.-H. Wu, G. T. Ho, C. H. Lam, and P. Koo, An Internet of Things (IoT)-based risk monitoring system for managing cold supply chain risks, *Industrial Management & Data Systems* 118 (7) (2018) 1432–1462.

[38] C. Verdouw, R. M. Robbemond, T. Verwaart, J. Wolfert, and A. J. Beulens, A reference architecture for IoT-based logistic information systems in agri-food supply chains, *Enterprise Information Systems* 12 (7) (2018) 755–779.

[39] L. Liu, B. Zhou, Z. Zou, S.-C. Yeh, and L. Zheng, A smart unstaffed retail shop based on artificial intelligence and IoT, in: *2018 IEEE 23rd International Workshop on Computer Aided Modeling and Design of Communication Links and Networks (CAMAD)*, IEEE, 2018, pp. 1–4.

[40] M. Mehra, S. Saxena, S. Sankaranarayanan, R. J. Tom, and M. Veeramanikandan, IoT based hydroponics system using deep neural networks, *Computers and Electronics in Agriculture* 155 (2018) 473–486.

[41] A. Goap, D. Sharma, A. Shukla, and C. R. Krishna, An IoT based smart irrigation management system using machine learning and open source technologies, *Computers and Electronics in Agriculture* 155 (2018) 41–49.

[42] C. Kamienski, J.-P. Soininen, M. Taumberger, *et al.*, An efficient collision power attack on AES encryption in edge computing, *IEEE Access* 7 (2019) 18734–18748.

[43] Z. A. Baig, S. Sanguanpong, S. N. Firdous, *et al.*, Averaged dependence estimators for dos attack detection in IoT networks, *Future Generation Computer Systems* 102 (2020) 198–209.

[44] L. Liu, Z. Ma, and W. Meng, Detection of multiple-mix-attack malicious nodes using perceptron-based trust in IoT networks, *Future Generation Computer Systems* 101 (2019) 865–868.

[45] A. K. Mishra, A. K. Tripathy, D. Puthal, and L. T. Yang, Analytical model for sybil attack phases in Internet of Things, *IEEE Internet of Things Journal* 6 (1) (2018) 379–387.

[46] M. L´opez, A. Peinado, and A. Ortiz, An extensive validation of a SIR epidemic model to study the propagation of jamming attacks against IoT wireless networks, *Computer Networks* 165 (2019) 106945.

[47] M. Almiani, A. AbuGhazleh, A. Al-Rahayfeh, S. Atiewi, and A. Razaque, Deep recurrent neural network for IoT intrusion detection system, *Simulation Modelling Practice and Theory* (2019) 102031.

[48] S. Deshmukh-Bhosale and S. S. Sonavane, A real-time intrusion detection system for wormhole attack in the RPL based Internet of Things, *Procedia Manufacturing* 32 (2019) 840–847.

[49] D. Yin, L. Zhang, and K. Yang, A DDoS attack detection and mitigation with software-defined Internet of Things framework, *IEEE Access* 6 (2018) 24694–24705.

[50] C. Li, Z. Qin, E. Novak, and Q. Li, Securing SDN infrastructure of IoT–fog networks from MitM attacks, *IEEE Internet of Things Journal* 4 (5) (2017) 1156–1164.

[51] H. Yi and Z. Nie, Side-channel security analysis of UOV signature for cloud based Internet of Things, *Future Generation Computer Systems* 86 (2018) 704–708.

[52] H. Yan, Y. Wang, C. Jia, J. Li, Y. Xiang, and W. Pedrycz, IoT-FBAC: function-based access control scheme using identity-based encryption in IoT, *Future Generation Computer Systems* 95 (2019) 344–353.

[53] S. Tu, M. Waqas, S. U. Rehman, *et al.*, Security in fog computing: a novel technique to tackle an impersonation attack, *IEEE Access* 6 (2018) 74993–75001.

[54] P. Zhang, S. G. Nagarajan, and I. Nevat, Secure Location of Things (SLOT): mitigating localization spoofing attacks in the Internet of Things, *IEEE Internet of Things Journal* 4 (6) (2017) 2199–2206.

[55] K. Zhang, X. Liang, R. Lu, and X. Shen, Sybil attacks and their defenses in the Internet of Things, *IEEE Internet of Things Journal* 1 (5) (2014) 372–383.

[56] N. Zhang, R. Wu, S. Yuan, C. Yuan, and D. Chen, RAV: relay aided vectorized secure transmission in physical layer security for Internet of Things under active attacks, *IEEE Internet of Things Journal* 6 (5) (2019) 8496–8506.

[57] Q. D. La, T. Q. Quek, J. Lee, S. Jin, and H. Zhu, Deceptive attack and defense game in honeypot-enabled networks for the Internet of Things, *IEEE Internet of Things Journal* 3 (6) (2016) 1025–1035.

[58] M. Alshamkhany, W. Alshamkhany, M. Mansour, S. Dhou, and F. Aloul, Botnet attack detection using machine learning, in: *Proceedings of the 2020 14th International Conference on Innovations in Information Technology (IIT)*, Abu Dhabi, United Arab Emirates, 16–17 November 2020; pp. 203–208. 10.1109/IIT50501.2020.9299061.

[59] G. Saroj and R. Patil, A survey paper on identity theft in the Internet, *International Journal of Trend in Scientific Research and Development* 3 (2019) 969–970. 10.31142/ijtsrd23966.

[60] A. O. Imaji, Ransomware Attacks: Critical Analysis, Threats, and Prevention Methods, Fort Hays State University (2019).

[61] A. B. Craig Beaman, S. H. Toluwalope David Akande, and Muhammad Khurram Khan, Ransomware: recent advances, analysis, challenges and future research directions, *Computers & Security* 111 (2021) 102490, ISSN 0167-4048, https://doi.org/10.1016/j.cose.2021.102490.

[62] V. Grover and Gagandeep, An efficient Brute Force attack handling techniques for server virtualization (March 30, 2020), in: *Proceedings of the International Conference on Innovative Computing & Communications (ICICC)*, 2020.

[63] N. Pecka, L. Ben Othmane, and A. Valani, Privilege escalation attack scenarios on the DevOps pipeline within a Kubernetes environment, in: *ICSSP/ICGSE*, 2022, pp. 45–49. 10.1145/3529320.3529325.

[64] I. Kotenko, I. Saenko, and A. Branitskiy, Framework for mobile Internet of Things security monitoring based on big data processing and machine learning, *IEEE Access* 6 (2018) 72714–72723.

[65] E. Hossain, I. Khan, F. Un-Noor, S. S. Sikander, and M. S. H. Sunny, Application of big data and machine learning in smart grid, and associated security concerns: a review, *IEEE Access* 7 (2019) 13960–13988.

[66] N. Chaabouni, M. Mosbah, A. Zemmari, C. Sauvignac, and P. Faruki, Network intrusion detection for IoT security based on learning techniques, *IEEE Communications Surveys & Tutorials* 21 (2019) 2671–2701.

[67] E. Anthi, L. Williams, M. S lowińska, G. Theodorakopoulos, and P. Burnap, A supervised intrusion detection system for smart home IoT devices, *IEEE Internet of Things Journal* 6 (2019) 9042–9053.

[68] A. Ghosh, D. Chakraborty, and A. Law, Artificial intelligence in Internet of Things, *CAAI Transactions on Intelligence Technology* 3 (4) (2018) 208–218.

[69] S. Wang and Z. Qiao, Robust pervasive detection for adversarial samples of artificial intelligence in IoT environments, *IEEE Access* 7 (2019) 88693–88704

[70] M. Zolotukhin and T. Hämäläinen, On artificial intelligent malware tolerant networking for IoT, in: *2018 IEEE Conference on Network Function Virtualization and Software Defined Networks (NFV-SDN)*, IEEE, 2018, pp. 1–6.

[71] S. S. Panda, B. K. Mohanta, U. Satapathy, D. Jena, D. Gountia, and T. K. Patra, Study of blockchain based decentralized consensus algorithms, in: *TENCON 2019–2019 IEEE Region 10 Conference (TENCON)*, IEEE, 2019, pp. 908–913.

[72] B. K. Mohanta, D. Jena, S. S. Panda, and S. Sobhanayak, Blockchain technology: a survey on applications and security privacy challenges, *Internet of Things* 8 (2019) 100107.

[73] M. Banerjee, J. Lee, and K.-K. R. Choo, A blockchain future for Internet of Things security: a position paper, *Digital Communications and Networks* 4 (3) (2018) 149–160.

[74] G. Verma, A. P. Shahi, and S. Prakash, A study towards recent trends, issues and research challenges of intelligent IoT healthcare techniques: IoMT and CIoMT, in: Kaiser, M.S., Bandyopadhyay, A., Ray, K., Singh, R., and

Nagar, V. (eds), *Proceedings of Trends in Electronics and Health Informatics. Lecture Notes in Networks and Systems*, vol. 376. Springer, Singapore, 2022. https://doi.org/10.1007/978-981-16-8826-3_16.

[75] S. P. Garima Verma, Design and implementation of modified unicode strategy for data security in IoT, *International Journal of Advanced Science and Technology* 29 (06) (2020) 6271–6294. Retrieved from http://sersc.org/journals/index.php/IJAST/article/view/19913.

[76] G. Verma and S. Prakash, Internet of Things for healthcare: research challenges and future prospects, in: Hura, G., Singh, A., and Siong Hoe, L. (eds), *Advances in Communication and Computational Technology. Lecture Notes in Electrical Engineering*, vol. 668. Springer, Singapore, 2021. https://doi.org/10.1007/978-981-15-5341-7_80.

[77] G. Verma and S. Prakash, S, Emerging security threats, countermeasures, issues, and future aspects on the Internet of Things (IoT): a systematic literature review, in: Kumar, N., Tibor, S., Sindhwani, R., Lee, J., and Srivastava, P. (eds), *Advances in Interdisciplinary Engineering. Lecture Notes in Mechanical Engineering*. Springer, Singapore, 2021. https://doi.org/10.1007/978-981-15-9956-9_6.

[78] G. Verma and S. Prakash, A study towards current trends, issues and challenges in Internet of Things (IoT) based system for intelligent energy management, in: *2019 4th International Conference on Information Systems and Computer Networks (ISCON)*, 2019, pp. 358–365, doi:10.1109/ISCON47742.2019.9036182.

[79] G. Verma and S. Prakash, A comparative study based on different energy saving mechanisms based on Green Internet of Things (GIoT), in: *2020 8th International Conference on Reliability, Infocom Technologies and Optimization (Trends and Future Directions) (ICRITO)*, 2020, pp. 659–666, doi: 10.1109/ICRITO48877.2020.9197848.

[80] G. Verma and S. Prakash, Pneumonia classification using deep learning in healthcare, *International Journal of Innovative Technology and Exploring Engineering (IJITEE)* 9(4) (2020) 1715–1723.

[81] D. Li, D. Han, T. H. Weng, *et al.*, Blockchain for federated learning toward secure distributed machine learning systems: a systemic survey, *Soft Computing* 26 (2022) 4423–4440. https://doi.org/10.1007/s00500-021-06496-5.

[82] R. Gaur and S. Prakash, Performance and parametric analysis of IoT's motes with different network topologies, in: Mekhilef, S., Favorskaya, M., Pandey, R.K., and Shaw, R.N. (eds), *Innovations in Electrical and Electronic Engineering. Lecture Notes in Electrical Engineering*, vol. 756. Springer, Singapore, 2021. https://doi.org/10.1007/978-981-16-0749-3_61

[83] S. A. Latif, F. B. Xian Wen, C. Iwendi, *et al.*, AI-empowered, blockchain and SDN integrated security architecture for IoT network of cyber physical systems, *Computer Communications* 181 (2022) 274–283, ISSN 0140-3664, https://doi.org/10.1016/j.comcom.2021.09.029.

Chapter 14

Alleviating 6G security and privacy issues using artificial intelligence

Lateef Adesola Akinyemi[1,2], Oluwagbemiga Omotayo Shoewu[1], Comfort Oluwaseyi Folorunso[3], Oluwafemi Ipinnimo[3], Abiodun Afis Ajasa[1,4], Quadri Ademola Mumuni[1] and Joseph Folorunsho Orimolade[5]

Abstract

The emergence of sixth-generation (6G) networks has brought about new security and privacy concerns, emphasising the need to address these issues promptly and comprehensively. This book chapter thoroughly examines the flaws and potential solutions in 6G networks, focussing on a multi-objective optimisation problem that considers the deployment of mobile users and prioritises energy efficiency, data integrity, and end-to-end encryption. A genetic algorithm scheme is employed to solve the optimisation problem, and the performance of several machine learning algorithms is evaluated using various metrics such as mean absolute error, mean square error, root mean square, and *R*2 score. It is important to note that all methods for the *R*2 score produced exact unity results for the LR, RF, KNN, and SVM with 1.000, 0.999, 0.993, and 0.993, respectively. However, the study uses artificially generated data to overcome the lack of available data on the energy efficiency, throughput, latency, and spectral efficiency of 6G networks. Additionally, the study highlights the security and privacy challenges posed by 6G networks and draws insights from previous technological advancements to identify potential solutions and prospects. This work is a comprehensive guide for researchers, practitioners,

[1]Department of Electronic and Computer Engineering, Faculty of Engineering, Lagos State University, Epe Campus, Nigeria
[2]Department of Electrical Engineering, Faculty of Engineering and the Built Environment, University of Cape Town, South Africa
[3]Department of Systems Engineering, Faculty of Engineering, University of Lagos, Nigeria
[4]School of Electrical Engineering, Faculty of Electrical Engineering, Universiti Teknologi Malaysia, Malaysia
[5]School of Computing and Information Sciences, Caritas Institute of Higher Education (CIHE), Hong Kong

and policymakers to navigate the emerging landscape of 6G networks while ensuring security, privacy, and optimal performance.

Keywords: 6G networks; Security scheme; Privacy; Optimisation; Artificial intelligence; Energy efficiency

14.1 Introduction

The fifth-generation (5G) network has recently undergone a significant transition to the sixth-generation (6G) network, and 6G, in particular, is seeing an increase in security and privacy problems. High data rates, throughput, and efficiency must be achieved before users can take advantage of them. As a result, artificial intelligence (AI) will significantly contribute to providing intelligent security and privacy in the 6G network. This study will present applications of AI or machine learning (ML) schemes in 6G security, 6G architecture security, 6G technology security, and 6G privacy. Some privacy, security, and trust issues will be highlighted and addressed. To provide a fully autonomous network, 6G uses AI. As a result, 6G will be impacted by cyberattacks on intelligent systems, particularly ML devices. Possible security concerns affecting ML systems include toxicity threats, data infiltration, data transformation, logic contamination, model avoidance, model translation, model harvest, and membership reasoning assaults [1]. AI systems can perform better thanks to the accumulation of more features. Privacy concerns arise due to data breaches and the unintentional use of personal information during data processing, which the users cannot see. Another crucial technology for unlocking the benefits of the 6G network is blockchain. Blockchain is appropriate for distributed spectrum sharing, service administration, and resource planning in extremely big and dispersed 6G networks [2]. The fact that 6G realises linked intelligence through AI-inspired functionalities, particularly with ML devices vulnerable to security assaults, is one of the security concerns with AI for 6G mentioned in the literature. The poisoning type of attack, which determines the learning stage of an ML scheme and thus results in the system's model learning erroneously, is a typical example of such an attack [3]. The potential solution for this problem that has been proposed thus far includes negative ML and changing target protection systems that can result in adaptable AI systems. The nature of huge data, which can be managed in terms of the speed of future systems and necessitates future networks like 6G, presents another hurdle in AI. AI can easily violate a user's privacy and confidentiality. Due to the magnitude of user data and the need to solve this issue, subscribers and users are unable to forecast how externally based systems will handle their data. Thus, edge-inspired deep learning or federated learning is used to protect private information by applying a realistic constraint that keeps data closer to the subscriber [4]. To accomplish the 6G vision, security must be ensured as it is a vital driver for the next mobile network generation and AI. AI-powered 6G security offers shrewd, reliable security measures. Software-defined network/network functions virtualisation (SDN/NFV) can be used to detect and counteract multidimensional attacks using deep reinforcement learning and deep neural network (DNN) [5]. Compared to many conventional strategies, they successfully protect against assaults such as host location hijacking, distributed denial-of-service (DDoS), Internet protocol spoofing, and flow table overloading. Due to their rapid processing and high accuracy, Decision Trees and Random Forest, two ML approaches, are effective in detecting DDoS attacks in SDN/NFV setups [6–9]. Demand-driven placement of virtual functions in the 6G network will be expected dynamically, making ML-based adaptive security techniques effective against SDN-related threats. The assaults also use advancements in AI algorithms that can find weaknesses in a widely dispersed network.

14.1.1 Contributions

In this chapter regarding alleviating 6G security and privacy issues utilising artificial intelligence, the following are the major contributions:

1. The thorough review presented herein gives up-to-date works on the 6G network and privacy issues, thereby serving as a hands-on for researchers who deem it fit to carry out further work in this area.
2. A few possible uses of AI for 6G networks with privacy and security issues are detailed in this work.
3. Based on data from recent studies and trends, this paper presents some educated assumptions of the effects of AI on 6G security and 6G privacy, even though 6G is still a work in progress.
4. Fresh privacy and security challenges necessitated by the majority of the innovations presented in this paper were identified, and the potential solutions were proffered, given the state of today's technology.
5. A possible area of application of blockchain technology to privacy and security in 6G communication was proposed in this research.
6. A scenario of a terrestrial-satellites-inspired network system to demonstrate the idea of network optimisation in 6G network systems using MATLAB® environment was formulated.
7. Contributions, lessons learned, and possible future directions or recommendations from this work were also concisely itemised.

14.1.2 Chapter organisation

Section 14.1 discusses the introductory section and contributions of the book chapter for 6G privacy and security issues by employing AI. Furthermore, Section 14.2 discusses the related work regarding privacy and security in the 6G network. A summary of the reviewed works is equally presented in a tabular form stating the strengths and weaknesses. The application of AI on a 6G network concerning privacy and security is discussed in Section 14.3. Section 14.4 briefly describes the 6G security and privacy challenges, while Section 14.5 discusses the solutions to 6G security and privacy challenges. Network optimisation in the 6G network, primarily on problem formulations and numerical simulation results, is presented and discussed in Section 4.7. More importantly, lessons that have been learned and summarily discussed with 6G privacy and security and presented in Section 14.8. Finally, Section 14.9 completes the book chapter by making conclusions regarding alleviating 6G privacy and security issues using artificial intelligence.

14.2 Related works

The innovation of PC frameworks that can do errands that would normally require human insight, for example, discourse acknowledgment, direction, and language interpretation, is alluded to as computerised reasoning (AI) [10].

To evaluate data, spot patterns, and base predictions or conclusions on the data, AI systems use statistical models and algorithms. The application of algorithms that can learn from data without being explicitly taught is known as ML, a subset of artificial intelligence. Image and speech recognition are two applications of deep learning, a type of ML that uses an artificial neural network [11].

Rule-based systems, decision trees, Bayesian networks, neural networks, and evolutionary computation are a few categories into which AI can be divided. In many industries, including finance, healthcare, transportation, manufacturing, and entertainment, AI offers a wide range of useful applications.

The administration and optimisation of network resources and the creation of new services and applications are all expected to be major applications of AI in the 6G network. AI can be used, for instance, to enhance network performance, forecast possible network faults, and dynamically distribute network resources based on demand.

The effectiveness, performance, and security of the network can all be improved by using AI's capabilities. Here are a few possible uses for AI on a 6G network [12–15].

1. Network optimisation: By evaluating network traffic and forecasting demand, AI can be utilised to improve the 6G network. AI can optimise network resources, lower latency, and enhance overall network performance by applying ML algorithms to predict network traffic.
2. Intelligent traffic routing: With AI, traffic may be intelligently routed based on the user's demands. By examining network traffic patterns, AI can choose the optimum path to reduce latency and assure high-speed data delivery.
3. Autonomous network management: Network management chores like network configuration and optimisation can be automated using AI. AI can increase network efficiency and lighten the stress on network administrators by automating certain processes.
4. Security and privacy: With AI, 6G network security and privacy can be improved. AI can provide early warning of security breaches by examining network data and spotting potential threats.
5. Smart edge devices: On a 6G network, AI can power smart edge devices. AI can lower latency and increase the effectiveness of edge devices by employing ML algorithms to handle data locally.

6. Improved user experience: By examining user behaviour and preferences, AI can be utilised to improve the user experience on a 6G network. AI can increase user pleasure and loyalty by offering tailored services and recommendations.
7. In general, adding AI to the 6G network can potentially enhance user experience, security, and network efficiency. We may anticipate additional cutting edge uses of AI in this field as the rollout of the 6G network advances.

A careful balance between security, privacy, and functionality will be necessary for developing the 6G network and using AI in them. The development of strong security and privacy mechanisms that can keep up with the changing threat landscape and ensure that user data is protected while utilising the full potential of AI to enable creative new services and applications will require collaboration between researchers and industry experts. The security and privacy issues surrounding crucial 6G technology are outlined in Table 14.1 [16].

Table 14.1 The security and privacy issues associated with 6G technology

Key technology	Security and privacy issues	Key technology contribution
AI	Access control	Processes for fine-grained control
	Malicious behaviour	Find network abnormalities and convey early cautions
	Authentication	A procedure for solo discovering that could be applied to the validation cycle to work on the actual layers' security
	Communication	An ML-based radio wire plan that may be applied to actual layer correspondence to stop data leaks
	Encryption	Quantum encryption techniques and ML
Molecular communication	Malicious behaviour	An enemy that interferes with molecular communication or its procedures.
	Encryption	A coding plan that could work on the security of data transmission
	Authentication	Gives guidance for creating new authentication methods
Quantum communication	Encryption	Protection techniques for quantum encryption keys
	Communication	Various quantum communication techniques
Blockchain	Authentication	Authorisation of mobile services using a new conceptual architecture
	Access control	A strategy for upgrading access conventions
	Communication	Hashing power is utilised to check exchanges.
THz (terahertz)	Authentication	the use of magnetic signatures for authentication
	Malicious behaviour	An eavesdropper can still receive a signal even while it is being broadcast via a narrow beam.
VLC (visible light communication)	Communication	A communication method that uses a secure protocol
	Malicious behaviour	Collaborating with listeners may weaken security

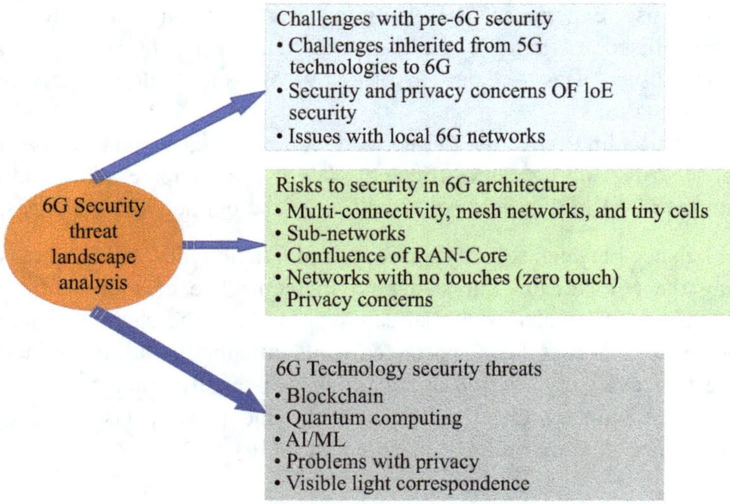

Figure 14.1 6G security threat landscape analysis [9]

The protection of our personal information is becoming more and more crucial as technology develops. To protect users, it is crucial to solving any security issues with the impending rollout of the 6G network. AI can be used as a remedy for these problems. AI may be used to identify security problems, take appropriate action, and safeguard data from online attacks [9]. The analysis of the 6G security threat landscape is shown in Figure 14.1.

AI can be applied in the following ways to address 6G security issues:

1. Threat detection: AI may be used to quickly identify security threats. AI can swiftly recognise and address possible attacks by examining network data and spotting strange behaviour.
2. Using predictive analytics, AI can foresee potential security issues before they materialise. AI can assist in uncovering possible weaknesses and thwart attacks by looking for patterns in data.
3. Access control: AI can be used to handle access control, ensuring that only authorised users have access to critical information. AI can identify possible unauthorised access attempts and stop them from happening by examining user behaviour and device profiles.
4. Behavioural analysis: AI can examine user behaviour to detect suspicious behaviour. AI can detect odd behaviour and notify users and administrators of potential security issues by keeping track of user activities.
5. AI can assist in a speedy and efficient reaction to security concerns. AI can lessen the effect of a security compromise and limit further damage by automating incident response procedures.
6. Management of vulnerabilities: AI can assist with managing vulnerabilities by checking the network for potential flaws and recommending remedies. AI can

aid in preventing security breaches by spotting vulnerabilities before they are used.

In general, AI can be quite helpful in addressing 6G security issues. AI may assist in ensuring that users are safe in the 6G age by identifying and responding to security threats, controlling access control, and managing vulnerabilities.

14.2.1 Summary of related works

Some of the strengths and weaknesses identified in the literature are highlighted in Table 14.2.

Table 14.2 Summary of the related works

Reference	Key technology	Strength	Weaknesses
[12]	Optical wireless technology and AI	They explained the state of the impending 6G wireless technology	It did not apply the concept of metaheuristic algorithms and expert systems to the addressed issue, specifically privacy and security in 6G
[13]	Intelligent RAN architecture	The study provides an intelligent RAN architecture for 6G and highlights the most recent network cloudification and intelligence developments	The study failed to address the issues of privacy and security in the 6G network
[15]	Internet of Things (IoT)	They examine the Internet and 6G network communication difficulties and technologies with the assistance of the IoT	This study did not consider security, privacy and trust as a challenge
[16]	• Intelligent edge computing in real-time • Distributed AI • Smart radio • 3D intercoms	They investigated a thorough analysis of the privacy and security concerns with the 6G network	The study failed to apply artificial intelligence or any metaheuristic algorithm in the considered 6G network
[9]	AI	They presented an overview of AI's involvement in 6G systems to understand the many opportunities and difficulties of having intelligent security and private provision	The paper failed to formulate a network optimisation problem and thereafter applied AI

14.3 Addressing 6G security and privacy issues using AI/ML

Essentially, there are four main elements of a 6G-inspired network: a decentralised AI system, real-time cognitive edge, three-dimensional (3D) intercommunications, and an intelligently based radio network. Because the 6G network consists of the most vital aspect of the 6G research being conducted, this book chapter focusses on these four identified areas. In addition, they are usually confronted with the most challenges of privacy and protection (security) [17]. Hence, the technological innovations or technologies employed in the prevalent investigation encompass Terahertz (THz) advancements, technologically based blockchain, molecularly inspired communications, visible light transmission (VLT), quantum-based communication systems, and expert systems.

In this case, the primary parts of privacy and security challenges are network management, authorisation and authentication, information transmission, and cryptography. Besides, 6G applications have critical deficiencies as well. The VLT and AI innovations are generally utilised by connected robotics and self-configured systems in which information transmission, criminal behaviour, and encryption can be regarded as problems. The multi-based sensory extended reality (MBSXR) application employs similar technology as a wireless brain based computer interfacing system; however, they have different challenges as regards privacy and security. The greatest deficiencies are encryption and maleficent behaviour.

14.3.1 The role of AI in 6G security

Foreseeing AI's precise effects on 6G security is challenging because 6G networks are still under development. But, using data from recent studies and trends, we can make some educated assumptions. The capacity to identify threats in real-time and take appropriate action is one possible advantage of AI in 6G security. AI systems might instantly scan vast volumes of data to discover and respond to security breaches faster and more accurately than human operators, thanks to the enhanced speed and capacity of the 6G network [18].

AI may also be used to create stronger encryption techniques that withstand attacks from quantum computers, which are anticipated to become more common in the future. AI, though, may potentially bring about fresh security threats. Hackers might, for instance, utilise AI algorithms to design more sophisticated and targeted attacks that are challenging to stop. Furthermore, when AI develops further, it may eventually become autonomous, which would allow it to operate independently of a human and present new security issues [9].

Future AI-powered wireless networks will succeed or fail based on their capacity to use AI and their 6G security. The enormous number of paying customers that 6G network operators serve (such as mobile users, businesses, and industries) may give them greater incentive to increase their security interest by implementing the most recent general AI developments. Overall

6G-AI-powered security solutions will thus profit from the development of general AI [18].

Numerous customary arrangements, like firewalls and interruption recognition frameworks, have been tasked with defending against security attacks, in any case, simulated intelligence (AI) makes such frameworks more competent and shrewder. While being widely utilised, conventional security systems (such as signature based intrusion detection) have shortcomings when it comes to tackling sophisticated attacks in a 6G context [18].

According to [18], many 6G enabling technologies at the network layer, support the use of AI to improve system performance. The architecture of an AI-enabled intelligent 6G network is shown in Figure 14.2. The benefits of massive data analysis and pattern recognition have led to the application of AI to many important technologies, including but not limited to the following:

1. Checking hub conduct for insider danger discovery in supporting confided-in network (in light of CNN/RBN).
2. Anticipating network attacks to reroute traffic, suggest smart adjustments to the network, and isolate questionable services in SD-WAN/SDN network (DRL-based).
3. vRAN/Open RAN radio and computation control policy optimisation (deep autoencoder-based).
4. Setting equipment recovery priorities and identifying failing VNFs (DRL-based).

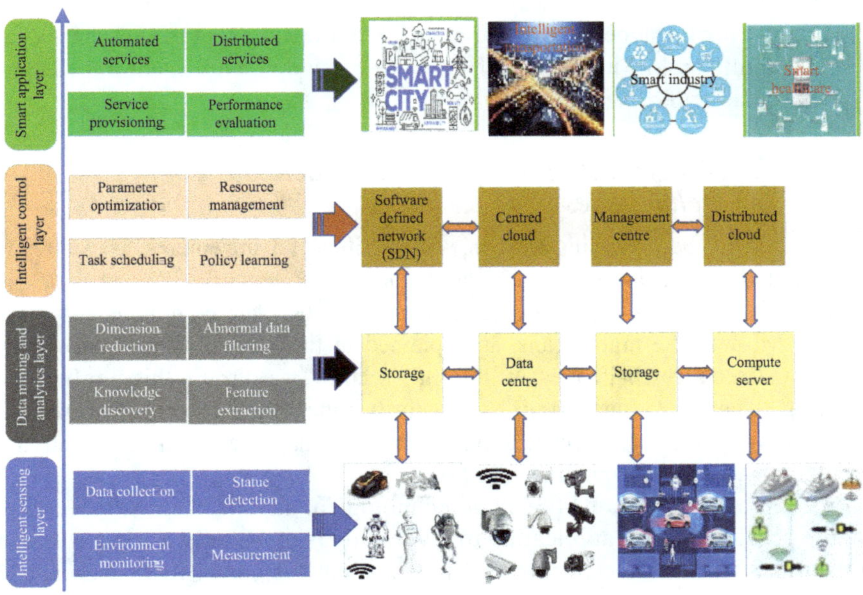

Figure 14.2 The design of an AI-based network

5. Analysing traffic and network access patterns to detect and block malignant traffic (RBM, CNN, DNN, DBN, autoencoder, LSTM).

14.3.2 The role of AI on 6G privacy

In the imagined time of 6G, security assurance is a key exhibition need and a significant component in remote networking, which presents three major challenges according to [8]:

1. The extraordinarily high volume of brief data transfers enabled by 6G could represent a higher risk to people's privacy, drawing considerable attention from governments and other commercial groups. In the 6G age, there is a larger risk that data collection and accessibility would compromise user privacy and complicate regulatory issues.
2. More advanced applications will operate on mobile devices, raising the possibility of assaults as intelligence moves to the edge of the network. It will be difficult to include privacy-protecting features in devices with limited resources.
3. It is important to strike a compromise between preserving the availability of highly accurate services and safeguarding user privacy. To actualise many smart applications, location data and IDs are required. This necessitates careful consideration of data ownership and access rights, management, and privacy protection laws.

The impact of AI and ML innovation on security is more grounded in two ways. In some ways, the appropriate utilisation of ML can protect security in 6G, however, in alternate ways, ML attacks might abuse security. The preparation (for instance, a harming assault) and testing stages are where protection assaults on ML models can happen (e.g., switch, enrolment impedance, ill-disposed assaults).

14.3.3 Challenges with security and confidentiality in 6G technologies

As indicated in the preceding section, some critical advancements have already demonstrated effectiveness in critical areas of the 6G network. They provide big dependability, minimal delay, secure, and effective 6G data communication capabilities. Most of the innovations, as explained in the section before, raise fresh privacy and security challenges. Within this section, we discuss this briefly. The following essential 6G innovations have security and privacy concerns having been raised in the literature: identification, identity management, criminal activity, cryptography, and telecommunication [16].

As depicted in Figure 14.3, each technological advancement substantiates notable possibilities for application in diverse areas of 6G network systems and applications with connected robotics, independent systems, MBSXR uses, and wireless brain-based computer interfacing systems, decentralised ledger innovations, and blockchain. Moreso, Figure 14.3 shows these essentialities as

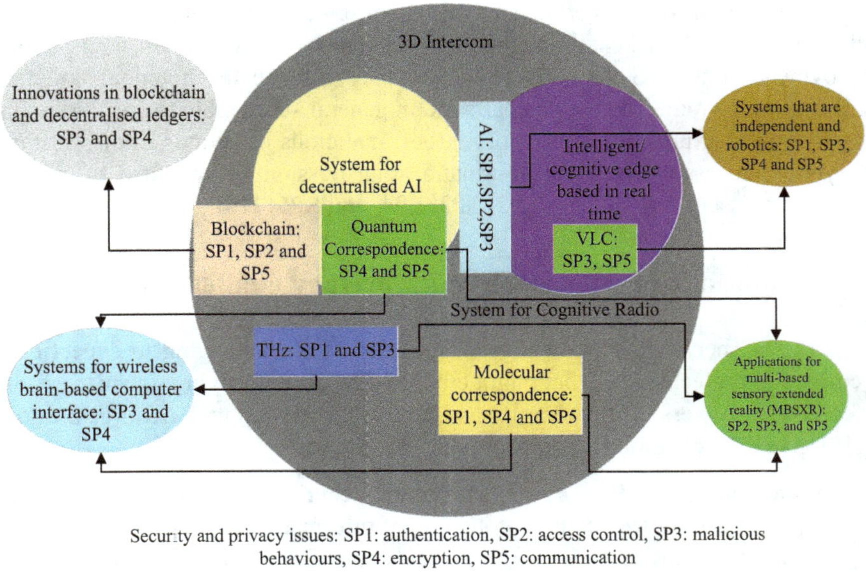

Figure 14.3 6G security and privacy challenges

clouds. The extremely first three central areas are decentralised AI system, real-time-based cognitive/intelligent edge and cognitive radio system – which incorporates the potentials spanning [16]. Supposing that AI would inspire the 6G network system and more importantly it houses at the link or juncture of all three of these fields. Then, it is depicted that the schematic diagram is accompanied by a listed summit five privacy and security challenges and issues.

A large number of the essential elements in Figure 14.3 are susceptibly sensitive to various identity management, maleficent behaviour, authentication, and identification. Nevertheless, certain advancements and innovative technologies are more susceptible than others to specific challenges. For illustrative purposes, the VLT or VLC is specifically susceptible to the data communication process and maleficent behaviour alongside the constant mental edge and shrewdly based radio framework. On the other hand, the atomic-based communication framework and THz innovation have capacities to help mental radio frameworks. The molecularly-based communication technology's encryption, authentication, identification and communication parts are analogously connected with the challenges in privacy and security in a dense 6G network. Therefore, the THz is essentially exposed to malicious behaviour and fraudulent acts. Hence, the technologically based blockchain and quantum-inspired communications system are connected to distributed AI and intelligently based radio networks [16].

The decentralised ledger advancements and blockchain which are fundamentally centred on blockchain technology are primarily secure as the ultimate deployment of the 6G network. However, they may be the potential target of malicious behaviour. These extra parts are in general susceptible to five dissimilar classifications of privacy and security issues: malicious behaviour, authentication, encryption, access control, and information transmission are regarded as security and privacy issues 1 to 5 (SP1, SP2, SP3, SP4, and SP5) [16].

14.4 Solutions to 6G security and privacy challenges

Since the 6G network is technology in progress, it is challenging to expect each security and protection risk that might emerge. The accompanying, in any case, are a few likely answers for the security and protection worries with 6G organisations [8,19] given the condition of innovation today:

1. End-to-end encryption (E2EE): A critical security part that can assist with safeguarding the secrecy and protection of information communicated through 6G organisations is start-to-finish encryption (E2EE). E2EE prevents unauthorised access and data interception by encrypting data from the point of origin to the point of destination [20].
2. Access control and authentication: Powerful access control and authentication procedures are crucial for ensuring that only authorised users may access the 6G network and services. To achieve this, complex access control methods such as multi-factor authentication (MFA), biometric authentication, and others can be used [21].
3. Blockchain-based security: Blockchain technology can add a layer of security to the 6G network by enabling secure and transparent data exchange, decentralised authentication, and tamper-proof data storage [22].
4. AI/ML-based threat detection: Using AI/ML approaches, real-time threat detection and response can be accomplished. Using ML algorithms, network traffic patterns can be examined, abnormalities can be discovered, and risks can be dealt with automatically.
5. Hardware-based security: Hardware-based security can add a layer of defence against hardware-based threats. Examples of such solutions are secure enclaves and trusted execution environments.
6. Privacy-preserving technologies: Data transported over a 6G network can be kept private by using tools like secure multi-party calculation, differential protection, and homomorphic encryption.

A multi-layered approach to security and privacy will be needed to overcome the problems the 6G network brings. A mix of technical solutions, regulatory frameworks, and industry standards will be required to ensure that the 6G network is secure, dependable, and trustworthy.

The summary of the solutions to 6G security and privacy challenges is shown in Figure 14.4.

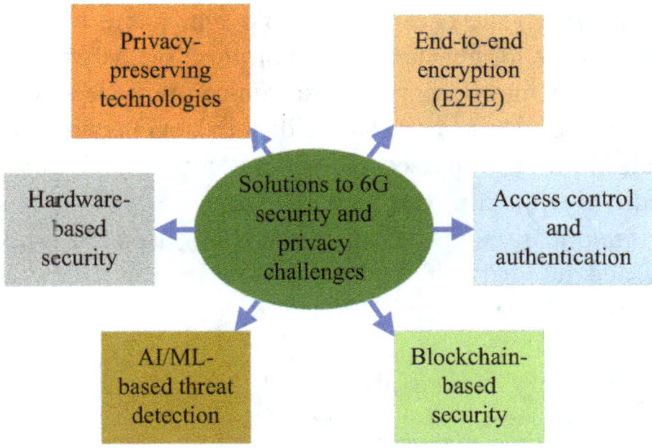

Figure 14.4 Summary of the solutions to 6G security and privacy challenges

14.5 Application of blockchain technology in alleviating security and privacy in 6G networks

The advent of the 6G network has brought about new challenges in terms of security and privacy. As this network continues to grow, there is a need for new solutions that can address the security and privacy concerns that arise. Blockchain technology has emerged as a promising solution to this problem.

Blockchain technology is a secure and transparent ledger that is decentralised, distributed, and utilised for recording transactions. It is a tamper-proof and immutable system that can be used to ensure data integrity and prevent unauthorised access. With its unique features, blockchain technology has the potential to address some of the key challenges of the 6G network, including data privacy, security, and authentication [23].

One of the key applications of blockchain technology in the 6G network is data privacy. Blockchain innovation can be utilised to store and oversee delicate information like individual data, monetary information, and clinical records. The data can be encrypted and stored in a decentralised manner, ensuring that it is secure and cannot be accessed by unauthorised parties [24].

Another important application of blockchain technology in the 6G network is security. Blockchain technology can create a secure and tamper-proof system that can prevent unauthorised access to network resources. It can also be used to authenticate users and devices, ensuring that only authorised users and devices are allowed to access the network.

Blockchain innovation can possibly further develop security and protection in different courses in 6G organisations [24]. Some possible applications of blockchain to security and privacy in the 6G network include [25]:

1. Secure authentication and access control: Blockchain can be used to securely authenticate and authorise access to the 6G network, preventing unauthorised

access or tampering. The decentralised and immutable nature of blockchain can make it difficult for attackers to compromise the network.
2. Secure data storage and sharing: Blockchain can provide a secure and decentralised platform for storing and sharing data in a 6G network. This can protect sensitive data from unauthorised access and ensure that data is not tampered with or modified without proper authorisation.
3. Secure communication: Blockchain can be used to ensure the confidentiality and integrity of communication in a 6G network, preventing eavesdropping and tampering. This can be achieved using encryption and digital signatures.
4. Decentralised identity management: Blockchain can provide a decentralised and secure identity management system for the 6G network, eliminating the need for centralised authorities and reducing identity theft risk.
5. Smart contract-based security: Blockchain-based smart contracts can be used to automatically enforce security policies and protocols in a 6G network, reducing the risk of human error and ensuring that security measures are consistently applied.
6. Threat detection and response: Blockchain-based threat detection and response systems can be used to detect and respond to security threats in real time, helping to prevent or mitigate attacks on the 6G network.

Overall, blockchain technology has the potential to significantly enhance the security and privacy of the 6G network, enabling the development of more secure and trustworthy communication systems [25]. The framework for the blockchain system in security is shown in Figure 14.5, while the schematic diagram indicating the flow of information is shown in Figure 14.6.

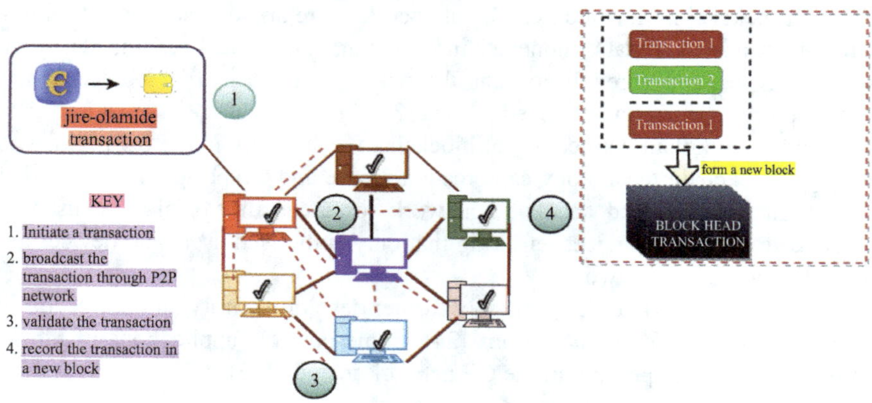

Figure 14.5 Security IoT framework with blockchain

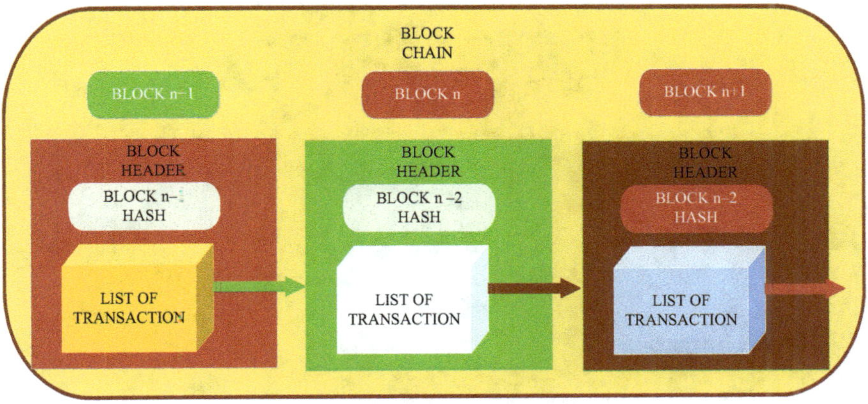

Figure 14.6 The flow of information in the blockchain system

14.6 Network optimisation in 6G network

Metaheuristic algorithms were used in the works of researchers in [26–30] to perform a wide range of network optimisations on 4G and 5G networks, respectively. Considering what is feasible, such research yielded startling results. However, the techniques discussed in previous studies have not been applied to 6G security and privacy concerns. This justifies the present metaheuristic or AI-based research geared toward network optimisation, 6G security, and privacy concerns. In the 6G network era, it is envisioned that more functionalities, services, and applications such as video, audio, realities, and so on will be available for all mobile users. To this end, more network resources such as power, channels, bandwidth, and so on will be used and allocated among mobile users be it primary or secondary or user equipment (UE) for seamless service delivery [31–34]. Therefore, in this book chapter's section, efforts are geared towards alleviating issues that might arise in 6G security and primarily using AI to formulate a network optimisation to address some of the growing challenges that will confront users while trying and exchange data communication with one another securely and confidentially i.e., end-to-end encryption. Emergency service notifications and the automatic updating of the computer's operating system and the software used, for instance, could be transmitted in multicast, broadcast, or unicast mode. In comparison, the IoT, vehicular use cases, the entertainment world, and heterogeneous modes are thought to be able to meet quality-of-service (QoS) requirements. Such QoS requirements must be fulfilled to guarantee the privacy and security of information transmitted from one user to another. In this section, let us assume an instance for the sake of demonstration to illuminate the idea of network optimisation in a 6G network when fully deployed. Assuming it is a multi-objective network optimisation problem inspired (MONOPI) using a pair of objective functions (OFs) or by breaking it into sub-problems for easy tractability. In addition, without any loss of generality, assuming all terrestrial-based devices/users will access the network resources in an

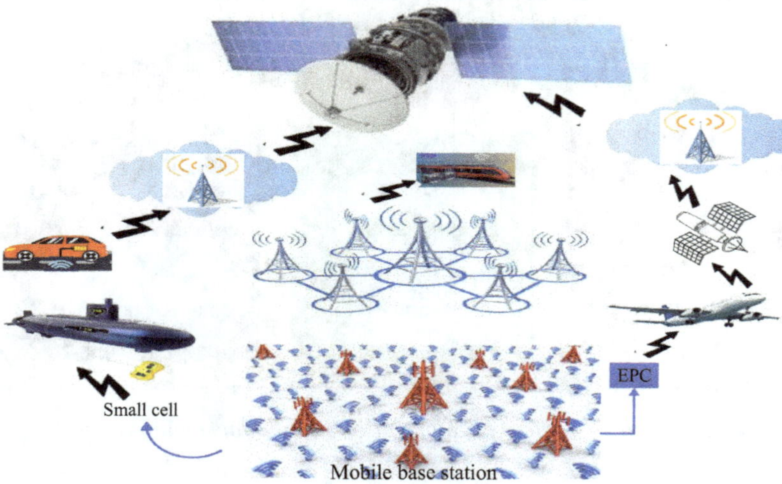

Figure 14.7 The terrestrial-satellite-inspired model for network optimisation

orthogonally based manner to avoid interference among users. For illustration in this part, let us use an example to illustrate the concept of network optimisation in a fully operational 6G network. Assuming, the optimisation problem has been divided into smaller issues for greater tractability or a multi-objective network optimisation-inspired (MONOPI) problem. In addition, without sacrificing generality, presumptively all terrestrial-based devices and users will connect to the network's resources via orthogonal bases to prevent user interference. Therefore, the terrestrial-satellite-inspired model for the network optimisation system model is depicted in Figure 14.7.

14.6.1 Problem formulations and method

Let θ_j and β_j represent the throughput and the delay in the propagation of information of the users in the sub-layer network γ_j to the terrestrial users' devices, correspondingly. Assume $\alpha_j, m = 1$ denotes that the terrestrial (ground mobile users) devices m is connected by the platform in layer j; else $\alpha_j, m = 0$ indicating no connection between the user's device and the layers j (note that j runs from terrestrial, air-, and space networks). Hence, the overall network multi-objective optimisation problem can be cast as follows:

$$\max_{\alpha} N_1 = \sum_{j\epsilon J, m\epsilon M}^{J,M} (\theta_j \alpha_j, m) \tag{14.1}$$

$$\min N_2 = \sum_{j\epsilon J}^{J} (\beta_j\, \alpha_j, m) \tag{14.2}$$

s.t.

$$C1 : \sum_{j\epsilon J}^{J} \alpha_j, m \geq 1, \forall m\epsilon M \tag{14.3}$$

$$C2 : \alpha_j, m\{0, 1\}, \forall m\epsilon M, \forall j\epsilon J \tag{14.4}$$

where N_1 in (14.1) represents the total spectral energy or energy efficiency of all the network and N_2 in (14.2) denotes the delay in the transmission encountered by terrestrial mobile users. C1 in (14.3) is the constraint that ensures that each terrestrial device is connected by at least one network (terrestrial mobile base station, air station, or space-based base station) to enable secure and private transmission of information. The C2 in (14.4) indicates the binary variable if connected to any platform, gives one otherwise zero. It is noted that the aim of the binary variable α_j, m is to compute the best way of combining the N_1 and N_2 concurrently resulting in a set of optimal solutions. This representation tells us that there is always a compromise/trade-off between the network spectral energy/efficiency and the delay associated with the transmission.

Let the transmitter (sender/source) be the mobile base station and the receiver (distribution mode) be the users (primary or secondary user). For every distribution, let $T = S$ be the broadcasting base station and receiving station and the recurring station (two modes) be mobile users MU represented in (14.5)

$$T = S = TBS^{DL}, \ R = D = MU^{DL}, \text{ for downlink} \tag{14.5}$$

Conversely, the uplink is given in (14.6) as

$$T = S = MU^{UL} \tag{14.6}$$

Hence, the sets TBS and MU are at times used in place of BS and MU, respectively, to mean base station and mobile user. Furthermore, by introducing a real non-negative decision variable called β_{il} which represents the power that the transmitting station (base station) $i\epsilon TBS$TBS employed to transmit data on channel $l\epsilon C$. Additionally, the binary decision variable α_{ijk} that represents which data or information can be transmitted by the transmitting station $i\epsilon TBS$ to the receiving station (mobile user) $j\epsilon MU$.

Where α_{ijl} in (14.7) indicates the binary variable indicating if user i is connected to a base station j using channel l

$$\alpha_{ijl} = \begin{Bmatrix} 1 \text{ if } i \in TBS \text{ can transmit to } j \in MU \text{ on } l \in C \\ 0, \text{otherwise.} \end{Bmatrix} \tag{14.7}$$

Also, let $\delta_{ij}\epsilon$ [0, 1] in (14.8) be binary indicator, $i\epsilon$ TBS, $j \in$MU which is a result of constraint that will serve as a binary decision variable or zero-one condition:

$$\delta_{ij} = \begin{Bmatrix} 1 \text{ } if \text{ } \alpha_{ijl} = 1 \text{ on any channel } i \in C \\ 0, \text{otherwise.} \end{Bmatrix} \tag{14.8}$$

To have a full understanding of the effective measurement of the signal, we introduce source with base station ideas. To start with, the path gain Ω_{ij} is defined to be properties of the power that the transmitting station $i \in TBS$ utilises that will act to the receiving station or mobile user (MU) $j \in$ MU. In this study, it is assumed that the path gain vector Ω of the whole network is known a prior by each transmitting base station (TBS) $i \in TBS$ and each receiving station $j \in$ MU as expressed in (14.9).

It then holds that

$$\Omega_{ij} \geq 0, i \in TBS, j \in \mathrm{MU} \tag{14.9}$$

Furthermore, the signal-to-interference-plus-noise ratio (SINR) is the ratio of the expected signal/desired signal over the unwanted noise or which for the transmitting base station (TBS) i, receiving base station (MU) j and channel l or c is expressed as expressed in (14.10):

$$SINR_{ijl} = \frac{\Omega_{ij}\beta_{ij}}{\sigma_j^2 + \sum_{m\epsilon TBS(i)} \Omega_{mj}\beta_{ml}}, \tag{14.10}$$

where $\sigma_j^2 > 0$ is the noise at the receiving station (MU) j which is assumed to be the known parameter again.

Hence, the channel capacity of user one (i, j) i.e., transmitting i and receiving j in channel l or c is expressed as in (14.11):

$$\theta_{ij=} W \log_2 (1 + SINR_{ijl}), i\epsilon TBS, j \in \mathrm{MU},\ i\mathrm{C} \text{ or } \mathrm{L}, \tag{14.11}$$

where W is the position scalar given by the size of the sequence; in our case, $W =$ is a known constant.

Moreover, the θ_{ijl} is related to the mobile user which is (i, j) which is receiving in the downlink or transmitting in uplink. Therefore, the total channel capacity of the mobile user (i, j) be cast compactly in (14.12) and (14.13)

$$\theta_{ij=} \sum_{i \in C} \theta_{ijl, i\epsilon TBS, J\epsilon MU} \tag{14.12}$$

$$\theta_{ij=W} \sum_{I\epsilon C} \log_2 \left(1 + \frac{\Omega_{ij}\beta_{ij}}{\sigma_j^2 + \sum\limits_{m\epsilon TBS(i)} \Omega_{mj}\beta_{ml}} \right) \tag{14.13}$$

Additionally, we shall introduce a utility function or cost function that aims for the excellent quality of the system for all mobile users in the network. In this study, the purpose is to "maximise the total channel capacity" θ_{ij} expressed in (14.13).

The following constraint needs to be stated, which assumes that the system network performs as expected.

(i) The first constraint guarantees that each mobile user (PU or SU) communicates precisely with one base station (transmitting base station in the case of downlinks and receiving station in the case of uplink) while each base station in simple terms can transmit to many mobile users concurrently.

(ii) The scored constraint ensures that each base station utilises each channel $l \epsilon C \; or \; l \epsilon L$ at most one time.

(iii) A power constraint on each transmitting base station (TBS) guarantees that the total power utilised is at most the upper threshold power consumption constant.

14.6.2 Power distribution and joint channel allocation for downlink and uplink in a system

Hence, the power distribution and joint channel allocation optimisation problem for the network for the downlink case is expressed in (14.14) and the respective constraints are given in (14.15)

$$x_{ijk} = \alpha_{ijl}; y_{ij} = \delta_{ij}; y_{ijk} = \Omega_{ijl}, \text{ Downlink signal flow: } \begin{Bmatrix} S \rightarrow D \\ TBS \rightarrow MU \\ BS \rightarrow M \\ B \rightarrow M \end{Bmatrix}$$

$$\underset{(\theta, \beta_{il,} \alpha_{ijl}, \delta_{ij})}{\text{Max}\theta} \tag{14.14}$$

Subject to (s.t):

$$\begin{aligned}
&(\text{C : i}) \; \frac{y_{ij}\theta}{\delta_{ij}\theta} \leq \sum_{l\epsilon L,\; k\epsilon C} \frac{x_{ijk}\theta_{ijl}}{\alpha_{ijl}\theta_{ijl}}, i\epsilon TBS,\; j\epsilon MU \\
&(\text{C : ii}) \sum_{i\epsilon TBS} \delta_{ij},\; j\epsilon MU \\
&(\text{C : iii}) \sum_{j\epsilon MU} \alpha_{ijl} \leq 1, i\epsilon TBS, i\epsilon L/l\epsilon C \\
&(\text{C : iv}) \sum_{i\epsilon C} \beta_{il \leq \beta^{max}_{TBS,i}, i\epsilon TBS} \\
&(\text{C : v}) \; \alpha_{ijl} \leq \Omega_{ij}, i\epsilon TBS, J\epsilon MU, i\epsilon C. \\
&(\text{C : vi}) \; 0 \; \leq \Omega_{ij} \leq 1 i\epsilon TBS, J\epsilon MU \\
&(\text{C : vii}) \; \alpha_{ijl}\epsilon\{0, 1\},\; i\epsilon TBS, J\epsilon MU, i\epsilon C. \\
&(\text{C : viii}) \; \beta_{il} \geq 0 i\epsilon TBS, i\epsilon C.
\end{aligned} \tag{14.15}$$

For the sake of clarity, we have used these notations: $S = TBS = BS$ (base station), $D = MU = M$ (mobile users)

If we let $S = MU = M$, $D = TBS = B$, MU $(S = M) \rightarrow TBS(BS) = D)$

Replacing constraints (C: iii) and (C: iv) in (14.15) with the following results in (14.16)

$$\sum_{j\epsilon TBS} \delta_{IJ} = 1, i\epsilon M, \quad \sum_{i\epsilon Mu} \alpha_{ijl} \leq 1, j\epsilon TBS \tag{14.16}$$

The power allocation scheme is defined by assuming that a realistic channel allocation is known prior i.e., in which both α_{ijk} and $i \epsilon TBS, j \epsilon MU, I \epsilon C$ and $\delta_{ij}, i \epsilon TBS, j \epsilon MU$ are equally known. Therefore, for each TBS (source) i, it is known that the destination or MU *j* in which the link it communicates with is established. This incites an innate extension of the notational system by introducing the idea of lines (connected (*i, j*)) where a line contains the TBS (source) $i \epsilon TBS$ and MU (destination), $j \epsilon MU$ that are communicating such that $\delta_{ij} = 1$. These are expressed in (14.17) and (14.18), respectively, for the link between TBS and MU and channel of connection is established:

$$L = \{(i, j) : Y_{ij} = 1,\ i \epsilon TBS,\ J \epsilon MU\} \ (\textit{the link between TBS and MU is established})$$
$$L_k = \{(i, j) \epsilon L : \ \alpha_{ijk} = 1\},\ c \epsilon C(\textit{when the link between i and j is established on channel } C) C_{ij} = \{i \epsilon C \ : \ \alpha_{ijk} = 1\},\ (i, j) \epsilon L \ (\textit{when the channel of connection is of } k \epsilon C, \textit{ then link established}) \tag{14.17}$$

More importantly, for the power allocation optimisation problem to be well behaved or formulated, the given channel allocated must connote that

$$L_k \neq \emptyset i \epsilon C \text{ and } C_{ij} \neq \varphi\ (i,j) \epsilon L \tag{14.18}$$

(the channel between link (*i, j*) is not zero)

Let the SINR as stated in (14.10) be recast for the assumption that existing links $(i,j) \epsilon L$ and the station capacity of a mobile user (primary user (PU) or secondary user (SU) on the link $(i,j) \epsilon L$ at channel *L* be expressed as in (14.19):

$$\theta_{ijl} = W \log_2 (1 + SINR_{ijl}),\ (i,j)L,\ i(i,j) \tag{14.19}$$

And the total channel capacity of a mobile user (PU/SU) *j* be stated as in (14.20):

$$\theta_{ij} = \sum_{l \epsilon C\_ij} \theta_{ijl}, \quad (i,j) \epsilon L \tag{14.20}$$

Worthy of note is the remaining links $(i,j) \notin L$ i.e., (i,j) no element of *L* is not employed or utilised. Therefore, their SINR and channel capacity need not be defined. Hence, the power allocation optimisation problem (PACOP) is expressed as in (14.21):

$$\left\{ \begin{array}{c} \text{Maximise} \quad \theta \\ \theta, \theta_{ij}, \beta_{il} \\ \text{S.E.} \\ (i)\ \theta \leq \theta_{ijl},\ (i,j) \epsilon L i \epsilon (i,j) \\ (ii)\ \sum_{L \epsilon C_i} \beta_{il} \leq \beta^{\max}_{TBS,i}\ i \epsilon TBS \\ (iii)\ \beta_{il} \geq 0, i \epsilon TBS, i \epsilon (i,j) \end{array} \right\} \tag{14.21}$$

where $\theta_{ij,}$ $(i,j)\in L$ stated in (14.19) and $\beta^{\max}_{\text{TBS,i}}$ is the same and maximum power consumption as expressed in (14.15 (C : iv)). By simplification via the introduction of an additional constraint on the channel allocation. The constraint is to only allow each link (i, j) ϵ L or use one single channel i.e., $|C_{ij}| = 1, (i,j)\in L$.

Let us regard this formulation as a SINGLE CHANNEL in each link power assignment optimisation problem expression in (14.22)

$$\left\{\begin{array}{c} \text{Maximise} \quad \theta \\ \theta, \theta_{ij}, \beta_{il} \\ \theta \le \theta_{ijl}, \quad (i,j)\epsilon Li\epsilon(i,j) \\ \sum_{L\epsilon C_i} \beta_{il} \le \beta^{\max}_{TBS,i,} i \in TBS \\ \beta_{il} \ge 0, i\epsilon TBS, i\epsilon(i,j) \end{array}\right\} \tag{14.22}$$

where $|C_{ij}| = 1, (i,j)\in L, \theta_{ijl} \in L, i \in (i,j)$

As one channel only occupies each link, the optimisation problem (14.22) in the uplink case can be decomposed into a separate optimal problem for each channel $l \in C$, since $|C_i| = |C_{ij}|, (i,j)\in L$ in the uplink.

14.6.3 Numerical simulation results

For this section, a scenario was created using a terrestrial-satellites-inspired network system to demonstrate the idea of network optimisation in a 6G system. Also, the numerical and performance analysis is carried out and achieved in a MATLAB® environment. Figures 14.8, 14.9, and 14.10 show the coverage area, deployment of users in a terrestrial-satellite-based scenario, and deployment of various mobile users

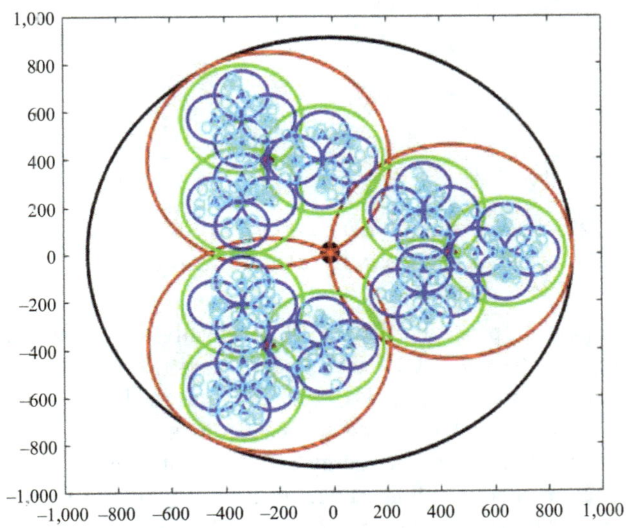

Figure 14.8 Coverage area for the terrestrial-satellite-based scenario

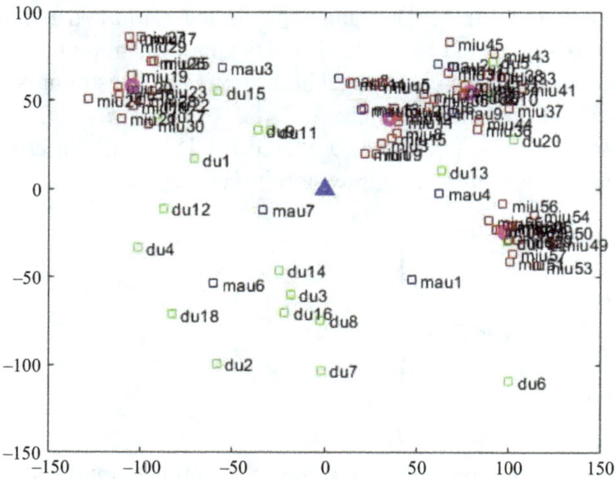

Figure 14.9 Grouping of users in terrestrial-satellite-based scenario

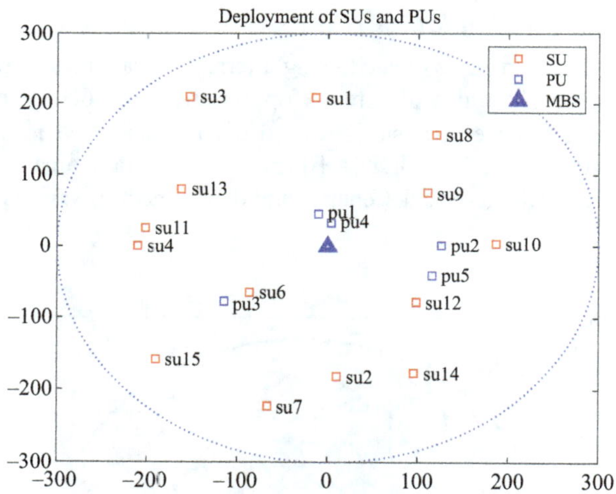

Figure 14.10 Distribution of different categories of users (PUs and SUs) in terrestrial-satellite-based scenario (cell-coverage)

within a base station. In particular, Figure 14.10 actually depicts how a base station serves both the primary users (PU) and secondary users (SU).

The total energy efficiency (EE) is displayed versus the iterations in Figure 14.8. The results of both the analytical and the GA comparisons of minimum energy efficiency (min-EE) and minimum energy efficiency utilising a genetic algorithm (min-EE-GA) are the same. The highest energy efficiency (max-EE) compared to the maximum energy efficiency utilising GA (max-EE-GA) is also plotted, and the startling findings are displayed in Figure 14.11. The median of both EE and EE-GA

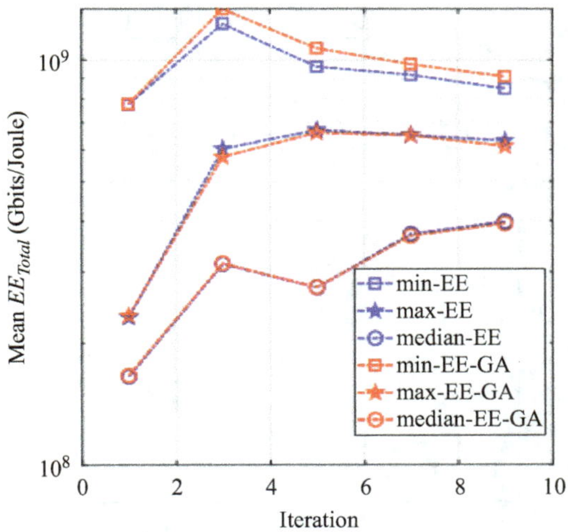

Figure 14.11 Mean energy efficiency versus iteration

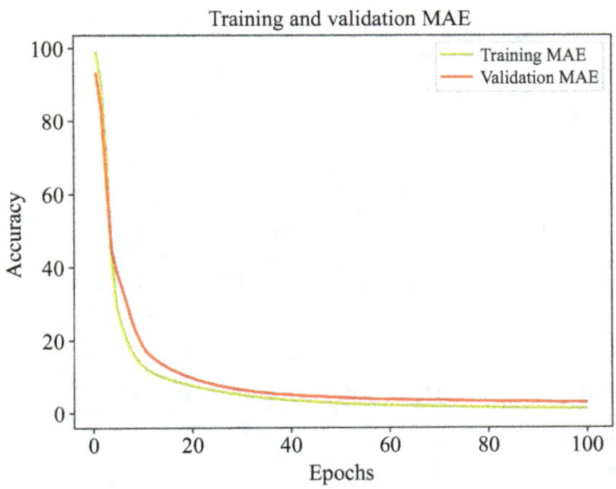

Figure 14.12 Training and validation of mean absolute error (MAE)

is then equally examined. A genetic algorithm scheme is used in this study as an example since it is exhaustive, provides the best solution, and guarantees that every constraint is met during the numerical simulation process [35].

The effectiveness of this study is also evaluated using a variety of performance indicators, including mean absolute error, mean square error, root mean square, and *R*2 score. Also, this study equally uses support vector machines (SVM), random forest (RF), the K-nearest neighbours (KNN) technique, and linear regression (LR). Nevertheless, as there is no information on the energy efficiency, throughput, latency, or spectrum efficiency of the 6G network, the data utilised for the training and testing of these algorithms is manufactured artificially or generated synthetically. Figures 14.12, 14.13, and 14.14, respectively, show the accuracy versus epoch, loss

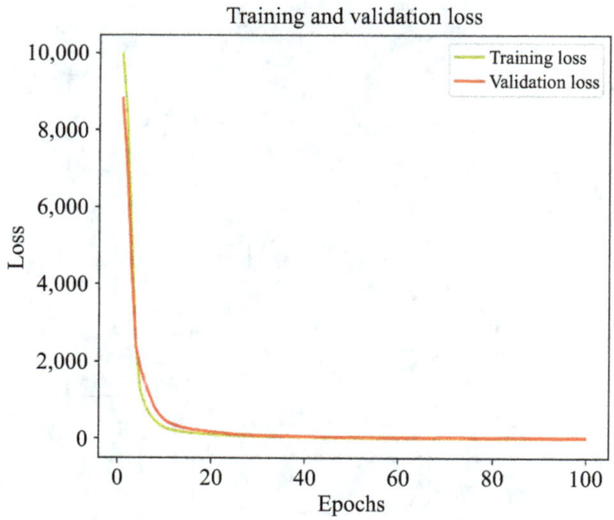

Figure 14.13 Training and validation loss

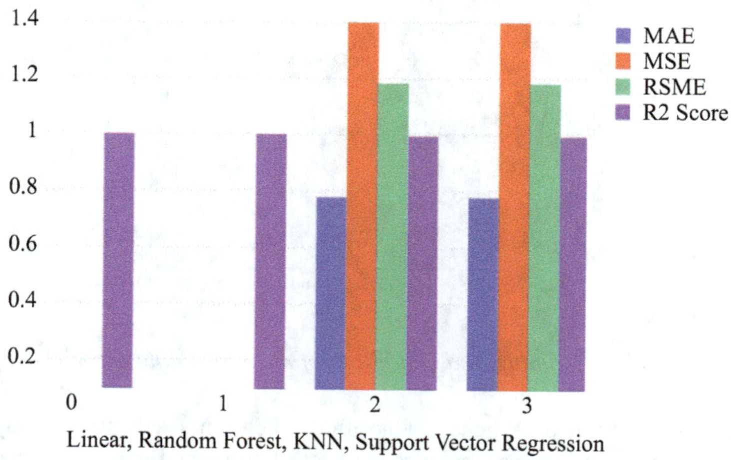

Figure 14.14 Bar chart for different algorithms with different performance metrics

versus epoch, and bar chart illustrating the level of performance of several algorithms. The *R*2 score is noteworthy, for which all algorithms provide an exact value of one in terms of numerical results for LR, RF, KNN, and SVM of 1.000, 0.999, 0.993, and 0.993, correspondingly shown in Table 14.3.

Table 14.3 Performance metrics used for the supervised ML

Performance metrics	Linear regression	Random forest	KNN	SVM
Mean absolute error	3.20752e−13	0.0049999	0.7784090	0.778409
Mean squared error	1.51784e−25	0.0023424	1.398068	1.398068
Root mean squared error	3.89545e−13	0.0483985	1.1823993	1.182399
*R*2 score	1.000000000	0.999989	0.9936443	0.993655

14.7 Lessons learned

In this study and previous studies, a lot of lessons have been learned regarding privacy and security for the 6G network. Some of the lessons learnt thus far are highlighted in the following sub-section.

14.7.1 Lessons learned from earlier wireless generations (1G–5G)

As we move towards the up-and-coming age of portable correspondence networks, 6G, it is essential to consider the privacy and security implications of technological advancements. With the consolidation of AI in 6G, there is a need to ensure that the systems are designed with privacy and security in mind to protect user data and prevent malicious activities. In this chapter, we will discuss some of the lessons learned from previous generations of the mobile telephone network and the future direction for 6G privacy and security issues using AI [36]. With the rise of 4G and 5G networks, we have witnessed numerous security threats that have impacted users' privacy and security [37,38]. Some of the lessons learned from these generations of the network include:

1. End-to-end encryption is crucial in protecting users' data from unauthorised access. In addition, the encryption of communication between devices and the network can prevent attackers from eavesdropping and intercepting data.
2. The importance of authentication – Authentication is essential to ensure that only authorised users can access the network. Implementing multi-factor authentication and biometric authentication can help prevent unauthorised access.
3. The need for secure software updates – Attackers can exploit security vulnerabilities in network software to gain unauthorised access. Regular and secure software updates can fix these vulnerabilities and prevent malicious activities [39].

While the following are the lessons learned from this work:

1. ML-based adaptive security strategies are efficient at thwarting SDN attacks. When AI approaches to get more advanced, they are also used in attacks to find holes in a frequently used network.
2. AI can be used to address 6G security challenges such as threat detection, attack prediction, access control, behavioural analysis, security compromise, and vulnerability management.

3. Because of the upsides of enormous information investigation and example acknowledgment, a few huge innovations, including CNN/RBN, SD-WAN/SDN organisations (DRL-based), vRAN/Open RAN radio, and VNFs (DRL-based), have started to utilise simulated intelligence.
4. Two factors have the effect of man-made intelligence and AI (ML) innovation on security being more grounded. In 6G, the right utilisation of ML can safeguard protection somehow or another, yet ML assaults may likewise abuse security in alternate ways.
5. The huge security parts that can assist with safeguarding the classification and protection of information sent employing 6G organisations are end-to-end encryption (E2EE), access control and authentication, blockchain-based security, ML-based threat detection, and privacy-preserving technologies.
6. The terrestrial-satellites-inspired network system can be utilised to demonstrate the idea of network optimisation in the 6G system.
7. Primary roadblocks and some research opportunities have been identified and highlighted for future studies.

14.7.2 Future directions

AI has the potential to revolutionise the mobile communication industry by improving network performance, reducing latency, and enhancing security. Some of the future directions for 6G privacy and security issues using AI include:

1. AI-based intrusion detection – computer-based intelligence (AI) can be utilised to identify and forestall malignant exercises in the organisation. AI-based intrusion detection systems can analyse network traffic and identify anomalies that may indicate an attack.
2. AI-powered encryption – AI can be used to improve encryption algorithms and make them more robust against attacks. AI can also generate and manage encryption keys to enhance security.
3. AI-based authentication – AI can be used to improve authentication systems by identifying patterns in user behaviour to authenticate users. This can help prevent unauthorised access by attackers who may have stolen user credentials.
4. AI-based threat intelligence – AI could be utilised to collect and discuss threat intelligence information through various sources to pinpoint emerging threats and prevent them before they cause harm. In conclusion, as we move towards 6G, we must consider privacy and security issues that may arise with technological advancements. With the incorporation of AI in the 6G network, we can enhance network security and privacy by leveraging AI-powered technologies. By learning from the lessons of previous generations and implementing robust privacy and security measures, we can ensure that the 6G network is safe and secure for users.

14.8 Conclusions

This book chapter focusses on using AI to address privacy and security issues in 6G communications. The chapter thoroughly reviews current works on the 6G network and privacy issues, which can serve as a helpful guide for researchers in the field. Additionally, the chapter details a few possible uses of AI for the 6G network. It provides educated assumptions on the effects of AI on 6G security and privacy based on recent studies and trends. The chapter also identifies new privacy and security challenges arising from innovations in the field and suggests potential solutions using current technology. Furthermore, the chapter proposes using blockchain technology to address privacy and security issues in 6G communications and formulates a network system scenario using MATLAB to demonstrate network optimisation. Finally, the chapter itemises the contributions, lessons learned, and possible future directions.

References

[1] C. Benzaid and T. Taleb, "AI for beyond 5G networks: a cyber-security defense or offense enabler?" *IEEE Netw.*, vol. 34, no. 6, pp. 140–147, 2020.

[2] T. Hewa, G. Gür, A. Kalla, M. Ylianttila, A. Bracken, and M. Liyanage, "The role of blockchain in 6G: challenges, opportunities and research directions," in *2nd 6G Wireless Summit (6G SUMMIT)*, 2020, pp. 1–5.

[3] T. S. Ajani, A. L. Imoize, and A. A. Atayero, "An overview of machine learning within embedded and mobile devices – optimizations and applications," *Sensors*, vol. 21, no. 13, p. 4412, 2021.

[4] Y. Sun, J. Liu, J. Wang, and Y. Cao, and N. Kato, "When machine learning meets privacy in 6G: a survey," *IEEE Commun. Surv. Tutorials*, vol. 22, no. 4, pp. 2694–2724, 2020.

[5] I. H. Abdulqadder, S. Zhou, D. Zou, I. T. Aziz, and S. M. A. Akber, "Multi-layered intrusion detection and prevention in the SDN/NFV enabled Cloud of 5G networks using AI-based defense mechanisms," *Comput. Networks*, vol. 179, p. 107364, 2020.

[6] R. Santos, D. Souza, W. Santo, A. Ribeiro, and E. Moreno, "Machine learning algorithms to detect DDoS attacks in SDN," *Concurr. Comput. Pract. Exp.*, vol. 32, no. 16, p. 5402, 2020.

[7] K. P. Trommler, "Security, Privacy and Trust in 6G Networks," *Bayern Innovativ*, 2021. Available: https://www.bayern-innovativ.de/de/seite/security-privacy-and-trust-in-6g-networks?fbclid=IwAR30qU5f_yCeCgZ7D4fcz91aobnxyKe4ogYUE9rcWgyGXl9LG0gPsj0_IAU. [Accessed: 20-Mar-2023].

[8] P. Porambage, G. Gür, D. P. M. Osorio, M. Livanage, and M. Ylianttila, "6G security challenges and potential solutions," in *Joint European Conference on Networks and Communications & 6G Summit (EuCNC/6G Summit)*, 2021, pp. 622–627.

[9] Y. Siriwardhana, P. Porambage, M. Liyanage, and M. Ylianttila, "AI and 6G security: opportunities and challenges," in *Joint European Conference on Networks and Communications & 6G Summit (EuCNC/6G Summit)*, 2021, pp. 616–621.

[10] L. Xiaojun, Z. Xiande, Z. Kexi, D. Zhenli, Z. Kai, and Wang Xi, "Study on the application fields and development prospects of artificial intelligence," in *2nd International Conference on Artificial Intelligence and Education (ICAIE)*, 2021, pp. 101–105.

[11] J. Harika, K. N. Palavadi Baleeshwar, and H. Shanmugasundaram, "A review on artificial intelligence with deep human reasoning," in *International Conference on Applied Artificial Intelligence and Computing (ICAAIC)*, 2022, pp. 81–84.

[12] S. M. Periannasamy, C. Thangavel, S. Latha, *et al.*, "Analysis of artificial intelligence enabled intelligent sixth generation (6G) wireless communication networks," in *IEEE International Conference on Data Science and Information System (ICDSIS)*, 2022, pp. 1–8.

[13] H. Liu, J. Zong, Q. Wang, Y. Liu, and F. Yang, "Cloud native based intelligent RAN architecture towards 6G programmable networking," in *7th International Conference on Computer and Communication Systems (ICCCS)*, 2022, pp. 623–627.

[14] W. Jiang, B. Han, M. A. Habibi, and H. D. Schotten, "The road towards 6G: a comprehensive survey," *IEEE Open J. Commun. Soc.*, vol. 2, pp. 334–366, 2021.

[15] K. K. Vaigandla, "Communication technologies and challenges on 6G networks for the Internet: Internet of Things (IoT) based analysis," in *2nd International Conference on Innovative Practices in Technology and Management (ICIPTM)*, 2022, pp. 27–31.

[16] M. Wang, T. Zhu, T. Zhang, J. Zhang, S. Yu, and W. Zhou, "Security and privacy in 6G networks: new areas and new challenges," *Digital Communications and Networks*, vol. 6, no. 3, pp. 281–291, 2022.

[17] A. L. Imoize, O. Adedeji, N. Tandiya, and S. Shetty, "6G enabled smart infrastructure for sustainable society: opportunities, challenges, and research roadmap," *Sensors*, vol. 21, no. 5, p. 1709, 2021.

[18] V.-L. Nguyen, P.-C. Lin, B.-C. Cheng, R.-H. Hwang, and Y.-D. Lin, "Security and privacy for 6G: a survey on prospective technologies and challenges," *IEEE Commun. Surv. Tutorials*, vol. 23, no. 4, pp. 2384–2428, 2021.

[19] M. B. Gracia, V. Malele, S. P. Ndlovu, T. E. Mathonsi, L. Maaka, and T. Muchenje, "6G security challenges and opportunities," in *IEEE 13th International Conference on Mechanical and Intelligent Manufacturing Technologies (ICMIMT)*, 2022, pp. 339–343.

[20] P. Porambage, G. Gur, D. P. M. Osorio, M. Liyanage, A. Gurtov, and M. Ylianttila, "The roadmap to 6G security and privacy," *IEEE Open J. Commun. Soc.*, vol. 2, pp. 1094–1122, 2021, doi:10.1109/OJCOMS.2021.3078081.

[21] R. D. Shirwaikar, A. M. Faisal, A. Singh, and D. D. Shanbhag, "A review on privacy and security in 6G networks," in *International Conference on Forensics, Analytics, Big Data, Security (FABS)*, 2021, vol. 1, pp. 1–6.

[22] Y. Zhang, S. Ma, J. Li, K. Li, S. Nepal, and D. Gu, "SMARTSHIELD: automatic smart contract protection made easy," in *IEEE 27th International Conference on Software Analysis, Evolution and Reengineering (SANER)*, 2020, pp. 23–34.

[23] Z. Li, J. Peng, L. Li, and K. Zhang, "Blockchain technology in 6G wireless networks: opportunities and challenges," *IEEE Wirel. Commun.*, vol. 27, no. 6, pp. 24–30, 2020.

[24] B. Zohrevandi and M. Shojafar, "Blockchain-enabled 6G networks: challenges and opportunities," *IEEE Commun. Mag.*, vol. 59, no. 4, pp. 18–24, 2021.

[25] I. Stojmenovic and W. Liao, "Blockchain technology for 6G wireless networks: opportunities, challenges, and solutions," *IEEE Netw.*, vol. 34, no. 6, pp. 56–61, 2020.

[26] O. O. Shoewu, L. A. Akinyemi, and R. Edozie, "UAV cellular communication in 5G new radio wireless standards," in *Unmanned Aerial Vehicle Cellular Communications* (pp. 25–45). Cham: Springer, 2023.

[27] S. O. Oladejo, S. O. Ekwe, and L. A. Akinyemi, "Multi-tier multi-tenant network slicing: a multi domain games approach," *ITU J. Futur. Evol. Technol.*, vol. 2, no. 6, pp. 1–26, 2021.

[28] S. O. Ekwe, S. O. Oladejo, L. A. Akinyemi, and N. Ventura, "A socially inspired energy-efficient resource allocation algorithm for future wireless network," in *16th International Computer Engineering Conference (ICENCO)*, 2020, pp. 168–173.

[29] A. T. Ajibare, D. Ramotsoela, L. A. Akinyemi, and S. O. Oladejo, "RF EMF radiation exposure assessment of 5G networks: analysis, computation and mitigation methods," in *IEEE AFRICON*, 2021, pp. 1–6.

[30] S. O. Ekwe, L. A. Akinyemi, S. O. Oladejo, and N. Ventura, "Social-aware joint uplink and downlink resource allocation scheme using genetic algorithm," in *IEEE AFRICON*, 2021, pp. 1–6.

[31] Y. O. Imam-Fulani, N. Faruk, O. A. Sowande, *et al.*, "5G frequency standardization, technologies, channel models, and network deployment: advances, challenges, and future directions," *Sustainability*, vol. 15, no. 6, p. 5173, 2023.

[32] A. E. Ibhaze, A. L. Imoize, and O. Okoyeigbo, "A brief overview of energy efficiency resources in emerging wireless communication systems," *Telecom, MDPI*, vol. 3, no. 2, pp. 281–300, 2022.

[33] A. L. Imoize, H. I. Obakhena, F. I. Anyasi, and S. N. Sur, "A review of energy efficiency and power control schemes in ultra-dense cell-free massive MIMO systems for sustainable 6G wireless communication," *Sustainability*, vol. 14, no. 7, p. 11100, 2022.

[34] R. L. Kumar, Y. Wang, T. Poongodi, A. L. Imoize, eds., *Internet of Things, Artificial Intelligence and Blockchain Technology*, Springer, Cham, 2021.

[35] J. O. Ogbebor, A. L. Imoize, and A. A.-A. Atayero, "Energy efficient design techniques in next-generation wireless communication networks: emerging

trends and future directions," *Wirel. Commun. Mob. Comput.*, pp. 1–19, 2020.

[36] F. Hu, X. Wang, and K. Li, "5G security and privacy: a review," *IEEE Access*, vol. 8, pp. 121646–121663, 2020.

[37] K. Wang, K. Lu, and K. Ren, "Privacy-preserving machine learning: threats and solutions," *IEEE Commun. Mag.*, vol. 57, no. 11, pp. 56–61, 2019.

[38] K. Kim and K. Chung, "Security and privacy challenges in 5G and beyond," *IEEE Commun. Mag.*, vol. 58, no. 2, pp. 80–87, 2020.

[39] Y. Zhang, C. Wang, J. Zhang, X. Wang, and Z. Zhu, "Secure and efficient data communication in 6G networks: challenges and solutions," *IEEE Commun. Mag.*, vol. 59, no. 2, pp. 118–123, 2021.

Chapter 15

Interference and phase noise in millimeter wave MIMO-NOMA and OFDM systems for beyond 5G networks

Udayakumar Easwaran[1] *and Krishnaveni Vellingiri*[2]

Abstract

The fifth-generation mobile communication (5G) is designed to support huge connectivity, great data rates, and excellent dependability as the number of wireless devices linked to the network approaches billions. Additionally, mobile users generate most of their data traffic from video streaming, demanding a higher bandwidth and lower latency. Current mobile communication networks must be upgraded to acceptable to meet these criteria. It is investigated that a multiuser environment will require multiple access methods, such as non-orthogonal multiple access (NOMA) and orthogonal frequency division multiplex (OFDM). Using millimeter-wave (mmWave) spectrum and non-orthogonal multiple access (NOMA) has helped to address inefficient power allocation phase noise issues and hybrid beamforming complications to meet the lowest rate necessities of the network user. At the transmitter base station, the user facts are superimposed in the power domain NOMA, after which the user end is subjected to phase noise cancellation. Due to insufficient elimination of the unwanted user's interference in the multiuser downlink, the desirable user's interference is vulnerable to unsatisfactory successive interference cancellation (SIC). The key goal of this survey is to decrease interference, i.e., phase noise in millimeter wave 5G systems using the parametric phase noise filtering method.

Keywords: NOMA; MIMO; Mmwave; Phase noise; Bit error rate; 6G networks; Security and Privacy; SNR; Beamforming and OFDM

15.1 Introduction

The Fourth Industrial Revolution is predicted to user in a period of technology advancement and digitization that will necessitate constant connectedness on the

[1]Department of ECE, KIT – KalaignarKarunanidhi Institute of Technology, India
[2]Department of ECE, PSG College of Technology, India

part of users. In order to meet the high connectivity requirements for systems that include great data rates, energy-efficient systems, extremely low latencies, etc. Therefore, the frequency domain is used to characterize phase noise. While a real oscillator's waveform displays "skirts" about the core or carrier frequency, an ideal oscillator's spectrum has the shape of an impulse. To determine the amount of phase noise present in a unit bandwidth at an offset from the carrier power, the noise power in this bandwidth is first multiplied by the carrier power [1].

A frequency synthesizer's internal local oscillator (LO) produces the carrier signal for both mixers. If possible, the target signal band should be convolved with an impulse and translated to a lower (and higher) frequency without changing its shape. However, if phase noise is present in the LO output, both the down converted and unconverted signals for the receive and transmit pathways will be harmed. On occasion, the intended signal may coexist in a channel with a significant interferer, and the local oscillator may display finite phase noise. The down converted band is formed up of two overlapping spectra when the two signals are combined with the LO output, which causes the intended signal to suffer from substantial noise [2].

However, NOMA is thought to be a multiple access method to expand the present wireless communication network. When no spatial separation is required in the downlink, NOMA enables the optimizes of multiuser process and allots the similar frequency properties to each user. Using a superposition approach, this is possible. It is suggested to employ the Fourier transform to pulse-shape NOMA signs [3] in order to reduce multiuser interference brought on by a flawed SIC. According to analytical findings, Fourier transform with NOMA outperforms other strategies in reducing SIC and thus minimizing the residual error brought on by a faulty SIC.

Most academics view orthogonal frequency division multiplex-based fast Fourier transform (OFDM-FFT) as a pulse shaping approach in downlink NOMA. In addition, because its side lobes are spectrally constrained, the wavelet-centered NOMA system performs better in symbol error rate (SER) even when the receiver's channel characteristics are not fully understood. The users with great channel gains must interpret all of the signals from the other users in single carrier NOMA, which increases complexity and decoding delay. The key to overcoming each interference suppression weakness of FFT-NOMA is, therefore, wavelet transform-based NOMA, which can be functional in a 5G mobile networks to meet rising connection difficulties [12].

The wavelet NOMA is evaluated under perfect SIC conditions in earlier research, which means that unwanted user signals are entirely removed from the wanted signal until it is convalesced. To get the desired user's data, however, one must delete all of the unwanted user signals after the complex signal if the unwanted user's Channel Information is insufficient. Due to the faulty SIC caused by this, in addition to the desired user signal, a residual error also results. Therefore, by SIC's shortcomings, some residual inaccuracy is generated [18]. Our work draws attention to the problem with NOMA systems that target more than two users, channel issues, and problematic SIC at the receiver. Using FFT and wavelet NOMA, we compare the residual error at users 2 and 3.

Addressing the vast security and privacy issues, interference, and phase noise in millimetre wave MIMO-NOMA and OFDM systems is particularly important to the envisioned 6G networks. While 6G networks, which are anticipated to offer even faster data rates, lower latency, and better connectivity, are still being developed, 5G networks are already in use. The utilization of millimetre wave frequencies, which are even higher than those utilized in 5G networks, is one of the primary technologies being investigated for 6G networks.

Because of interference and phase noise's great sensitivity at millimetre wave frequencies, signal quality and data throughput can be severely impacted. This is especially true for MIMO-NOMA and OFDM systems, which depend on sophisticated signal processing methods to boost spectral efficiency and enhance general network performance. Therefore, the successful implementation of 6G networks depends on knowing and mitigating the impacts of interference and phase noise in these systems.

As 6G network development advances, research into interference and phase noise in millimetre wave MIMO-NOMA and OFDM systems for 5G networks is expected to continue and change. To enhance the performance of these systems at higher frequencies, new methods for interference cancellation, phase noise correction, and other signal processing methods might be created. In addition, new protocols and standards might be created to guarantee the seamless integration of 6G networks with current 5G networks.

15.1.1 Key contributions of the chapter

The results are the important contributions of this chapter:

(i) The chapter surveys the interference that exists in millimeter-wave MIMO-NOMA and OFDM schemes.
(ii) The significant challenges and the prospects of phase noise in millimeter wave 5G communication schemes.
(iii) For the reduction of interference and phase noise, use phase noise filtering method to reduce together inter-carrier and inter-symbol interference in MIMO-NOMA systems.

15.1.2 Chapter organization

Section 15.2 shows the related works on the MIMO-NOMA systems. Section 15.3 shows the system model of FFT-NOMA. Section 15.4 presents the uplink and downlink NOMA network. Section 15.5 discusses the MIMO-NOMA systems, including resource allocation, user clustering, monotonic optimization, combinatorial relaxation, power allocation in NOMA, and security and privacy in 5G systems. Section 15.6 shows the results and discussion. Section 15.7 concludes the chapter with future direction.

15.2 Related work

Systems for mobile wireless communication have become a need in contemporary life. However, as devices become more numerous and diverse, it becomes necessary for numerous users and/or applications to share the same radio spectrum. The mandate for the Internet of Things also brings the requirement to connect every person and thing through it. However, the stringent restrictions of the current communication networks prevent any system adjustments or enhancements that would match these expectations. Certain academics and businesses have proposed future technologies to meet the aforementioned stringent standards and take on the issues facing future generations [9].

Numerous users can use non-orthogonal properties simultaneously in NOMA, which produces excellent spectral efficiency. System ideas based on non-orthogonality have recently drawn a lot of attention from researchers and are being employed in communication networks. Since it does not require significant network changes, power-domain multiplexing is simpler to implement. Additionally, expanding bandwidth is not necessary to increase spectral efficiency. In an ideal [15] environment, orthogonal multiple access (OMA) approaches can achieve an acceptable system routine even with basic receivers since there is no mutual interference between users, but they are still unable to handle the new problems brought on by the rising needs for 5G systems and beyond.

The fixed power allocation technique can realize the different QoS needs. A power allocation approach influenced by cognitive radio ensures that the user's QoS needs are met right away. Additionally, the system OP's exact and asymptotic formulas have been developed. The downlink MIMO-NOMA network power minimization problem under full channel state information (CSI) and channel distribution information. The linear precoders that offer a higher total throughput while also enhancing the throughput of the user with a low-quality channel were constructed while also meeting the requirements of the QoS standard. It is also demonstrated that higher distinctive channel gains are required to reach the extreme number of users per cluster to realize an advanced NOMA presentation [16].

Since multicast beamforming provides improved sum capacity presentation uniform for numerous users, it can also be suggested as a technology for MIMO schemes. However, there are numerous uses for it. One approach relies on a sole beam that can be utilized by all users and produces a common signal that everyone picks up. Another tactic is to deploy a large number of beams that can be used by different user groups, resulting in each group receiving a distinct signal [6]. In particular, two users can share one beam. Since just two users with different channel quality can share the proposed beam, it is likely to be simple to apply clustering and power allocation methods to boost the overall capacity and minimize interferences between clusters and users. The comparison between bandwidth for various generations [38] is shown in Table 15.1.

Every receiver in SIC looks for more robust signals than the signal the user is attempting to receive first. The received signal is further isolated from the other signals when just the linked user's individual signal is left. Then, by considering users with lesser power quantities as noise, separately user decodes its specific signal. According to each uplink NOMA network to the BS, a mobile user sends their signal. The signals of mobile users at the BS are located using SIC iterations [17].

Table 15.1 Comparison between bandwidth for various generations

Generation	Frequency (GHz)	Wavelength (m)	Bandwidth (GHz)
5G (millimeter wave)	28	0.0107	1.3
	38	0.0079	1.4
	46	0.0065	1.4
	73	0.0041	5.0
	83	0.0036	5.0
4G	2.1	0.1429	1.1174
	2.6	0.1154	
3G	1.8	0.1667	
2G	0.8	0.3750	
	0.9	0.3333	

Light cannot pass through walls, hence visible light communication (VLC) by nature provides a greater level of data protection. Additionally, VLC is mainly helpful in some delicate settings, like aeroplane cabins and hospitals, here interference to current radio frequency (RF) schemes is a concern by nature. The use of NOMA to the downlink VLC situation has the possible to significantly improve the presentation of VLC systems without harming the quality of light emitting diode (LED) lighting, despite the significant disparities among the VLC channel and the RF channel. The NOMA-enabled VLC functional in many contexts and circumstances, such as NOMA in VLC, more study must be done. Resource distribution in NOMA-supported VLC is still a concern [10].

In a heterogeneous network (HetNet) that employs OMA, different users are assigned separate time-frequency resources. OMA enables users to access the network by using orthogonal channels, ensuring that each user's transmission does not interfere with others. When comparing OMA-enabled HetNet with the two-tier NOMA configuration, it has been determined that NOMA may have an impact on the performance of users at the cell edge.

A bandwidth-efficient data transmission technique, MIMO-OFDM, is utilized in several of the most recent broadband technologies, including WLAN, WiMax, LTE, and others. Phase noise, IQ imbalance, and other issues are the main downsides of analogue circuits. Here, data symbols are sent over significant low-rate subcarriers using OFDM. Phase noise is exceedingly challenging to measure and reduce [37]. The MIMO technology has been used to reduce the transmission nonideals.

15.3 System model of FFT-NOMA

The complexity of the system and other communication-related issues like channel dispersion are significantly influenced by the choice of an appropriate pulse-shape. Numerous characteristics of communication structures based on the OFDM method include receiver designs that are less complex, toughness to multipath delay spread, interoperability with MIMO systems, and straight forward equalizers. A usual NOMA scheme model with two users, with user 1 being the far user and user 2 being the close user to the base station (BS). This study takes into account a three-user NOMA system model with a transitional user in adding to the far and close users. Separately user's data is divided into distinct power levels and turned into a particular data stream using a superimposed technique, enabling non-orthogonal multiplexing of all three users over the wireless communication channel [25].

This study focuses on how interference between users 2 and 3 affects the communication error rate between them. Imperfect SIC at the receiver will produce a residual inaccuracy if user 3 and user 1 channel estimation are inaccurate. In our system model, wavelet filter banks are suggested to lessen the impact of incomplete SIC. At the transmitter side, baseband modulation and source coding are functional to the input data symbol. For each of the three users, variable power is assigned based on the channel circumstances. The ideal oscillator output becomes

$$m(t) = A_0 \cos\,(2\pi f_c t + \theta) \tag{15.1}$$

Here, A_0 is the amplitude, f_c is the carrier frequency, θ is the phase offset, $\mu(t)$ is the frequency deviation of an oscillator and power spectral density $N_0 = v/\pi$, and v is the oscillator linewidth.

Various minimal symbol errors using multicarrier modulation, which employs FFT for NOMA and the same wavelet technique for wavelet NOMA, data streams can be sent over non-orthogonal subcarriers [30]. The MMSE equalization on the receiver side, DFT filter banks are used to alter the existing data. After baseband demodulation and decoding, user 1 recovers each user's interference, and user 3 and user 2's interference is eliminated using SIC. User 1 considers each user's interference as noise and interprets the interference as noise. The impact of user interference is because of the wavelet NOMA that was recommended by this study attempt as a way to lessen user interference. The block of phase noise in OFDM systems [23] is shown in Figure 15.1.

In order to evaluate the SER performance of close, intermediate, and far users in the downlink NOMA system, considering both ideal and idealized CSI values. Based on the channel gains, the power issues are assigned to for each user individually in the instance of QPSK modulation. When the expected signal is demodulated, it is assumed that the channel is defective, which causes the SIC to be imperfect and leaves residual error [8]. Users are thought to be equally spaced apart. Both FFT-NOMA and

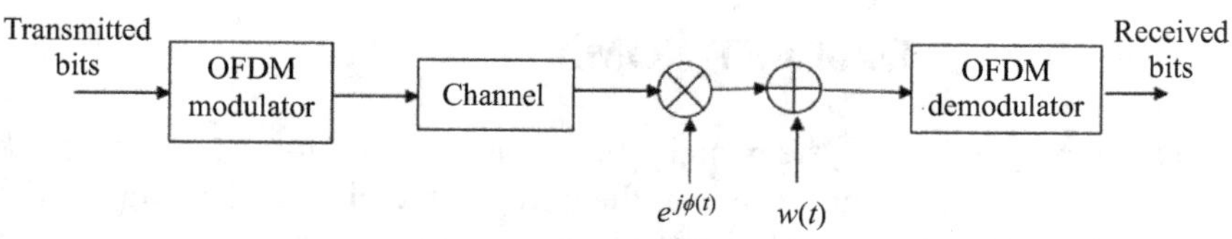

Figure 15.1 Block diagram of phase noise in OFDM systems

wavelet-NOMA assume additive white Gaussian noise (AWGN) channel environments, hence channel equalization is not taken into account [28]. However, the wavelet-transform-related NOMA method that combines NOMA as a multiple access method and wavelet as a pulse shaping method is proposed in this study. The flow diagram of SIC in millimeter wave NOMA systems is shown in Figure 15.2.

The interference caused by user 2's data on user 1 and user 3 is the primary subject of the research endeavor. The signal space for three users describes how users with high transmit power affect users with relatively low transmit power who are carrying out SIC. The Constellation opinions are rotated 45 degrees for the third user to make it easier to understand user 1's interference. User 2's constellation is centered on user 3, which is surrounded by user 3's constellation [27]. So that user 1's constellation is tilted 45 degrees toward user 2. This separation of each user from the others allows for easy identification of all users. The scenario given in this research effort has each user interfering with the other two participants. The circle in the middle represents user 3, which affects the received superposed signal in the same way as the added two high-power users. Where cross symbol indicates user 2, which will only affect user 1.

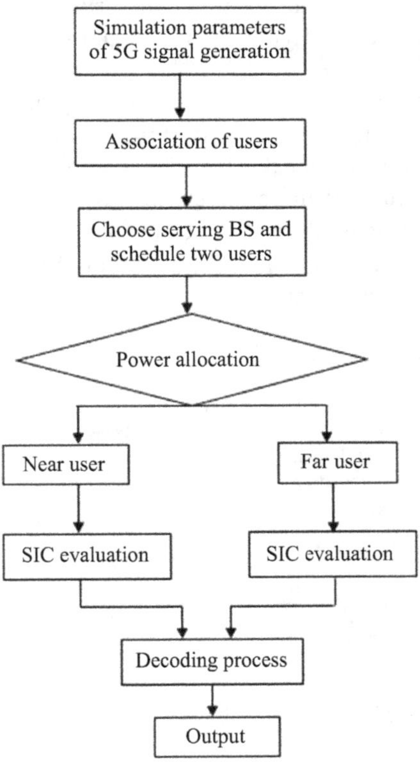

Figure 15.2 Flow diagram of SIC in millimeter wave NOMA systems

15.4 Uplink and downlink NOMA network

However, NOMA may simultaneously serve several users on distinct channels, which improves user fairness and allows for lower latency and greater large connectivity. Since the NOMA does not necessitate significant changes to the current architecture, it is also well-matched with present and forthcoming communication technologies [20]. In LTE-A systems, MUST makes advantage of superposition coding idea for a multiuser communication. The deployment situations, assessment methodology, and potential NOMA arrangement for the 3GPP radio access network (RAN) have all been looked into, respectively, while employing MUST. For any output sinusoid with noisy can be

$$c(t) = e^{j2\pi fct + j\theta(t)} \tag{15.2}$$

Where $\theta(t)$ is the phase noise random process over power spectral density, fc is the carrier frequency. To put it another way, OMA's poor performance renders it inappropriate and unable to provide the functionality essential by upcoming generations of wireless communication systems. Therefore, experts believe that NOMA is an excellent option for the next generation of MA techniques. Even while NOMA contains many characteristics that could benefit future generations, nearby are several problems that need to be resolved earlier it can be used to its full probable. The block diagram of conventional NOMA systems [31] is shown in Figure 15.3.

When linked to OMA, the computational difficulty of the receiver will rise in NOMA since individually user must first interpret the signals of nearly other users earlier decoding its individual signal. This will cause a longer delay. Additionally, the BS should be informed of the channel gains of all users, although doing so incurs a sizable channel state information (CSI) opinion upstairs. Furthermore, the error chance of subsequent decoding will grow if any SIC procedures at any user encounter faults. To prevent such error spread, the number of users should be decreased.

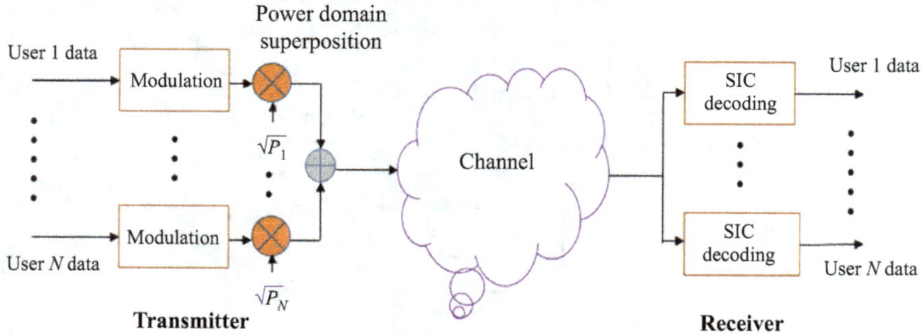

Figure 15.3 Block diagram of conventional NOMA systems

It is predicted that till the user's interference is recovered, each user's receiver will go through the SIC process in turn. Inversely related to the channel conditions is the distribution of user power coefficients. More transmission power is given to the user with a poor channel situation than to the user with a good channel situation. The user with the maximum broadcast power, as a result, immediately improves its signal lacking going through any SIC processes because it perceives the indicators of other users as noise. However, other users are required to do out SIC operations [36].

15.5 MIMO-NOMA systems

Using MIMO technology, wireless communication networks' capacity and error probability can both be significantly increased. The factors prevent the optimal power allocation from being achieved. Technique with a reduced level of complexity is necessary to maximize throughput. Conditions ergodic capacity is proposed. To reconcile the highest sum of mobile users with the best sum rate that MIMO-NOMA schemes are capable of achieving, entirety rate has modeled in two different ways. Power distribution among user clusters is the goal of the first strategy. Another tactic is to partition the users into several clusters, allowing each cluster to get orthogonal spectrum resources in accordance with the preferred user grouping approach. The efficiency of MIMO-NOMA and MIMO-OMA approaches for two users per cluster across Rayleigh fading channels has also been investigated [19].

SINR thresholds even though the total number of people accepted and the sum rate are the highest when they are equal, good results are still obtained despite the fact that each user's SINR threshold varies. The proposed method's low level of complexity increases linearly as user density in each cluster increases. A downlink MIMO-NOMA network's performance is introduced for the straightforward situation of dual users or one cluster. When comparing the sum rate and ergodic sum rate between MIMO-NOMA and MIMO-OMA, MIMO-NOMA performs better [22]. The sentence also mentions that this result holds true when using zero-forcing (ZF) precoding and signal arrangement together in a more realistic scenario with multiple users grouped into clusters and sharing a transmit beamforming path.

To prevent the negative impacts of using many antennas at once, antenna range methods have also been recognized as a potent remedy that can be practical to MIMO schemes. These side consequences include expensiveness, redundant power use, and hardware complexity. The benefits of variety that MIMO systems that can provide are still present. As they were created for MIMO-OMA systems, some works use antenna selection methods in MIMO-NOMA. The significant inter-user interference in MIMO-NOMA schemes, unlike in MIMO-OMA networks where information transfer occurs without hindrance, the improvements achieved in MIMO-NOMA cannot be easily replicated [26]. Consequently, some works have raised concerns and questioned the issue of selecting an antenna. The huge MIMO-NOMA system for two users employs the user scheduling approach that is employed in addition to a successful TAS scheme to exploit the sum rate for two situations. An effective search method is recommended for the first circumstance. This technique seeks to select the antennas with the maximum channel gains while restricting the search to a finite candidate set of suitable antennas for the users who are interested. The phase noise for a free-running oscillator is

$$\phi(t) = \sqrt{cB(t)} \tag{15.3}$$

where $B(t)$ is the Brownian motion and c is the diffusion rate. The standard Brownian motion (wiener process) is defined as the random process.

The antenna–user pair that contributes the greatest to the overall channel gain is picked because this algorithm precisely alters the percentage of the channel gain defined by a particular antenna–user pair to the overall channel gain. In addition, a hybrid antenna and user influence algorithm does not offer the greatest trade-off among system presentation and difficulty. Sadly, neither the writers of nor the authors of have conducted an analytical study of the system performance. However, the max–min–max antenna selection strategy enhances the user's immediate channel gain when the channel situation is low, while the max–max–max method offers the optimal choice when the channel state is good, state with a good channel condition. Additionally, both proposed algorithms' asymptotic closed-form formulations of the regular entirety rates are assessed [9].

15.5.1 Resource allocation

The effectiveness of multicast beamforming is evaluated for a downlink MISO-NOMA system in a humble model with two users using superposition coding. By sharing a signal, multicast beamforming makes use of the beam to serve numerous users within a cluster. BS's transmitter primarily has many antennas and bases its information stream on the multi-resolution broadcast idea, which only sends low priority signals to users who are distant after the BS, or users have poor channel value. Users with virtuous channel quality who are close to the BS receive together high priority and low priority signals. A least power beamforming issue has also been established using superposition coding to discover the beamforming vectors and powers for both users. A downlink MIMO-NOMA network's BS performs random beamforming [39]. According to the system classical, each beam is expected to be utilized by every user within a single cluster and to have an equal distribution of transmission power. To reduce inter-cluster interference, it is advisable to employ the zero-forcing beamforming technique, especially when different channel quality users are anticipated. Additionally, user–cluster algorithms and dynamic power allocation have been suggested in order to maximize throughput while minimizing interference.

Numerous studies have looked at the case of a perfect CSI, there is a resource allocation issues related to maximizing the sum rate. Two users in a single cluster with two dissimilar precoder implementations are involved in this challenge. It investigates the downlink MISO-NOMA system sum rate maximization problem [21]. However, a complex vector is used to weight each mobile user's sent signal. Moreover, the minorization–maximization method is recommended as an approximation in order to avoid the high computing complexity associated with non-convex optimization problems. The main goal of the minorization–maximization technique, which assumes perfect CSI, is to create complex weighting vectors that maximize system throughput for a given order of users. A downlink MIMO-NOMA scheme with various beams and perfect CSI existing at all nodes, BS transmissions precoded signs to all mobile users, meaning that each beam serves a number of users [29].

15.5.2 User clustering

Since large MIMO technology may guarantee substantial antenna diversity at a lesser cost, their performance over NOMA has also drawn a lot of interest. For instance, study a massive MIMO-NOMA system with little feedback, in which the BS has many more broadcast antennas than the customers' receive antennas. As a result, the superimposed code maximization strategy performs better than the orthogonal one when there are more mobile users and higher levels of mobility. Routine of massive access MIMO schemes, in contrast to massive MIMO, when the number of users surpasses the quantity of antennas engaged at the BS [23]. An iterative detection algorithm is commonly used in mmWave systems to handle multiuser detection with low complexity. This algorithm aims to minimize the mean square error and achieves fast convergence in terms of both mean and variance.

In order to lessen interbeam interference, a precoding approach based on the zero-forcing (ZF) principle has also been proposed. Additionally, a dynamic power allocation method and iterative optimization approach are suggested to increase sum rate while reducing complexity. One of the user scheduling strategies effectively handles multiple interference, while the other maximizes multi-collinearity among users, depending on how the signal space alignment is set up. Hereafter, change the non-convex constraints into their approximate inner convex forms. Additionally, it is demonstrated that using a maximum multicollinearity scheme results in a greater sum rate of centre users and a higher energy efficiency when less power is sent.

The mmWave-NOMA transmission scheme's outage sum rate [28] can be

$$R_{sum}^{NOMA} = \mathrm{P(M=1)}\left(1 - R_{OMA}^{1|M}\right)R_1 + \sum_{k=2}^{\infty} P(M=k) \times \left(\left(1 - P_{i|M}^{o}\right)R_i + \left(1 - P_{j|M}^{o}\right)R_j\right) \tag{15.4}$$

By using mmWave-OMA, the sum rate may be expressed as

$$R_{sum}^{OMA} = \mathrm{P(M=1)}\left(1 - R_{OMA}^{1|M}\right)R_1 + \sum_{k=2}^{\infty} P(M=k) \times \left(\left(1 - P_{OMA}^{i|M}\right)R_i + \left(1 - P_{OMA}^{j|M}\right)R_j\right) \tag{15.5}$$

Conditional outage probability when OMA is applied is indicated by $R_{OMA}^{1|M}$.

The number of users in the disc is Poisson distributed, therefore user *i* will treat message as noise and immediately decode its information (M users in D).

As was already established, NOMA capacities to meet the IoT's demand for quick service for small packet communications from a large quantity of users. As a result, research on MIMO-NOMA performance for IoT is a common theme in the literature. In a MIMO-NOMA downlink system, one transmitter is assumed to be delivering information to two users. The second user has a larger data rate whereas the first user has a little data rate, or minor packet transmission. Investigations focus in particular on outage performance when precoding and power allocation methods are used [11]. Additionally, it is demonstrated that NOMA still has promise even when channel and user attributes are similar.

15.5.3 Monotonic optimization

Due to its inherently combinatorial character, user clustering is generally a challenging problem. In fact, it has been demonstrated for downlink that allocating. Connecting users in NOMA to orthogonal resource blocks, such as subcarriers and subchannels, is a challenging task known to be NP-hard. To address this issue, greedy user clustering algorithms have been proposed, leveraging positive correlation for effective user grouping. It has been widely used to compare the difference between channel gain and performance gain [32]. But as a result, the performance advantage may not be distributed equally throughout the numerous clusters. According to the users, it should initially be separated into two collections based on their channel gains to overcome this issue. Then, *I* is matched with the user in group one with the highest channel gain that is equivalent to group two, and so on. Furthermore, it has been demonstrated that using the right power allocation in conjunction with this user clustering strategy can produce the best result. This is only applicable to channels with flat frequency. These user clustering techniques based on channel gain are simple. However, because they are heuristic methods, their performance could be unstable. Systematic frameworks should be used to establish a balance between complexity and efficacy [30].

Due to intra-cluster interference, the resource distribution difficult in NOMA systems is typically non-convex. As a result, utilizing the convex optimization theory to find the best solution is quite difficult. Convexity should therefore not be the only property used. Among these, monotonicity is a crucial characteristic that can be applied to solve non-convex issues. It provides an ideal solution to the combined power and subcarrier distribution that is difficult via monotonic optimization [29].

15.5.4 Combinatorial relaxation

The binary variable that links a person to their associated cluster presents the fundamental difficulty in user clustering. The unique NP-hard difficult may frequently be converted into a convex difficult by soothing this binary variable into a uninterrupted one. Convex optimization can then be used to find the best solution.

Keep in mind that methods like rounding should be used to retrieve the binary variables [14]. This kind of unwinding and healing, nevertheless, frequently outcomes presentation break among the unique issue and the calm one. By matching theory and game theory in NOMA systems, user clustering has recently seen a widespread use of game theory. It should be noted that the conventional adopts coalition game, in which each user's objective is to maximize its personal utility rather than enhancing system performance.

The particle swarm optimization technique, which modifies the effectiveness purpose for each user in the direction of a global optimal solution, is used to improve the coalition game that is planned in as a way to combat this. For a mmWave NOMA system, a Stackelberg game strategy is suggested by taking user clustering as the leader and power distribution as the supporter, with the exception of the coalition game. The unilateral equilibrium deviation and distributed implementation are the main drawbacks of game theory-based strategies [23].

The aforementioned methods' computational complexity might be too great for actual execution. To solve this problem, it is a good idea to refrain from comparing candidate user pairs needlessly for user collections that are not suitable for NOMA multiplexing. The prior user clustering techniques also do not take into account the possibility that there might not be enough strong. After each robust user is balancing with its partner in this condition, there can be still weak users left over. A hybrid strategy that uses OMA to reach the remaining users can be used to handle this. However, this denies these consumers the benefits that NOMA offers. The perception of virtual user clustering, in which a strong user, two weak users, and a third user share a frequency band, can be used as an alternative to implement NOMA. The bandwidth is split equally between the strong and one weak user and third weak user and the strong user [24].

15.5.5 Power allocation in NOMA

Due to user multiplexing in the power field, PA is crucial in NOMA. The control of interference, rate circulation, and even user charge are all directly jammed. A bad PA could cause a system breakdown due to SIC failure as well as an unjust rate allocation among consumers. Users' channel circumstances, CSI availability, QoS supplies, total power constraints, system objects, and other factors need to be taken into account while creating PA methods. Thus, the objectives of PA in NOMA are to either increase the amount of admitted users, increase the total rate, or increase energy efficiency (EE), or justice that is balanced while using the least amount of energy. A classification of the numerous PA techniques that have been put out in the works to address dissimilar facets of PA in NOMA. The two subsections that follow focus on single-carrier (SC) systems and multiple-carrier (MC) systems, respectively, as we introduce PA [5].

To exploit the total rate, three proposed techniques are integrated, though. In the first, weighted sum rate maximization suggests creating a unique beamforming matrix for each beam that takes advantage of all CSI at the BSs. In the second approach, each mobile user's super SIC is intended to be at the receiver. In order to fully profit from SIC, channel gain differences between clusters must be strong, and channel correlation among mobile users must be high [34]. The fourth one aims for the optimization with fixed power allocation, providing a greater sum rate as well as practical presentation for the user with poor channel quality. It is researched how much transmission power can be used by each mobile user [44–46].

Due to resource constraints, spectral efficiency (SE) and energy efficiency (EE) are two essential presentation indicators in 5G networks. The NOMA allocates the same resource blocks to different users based on their power level, it is remarkably effective in terms of both spectrum and energy. The rate of the total capacity over the BS's overall power consumption [29] is known as

$$\text{Energy efficiency} = \frac{R}{P_{total}} = \text{SE}\frac{B_T}{P_{total}} \text{(bits/Joule)} \tag{15.6}$$

where B_T is the transmission bandwidth, P_{total} is the total signal power used by the BS, and R is the sum capacity.

In terms of bps/Hz, SE can be represented as R_T/W. Like in the majority of wireless networks, SE and EE cannot be optimized simultaneously in NOMA networks.

15.5.6 Security and privacy in 5G systems

The major drawback of wireless communication systems is interference. The foundation for the parametric phase noise reduction strategy is the security in MIMO-NOMA systems utilizing parametric amplifier working in its instability area in a non-autonomous feedback loop connected at the output of a noisy oscillator. A system can function at particular regions where it behaves like a parametrically driven Duffing resonator, effectively becoming resistant to the phase variations that affect the oscillator output signal. The Parametric Phase Noise Filtering (PFIL) prototype is used in application at very high frequencies. The PFIL prototype has successfully reduced the phase noise at the output of a commercial signal generator. This phase noise suppression technique has been demonstrated, indicating the potential for implementing passive, low-cost phase noise cancellation circuits [33].

The majority of recent MIMO-NOMA research focuses on capacity and sum rate optimization issues. However, wireless communication scheme presentation in terms of symbol error rate (SER) is also very important. The SER performance in MIMO-NOMA networks is examined using the least Euclidean distance precoding method. Two-user MIMO-NOMA is examined for a straightforward broadcast scenario. However, two-user pairing techniques are used to make the practical example of a multiuser MIMO-NOMA system more easily realizable. Time, frequency, space, code, and power are the key resources used by wireless communication systems [13]. A NOMA cluster is created by the multiple users that can be accommodated by each resource block (RB) in NOMA systems. It is crucial to consider how to distribute power and divide the users into NOMA clusters [41–43].

CPE and ICI serve to describe phase noise distortions. When the subcarrier spacing is increased, the phase noise that causes ICI can be decreased. CPE and ICI are the two impacts that the phase noise causes. The phase shift of each subcarrier is the cause of the common phase error [11]. Inter-carrier interference is the loss of orthogonality all neighboring subcarriers. By employing scattered pilots and data detection, common phase error can be eliminated. Gaussian noise is how ICI behaves [12]. The two impacts will harm SINR and synchronization. Each and every subcarrier shares CPE. The CPE's time-varying component is referred to as its frequency-dependent component. This will produce the unwanted and damaging ICI.

15.6 Results and discussions

Any local oscillator will experience phase noise. The oscillation's power is dispersed over neighboring frequencies to create the noise sidebands. The phase noise is enhanced by the factor referred as 20 log(N) dB if the wave's frequency is amplified by a factor of N. Any wireless communication system's system performance will be harmed by phase noise because it lowers signal quality and raises bit error rate (BER). The local oscillator's random variables are known as phase noise. The phase noise has increased as a result of the oscillators' flaws, and this has led to unpredictable fluctuations in the oscillator output. Both the transmitter side and the receiver side experience the phase noise, which results from the oscillator jitter.

In the context of the Fourth Industrial Revolution, robots are taking over a number of human jobs. It aspires to produce more effective mobility based on a variety of factors, including self-driving automobiles, intelligent industries, smart cities, legal and medical advice, and the employment of smart drones for a variety of purposes, with military. The Fourth Industrial Revolution is enabled by 5G communications, which include beamforming, huge MIMO, mm-wave communications, and device-to-device communication. The end result of using NOMA is to increase spectrum efficiency while maintaining appropriate user fairness [35].

It has the potential to completely alter how future radio access solutions are developed. Numerous cutting-edge communication methods that will be utilized in 5G and beyond are compatible with NOMA. The description about NOMA concept without performing a detailed scientific analysis [40]. Instead, the presentation focuses on conventional NOMA with MIMO considering the zero-forcing receiver in the downlink transmission. The simulation includes two NOMA users with power allocation values of (1, 0.5), where the first number represents the power of the reference user and the second number represents the power of the interfering user (3 dB below). During the downlink transmission, the BS simultaneously provides signals to both users at different strengths. In NOMA, users with poorer channel conditions receive more power, indicating a power allocation strategy based on channel quality. Users with worse channel conditions receive more power under NOMA. The practical estimator and the actual channel of OFDM systems are shown in Figure 15.4.

The BS provided the two user signals concurrently and at two different strengths during the downlink transmission. Users with worse channel conditions

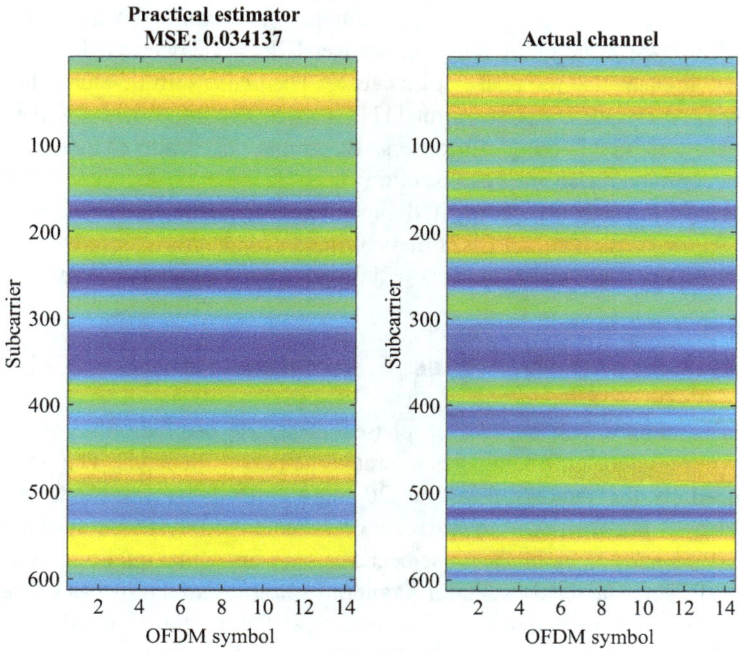

Figure 15.4 Practical estimator and actual channel of OFDM systems

receive more power under NOMA. Additionally, users in the cell's edge are more vulnerable to inter-cell interference, necessitating more power to maintain a reasonable signal-to-noise ratio. These power scales can naturally alter due to fading effects and the power management employed to counteract them [38]. By measuring the BER as a function of Eb/N0, where Eb is the energy of the transmitted bits and N0 is the one-sided power spectral density of the noise, using the parametric phase noise filtering approach, the performance was assessed using Monte Carlo simulations. The SC-FDE and perfect channel estimations were presummated with QAM modulation and a block length of N = 256 symbols (equivalent findings for various values of N, provided that N). The comparison of transmit power and BER in NOMA systems is shown in Figure 15.5.

The BS provided the two user signals concurrently and at two different strengths during the downlink transmission. Users with worse channel conditions receive more power under NOMA. Additionally, users in the cell's edge are more vulnerable to inter-cell interference, necessitating more power to maintain a reasonable signal-to-noise ratio. These power scales can naturally alter due to fading effects and the power management is employed to counteract them [38]. By measuring the BER as a function of Eb/N0, where Eb is the energy of the transmitted bits and N0 is the one-sided power spectral density of the noise, using the parametric phase noise filtering approach, the performance was assessed using Monte Carlo simulations.

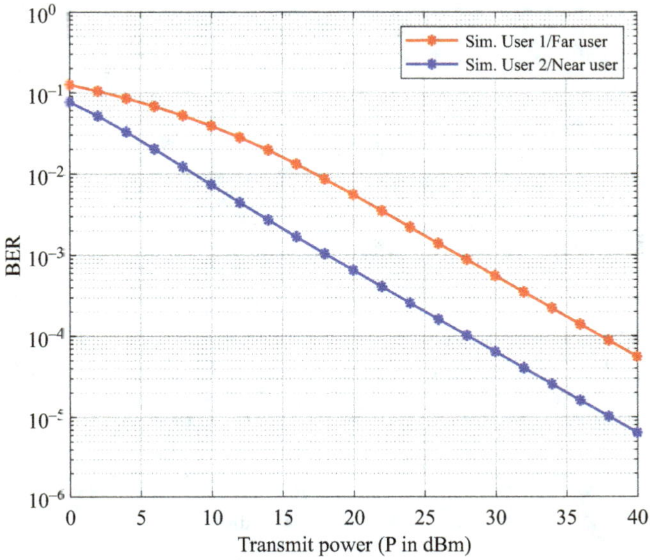

Figure 15.5 Comparison of transmit power and BER in NOMA systems

The heterogeneous networks (HetNet) are a good way to increase capacity while maintaining low costs and smart resource management in the setting of 5G needs. As opposite to a homogeneous network, which only has one BS in individually cell, HetNets have additional small BSs named micro-BS and pico-BS, in addition to the major BS. To improve spectral efficiency and sum of supplied users, these lesser BSs with fewer transmit powers and lesser coverages are connected inside the macrocell's coverage area. Maintaining user fairness is one of the major issues of HetNet's. The comparison of transmit power and outage probability in NOMA systems is shown in Figure 15.6.

Due to its enormous bandwidth resource, mmWave communication stands out as one of the most favorable technologies for 5G and elsewhere. IoT and cloud-assisted vehicle networks are two situations where it can be applied successfully. However, because mmWave systems transmission is very directed and users' channels are highly associated, their performance is significantly diminished. On the other hand, such a high correlation is a favorable circumstance for the use of NOMA. Therefore, mmWave and NOMA integration is ideally suited to offer a huge connection in dense nets. The use of NOMA in mmWave has been examined in a variety of contexts and circumstances, and NOMA has been demonstrated to be more effective than OMA. The NOMA-based mmWave networks research is still in its infancy. To further improve the system performance, multidimensional research are needed. The comparison of transmit power and achievable capacity in NOMA systems is shown in Figure 15.7.

By adopting layered transmission, each mobile user decodes signals throughout SIC sequence by sequence, which results in a significantly lower level of decoding difficult than in the case of non-layered transmission. It uses beamforming and user

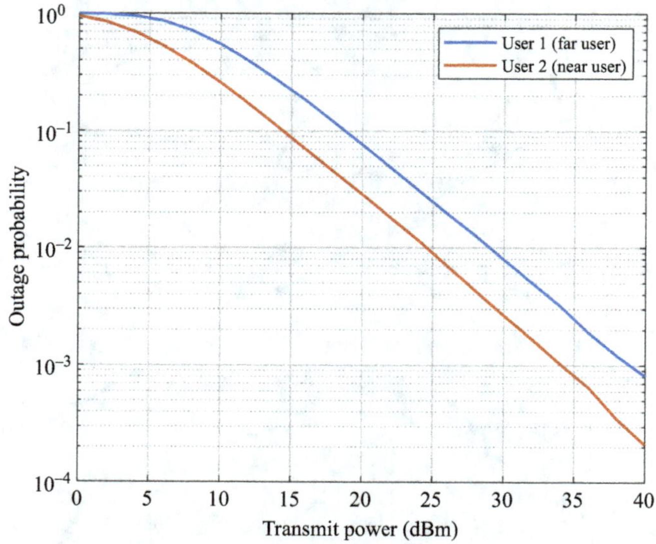

Figure 15.6 Comparison of transmit power and outage probability in NOMA systems

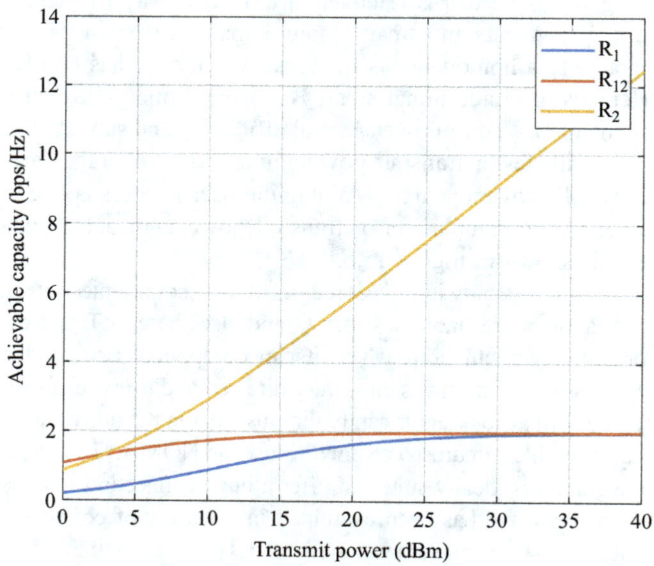

Figure 15.7 Comparison of transmit power and achievable capacity in NOMA systems

selection. The outage shows for both uplink and downlink systems in a MIMO-NOMA architecture with single-cell and multicell. Furthermore, by utilizing two power allocation methodologies, a suitable trade-off between fairness and throughput has been made.

15.7 Conclusions and future scope

5G is a key user of the Fourth Industrial Revolution since it supports point-to-point connectivity, greater communication speeds, and reduced latency than 4G. The NOMA is a capable multiple access for upcoming 5G releases, and it surveys particularly well for mMTC. NOMA has acknowledged as a robust choice among all MA procedures since it possesses key characteristics to get beyond OMA's shortcomings and fulfil the requirements. NOMA is superior to OMA: (i) spectrum efficiency and throughput OMA, like OFDMA, assign a precise frequency resource to individual user regardless of that user who is experiencing excellent or corrupt channel situations. However, the entire system has low spectral efficiency and throughput. Contrarily, with NOMA, several mobile users simultaneously share a single frequency resource with together good and terrible channel conditions. The resource that was granted to the weak user is used by strong user, and SIC at receivers can lessen interference. Massive connectivity cannot be supported by this strategy. NOMA was briefly demonstrated. This survey clearly shows interference reduced in millimeter wave NOMA systems compared with OFDM systems using parametric phase noise filtering method. In future, we are going to implement 6G mmWave systems in massive MIMO-NOMA and OMA systems.

References

[1] S. Islam, M. Zeng, O. A. Dobre, and K. S. Kwak, "Resource allocation for downlink NOMA systems key techniques and open issues," *IEEE Wireless Communication*, vol. 25, no. 2, pp. 40–47, 2018.

[2] L. Zhu, J. Zhang, Z. Xiao, X. Cao, D. O. Wu, and X. Xia, "Millimeter-wave NOMA with user grouping, power allocation and hybrid beamforming," *IEEE Transactions on Wireless Communications*, vol. 18, no. 11, pp. 5065–5079, 2019

[3] L. Zhu, Z. Xiao, X.-G. Xia, and D. O. Wu, "Millimeter-wave communications with non-orthogonal multiple access for B5G/6G," *IEEE Access*, vol. 7, pp. 116123–32, 2019.

[4] Z. Xiao, L. Zhu, J. Choi, P. Xia, and X. Xia, "Joint power allocation and beamforming for non-orthogonal multiple access (NOMA) in 5G millimeter wave communications," *IEEE Transactions on Wireless Communications*, vol. 17, no. 5, pp. 2961–2974, 2018.

[5] H. Marshoud, V. M. Kapinas, G. K. Karagiannidis, and S. Muhaidat, "Non-orthogonal multiple access for visible light communications," *IEEE Photonics Technology Letters*, vol. 28, no. 1, pp. 51–54, 2016.

[6] A. Boulogeorgos, N. D. Chatzidiamantis, and G. K. Karagiannidis, "Non-orthogonal multiple access in the presence of phase noise," *IEEE Communications Letters*, vol. 24, no. 5, pp. 1133–1137, 2020.

[7] W. Xu, X. Li, C.-H. Lee, M. Pan, and Z. Feng, "Joint sensing duration adaptation, user matching, and power allocation for cognitive OFDM-NOMA systems," *IEEE Transactions on Wireless Communications*, vol. 17, no. 2, pp. 1269–1282, 2018.

[8] F. Kara and H. Kaya, "BER performances of downlink and uplink NOMA in the presence of SIC errors over fading channels," *IET Communications*, vol. 12, no. 15, pp. 1834–1844, 2018.

[9] K. Higuchi and A. Benjebbour, "Non-orthogonal multiple access (NOMA) with successive interference cancellation for future radio access," *IEICE Transactions on Communications*, vol. E98.B, pp. 403–414, 2015.

[10] Q. He, Y. Hu, and A. Schmeink, "Closed-form symbol error rate expressions for non-orthogonal multiple access systems," *IEEE Transactions on Vehicular Technology*, vol. 68, no. 7, pp. 6775–6789, 2019.

[11] L. Dai, B. Wang, Y. Yuan, S. Han, I. C-L, and Z. Wang, "Non-orthogonal multiple access for 5G solutions, challenges, opportunities, and future research trends," *IEEE Communications Magazine*, vol. 53, no. 9, pp. 74–81, 2015.

[12] Y. Wu, E. Attang, and G. E. Atkin, "A novel NOMA design based on Steiner system," in 2018 *IEEE International Conference on Electro/Information Technology (EIT)*, Rochester, MI, 2018, pp. 846–850.

[13] Q. Wang, Y. Liu, X. Yan, and H.-C. Wu, "An innovative pulse-shaping scheme using multiwavelets for non-orthogonal multiple-access," *IEEE Communications Letters*, vol. 23, no. 12, pp. 2376–2380, 2019.

[14] R. Kizilirmak, "Non-orthogonal multiple access (NOMA) for 5G networks," in *Towards 5G Wireless Networks – A Physical Layer Perspective*, 2016, pp. 83–98.

[15] Z. Ding, F. Adachi, and H. V. Poor, "The application of MIMO to non-orthogonal multiple access," *IEEE Transactions on Wireless Communications*, vol. 15, no. 1, pp. 537–552, 2016.

[16] Q. Sun, S. Han, I. Chin-Lin, and Z. Pan, "On the ergodic capacity of MIMO NOMA systems," *IEEE Wireless Communications Letters*, vol. 4, no. 4, pp. 405–408, 2015.

[17] M. Zeng, A. Yadav, O. A. Dobre, G. I. Tsiropoulos, and H. V. Poor, "On the sum rate of MIMO-NOMA and MIMO-OMA systems," *IEEE Wireless Communications Letters*, vol. 6, no. 4, pp. 534–537, 2017.

[18] E. Udayakumar and V. Krishnaveni, "A review on interference management in millimeter-wave MIMO systems for future 5G networks," in *Proceedings of 4th International Conference on Innovations in Electrical and Electronics Engineering (ICIEEE 2019),* Guru Nanak Institute of Technology, Hyderabad, pp. 715–721, 2019.

[19] M. S. Ali, E. Hossain, and D. I. Kim, "Non-Orthogonal Multiple Access (NOMA) for downlink multiuser MIMO systems user clustering, beam-forming and power allocation," *IEEE Access*, vol. 5, pp. 565–577, 2017.

[20] E. Udayakumar and V. Krishnaveni, "Performance evaluation of phase noise reduction for MIMO FBMC-OQAM systems," *Solid State Technology*, vol. 63, no. 5, pp. 5668–5675, 2020.

[21] J. Choi, "On the power allocation for MIMO-NOMA systems with layered transmissions," *IEEE Transactions on Wireless Communications*, vol. 15, no. 5, pp. 3226–3237, 2016.

[22] M. F. Hanif, Z. Ding, T. Ratnarajah, and G. K. Karagiannidis, "A minorization-maximization method for optimizing sum rate in the downlink of non-orthogonal multiple access systems," *IEEE Transactions on Signal Processing*, vol. 64, no. 1, pp. 76–88, 2016.

[23] E. Udayakumar and V. Krishnaveni, "Analysis of phase noise issues in millimeter wave systems for 5G communications," in *Wireless Personal Communications*, Springer, vol. 126, pp. 1601–1619, 2022.

[24] Y. Liu, G. Pan, H. Zhang, and M. Song, "On the capacity comparison between MIMO-NOMA and MIMO-OMA," *IEEE Access*, vol. 4, pp. 2123–2129, 2016.

[25] M. Zeng, A. Yadav, O. A. Dobre, G. I. Tsiropoulos, and H. V. Poor, "Capacity comparison between MIMO-NOMA and MIMO-OMA with multiple users in a cluster," *IEEE Journal on Selected Areas in Communications*, vol. 35, no. 10, pp. 2413–2424, 2017.

[26] E. Udayakumar and V. Krishnaveni, "Phase noise effect on millimeter wave MIMO-OFDM and FBMC-OQAM systems," in *AIP Conference Proceedings of International Conference on Applied Data Science and Smart Systems*, Chitkara University, Punjab, 4th and 5th November, 2022.

[27] B. Kimy, S. Lim, H. Kim, *et al.*, "Non-orthogonal multiple access in a downlink multiuser beamforming system," in *Proceedings of the IEEE Military Communications Conference (MILCOM)*, San Diego, CA, pp. 1278–1283, 2013.

[28] Z. Ding, P. Fan and H. V. Poor, "Random beamforming in millimeter-wave NOMA networks," *IEEE Access*, vol. 5, pp. 7667–7681, 2017.

[29] U. Ozmat, O. Ulgen, and E. Gunaydin, "Bit error rate analysis of non-orthogonal multiple access (NOMA) technique in 5G with different power and user scenarios," in *2018 Advances in Wireless and Optical Communications (RTUWO)*, Riga, Latvia, 2018, pp. 45–49.

[30] M. Zeng, W. Hao, O. A. Dobre, and H. V. Poor, "Energy-efficient power allocation in uplink mmWave massive MIMO with NOMA," *IEEE Transactions on Vehicular Technology*, vol. 68, no. 3, pp. 3000–3004, 2019.

[31] M. Abdelmoniem, S. M. Gasser, M. S. El-Mahallawy, *et al.*, "Enhanced NOMA system using adaptive coding and modulation based on LSTM neural network channel estimation," *Applied Science*, vol. 9, no. 15, p. 3302, 2019.

[32] Z. Ding, P. Fan, and H. V. Poor, "Impact of user pairing on 5G non-orthogonal multiple access downlink transmissions," *IEEE Transactions on Vehicular Technology*, vol. 65, no. 8, pp. 6010–6023, 2016.

[33] C. Cassella, "Phase noise suppression through parametric filtering," *Applied Physics Letters*, vol. 110, no. 6, 063503, 2017.

[34] S. M. Pishvaei, F. T. Miandoab, and B. M. Tazehkand, "Outage analysis of mmWave-NOMA transmission in the presence of LOS and NLOS paths," *Future Generation Computer Systems*, vol. 128, pp. 88–101, 2022.

[35] R. Adeogun, G. Berardinelli, and P. E. Mogensen, "Enhanced interference management for 6G in-X subnetworks," *IEEE Access*, vol. 10, pp. 45784–45798, 2022.

[36] J. Arellano, C. D. Altamirano, and H. R. C. Mora, "On the interference reduction factor in massive MIMO system over Rician fading channels," in 2022 *IEEE Sixth Ecuador Technical Chapters Meeting (ETCM)*, 2022, pp. 1–5.

[37] I. Budhiraja, N. Kumar, and S. Tyagi, "ISHU: interference reduction scheme for D2D mobile groups using uplink NOMA," *IEEE Transactions on Mobile Computing*, vol. 21, no. 9, pp. 3208–3224, 2022.

[38] E. Udayakumar and V. Krishnaveni, "Analysis of various interference in millimeter- wave communication systems: a survey," in *Proceedings of 10th IEEE International Conference on Computing, Communication and Networking Technologies (ICCCNT 2019)*, Indian Institute of Technology Kanpur, Uttar Pradesh, 2019, pp. 1–5.

[39] N. Xie, Y. Xu, J. Zhang, and J. Chen, "Joint estimation of channel responses and phase noises in asynchronous MIMO systems with intentional timing offset," *IEEE Transactions on Communications*, vol. 71, no. 1, pp. 412–426, 2023.

[40] D. Yang, J. Xu, W. Xu, Y. Huang, and Z. Lu, "Secure communication for spatially correlated RIS-aided multiuser massive MIMO systems: analysis and optimization," *IEEE Communications Letters*, vol. 27, pp. 797–801, 2023.

[41] H. I. Obakhena, A. L. Imoize, F. I. Anyasi, *et al*., "Application of cell-free massive MIMO in 5G and beyond 5G wireless networks: a survey," *Journal of Engineering and Applied Science* vol. 68, p. 13, 2021.

[42] M.-S. Van Nguyen, D.-T. Do, V.-D. Phan, W. U. Khan, A. L. Imoize, and M. M. Fouda, "Ergodic performance analysis of double intelligent reflecting surfaces-aided NOMA–UAV systems with hardware impairment," *Drones* vol. 6, p. 408, 2022.

[43] A. Olumide, A. Gbenga-Ilori, M. Adelabu, A. L. Imoize, and O. Ladipo, "Energy efficiency techniques in ultra-dense wireless heterogeneous networks: an overview and outlook," *Engineering Science and Technology, An International Journal*, vol. 23, no. 6, pp. 1308–1326, 2020.

[44] V. B. Kumaravelu, A. L. Imoize, F. R. Castillo Soria, and P. G. S. Velmurugan, "Outage probability analysis and transmit power optimization for blind-reconfigurable intelligent surface-assisted non-orthogonal multiple access uplink," *Sustainability* 14, no. 20, p. 13188, 2022.

[45] A. L. Imoize, H. I. Obakhena, F. I. Anyasi, and S. N. Sur, "A review of energy efficiency and power control schemes in ultra-dense cell-free massive MIMO systems for sustainable 6G wireless communication," *Sustainability*, vol. 14, no. 17, p. 11100, 2022.

[46] A. L. Imoize, O. Adedeji, N. Tandiya, and S. Shetty, "6G Enabled smart infrastructure for sustainable society: opportunities, challenges, and research roadmap," *Sensors*, vol. 21, no. 5, p. 1709, 2021.

Chapter 16

A generative adversarial network-based approach for mitigating inference attacks in emerging wireless networks

Olakunle Ibitoye[1], Ashraf Matrawy[1], Omair Shafiq[1] and Agbotiname Lucky Imoize[2,3]

Abstract

The proliferation of smart, connected, always-listening devices has introduced significant privacy risks to users in wireless networks comprising dense massive devices. Beyond the notable risk of eavesdropping, intruders can adopt machine learning techniques to infer sensitive information from audio recordings on these devices, resulting in a new dimension of privacy concerns and attack variables for wireless network users. Techniques such as sound masking and microphone jamming have effectively prevented eavesdroppers from listening to private conversations. In this study, we explore the problem of adversaries spying on wireless network users to infer sensitive information with machine learning techniques. We then analyze the role of randomness in the effectiveness of sound masking for mitigating sensitive information leakage. We propose a generative adversarial network (GAN)-based approach for privacy preservation in the network, which generates random noise to distort the unwanted machine learning-based inference. Our experimental results demonstrate that GANs can be used to generate more effective sound masking noise signals which exhibit more randomness and effectively mitigate deep learning-based inference attacks while preserving the semantics of the audio samples in wireless networks. The GANs would find useful applications in addressing the proliferating privacy and security concerns in 5G and the envisioned 6G wireless networks.

Keywords: 5G wireless networks security; Privacy-preservation; Dense massive devices; Generative adversarial networks; Information leakage

[1]School of Information Technology, Carleton University, Canada
[2]Department of Electrical Engineering and Information Technology, Institute of Digital Communication, Ruhr University, Germany
[3]Department of Electrical and Electronics Engineering, Faculty of Engineering, University of Lagos, Nigeria

16.1 Introduction

The presence of "always listening" devices in a user's environment poses significant privacy concerns, as an adversary may leverage these devices to eavesdrop on a user's private conversations. With the proliferation of Internet of Things (IoT) smart home devices, the number of microphone-bearing devices in residential homes has increased exponentially over the past decade resulting in an increased attack surface for adversaries. A typical case study in [1] showed that researchers, with minimal coding effort, converted an Amazon Echo smart speaker into a spy device for eavesdropping on homeowners. Similar studies [2,3] have suggested a wide range of adversaries, including IoT platform owners, app developers, device manufacturers, and solution providers.

The ability of generative adversarial networks (GANs) to create high-dimensional data has been researched [4]. This study seeks to correlate the relationship between high-dimensional data in GAN-generated audio samples and increased randomness. The use of sound masking noise signals has been an important technique in protecting user privacy against eavesdroppers who attempt to gain unauthorized access to users' private conversations. One recommended approach for protecting user privacy in smart homes from always listening devices is using sound masking by adding white noise to audio signals [5]. White noise includes all frequencies at equal energy. However, generating audio signals that sound more comfortable to listeners is more desirable. As such, only the specific frequency spectrum required to increase privacy is produced with minimal distraction.

Providers and vendors of smart wireless devices have argued that connected, always-listening devices such as Amazon Echo speakers implement a temporary buffer [6] that prevents the device from continuously recording user conversations. In addition, users can review and delete their voice recordings through the app or voice commands. While these techniques are a reasonable proposition, their effectiveness is not yet proven in preserving user privacy, especially since the attack surface increases with Internet connectivity and accessibility to various third-party apps. Moreover, this requires complete trust in several providers (hardware, software, etc.) and the insiders within the organization, which may not always be guaranteed.

This study explores risks beyond eavesdropping and considers information leakage in smart home environments. An information leakage attack provides a larger attack surface because the adversary can deduce or infer sensitive information with the aid of computation and machine learning techniques. For example, an adversary can infer that there is an infant child in the home, can infer the race and gender of the occupants, or activities being performed in a home by merely running inference attacks on the smart home devices [7,8].

Sound-based inference attacks may provide greater incentives to adversaries. For example, they can get thousands of users to download an app and infer certain sensitive information, such as behavioral patterns for a large number of people, which can then be used for commercial purposes such as advertising and sales targeting [9]. An adversary may also use such information for more malicious purposes, which could jeopardize the safety of the smart home occupants, such as inferring home occupancy and planning a robbery attack [10].

Our contributions from this study are twofold. In our first contribution, we demonstrate that GAN-generated noise results in better performance in mitigating machine learning-based information leakage inference in smart home environments due to the increased randomness in the GAN noise. We show the relationship between randomness in audio signals and the effectiveness of a sound-masking noise signal in preventing sensitive information leakage.

For our second contribution, we introduce a novel GAN structure for producing sound-masking noise signals that are proven to be truly random. We adopt existing frameworks for measuring the randomness element in discrete signals and demonstrate that the GAN-based audio noise signals have more entropy-based randomness compared to digitally generated white noise signals.

The novelty of our research is demonstrated in the following ways. To the best of our knowledge, this is the first study to investigate the use of GAN-based noise for mitigating sound-based privacy leakage inference attacks targeted against smart home environments.

Also, this is the first study, to the best of our knowledge, to investigate the effect of randomness on the ability of a sound masking noise signal to mitigate sensitive information leakage. Our findings show that information leakage mitigation is strongly correlated with the randomness element in the sound masking audio signal.

We further demonstrate that GANs can generate noise signals which can effectively mitigate sound-based privacy inference attacks while maintaining the semantics of the audio signal, as shown in Section 16.6.3.

The presented GAN-based approach for privacy preservation in the wireless network would find a useful application in emerging wireless networks such as the beyond 5G and 6G networks. Specifically, the GAN-based technique would provide robust security against sophisticated attacks on open wireless communication channels. This will help to protect sensitive user data on the wireless edge and guarantee the confidentiality of critical user information over the channels. Additionally, the security framework will enhance trust and safety among all parties in the wireless ecosystem.

16.2 Related work

Existing research for privacy preservation/information leakage prevention with noise distortion has focused on signal jamming – to distort the signal and prevent an eavesdropper from listening. No existing solution has utilized generative adversarial networks to create noise distortion. Similar work [11] has also used ultrasonic transmission to jam nearby microphones.

Lei *et al.* [12] proposed a physical presence-based access control mechanism that ensures that physical presence is detected before activating the "wake word" in voice assistants as a security measure or before accepting voice commands from a voice-activated digital assistant. While this technique is effective, carefully crafted malware can effectively fool this safeguard [13]. In addition, the presence of a home occupant is not a deterrent for an intruder who is deploying and executing malware remotely since the malicious app will most likely operate saliently and quietly. The authors in [14] proposed a Doppler radar-based liveliness detector to prevent spoofing attacks on voice assistants and ensure a human is present before accepting voice commands. The work of [15] proposes a framework that implements a solution that jams the device microphone until the user issues a voice command.

Authors in [16,17] discussed the limitations of GANs in that it learns the data distribution from the dataset and tends to remember the training samples, which can be used to infer sensitive information from the dataset. The authors thus propose approaches to incorporate privacy preservation techniques into the structure of the GAN.

Researchers have investigated the use of noise audio signals for privacy preservation in smart environments, such as in [18], where a noise generator was proposed for preserving privacy in smart tactical platforms. In [19], researchers exploited audio masking to prevent sensitive information leakage in smartphones.

The feasibility of inferring sensitive information in smart homes has been studied in numerous contexts. In [20], user activities such as walking and sleeping could be inferred by observing the network traffic in a smart home. Even when such traffic is encrypted as in [21], an adversary can still perform information leakage attacks on the smart home, compromising user privacy, and confidentiality.

Several attempts have been made to explore the usage of GANs in network security. The authors of [22] proposed the use of GANs for defending against adversarial attacks in network security.

From our literature review, we observed that no published work had explored the use of GANs to generate audio noise signals to mitigate audio inference attacks in smart home environments. Our research, therefore, seeks to close this gap.

16.3 Problem statement and proposed solution

Several users have installed various IoT devices to make their homes smarter. These devices are always connected, measuring and collecting data about the environment. An adversary can use the information from those sensors to infer sensitive information about the occupants of the home. This raises significant privacy concerns. For example, researchers have been able to infer the TV content of home users by listening to the sound from the TV [23].

While it is easy for someone familiar with the movie to tell just by listening to the audio sound if the person is in close proximity to the home, the proliferation of smart, connected devices that are always listening creates a larger attack surface. This means that IoT devices or smartphones could be accessed remotely without the owner's authorization or consent to deduce and infer such sensitive content. We term this for the scope of this study as an inference attack.

As machine learning algorithms become more sophisticated, adversaries will utilize machine learning and deep learning techniques to compromise the privacy of smart home users. In one of such attack variations, an adversary could seek to intercept digital voice assistants, which are very common in many smart home environments and are incorporated into various devices, such as smart speakers, smart refrigerators, and smartphones. In order to prevent sensitive information leakage from digital voice assistants, which are heavily integrated into smart home devices, we need to understand how an adversary can achieve such information leakage and the risk associated with it as well as the consequences of such leakage.

16.3.1 What is an inference attack?

In the context of this study, an inference attack occurs when an external party infers sensitive information from data that they have access to [20]. Deep neural networks (DNNs) are widely used for various audio processing tasks, which fall under two broad categories: audio analysis and audio synthesis/transformation [24]. Our study represents a borderline between these two categories where we draw a distinction between the two categories and differentiate between audio recognition and audio inference. In this study, we focus on audio inference aspects whereby our target is not to recognize what was said but what could be inferred from what was said. Figure 16.1 illustrates the difference between eavesdropping and inference.

Consider a similar case scenario in which a user downloads a malicious app that exploits the "always listening" capability of a smart device and then runs a script to infer the user's movie preferences. This is also an example of an information leakage attack.

16.3.2 MaskGAN: our proposed solution

In our proposed solution, we utilize GANs [25] to generate sound masking audio noise to mitigate the information leakage as a result of the machine learning-based inference. More details about the MaskGAN structure is provided in Section 16.4.5. The advantage of our proposed solution is twofold. First, GANs, due to the lack of a deterministic bias [25] can generate synthetic data samples that are truly random. Our objective in this study is to investigate if the noise generated by MaskGAN can mitigate information leakage while preserving the semantics of the audio, as shown

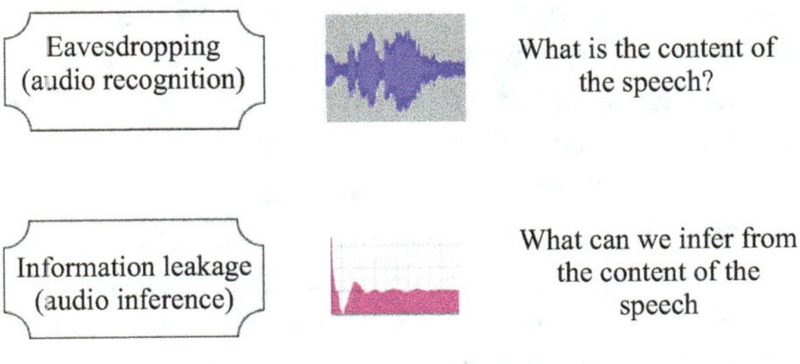

Figure 16.1 Eavesdropping vs. inference

in the results (Section 16.6). The second advantage is that our solution is independent of the smart home device manufacturer, vendor, or solution provider and is completely within the control of the user. We ensure that the noise generated by the MaskGAN does not exceed a sound intensity of 45 dB, which is within the comfort zone for human hearing [26].

In this study, we seek to understand the role that randomness plays in sound masking privacy preservation. We conduct experiments to determine if our GAN-based approach for generating sound masking noise signals can produce audio noise signals that exhibit more randomness compared to white noise.

16.3.3 Research questions

Our study seeks to answer the following research questions:

1. Are GAN-generated noise samples effective for information leakage prevention in smart home environments to deter various adversaries from inferring sensitive information from user conversations?
2. Can GAN-generated noise samples be used to deter adversaries from inferring sensitive information from smart home devices while maintaining the semantics of the audio samples?
3. Are GAN-generated noise samples more random compared to white noise? What role does randomness play in improving the ability of privacy-preserving sound masking techniques to prevent the risk of inferring sensitive information from "always listening" smart home devices?

Our findings to these three research questions are reported in the results (Section 16.6).

16.4 Threat model

The threat model in our study assumes an information leakage scenario in which an adversary accesses audio files from an "always listening" connected device in a smart home and infers sensitive information such as user demographics or activities of the home occupants. Figure 16.2 illustrates our threat model.

We assume the adversary is anyone other than the legitimate smart home device's data owner. The adversary could be a device manufacturer, an insider, an authorized third-party app developer, an unauthorized intruder such as one who deploys a malicious app, or a possible state actor, as discussed in [27]. The adversaries have different capabilities, but it is assumed that all adversaries can access the smart home device either physically or remotely. The threat model illustrated in Figure 16.2 illustrates an unauthorized third-party adversary who deploys a malicious app onto the smart device through physical access or a phishing attack. The malicious app compromises any existing protection, e.g., the temporary buffer which prevents the smart device from continuously recording conversations [6]. Different adversaries follow the same pattern of attack against the end-user with the same end goal – in which sensitive information the smart

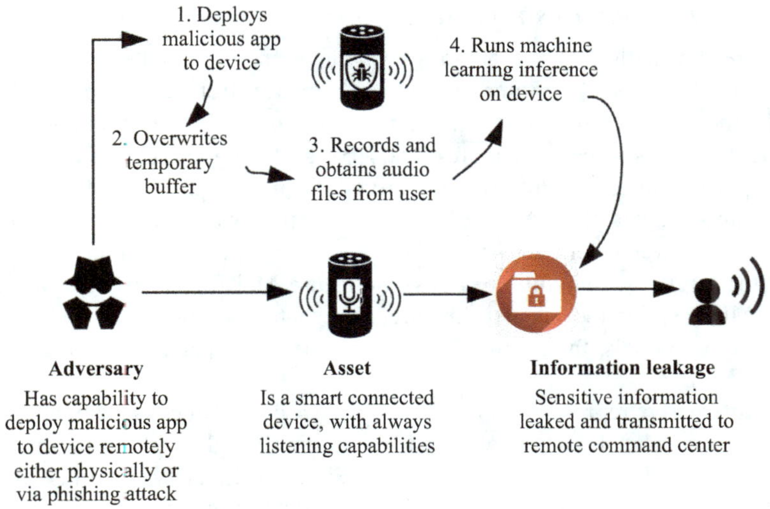

Figure 16.2 Threat model

device user has not given consent to is inferred and ultimately used for the adversary's gain. We note that there are various capabilities for the different adversaries and various types of attacks that each of the adversaries can launch based on their capabilities, such as eavesdropping, inference attacks, or other malicious purposes. For the scope of this study, however, we focus only on inference attacks, in which the adversary applies machine learning techniques to infer sensitive information from the audio recordings from those devices.

It is also assumed that the adversary has full knowledge of the model in what could be referred to as a white box attack. The adversary gains access to the recordings and carry out an inference attack on the recorded conversation. In our experimental approach, three different DNN models, namely the convolutional neural network (CNN), recurrent neural network (RNN), and the convolutional RNN (CRNN), were used to infer sensitive information from the audio recordings.

16.4.1 Solution overview

GANs [25] belong to the set of unsupervised deep learning algorithms known as generative models, which learn the underlying hidden structure of given data without specifying a target value. Generative models typically generate synthetic inputs x', given input data x, by learning the intrinsic distribution function $p(x)$ of the input data, in contrast to discriminative models, which tend to model the conditional probability distribution function $p(y|x)$, for a given function $y(x)$, generative models are direct density implicit models which model $p(x)$ without attributing the probability distribution function.

16.4.2 Audio features representation

Feature representation of the audio signal plays an important role in the deep learning model's ability to infer sensitive information from an audio sample. We consider the task of feature representation for this study different from that of audio classification tasks since the features that serve best for audio classification might not adequately suffice for inferring sensitive information [28]. As a basic foundation, the upper layers of a DNN are best suited for performing feature extraction. In contrast, the lower layers are established to perform class discrimination [29] to output the target class. While it is possible to use Mel frequency cepstral coefficients (MFCCs) for the acoustic feature representation, since our study utilizes deep learning models, this approach is ignored because spatial information is lost from the MFCC.

An alternative representation known as the spectrogram consists of a temporal sequence of spectra. It can be obtained by omitting the discrete cosine transform (DCT) to yield the log-mel spectrum [24]. Figure 16.3 shows an illustration of spectrogram images for an audio sample in our dataset and a white noise sample generated for our experiment.

Even though the spectrograms are similar to images, the approach for audio processing using DNNs is considered to be different from image classification due to the variation in value distribution for audio samples as compared to image samples.

We desist from using the time-domain waveform samples of the audio representation since they do not capture sufficient spatial information, which is crucial for our machine learning model and technique.

16.4.3 Neural network models

In the past, it was common practice to model and analyzed audio signals using Gaussian mixture due to their mathematical elegance [24]. However, in recent times, DNNs have been shown to be more accurate for audio processing, and classification tasks [30]. In this study, we examine the performance of three types of neural network models, namely – CNN, RNN, and CRNN, for our task of inferring sensitive information from audio samples.

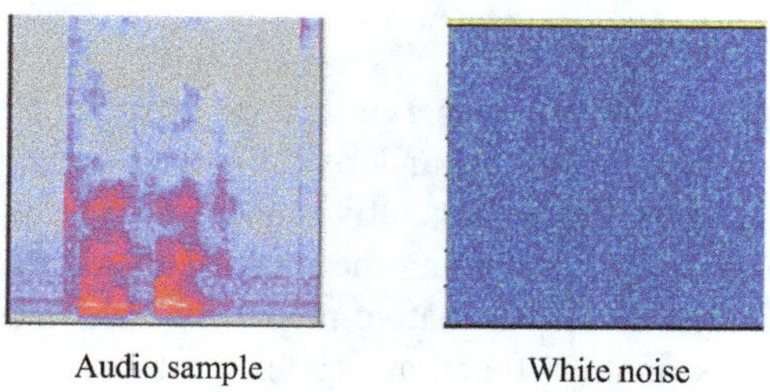

Figure 16.3 Mel spectrogram of audio sample vs. white noise

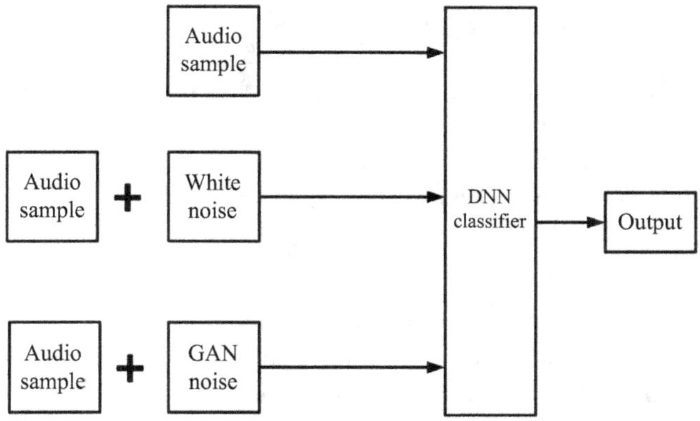

Figure 16.4 Solution architecture

A CNN consists of a series of convolutional layers passed through pooling layers, followed by one or more dense layers. Since our study is based on spectral input features from the Mel spectrogram, a two-dimensional time-frequency convolution is used for computing the feature maps, which are further downsampled by the pooling layers. The optimal parameters for the CNN are obtained experimentally based on the validation error observed during the training process. Recurrent neural networks (RNNs) are well suited for sequence modeling tasks such as audio processing [31] due to the fact that they intrinsically model the temporal dependency in the input features. A CRNN [32] is an extension of the CNN in which an RNN is implemented to process the output of a CNN. While the purpose of the convolutional layers is to perform feature extraction, the recurrent layers enable the model to make sense of the longer temporal context.

The audio samples are all processed into 16-bit, 48 kHz wave format before being converted into spectrogram images. After the audio samples are pre-processed into spectrogram images, spectral feature extraction is carried out, and the input is then fed into the DNN classifiers. Figure 16.4 shows a diagrammatic representation of the solution architecture.

16.4.4 Noise generation methodology

Our solution is based on the premise that GAN-generated noise, when combined with audio recordings from the smart home device, reduces the effectiveness of machine learning-based inference from the audio recordings. This enhances smart home user privacy from various forms of adversaries with varying capabilities discussed in the preceding paragraph. Our results from Section 16.6 highlight more details on this.

The GAN noise signal is generated by an external device that is permanently in the smart home user's environment and constantly producing noise signals which when combined with audio recordings from the smart home device prevents an adversary from inferring sensitive information from the audio recordings. The noise amplitude of the external noise generator is audible for human perception, but it should not exceed the acceptable noise threshold for human comfort.

In this study, we evaluate the effectiveness of the GAN noise with white noise. The white noise is generated with a python script using the same hardware for generating the GAN noise. In our evaluation, both noise samples are produced at the same amplitude to ensure consistency in the results.

16.4.5 MaskGAN overview

In the original GAN setup introduced by [25], GANs were used to generate synthetic data samples by taking as input, statistically independent noise samples. To the best of our knowledge, GANs have not been used to generate random noise signals. We choose to implement GANs in our approach to create audio samples as against other generative models such as variable autoencoders (VAE) because GANs do not introduce any deterministic bias and work better with discrete latent variables [25].

Our solution which we refer to as MaskGAN is an adaptation of deep convolutional GANs (DCGANs) [33]. DCGANs are a notable architecture for adversarial image generation in which a transposed convolution operation is implemented for creating high-resolution images from low-resolution feature maps. Since DCGAN outputs 64 × 64 pixel images, we add two additional layers to produce 2 s of audio at 16 kHz. Furthermore, the two-dimensional convolutions are flattened into one-dimensional with the stride factors increased twofold.

Our proposed MaskGAN structure consists of two models as shown in Figure 16.5. The first model known as the generator tries to generate new and synthetic audio samples that are identical to the target white noise audio sample. The second model known as the discriminator performs an adversary role by trying to detect if the synthetic audio sample is real or fake, hence helping to improve the knowledge of the generator until the generator eventually succeeds in creating

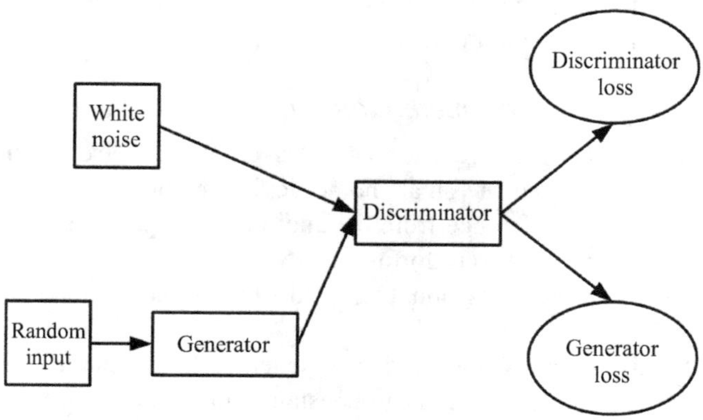

Figure 16.5 MaskGAN structure

some synthetic audio sample that is realistic and as indistinguishable from the actual white noise audio sample as possible.

The generator model is represented as $G(z, \theta_g)$ while the discriminator model is represented as $D(x, \theta_d)$ where x represents the input audio samples and z represents the generated synthetic samples. The weights of the neural network also known as parameters are represented as θ. The parameters of the generator θ_g are updated to maximize the probability that the synthetic audio is classified as the real audio dataset. The loss function of the generator network seeks to maximize $D(G(z))$. With regard to the discriminator, the parameters are optimized to maximize the probability that the synthetic noise audio samples are classified as real audio samples. Hence, the loss function of the discriminator seeks to maximize the function $D(x)$ while minimizing the function $D(G(z))$.

The minmax game between the generator and the discriminator is represented as a value function $V(G, D)$, whereby the generator seeks to maximize the probability that its output is classified as real. In contrast, the objective of the discriminator is to minimize this probability.

The input to the MaskGAN model is a random seed and the target output is a white noise signal that has been generated from our python code. After several iterations, the generative model finally arrives at a synthetic noise signal that is indistinguishable from the digital white noise signal, based on the assessment of the discriminative model.

16.4.6 Dataset, developmental tools, hardware, and software

For our experimental work, we used a standalone desktop PC running windows 10 Education OS. The hardware components consist of an AMD Ryzen 7 2700X processor at 3.70 GHz with 32 GB of RAM and 1TB SSD storage. The graphics card is a standalone GPU – NVIDIA GeForce RTX 2080 TI with 11 GB RAM.

All software development was carried out using publicly available and open-source tools. The software code was written in Python programming language using the Spyder Integrated Development Environment (IDE), which is part of the "Anaconda software distribution." For the deep learning framework, we used the Google TensorFlow v2 deep learning framework.

The three datasets we used represent the three inference attack case scenarios that were explored in this study, namely music genre inference (MGI), user demographics inference (UDI), and speech emotion inference (SEI). The datasets used are publicly available, and details of each dataset are further discussed in Section 16.6.1.

16.5 Experimental approach

Assuming the smart home has devices that are equipped with always-listening capabilities. As discussed in Section 16.4 above, these devices could be harnessed by an adversary to leak sensitive information from the occupants of the home. Our experiments seek to deter such leakage inference attacks using truly random noise generated by a GAN neural network model which we term as MaskGAN. The first subsection describes our approach for generating audio noise with increased randomness using our GAN solution. Section 16.5.2 describes our approach for measuring the randomness of the GAN-generated noise and performing a comparison to the white noise using two different runs tests methods. In Section 16.5.3, we describe how we perform the inference attacks for three different scenarios using three different datasets, and, for each dataset, three different neural network models are utilized. In this step, the original audio dataset is used without adding any form of noise mitigation. In Section 16.5.4, we discuss our approach to mitigate the leakage of sensitive information via inference attacks with the use of the noise generated by the GAN and white noise. In Section 16.5.5, we discuss the different metrics we utilize for evaluating our methodology and results.

Since this paper focuses on information leakage from smart homes rather than eavesdropping, our case study scenarios and dataset selection best reflect this context. For example, rather than selecting datasets for automatic speech recognition such as [34], we instead select datasets in which information inference is sought from the audio samples. In our "semantic preservation factor" evaluation metrics in Section 16.5.5.2, we discuss our approach to experientially highlight the difference between both contexts and report our results in Section 16.6.3. For our case study, we consider the possibility of an adversary seeking to infer what genre of music the occupants of a home prefers to listen to and therefore provide targeted ads to the user. The second case scenario demonstrates an adversary who infers the user demographics such as race and gender of the home occupants, while the third case scenario discusses an adversary who seeks to infer the emotion of the home occupants. The adversary achieves this sensitive information leakage or inference attacks using machine learning or deep learning techniques applied to the audio recordings. Other possible adversary scenarios may include the possibility to allow speech recognition while blocking out contextual information leakage.

16.5.1 Generate noise signals with GAN

The first step in our experimental approach entails using the GAN structure described in Section 16.4.5 to create noise samples using white noise as the target output. As illustrated in Figure 16.5, the generative model produces audio samples from a random seed and learns to improve as the discriminator determines how close the audio sample is to the white noise signal. The amplitude of the generated GAN noise does not exceed 45 db in order to remain within the human comfort level as specified in [26].

16.5.2 Measuring the degree of randomness in noise signals

In the second step of our experimental approach, we compute the degree of randomness of the original sample, the white noise, and the GAN noise. In this section, we use two different non-parametric approaches in determining the degree of randomness of the audio samples.

Each audio sample is represented as a matrix of integers, with the shape representing the dimensions. For the scope of our study, we focus on notable runs tests in which upward and downward run counts are carried out for a sequence of variables, by floating the integers of the audio samples represented as an integer matrix.

Two measures of randomness namely the Wald-Wolfowitz runs test [35] and the Cox–Stuart test [36] are used to measure and compare the degree of randomness between the three audio signals. The results are reported in Sections 16.5.2.1 and 16.5.2.2.

16.5.2.1 Wald–Wolfowitz runs tests

The Wald–Wolfowitz runs tests [35] consider each integer in the integer matrix representation of the audio sample as n observations with a median value. A measure of the expected runs $E(R) = \frac{2n_1n_2}{n} + 1$ and the variance $V(R) = \frac{2n_1n_2(2n_1n_2-n)}{n^2(n-1)}$ are computed respectively below to establish the statistical ratio. The equation below from [35]

$$Z_R = \frac{(R - E(R))}{V(R)} \tag{16.1}$$

which represents the number of runs in the representation of the audio file corresponding to its size.

16.5.2.2 Cox–Stuart test

The Cox–Stuart test [36] focuses on randomness based on negative or positive tests in data. Taking into consideration the sum of positive signs for an integer matrix representing each audio sample, a p-value is taken as a cumulative probability function for a binomial distribution of the dataset. The integer matrix representation of the dataset is grouped into pairs with the sign computed. The sign test in the equation below is used to determine if there is a trend in randomness as observed in the integer matrix representation of the audio sample. The equation below from [36]

$$sign(x_i, x_i + c) = \begin{cases} + & \text{if } X_i = X_{i+c} \\ 0 & \text{if } X_i \leq X_{i+c} \\ - & \text{if } X_i \geq X_{i+c} \end{cases} \tag{16.2}$$

Thus, the p-value with a count of the positive comparisons forms the statistical ratio for the degree of randomness.

Outcome of the runs test for randomness: For each runs test, we compute the average across the entire dataset for each inference attack case scenario mentioned in Section 16.4.6. We repeat the process of the run test computation for each of the datasets with the white noise added and also with the GAN noise added. First, we compute the degree of randomness in the original audio sample. We then compare the degree of randomness with the audio sample superimposed with the white noise sample as well as the audio sample superimposed with the GAN noise. Figures 16.6, 16.7, and 16.8 show the results of the randomness tests. The results

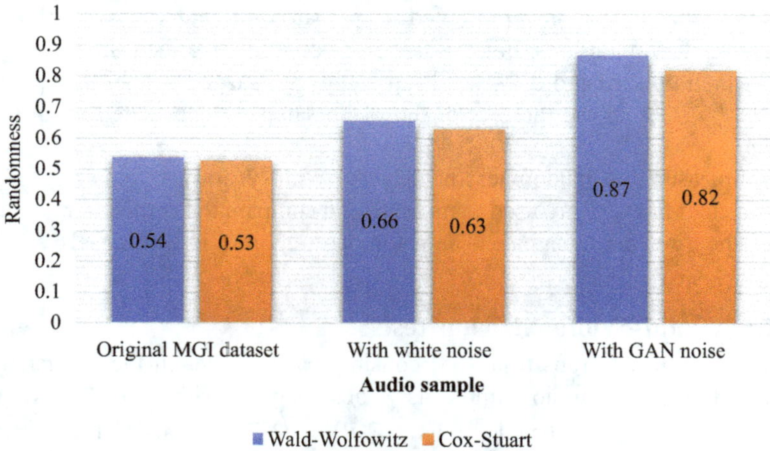

Figure 16.6 Randomness tests with MGI dataset

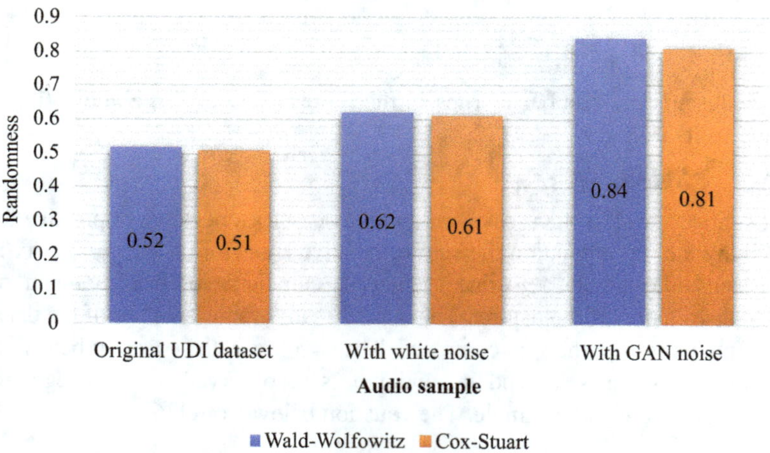

Figure 16.7 Randomness tests with UDI dataset

show that on a scale of 0–1, the audio sample overlaid with the GAN noise shows more randomness based on both runs test compared to the original dataset as well as the dataset with the white noise.

16.5.3 Perform inference attacks on original audio samples

Machine learning-based audio profiling of voice recordings from always-listening devices can be used to infer sensitive information from a smart home user's environment. We experiment with a total of three publicly available datasets, to explore three types of inference attacks to leak out sensitive information about the occupants of a home.

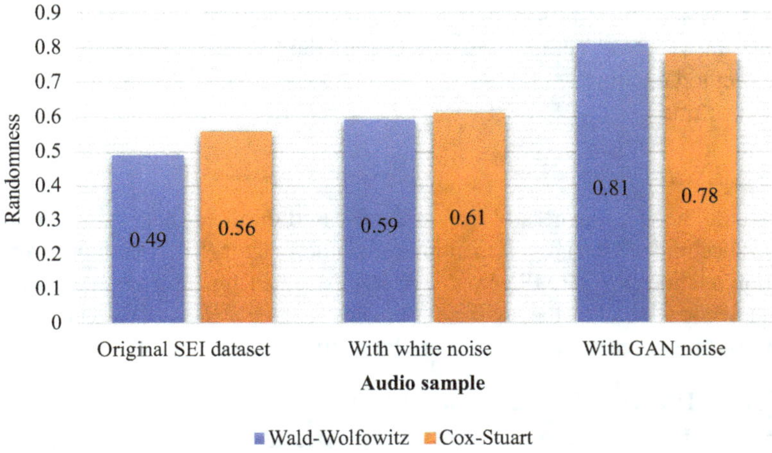

Figure 16.8 Randomness tests with SEI dataset

16.5.3.1 Inferring user music listening preferences from smart listening devices

We use the Free Music Archive Dataset [37] which is an open and easily accessible dataset suitable for evaluating several tasks in music information retrieval (MIR). It consists of full-length and high-quality audio which includes metadata and tags. The dataset consists of 106,574 tracks from 16,341 artists and 14,854 albums, arranged in a hierarchical taxonomy of 161 genres.

16.5.3.2 Inferring user demographics from smart listening devices

This task involves inferring three basic user demographics contexts from audio files, namely age, gender, and race. The dataset used is the Mozilla common voice dataset [38], which consists of about 51,000 voice recording samples. We use all three DNN architectures to perform multi-class, multi-label classification.

16.5.3.3 Inferring emotional content from smart listening devices

In the third step of our experimental approach, we explore the feasibility of an adversary to infer emotional context from a user's private conversations. As earlier discussed, monetary motives such as targeted advertisements may be a factor for such an adversary in implementing this form of inference attack. For this case scenario, the Ryerson Audio–Visual Database of Emotional Speech and Song (RAVDESS) [39] is used.

16.5.4 Mitigate sound inference attacks

Our methodology entails the superimposition of the original audio samples with some form of external audio noise to prevent an adversary from inferring unwarranted sensitive information from audio recordings. The external noise is generated in with two methods. For the first method, white noise is generated. The second method entails the use of a GAN architecture, which forms the basis of our proposed solution.

16.5.5 Evaluation

We evaluate the effectiveness of our solution based on three metrics. The mitigated inference accuracy (MIA), the semantic preservation factor (SPF), and randomness to mitigation relationship (RTMR). First, we establish a benchmark assessment which we report in Section 16.6.1 as the Baseline Inference Accuracy (BIA). We then proceed to evaluate our proposed solution based on the metrics described below.

16.5.5.1 Mitigated inference accuracy

The mitigated inference accuracy (MIA) denotes the prediction accuracy of the DNN model in inferring sensitive information from an audio dataset when the sound masking noise has been applied. We report this metric for all inference scenarios using the three different DNN architectures described in the study.

16.5.5.2 Semantic preservation factor

The semantic preservation factor (SPF) represents the attribute of the sound masking signal to preserve the semantics of the audio content. We use a different dataset for this experimental setup with the three DNN models to compare the SPF of both the white noise and the GAN noise.

16.5.5.3 Randomness to mitigation relationship

The third evaluation metric compares the randomness in the GAN noise and white noise with the mitigation inference accuracy. For both the white noise and the GAN noise, we calculate the element of randomness in the audio dataset when each noise sample is added, compared to the effect of the inference mitigation that was achieved.

16.6 Results

In this section, we report our experimental findings based on our three evaluation criteria discussed in Section 16.5.5. In Section 16.6.1, we establish the effectiveness of the inference attack on all three datasets. In Section 16.6.2, we compare the effect of both the GAN noise as well as the white noise in mitigating inference attacks for all three case scenarios. In Section 16.6.3, we show results that demonstrate that the GAN noise is more effective in preserving the semantics of the audio compared to the white noise. In Section 16.6.4, we show how the randomness for both the white noise and the GAN noise correlates with the mitigated inference accuracy for all three case scenarios.

16.6.1 Baseline inference accuracy

In the first experiment, we conduct a baseline assessment of the inference attacks against all three datasets for the three case scenarios we considered. All three

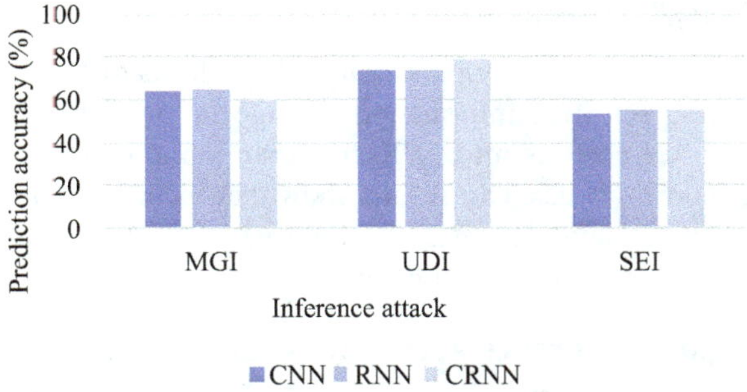

Figure 16.9 Baseline inference accuracy

machine learning techniques were effective in inferring information from the audio dataset with the highest achievable inference of 82% from the CRNN model for the user demographic inference (UDI). as shown in Figure 16.9. The figures reported are the best results achieved based on K-fold cross-validation which was used to determine the optimal parameter settings of the neural network models.

16.6.1.1 Music genre inference (MGI)

As part of privacy considerations and in the context of user privacy, a user's preferences for music listening may be chosen not to be shared with external parties without their consent. An adversary may however want to infer this information for example for monetary purposes such as targeted advertising without the user's consent. The ability of an adversary to infer this information is demonstrated using the Free Music Archive dataset [37]. We confirm using three different DNN architectures that the music genre can be correctly inferred with an accuracy of up to 67%.

16.6.1.2 User demographics inference (UDI)

We explore the feasibility of an adversary to infer user demographic data such as age, accent, and gender from the audio dataset using the Mozilla Common Voice dataset [38]. The dataset consists of about 51,000 voice recording samples of humans in 18 different languages. Our DNN models identify the demographic qualities of the speaker with an accuracy of 74%, 71%, and 89%, respectively. When all demographic properties are combined, an accuracy of 82% is achieved.

16.6.1.3 Speech emotion inference (SEI)

In our third privacy inference case scenario, we examine the ability of an adversary to infer the emotion of users from a given dataset. We use the Ryerson Audio-Visual Database of Emotional Speech and Song (RAVDESS) [39] which consists of 7,356 audio samples of 12 female and 12 male professional actors. Each of the actors is tasked with speaking out two lexically matched statements using a neutral North American accent. The dataset is labeled to distinguish a total of seven different emotions including calm, happy, sad, anger, fearful, surprise, and disgusted expressions.

16.6.2 Mitigated inference accuracy

In this section, we test the privacy preservation hypothesis of the GAN noise capability in preventing sensitive information leakage in smart homes. We seek to determine if the GAN noise is more effective than white noise in preserving the privacy of smart home devices. Our results show that the noise generated by GAN results in over 45% reduction in sensitive information leakage from smart home devices while maintaining the semantics of the audio.

16.6.2.1 Mitigated inference accuracy (white noise)

In our next experiment, we tested the ability of the DNN to correctly infer sensitive information from the dataset for the case scenarios discussed above. We notice very little difference in the ability of the white noise when combined with the original audio to mitigate against information leakage. When compared to the BIA results in Section 16.6.1, the maximum decrease in the inference that was observed when the white noise was added was less than 11% as shown in Figure 16.10.

16.6.2.2 Mitigated inference accuracy (GAN noise)

In our third experiment, we combine the GAN-generated noise with the original audio and repeat the inference attack using the three DNN models. When compared to the BIA results in Section 16.6.1, we observe up to a maximum of 45% decrease in inference accuracy when the GAN-generated noise is added to the original audio sample as shown in Figure 16.11.

16.6.3 Semantic preservation factor

In our fourth experiment, we demonstrate the ability of the GAN noise to preserve the semantics of the audio while effectively mitigating information leakage attacks. A visual representation of the results is shown in Figure 16.12 We performed speech recognition classification using the google speech commands dataset [34]. This fourth dataset was selected since the dataset was collected and labeled for recognizing the content of the speech. Unlike the other three datasets which were used in the inference attack discussed in Section 16.5.3, this dataset is more

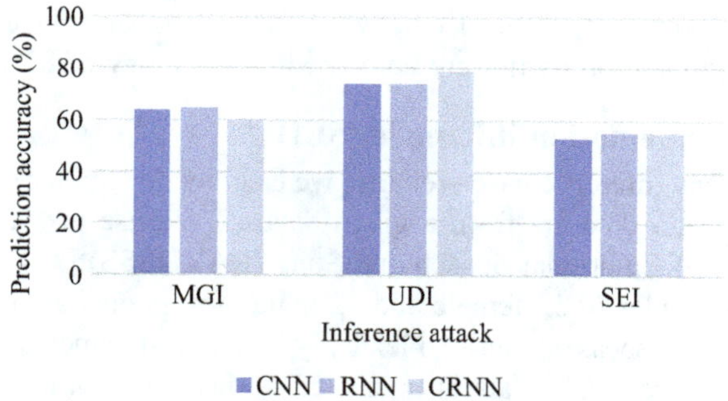

Figure 16.10 Mitigated inference accuracy (white noise)

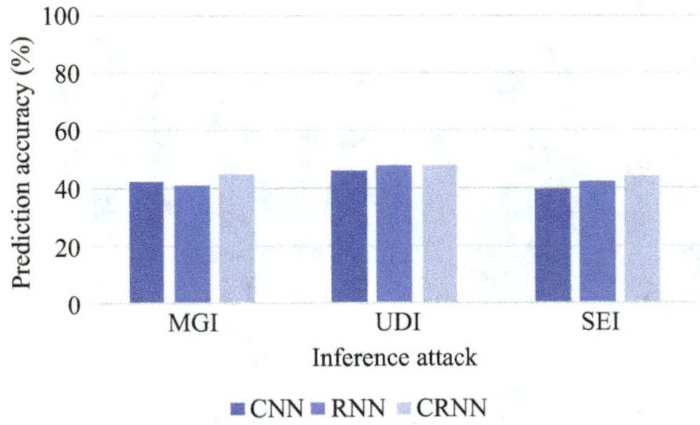

Figure 16.11 Mitigated inference accuracy (GAN noise)

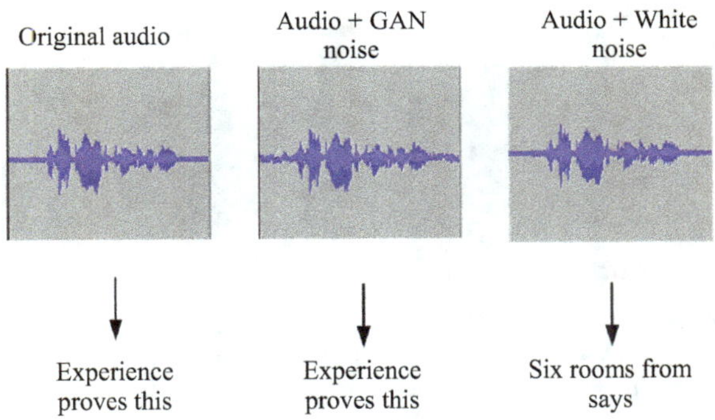

Figure 16.12 Illustration – semantic preservation factor

appropriate for deducing the content of the speech. The other three datasets were not used for the semantic experiment since they were not collected and labeled for speech classification tasks. The result of the experiment shows that the GAN noise has less impact on the DNN speech recognition classifier compared to the white noise as shown in Figure 16.13. Hence, we confirm that the GAN noise does indeed preserve the utility of the device by deterring inference attacks yet, maintaining the semantics of the conversation.

16.6.4 Randomness to mitigation relationship

We evaluated the relationship between randomness and the MIA in each inference attack case scenario. The result as illustrated in Figure 16.14 shows that the higher the degree of randomness, the higher the mitigation effect that the noise exhibits in deterring the inference attack. The mitigation achieved is calculated as the difference in the MIA for each case scenario with the white noise as well as the GAN noise. The GAN noise, having more randomness compared to the white noise, is proven to have a higher mitigation effect.

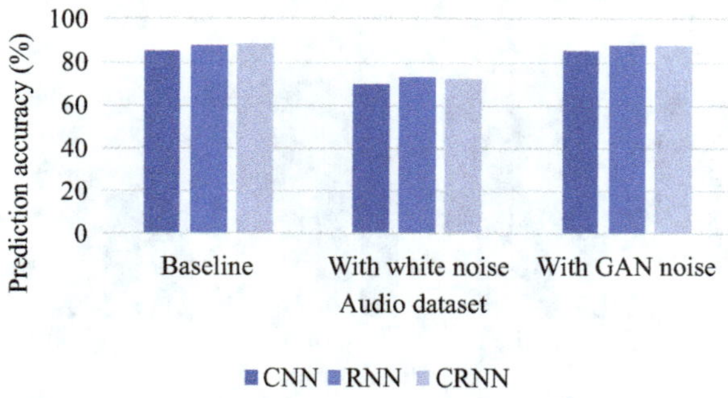

Figure 16.13 Semantic preservation factor

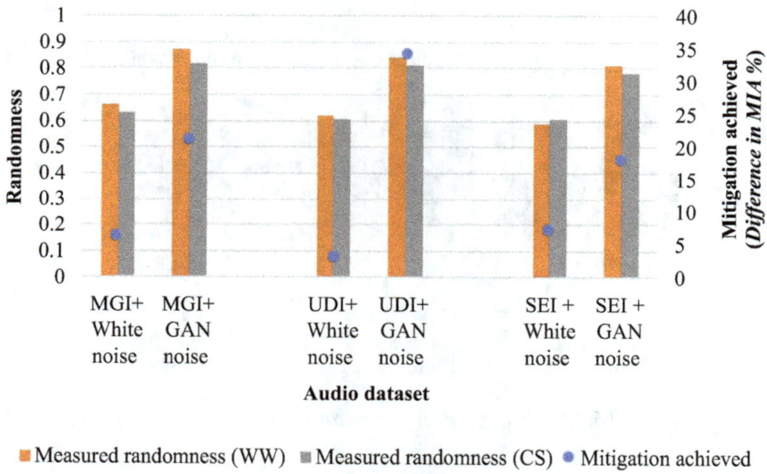

Figure 16.14 Randomness to mitigation relationship

16.7 Discussion

16.7.1 White noise and randomness

White noise is known to exhibit statistical characteristics that are similar to randomly generated numbers. To be considered as truly random, we expect the entity to be in fact unpredictable; but the possibility of a white noise generator exhibiting true randomness in the sense of unpredictability is questionable. Speicher *et al.* [40] noted that patterns can be noted in pseudo-random generated white noise, based on the fact that is contains possibly predictable elements for example, a linear congruential random generator, which is a typical algorithm used for producing digital noise output. Tzeng *et al.* [41] argued that it is best to treat randomness as a property of the process that generates the signal of the white noise, not of the white noise itself.

Hence, we establish that white noise with its pseudorandom property is thus limited in its ability as an effective measure in privacy preservation for use in audio masking for preventing sensitive information leakage attacks.

The ability of generative adversarial networks to create high-dimensional data has been researched [4] and the relationship of this high dimensionality to randomness is an object of interest. As the generator model in the GAN continuously learns to produce data samples that the adversary (discriminator) cannot predict, the randomness element in its output improves, as demonstrated in our results. Our solution entails the use of a generative model that has learned to produce realistic noise samples of a given dataset from low-dimensional, random latent vectors.

Several recent efforts have been made to generate sound using GANs including the use of CycleGAN by [42]. Their approach augments an existing audio sample with emotions and can also convert speeches between emotional variations e.g. convert an angry speech into a sad speech.

We differentiate our work from other studies such as [43] which use adversarial attacks to mitigate speech recognition i.e. the use of machine learning systems to determine the identity of the speaker. In this specific study, a state-of-the-art DNN known as X-vector was tested. By adding a carefully crafted inconspicuous noise to the original audio, their attack method was successful in fooling the DNN into making false predictions. The solution goes further to incorporate room impulse response (RIR) estimates while training the adversarial examples to demonstrate the effectiveness for both digital attacks as well as over-the-air attacks.

16.7.2 Mitigating privacy inference leakage in digital space vs. physical space

Sound masking in the context of this study occurs in the physical space and is more practical-oriented. Factors such as the room impulse ratio are considered in deploying real and tangible audio signals to mitigate unwanted sensitive information leakage due to machine learning inference. Traditionally, there have been several ways of attacking speech recognition systems. Adversarial examples, for instance, the work of Carlini and Wagner [44] could impact a speech recognition system to misclassify by adding a carefully crafted perturbation. This adversarial attack method was tested against speech recognition only, but not tested against speech inference. Furthermore, their attack was proven to be effective in the digital space. Similar studies [44,45] have proposed solutions mostly against automatic speaker recognition in the digital space. In our threat model, we considered various adversaries, including the manufacturer or solution developer who controls the digital space, and therefore, implementing a solution within the digital space will be ineffective against such adversaries.

Also, the work of Fuxun Yu *et al.* [46] introduced "MASKER," a solution that introduces human imperceptible adversarial perturbation into real-time audio signals with a significant increase in the word error rate (WER). Their work focused on mobile platforms and was not tested to work against digital voice assistants or smart home environments. Also, the solution is primarily effective only in the digital space and was not tested as an over-the-air solution.

Our method focuses on a solution that is implemented within the physical space. Since the user can perceive the sound generated, we ensure that this sound is within the audible comfort zone for human hearing of less than 45 db [26]. We tested with various amplitudes of the audio signal and determined that as the amplitude increases, the effectiveness also increases. However, compared to the original audio as well as the white noise, our GAN-generated noise achieves better mitigation based on several metrics.

We show that speech recognition and sound-based inference have varying and different unique characteristics and adversarial noise techniques should be considered differently. Refer to Figure 16.1 where we illustrate the difference between eavesdropping and inference attacks.

16.8 Conclusion

We proposed a novel method for mitigating sensitive information leakage in smart wireless networks. We highlighted a threat model whereby an adversary deploys a machine learning-based inference attack on connected, always-listening devices such as smartphones or smart speakers.

Our solution is based on a generative deep learning model known as the GAN. We established from our experiments that GAN-based audio samples have increased randomness compared to white noise. Also, when used for sound masking purposes, GAN-generated noise can effectively mitigate machine learning-based inference attacks in smart wireless networks while preserving the semantics of the audio conversation. The projected approach would find useful applications in future wireless communication systems such as beyond 5G and 6G networks. Future work would focus on enhancing the robustness of the security architecture to mitigate sophisticated security attacks in dense wireless networks.

Acknowledgment

This work was supported by the Natural Sciences and Engineering Research Council of Canada (NSERC) through the NSERC Discovery Grant program. The work of Agbotiname Lucky Imoize is supported in part by the Nigerian Petroleum Technology Development Fund (PTDF) and in part by the German Academic Exchange Service (DAAD) through the Nigerian-German Postgraduate Program under Grant 57473408.

References

[1] Checkmarx. Amazon Echo – Alexa Leveraged as a Silent Eavesdropper, 2018. Available from: https://developer.amazon.com/alexa-skills-kit https://developer.amazon.com/alexa-skills-kit%0Ahttps://info.checkmarx.com/hubfs/Amazon_Echo_Research.pdf.

[2] Lau J, Zimmerman B, and Schaub F. Alexa, are you listening? privacy perceptions, concerns and privacy-seeking behaviors with smart speakers. *Proceedings of the ACM on Human–Computer Interaction*. 2018;2 (CSCW):1–31.

[3] Zeng E, Mare S, and Roesner F. End user security and privacy concerns with smart homes. In: *Thirteenth Symposium on Usable Privacy and Security ({SOUPS} 2017)*, 2017. p. 65–80.

[4] De Bernardi M, Khouzani M, and Malacaria P. Pseudo-random number generation using generative adversarial networks. In: *Joint European Conference on Machine Learning and Knowledge Discovery in Databases*. Springer, 2018. p. 191–200.

[5] Jiang J, Li Y, Ma X, *et al.* Research on noise quality in anti-eavesdropping system based on acoustic masking. In: *2017 International Conference on Wireless Communications, Signal Processing and Networking (WiSPNET)*. IEEE, 2017. p. 823–827.

[6] Godwin S, Glendenning B, and Gagneja K. Future security of smart speaker and IoT smart home devices. In: *2019 Fifth Conference on Mobile and Secure Services (MobiSecServ)*. IEEE, 2019. p. 1–6.

[7] Bugeja J, Jacobsson A, and Davidsson P. On privacy and security challenges in smart connected homes. In: *2016 European Intelligence and Security Informatics Conference (EISIC)*. IEEE, 2016. p. 172–175.

[8] Apthorpe N, Huang DY, Reisman D, *et al.* Keeping the smart home private with smart (er) IoT traffic shaping. *Proceedings on Privacy Enhancing Technologies*. 2019;2019(3):128–148.

[9] Huang DY, Apthorpe N, Acar G, *et al.* IoT Inspector: Crowdsourcing Labeled Network Traffic from Smart Home Devices at Scale, 2019. arXiv preprint arXiv:190909848.

[10] Bastos D, Shackleton M, and El-Moussa F. Internet of Things: a survey of technologies and security risks in smart home and city environments. In: *Living in the Internet of Things: Cybersecurity of the IoT – 2018*, 2018.

[11] Chen Y, Li H, Teng SY, *et al.* Wearable microphone jamming. In: *Proceedings of the 2020 CHI Conference on Human Factors in Computing Systems*, 2020. p. 1–12.

[12] Lei X, Tu GH, Liu AX, *et al.* The Insecurity of Home Digital Voice Assistants – Amazon Alexa as a Case Study, 2017. arXiv preprint arXiv:171203327.

[13] An N, Duff A, Noorani M, *et al.* Malware anomaly detection on virtual assistants. In: *2018 13th International Conference on Malicious and Unwanted Software (MALWARE)*. IEEE, 2018. p. 124–131.

[14] Zhang L, Tan S, and Yang J. Hearing your voice is not enough: an articulatory gesture based liveness detection for voice authentication. In: *Proceedings of the 2017 ACM SIGSAC Conference on Computer and Communications Security*. CCS'17. New York, NY: Association for Computing Machinery, 2017. p. 57–71. Available from: https://doi.org/10.1145/3133956.3133962.

[15] Gao C, Chandrasekaran V, Fawaz K, *et al.* Traversing the quagmire that is privacy in your smart home. In: *Proceedings of the 2018 Workshop on IoT Security and Privacy*, 2018. p. 22–28.

[16] Liu Y, Peng J, Yu JJ, *et al.* Ppgan: Privacy-Preserving Generative Adversarial Network, 2019. arXiv preprint arXiv:191002007.

[17] Frigerio L, de Oliveira AS, Gomez L, *et al.* Differentially private generative adversarial networks for time series, continuous, and discrete open data. In: *IFIP International Conference on ICT Systems Security and Privacy Protection*. Springer, 2019. p. 151–164.

[18] Asatilla A and Kim DS. Information protection by noise generator for tactical smart platforms. In: *2018 3rd International Conference On Internet of Things: Smart Innovation and Usages (IoT-SIU)*. IEEE, 2018. p. 1–6.

[19] Tung YC and Shin KG. Exploiting sound masking for audio privacy in smartphones. In: *Proceedings of the 2019 ACM Asia Conference on Computer and Communications Security*, 2019. p. 257–268.

[20] Copos B, Levitt K, Bishop M, *et al.* Is anybody home? Inferring activity from smart home network traffic. In: *2016 IEEE Security and Privacy Workshops (SPW)*. IEEE, 2016. p. 245–251.

[21] Apthorpe N, Reisman D, Sundaresan S, *et al.* Spying on the Smart Home: Privacy Attacks and Defenses on Encrypted IoT Traffic, 2017. arXiv preprint arXiv:170805044.

[22] Usama M, Asim M, Latif S, *et al.* Generative adversarial networks for launching and thwarting adversarial attacks on network intrusion detection systems. In: *2019 15th International Wireless Communications & Mobile Computing Conference (IWCMC)*. IEEE, 2019. p. 78–83.

[23] Xu Y, Frahm JM, and Monrose F. Watching the watchers: automatically inferring tv content from outdoor light effusions. In: *Proceedings of the 2014 ACM SIGSAC Conference on Computer and Communications Security*, 2014. p. 418–428.

[24] Purwins H, Li B, Virtanen T, *et al.* Deep learning for audio signal processing. *IEEE Journal of Selected Topics in Signal Processing*. 2019;13 (2):206–219.

[25] Goodfellow I, Pouget-Abadie J, Mirza M, *et al.* Generative adversarial nets. In: *Advances in Neural Information Processing Systems*, 2014. p. 2672–2680.

[26] Fincher W and Boduch M. Standards of human comfort: relative and absolute. In: *UTSoA – Seminar in Sustainable Architecture*. University of Texas at Austin School of Architecture, 2009.

[27] Cauley L. NSA has massive database of Americans' phone calls. *USA Today*. 2006;11(06).

[28] Avci U, Akkurt G, and Unay D. A pattern mining approach in feature extraction for emotion recognition from speech. In: *International Conference on Speech and Computer*. Springer, 2019. p. 54–63.

[29] Mohamed Ar, Hinton G, and Penn G. Understanding how deep belief networks perform acoustic modelling. In: *2012 IEEE International Conference*

on Acoustics, Speech and Signal Processing (ICASSP). IEEE, 2012. p. 4273–4276.

[30] Hinton G, Deng L, Yu D, *et al.* Deep neural networks for acoustic modeling in speech recognition: the shared views of four research groups. *IEEE Signal Processing Magazine.* 2012;29(6):82–97.

[31] Lipton ZC, Berkowitz J, and Elkan C. A Critical Review of Recurrent Neural Networks for Sequence Learning, 2015. arXiv preprint arXiv:150600019.

[32] Choi K, Fazekas G, Sandler M, *et al.* Convolutional recurrent neural networks for music classification. In: *2017 IEEE International Conference on Acoustics, Speech and Signal Processing (ICASSP)*. IEEE, 2017. p. 2392–2396.

[33] Radford A, Metz L, and Chintala S. Unsupervised representation learning with deep convolutional generative adversarial networks. In: Bengio Y and LeCun Y, editors, *4th International Conference on Learning Representations, ICLR 2016*, San Juan, Puerto Rico, May 2–4, 2016, Conference Track Proceedings, 2016. Available from: http://arxiv.org/abs/1511.06434.

[34] Warden P. Speech Commands: A Dataset for Limited-Vocabulary Speech Recognition, 2018. ArXiv e-prints. Available from: https://arxiv.org/abs/1804.03209.

[35] Wald A and Wolfowitz J. On a test whether two samples are from the same population. *The Annals of Mathematical Statistics.* 1940;11(2):147–162.

[36] Cox DR and Stuart A. Some quick sign tests for trend in location and dispersion. *Biometrika*. 1955;42(1/2):80–95.

[37] Defferrard M, Benzi K, Vandergheynst P, *et al.* Fma: A Dataset for Music Analysis, 2016. arXiv preprint arXiv:161201840.

[38] Ardila R, Branson M, Davis K, *et al.* Common Voice: A Massively-Multilingual Speech Corpus, 2019. arXiv preprint arXiv:191206670.

[39] Livingstone SR and Russo FA. The Ryerson Audio-Visual Database of Emotional Speech and Song (RAVDESS): a dynamic, multimodal set of facial and vocal expressions in North American English. *PLoS ONE.* 2018;13(5):e0196391.

[40] Speicher R. A new example of 'independence' and 'white noise'. *Probability Theory and Related Fields*. 1990;84(2):141–159.

[41] Tzeng S and Wei LY. Parallel white noise generation on a GPU via cryptographic hash. In: *Proceedings of the 2008 Symposium on Interactive 3D Graphics and Games*, 2008. p. 79–87.

[42] Asakura T, Akama S, Shimokawara E, *et al.* Emotional speech generator by using generative adversarial networks. In: *Proceedings of the Tenth International Symposium on Information and Communication Technology*, 2019. p. 9–14.

[43] Li Z, Shi C, Xie Y, *et al.* Practical adversarial attacks against speaker recognition systems. In: *Proceedings of the 21st International Workshop on Mobile Computing Systems and Applications*, 2020. p. 9–14.

[44] Carlini N and Wagner D. Audio adversarial examples: targeted attacks on speech-to-text. In: *2018 IEEE Security and Privacy Workshops (SPW)*. IEEE, 2018. p. 1–7.

[45] Alzantot M, Balaji B, and Srivastava M. Did You Hear That? Adversarial Examples Against Automatic Speech Recognition, 2018. arXiv preprint arXiv:180100554.

[46] Yu F, Xu Z, Liu C, *et al.* MASKER: adaptive mobile security enhancement against automatic speech recognition in eavesdropping. In: *Proceedings of the 56th Annual Design Automation Conference 2019*, 2019. p. 1–6.

Chapter 17

Adversarial resilience of self-normalizing convolutional neural networks for deep learning-based intrusion detection systems

Olakunle Ibitoye[1], Ashraf Matrawy[1], Omair Shafiq[1] and Agbotiname Lucky Imoize[2,3]

Abstract

Deep learning-based intrusion detection systems (IDSs) can easily be fooled by the presence of adversarial examples, thus, limiting their usefulness in security-critical applications such as in 5G and 6G wireless networks. The cause for the adversarial vulnerability of the neural network requires an elaborate investigation. Still, some researchers have proposed that regularization and normalization techniques applied to the neural network models play a significant role. In this paper, we evaluate the role of self-normalization in the adversarial vulnerability of neural network models within the context of IDSs for application in the envisioned 6G wireless networks. We propose the design and implementation of a deep learning-based IDS for botnet traffic and subject it to various forms of adversarial attacks. We then investigate the impact of self-normalization on the adversarial resilience of our deep learning-based IDS and compare our findings with that of image classification neural network models. Our study proposes a customized convolutional neural network (CNN) model that utilizes self-normalizing activation in the fully connected layers. Our results show that self-normalization of the deep learning-based IDS using scaled exponential linear unit (SELU) results in greater resilience to various adversarial samples. The projected adversarial resilience of self-normalizing CNNs for deep learning-based IDSs would be useful in future wireless communication systems such as 6G networks.

Keywords: 6G wireless networks; Security and privacy; Intrusion detection system (IDS); Adversarial samples; Resilience; Self-normalizing convolutional neural networks (SCNN)

[1]School of Information Technology, Carleton University, Ottawa, Canada
[2]Department of Electrical and Electronics Engineering, Faculty of Engineering, University of Lagos, Nigeria
[3]Department of Electrical Engineering and Information Technology, Institute of Digital Communication, Ruhr University, Germany

17.1 Introduction

As cyberattacks become more sophisticated and more difficult to overcome using standard techniques adopted by security operations centers, the role of artificial intelligence and machine learning in preventing cyberattacks has become more significant. Intrusion detection systems (IDS), which are a primary defense layer in any security operative program, have progressively utilized machine learning and deep learning techniques.

Deep learning-based IDS have an advantage over conventional anomaly-based IDS because they help overcome the challenge of proper feature selections [1]. However, two major challenges of deep learning in security applications are the lack of transparency of the deep learning models [2] and the vulnerability of the deep learning models to adversarial attacks [3]. For the scope of this study, we focus on the adversarial vulnerability of deep learning models.

An adversarial attack occurs when an adversarial example is fed as an input to a machine-learning model. An adversarial example is an instance of the input in which some feature has been intentionally perturbed to confuse a machine learning model to produce a wrong prediction. Szegedy *et al.* [3] demonstrated how a deep learning model for image recognition could be confused into making wrong predictions by introducing a tiny perturbation to the image. Other researchers [4,5] have also proved that adversarial attacks are equally effective against deep learning models in network security applications such as malware detection and IDSs.

Normalization is used in neural networks to dampen the oscillations that occur in the distribution of activations at the output of each neuron or node. When normalization is applied, a neural network model's ability to generalize is significantly improved, and the training time is reduced [6] as a result of normalization during the backpropagation process. Also, the neural network becomes less prone to vanishing and exploding gradients due to the normalization process.

A self-normalizing neural network (SNN) [7] is a type of deep learning model that maintains the stability of the network during the gradient descent process. A SNN is built by replacing the standard activation functions in a typical neural network with a specific activation function known as the SELU.

Researchers have shown that self-normalization increases the resilience of deep neural networks against exploding or vanishing gradients. Still, no publication, to the best of our knowledge, has demonstrated the impact of self-normalization on the adversarial resilience of convolutional neural networks (CNNs) in the context of IDSs. In our previous work [8], we studied how deep learning-based IDSs for IoT networks could be more resilient against adversarial samples. In our study referred to above, we evaluated both feed-forward neural networks (FNN) and SNN models.

This paper is an extension of our previous work [9] where we now investigate the adversarial resilience of self-normalizing CNNs (SCNN). We focus our study on the context of IDSs for botnet traffic as well as computer vision. Our contributions in this paper are as follows: For our first contribution, we created a novel deep learning-based IDS which is a variant of the CNN model. We call our method the SCNN-IDS based on the self-normalizing properties of the activation function employed within the dense layer of the neural network model. We evaluate the performance of our model with a standard CNN-based IDS (CNN-IDS) and report our results in Section 8.1.

In our second contribution, we studied the adversarial resilience of the SCNN-IDS within the context of botnet network traffic classification. We demonstrate that the self-normalizing properties of the SCNN-IDS reduce its vulnerability to adversarial examples. The results are shown in Section 8.2.

In our third and final contribution, we show that our results on the adversarial resilience of the SCNN-IDS are applicable to other domains, such as computer vision. As reported in Section 8.3, we repeat our experiments with an image dataset and show that the adversarial resilience exhibited by the SCNN-IDS is consistent when the experiments are repeated with an SCNN-image classifier.

In another contribution, we explore how adversarial samples vary in network security compared to computer vision. We explore why common defenses used against adversarial samples in computer vision may be inadequate for network security.

It is worthy of note that the adversarial resilience of self-normalizing CNNs for deep learning-based IDSs presented in this study would find useful applications in emerging wireless communication systems, specifically beyond 5G and 6G wireless networks. Compared to the available techniques, the projected system helps to detect network intrusion and other vulnerabilities in real-time, thereby mitigating network failures resulting from complex security attacks and enhancing the privacy-preservation and confidentiality of critical user information.

17.2 Related work

We compare our research with previous studies that have attempted to improve the resilience of CNN-based IDS to adversarial samples. CNN-based IDS have a unique peculiarity in their feature representation that utilizes high-grade abstraction in data through the complex structure to produce higher detection rates. These peculiarities would thus require finding a solution that takes into consideration the feature representation attributes of CNN models. Zhang *et al.* [10] proposed a tri-fold solution that incorporates model voting ensembling, ensembling adversarial training, and query detection. Their experimental study showed significant improvement to the robustness of the CNN-based IDS, bringing the detection rates close to 100%. Even though this is an effective solution, it proves to be very complex to implement as discussed by the authors.

A similar study by Abou-Khamis *et al.* [11] evaluated the use of min–max optimization to defend against adversarial samples in deep learning-based, which showed low results of less than 80% effectiveness for mitigating adversarial samples in CNN-based IDS.

Based on our literature review findings, no publication has evaluated the impact of self-normalization on the adversarial robustness (AR) of a CNN-based IDS. Hence, our study is novel and offers a useful contribution in understanding the security of machine learning and artificial intelligence in network security.

17.3 Background – adversarial machine learning

Deep learning is vulnerable to well-crafted input samples which are designed to fool the deep learning model. These well-crafted inputs are known as *Adversarial Samples*.

17.3.1 Adversarial taxonomy

Below is a brief taxonomy on adversarial machine learning based on six dimensions. Figure 17.1 illustrates this taxonomy.

17.3.1.1 Knowledge

White box attacks in which the attacker has full knowledge of the machine learning model or algorithm. Black box attacks are attacks in which the attacker has limited or no knowledge of the machine learning model or algorithm.

17.3.1.2 Falsification

False-positive attacks seek to misclassify a negative sample as a positive one. In the context of IDSs, botnet traffic will be misclassified as benign network traffic based on adversarial manipulation. False-negative attacks seek to misclassify a positive sample as a negative one. For example, benign network traffic is misclassified as botnet traffic.

17.3.1.3 Perturbation scope

Universal attacks create a universal perturbation for the entire dataset while individual attacks generate unique perturbations for each input sample in the dataset.

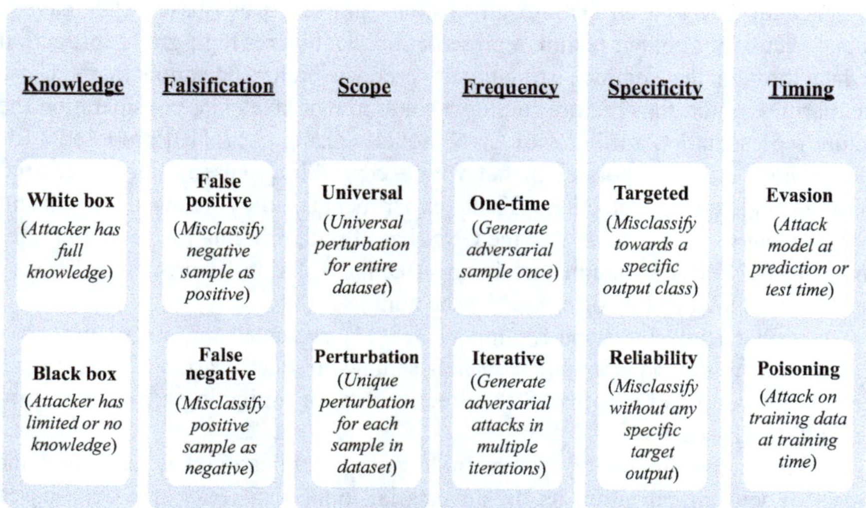

Figure 17.1 Adversarial taxonomy

17.3.1.4 Attack frequency

One-time attacks generate the adversarial sample only once without iteration. This is often most feasible in computationally intensive tasks such as in reinforcement learning. Iterative attacks, as the name suggests, create the adversarial sample over multiple iterations, with each iteration based upon interaction with the target classifier.

17.3.1.5 Specificity

Targeted attacks create a misclassification towards a specific class while Reliability attacks only seek to mislead the model without any specific output class in mind.

17.3.1.6 Timing

Evasion attacks create a misclassification during the prediction phase of machine learning by attacking the model. Poisoning attacks attack the training data and are launched during the training phase of the machine learning model.

Other taxonomy classifications such as perturbation limitation and measurement are exempted from this taxonomy since they are more applicable to the computer vision domain than to network security.

17.3.2 Generating adversarial samples

Most approaches for generating adversarial examples aim to alter a target prediction while minimizing the distance between the adversarial sample and the target instance. If the attacker has access to the model gradient, then gradient-based models such as neural networks are vulnerable. However, model-agnostic adversarial crafting methods do not require access to the model gradient but only the prediction function. To understand how adversarial examples are created, first, we establish that deep learning classifiers aim to optimize parameters denoted by θ while minimizing the average loss over a training sample $\{x_i \in X_i, y_i \in \mathbf{Z}\}, i= 1, ..., m$. This optimization problem is depicted by the equation:

$$_{\theta}\frac{1}{m}\sum l(h_{\theta}(x_i), y_i) \tag{17.1}$$

is often solved by some optimization algorithm such as stochastic gradient descent [12]. The gradient of the loss is then computed with respect to parameters θ while θ is continuously modified in the negative direction until convergence is reached.

$$\theta = \theta - \frac{\alpha}{|\beta|}\sum_{i\in\beta} \Delta_{\theta} l(h_{\theta}(x_i), y_i) \tag{17.2}$$

The gradient $\Delta_{\theta} l(h_{\theta}(x_i), y_i)$ in deep neural networks is typically computed using backpropagation, based on automatic differentiation, which determines how small changes to each input affect the loss function. However, to create the adversarial samples, the adversary rather than modifying the input sample to minimize the loss, modifies the input sample to maximize the loss:

$$\widehat{x}l(h_\theta(\widehat{x}), y_i) \tag{17.3}$$

where $\widehat{x}$ represents the adversarial example [13].

Several attack techniques and methods for creating adversarial samples exist, and for this study, we choose five of those techniques. We select attack methods that are generally applicable to both image classification as well as other domains, such as network security. Methods such as one-pixel attack [14] are restricted to only image classification domains and were not included in this study. We also focus on evasion attack methods, consistent with our threat method in Section 4.3. We like to note below, based on the selected methods, that the robustness of a machine learning model depends on the ability of an attacker to find an adversarial sample that is as close as possible to the original input. For the scope of this study, we focus on five different adversarial attack methods as illustrated in Figure 17.2. The five attacks are briefly described in the section below.

17.3.2.1 Fast gradient sign method (FGSM) attack

In the FGSM method proposed by Goodfellow *et al.* [13], adversarial samples are created by finding the maximal direction of positive change in the loss. In this method, a one-step gradient update is performed along the direction of the sign gradient at each level. The FGSM attack is highly efficient in its computational requirements.

17.3.2.2 Carlini and Wagner (C&W) attack

Carlini *et al.* [15] developed a targeted attack specifically for existing adversarial defense methods. In this study, we deploy the untargeted version of the attack by

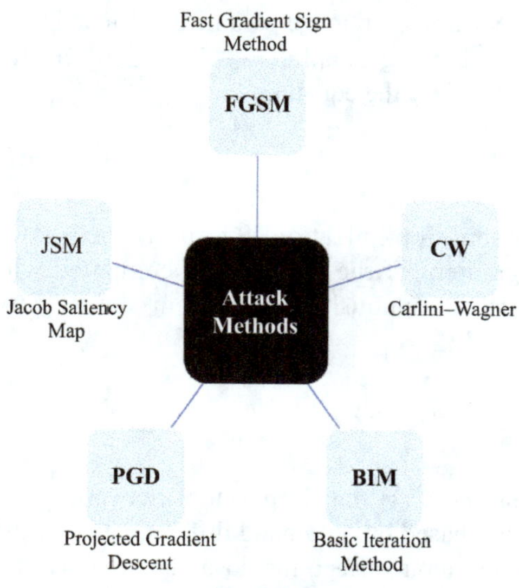

Figure 17.2 Adversarial attack methods

modifying the objective function. Carlini and Wagner attack is more computationally intensive than the FGSM attack and it was discovered that defenses such as defensive distillation [16] were ineffective towards the Carlini and Wagner attack [17].

17.3.2.3 Basic iterative method (BIM) attack

The FGSM is based on the assumption that the adversarial samples could be fed directly into the model. However, this is not always practical. The basic iterative method proposed in [18] overcomes this limitation by running the gradient update in multiple iterations.

17.3.2.4 Post gradient descent (PGD) attack

PGD [19] seeks to solve a constrained optimization problem. The perturbation which maximizes the loss of a model is computed while ensuring the perturbation amount is constrained to a set value of *epsilon*. This is achieved by projecting the attack result back to the original input during each iteration.

17.3.2.5 Jacobian-based saliency map (JSM) attack

The JSM attack is proposed by Papernot *et al.* [4], in which the Jacobian matrix of input sample x is computed to determine which input features most significantly affect the output. As such, a corresponding perturbation to the input feature is computed to generate the adversarial attack.

17.4 Problem definition and proposed study

17.4.1 Problem definition

Adversarial attacks were introduced in 2014 as an inherent weakness of image classification deep learning models [3]. As artificial intelligence and machine learning became mainstream in network security, researchers have begun to demonstrate that adversarial machine learning is also a significant threat against machine learning applications in network security. Zheng [5] demonstrated that a deep learning-based IDS that could correctly identify Denial of Service (DoS) attacks with an accuracy of 93% could have its performance degraded to as low as 24% with adversarial samples. The deep learning model used in the study was a feed-forward artificial neural network (ANN).

There has been much interest in finding ways to improve the AR of deep learning models [20]. Various adversarial defenses have been proposed for attaining adversarial resilience, most notably adversarial training [3]. In adversarial training, the resilience of the deep learning model is achieved by enhancing the training data with adversarial samples. Due to the limitation of this method, Tramer *et al.* [21] proposed ensemble adversarial training. This technique further improves the adversarial training technique by enhancing the training data with perturbations imported from a separate model.

Regularization has also been proposed as a possible technique for mitigating adversarial attacks. The relationship between adversarial vulnerability and input gradient regularization was explored by Ross *et al.* [22], who demonstrated that by regularizing the input gradient, the robustness of the neural network to adversarial perturbations was increased as much as by adversarial training. In addition to increasing the AR, the interpretability of the neural network was also improved with input gradient regularization.

Normalization is another technique primarily designed to dampen excessive oscillations in the distribution of activations in the output of a neuron. [6]. In the context of adversarial resilience, Farnia *et al.* [23] proposed the use of spectral normalization for improving the AR of deep learning models.

Despite several proposed methodologies and ideas some of which have been highlighted in the paragraphs above, a generalized solution for adversarial machine learning in network security does not exist. The majority of adversarial defense techniques have been strictly implemented within the context of computer vision, and their applicability to other domains such as network security is either irrelevant or unproven. The goal of this study is to attempt to investigate techniques to totally eliminate or at least reduce the vulnerability of deep learning-based IDSs to adversarial samples.

17.4.2 Proposed study

In this study, we investigate the effect of self-normalization on the adversarial resilience of deep learning-based IDSs. Self-normalization is a technique proposed by Klambauer *et al.* [7] for preventing exploding gradient problems in deep neural networks.

For our study, we utilize the CNN which is a type of feed-forward ANN. We propose to investigate if self-normalization applied to CNNs can result in IDSs that are more resilient to adversarial samples. Furthermore, we explore if the proposed technique can be applied to other domains such as computer vision by comparing our results with that of an SCNN-based image classification model.

The goal of this study is to attempt to answer the following research questions: (1) Does self-normalization reduce the vulnerability of CNN-based IDSs to adversarial samples? (2) Can the effect of self-normalization on adversarial samples within the context of IDSs be applied to other domains such as computer vision?

17.4.3 Threat model

Figure 17.3 depicts the threat model for this study with respect to the taxonomy in Section 3.1. Our study considers the following assumptions. First, we limit our study on adversarial attacks to evasion attacks which are launched during the prediction phase of the deep learning model. We consider that the attacker has complete knowledge of the deep learning model, dataset, and deep learning algorithm, hence resulting in a white-box attack. For the third assumption, in this study, the attacker does not target any specific prediction outcome, rather the attacker seeks to mislead the deep learning classifier to make a mistake and produce a misclassification; hence a reliability attack. The expected outcome is to degrade the performance of the deep

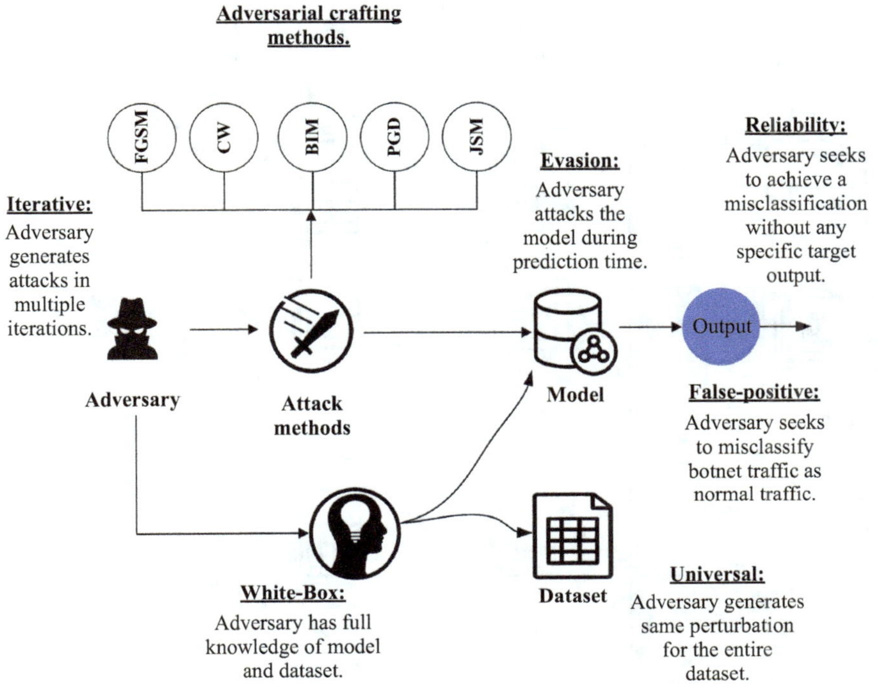

Figure 17.3 Threat model

learning classifier, as measured by various performance metrics. For each adversarial crafting method, the same perturbation is generated for the entire dataset, so we have universal perturbation attacks. The objective of the attack is to allow botnet traffic to go undetected by the deep learning classifier; hence we propose false-positive adversarial attacks. Finally, in the context of this study, for each adversarial attack method, the adversarial samples are crafted in multiple iterations, resulting in iterative attacks.

17.5 Experimental approach

In carrying out this study, we implement two deep learning-based IDS classifiers using a CNN and a variation of the CNN which we call the self-normalizing CNN (SCNN). The variation of the CNN is achieved by implementing a SELU activation function in the fully connected layers of the CNN. The four steps described below detail our approach to conducting the experiment. An illustration of the experimental setup is shown in Figure 17.4.

First, using a botnet network traffic dataset (CTU-13) [24], we train both IDS's and compare their performance accuracy as well as other classification metrics such as precision, recall, F1-score, and support.

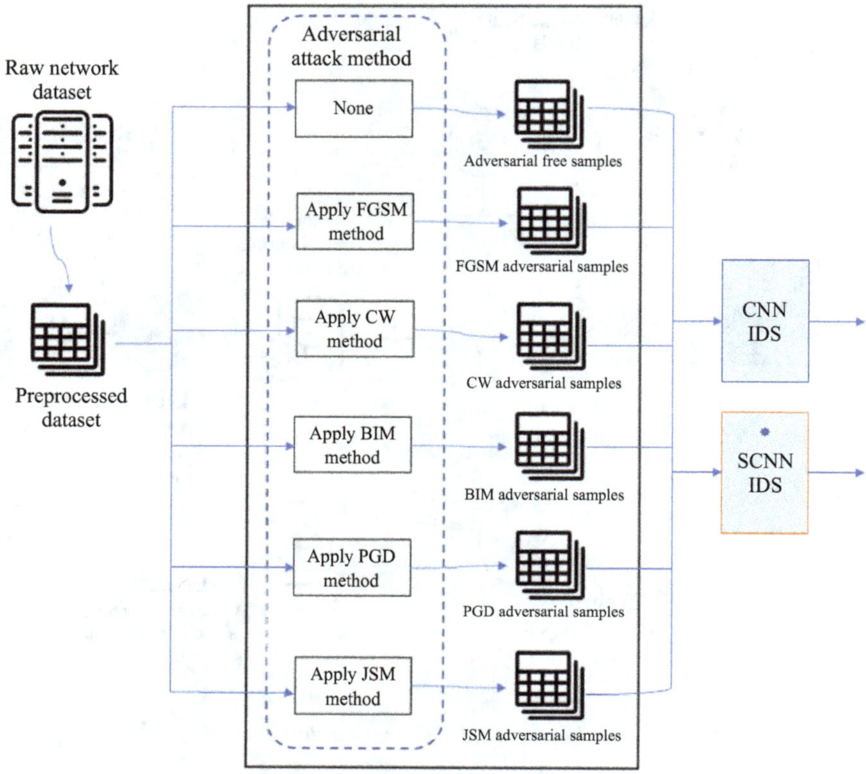

Figure 17.4 Solution overview architecture

In the second step, we process the CTU-13 intrusion detection dataset to create adversarial samples using five different adversarial attack methods. The methods used in crafting the adversarial samples for this study are the FGSM [13], C&W method [15], basic iteration method (BIM) [18], the projected gradient descent (PGD) method [19], and the Jacob-saliency map (JSM) [4] method.

In the third step, we test the effectiveness of both IDSs by subjecting them to adversarial samples. We recall that both IDSs were trained using adversarial free sample datasets. We evaluate and compare to determine if both IDSs respond to the adversarial samples in the same manner or if the SCNN-IDS is more resilient to adversarial samples based on the self-normalization techniques.

In the last step, we repeat the experiments using an image dataset – the Sokoto Coventry Fingerprint Biometric Dataset (SocoFing) and seek to determine if the results of the IDS dataset are consistent with that of an image dataset. Asides from the dataset, we keep every other factor kept constant in the experimental setup including the model parameter and hyperparameter configurations when conducting the image and IDS dataset comparison.

17.6 Solution description

In this section, we provide a detailed description of the self-normalizing CNN (SCNN) architecture. We describe in detail how the SCNN architecture differs from the basic CNN architecture that was used as the evaluation and comparison basis for this study.

17.6.1 SCNN

A SCNN is a variation of the CNN that incorporates an activation function with self-normalizing properties in the fully connected layers. CNNs generally learn the relationship between inputs and outputs, and the learned experience is stored in the filter weights. A CNN is a type of FNN. FNNs are known to be excellent universal approximators in the sense that they possess the capability to approximate a certain function *f* to any desired accuracy.

An illustration of the SCNN is shown in Figure 17.5. The convolution layer combined with a pooling layer together forms the convolution phase of the model. A convolution layer carries out convolutional operations on the input to detect patterns. A pooling layer provides spatial-based dimension reduction. A non-linear activation such as the rectified linear unit (ReLU) is applied after each convolution phase. The role of the ReLU layer is to clip the negative values to zero while maintaining the positive values. The ReLU layer performs a significant role in the forward propagation process since the system performance is noticeably degraded without the ReLU layer. The author of [25] highlights the necessity of a non-linear activation function in all intermediate layers for a CNN. The output of the pooling layer is flattened before it is fed into a fully connected layer.

Since the convolutional operation is a linear operation, applying the ReLU activation immediately after each convolution phase and before the next means a non-linear operation in between two linear operations.

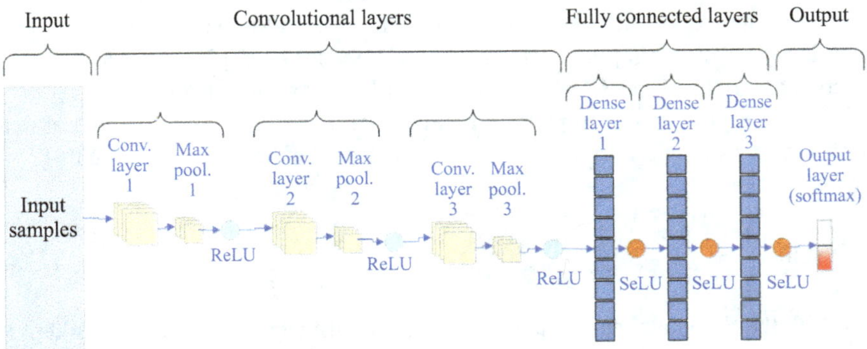

Figure 17.5 SCNN architecture: the SCNN. The convolutional layers perform feature extraction on the dataset while the fully-connected layers perform class discriminatory tasks to determine the output class.

The convolutional layer has its filter weights initialized and subsequently continuously updated using backpropagation to minimize a cost function. This occurs during the training phase while during the testing phase, the weights remain fixed.

We note that a typical CNN model will combine convolutional layers with fully connected layers. The purpose of the convolutional layers is to bring out spectral correlation while reducing spectral variation. The fully connected layers on the other hand receive the learned information from the convolutional layers as input for performing class discrimination. Hence, convolution neural networks (CNNs) have an advantage over multi-layer FNNs.

In each neural network architecture, we implement fully connected (dense) layers as provided by the Keras deep learning framework [26]. The dense layer in a Keras neural network implements the equation:

$$output = activation\ function * (input, kernel) + bias$$

This refers to the dot product of the input tensor and the weight kernel matrix, plus a bias vector. The output value is then passed through an activation function. Since the dense layer possesses an input with a rank of 2 or higher, the input is flattened before the dot product is calculated.

17.6.2 Activation functions

Activation functions in the context of ANNs symbolize an abstract representation of the firing activity of an artificial neuron. The major task of the activation function is to map the input of the neuron to the appropriate response variables or class labels. The neural network's ability to learn and perform complex tasks is dependent on the non-linear transformation of the input by the activation function since it would otherwise result in a non-linear combination of the inputs. Hence, the role of the activation function is to add non-linearity to the network. Due to the non-linearity, we can produce a neural network model having a target variable that has a non-linear relationship with the explanatory variables or features.

Neural networks with deep architectures have been known to experience gradient decay, resulting in poor performance. Klambauer *et al.* [7] proposed the SNN which is a variant of the ANN that uses a SELU activation function.

For the SNNs, a custom method for initializing the weights in the neural network known as Lecun normal is utilized. The scaled exponential linear unit [7] is shown as:

$$\text{selu}(x) = \lambda \begin{cases} x & \text{if } x \geq 0 \\ \alpha e^x - \alpha & \text{if } x \leq 0 \end{cases} \tag{17.4}$$

With the SeLU activation function, the mean of the activation output is kept at zero and the variance is kept at one. This allows the deeper neural network architectures to be trained without suffering significant gradient decay.

During stochastic gradient descent, the distribution of the weights W in the neural network as well as the outputs x of each layer are known to vary significantly for every iteration of the stochastic gradient descent process. As a result of the variations, the training process becomes very unstable, hence resulting in saturated activations and consequently introducing the problem of vanishing gradients.

The initial value for each layer in the neural network is provided with an initial value denoted by the initializer parameter. In our neural network architecture, the initial values for the weight matrix and the bias vector are specified using the kernel initializer parameter. For the bias, the default is the zero initializers which set the value of the bias vector to all zeros.

The default kernel weight matrix is the Glorot uniform [27] initializer which takes its values from a uniform distribution using the equation

$$limit = sqrt(6/(fan\_in + fan\_out))$$

Where the uniform distribution will fall uniformly between [−limit, limit]; Fan_in and fan_out are the units in the input and output tensors, respectively.

The regularization hyperparameters are unused by default. It is usually applied to either the weight matrix, the bias vector, or the entire layer output after activation. The effects include achieving a weight matrix close to zero or making the network sparse.

Our SCNN model in Figure 17.5 is divided into two sub-networks. The first subnetwork is the feature extraction subnet which consists of three convolutional layers, each followed by a max pooling layer and a ReLU activation function.

The other subnetwork is the class discriminatory subnet which determines the output class for the model. This is made up of three dense layers, each followed by a SeLU activation function.

17.6.3 Weight initialization

The convolutional layers in the SCNN have their weights initialized with the Glorot uniform initialization technique. The Glorot uniform initialization takes its samples around zero while the standard deviation is calculated following the formula below:

$$stddev = \sqrt{\frac{2}{(in + out)}}$$

For the fully connected layers, the weight initialization is performed using LeCun normal initialization. Similarly, the samples drawn are centered around zero while the standard deviation is calculated as:

$$stddev = \sqrt{\frac{1}{in}}$$

where *in* corresponds to the number of nodes in the previous layer and *out* corresponds to the number of nodes in the current layer.

In a standard CNN, the weight initialization in both the convolutional layers and dense layers is carried out using the Glorot uniform initialization.

17.6.4 Dropout

Dropout technique [28] is extensively used with the ReLU activation function to set the weights of randomly selected neurons to zero with the purpose of preventing overfitting. For the SCNN model, the dropout technique produces an undesirable effect since the weights of zero can be attributed to having low variance.

We recall from [7] that

$$lim_{x \to \infty} selu(x) = -\lambda\alpha = \alpha.$$

As proposed in [7] AlphaDropout is used instead of standard Dropout. AlphaDropout randomly sets the input values to a predefined value for α instead of zero. As a result, the self-normalizing properties of the SELU activation function will be retained in our SCNN model. In the convolutional layers, however, we did not implement the SELU activation functions, we do not need to use the AlphaDropout. Instead, the standard dropout is utilized.

17.7 Experimental setup

The basic architecture for the SCNN model which was used for this study consists of one input layer, followed by three convolutional layers each consisting of a convolutional layer and a max-pooling layer. The convolutional layers are followed by three fully connected (Dense) layers with 512, 256, and 128 neurons, respectively. The output of the fully connected layers is fed into the output layer which utilizes a softmax activation function. For the CNNs, three additional convolutional layers and three max-pooling layers are added. Specific details for each experiment are included in the following sections.

The parameters of the SCNN model that affects the quality of the model include the parameters and the hyper-parameters. The SCNN parameters are configurations that are internal to the model and include the weight and the biases for training the model. The SCNN parameters are determined during the training process and are estimated using the ADAM optimization algorithm [29].

Conversely, the hyper-parameters of the SCNN model are external configurations that are manually set while creating the model. The number and size of hidden layers is one example of the model hyper-parameters. For our study, we arrived at an optimal size of three convolutional and max pooling areas after experimenting with a variation of sizes. Other hyper-parameters such as dropout rate, learning rate, and weight initialization scheme were manually determined by trial and error means.

For the SNN and SCNN, we use a SELU activation function while for the ANN and CNN, we use a ReLU in the Dense layers. The ANN and CNN use basic dropout in the Dense layers to prevent overfitting and ensure better stability in the network during the learning phases while the SNN and SCNN use the AlphaDropout layer to retain the mean and variance at 0 and 1, respectively.

For initializing the weights, we select Glorot Uniform initializer [27] for the ANN and SNN while we use a Lecun Uniform Initializer [30] for the SNN.

For all the neural network models, we select the adaptive moment estimation (Adam) optimization algorithm to optimize the loss. Compared to other optimization methods, Adam calculates adaptive learning rates during the network learning process for each of the parameters.

In order to accurately calculate the perturbation timing factor, we run the experiment for each model at a single time to maintain consistency. Each of the neural network models is trained for a total of 100 epochs. We implemented an early-stop strategy to terminate the model training if the validation loss remains consistent for 10 consecutive epochs. A batch size of 1,024 is selected for the experiment.

17.7.1 Hardware platform

The experiments were carried out on a virtualized desktop computing environment with processor details – Intel core processor (Haswell) 32 cores, @2.85 GHz, 32 GiB DIMM random access memory (RAM), and 64 Gb SSD storage. For the graphics processing unit (GPU), a Tesla V100 PCIe 16Gb was utilized which was also attached to the virtualized desktop computing environment.

17.7.2 Development platform and tools

For the software, the code was written in Python 3.7 using the Spyder integrated development Environment prepackaged with the Anaconda data science distribution. The deep learning model was implemented using Tensorflow machine learning framework [31] with the specific version Tensorflow-gpu v1.15. We created a virtual environment for our experiments with specific versions of python software packages that are compatible with our code.

The adversarial samples were generated using the IBM AR Toolbox (ART) framework [32]. The ART framework is an open-source python library for deploying adversarial attacks and defenses using various methods and techniques. Several machine learning and deep learning frameworks including Keras, PyTorch, TensorFlow, and Scikit-learn are supported in the ART framework. In this study, we used the TensorFlow deep learning framework.

17.7.3 Dataset description

We use two types of datasets for our study: an IDS dataset and an image dataset. For the intrusion detection dataset, the CTU-13 dataset [24] is used which is a network traffic dataset consisting of real botnet traffic combined with background traffic and normal traffic. The dataset is captured as network flows. A network flow is defined as a network connection sharing similar characteristics. The original dataset consists of 13 different scenarios with each scenario representing various specific malware, protocols, and intrusion activities.

17.7.3.1 CTU-13 intrusion detection dataset

The original CTU-13 dataset contains 16 features out of which we use six features that are considered most relevant to the study. The remaining features were omitted

Table 17.1 Dataset characteristics

Duration	19.5 h
No. of packets	155,207,799
No. of netflows	2,954,231
Bot type	Murlo

Table 17.2 Traffic distribution

Description	No. of pkts	% of pkts
Total traffic	2,954,231	100
Botnet traffic	5,052	0.17
Normal traffic	72,822	2.46
C&C traffic	1,074	2.4
Background traffic	2,875,282	97.32

since they had less impact on the output and were likely to result in overfitting the model. The dataset was standardized by scaling them to values between 0 and 1. Also, the target variable was encoded from a string object into numerical values using LabelEncoder() class from Scikit Learn. The dataset was split into training and test values using a test size of 33% of the entire dataset sample.

Scenario 8 of the CTU-13 dataset has the following characteristics as shown in Table 17.1. Traffic distribution is shown in Table 17.2.

17.7.3.2 MNIST digits classification dataset

The MNIST digits classification dataset is an image dataset of handwritten digits that contains 60,000 training images and 10,000 test images. Each of the 70,000 images consists of 28 × 28 grayscale images of the 10 digits from zero to nine.

17.7.4 Dataset preparation

To prepare the CTU-13 IDS dataset for the convolution layer, the dataset was reshaped into 3 dimensions to represent a shape of (28,28,1). We perform feature selection for the CTU-13 IDS dataset in order to reduce the number of features. This helps to reduce the training time of the neural network model and helps to prevent overfitting problems. Ports 80 (HTTP) and 443 (HTTPS) are mostly utilized by the majority of network hosts for network traffic and carry HTTP-based botnet traffic. These two features were omitted from the study. Protocols TCP and UDP are mostly common. Protocols that are rarely used will create an outlier in the prediction model and impact the accuracy of the model prediction. These features were omitted from the study. The MNIST dataset is reshaped, rescaled to pixel values between 0.0 and 1.0, and one-hot encoded to categorical features.

For both the CTU-13 IDS dataset and the MNIST dataset, we select only the first 100 test samples for our experiment to speed up the prediction time and the adversarial attack time. All training samples are, however, used to ensure a more accurate model build.

17.7.4.1 Addressing imbalanced dataset for CTU-13

The CTU-13 IDS dataset is highly imbalanced with botnet traffic representing only 0.05% of the entire dataset. To address this imbalance in the dataset, the classifier apportions heavier weights to the few botnet samples available. We implement this by passing Keras weights as a parameter for both classes in the dataset – normal traffic and botnet traffic. This enables the deep learning model to focus more on the botnet samples during the model training process.

17.7.5 Generating the adversarial samples

The adversarial samples for this study were generated using the AR Toolbox (ART) [32] framework which is provided by IBM and is available for public use.

The first method used in generating the adversarial examples is the FGSM. The FGSM method performs a one-step gradient update along the direction of the sign of gradient for every input in the dataset [13]. The second method is the BIM which runs a finer optimization of the FGSM with minimal smaller changes for multiple iterations [18]. In each iteration, each feature of the input values is clipped to avoid too large a change on each feature. The third method is the projected gradient descent (PGD) which is also a variation of the FGSM attack but omits the random start feature of the FGSM [19]. The first three methods are model-dependent methods [33] and rely on the model gradient. In addition to the model-dependent methods, we also utilize two other model-agnostic adversarial attack methods namely the C&W attack as well as the JSM attack.

In our experiments, the adversarial samples are crafted with the intent of misleading the classifier without any specific target labels being specified. A complete knowledge of the deep learning model and algorithm is also assumed.

17.7.6 Evaluation metrics

Our evaluation compares the effectiveness of a CNN model with a SCNN model for intrusion detection using two main metrics: classification accuracy (CA) metrics and AR metrics. The CA metrics compare the performance of both models in classifying botnet traffic from normal network traffic using adversarial-free samples. AR metrics compare the effectiveness of both models using adversarial samples.

17.7.6.1 Classification accuracy (CA) metrics

The CA metrics include prediction accuracy, F1 score, recall, precision and support. The prediction accuracy measures the correlation of the actual prediction of the deep-learning model with the actual score. F1-score measures the harmonic mean between the precision and the recall. Recall represents the number of true positives that were correctly classified. Precision represents a measure of how exact the classifier is with a high precision representing a low number of false positives. Support represents how many actual occurrences of the class were present in the dataset.

17.7.6.2 AR metrics

The AR metrics compare the robustness of both the CNN and SCNN models to adversarial samples which were generated using the five adversarial crafting methods discussed in Section 17.7.5. The AR metrics are first evaluated in the context of intrusion detection. A similar evaluation in the context of image classification is performed and the observations are discussed in Section 17.9.

17.8 Results

In the first two results, we compare the performance of both the CNN and SCNN using the CA metrics as well as the AR metrics within the context of IDSs. In the following sub-sections, we evaluate using the same metrics in the context of computer vision/image classification. Our intent is to find out if AR in network security and computer vision follows the same pattern of performance.

17.8.1 Classification accuracy of CNN vs. SCNN for IDSs

Figure 17.6 shows the prediction accuracy of CNN vs. SCNN for intrusion detection. Both models accurately perform the task of classifying normal traffic from botnet traffic, based on the CTU-13 dataset. The SCNN-based IDS model achieves a prediction accuracy of 95% while the CNN-based IDS model achieves a 96% prediction accuracy.

We evaluate four different classification metrics namely the F1-score, recall, precision, and support. These classification metrics provide a more holistic report on the performance between the two IDS classifiers by taking into consideration

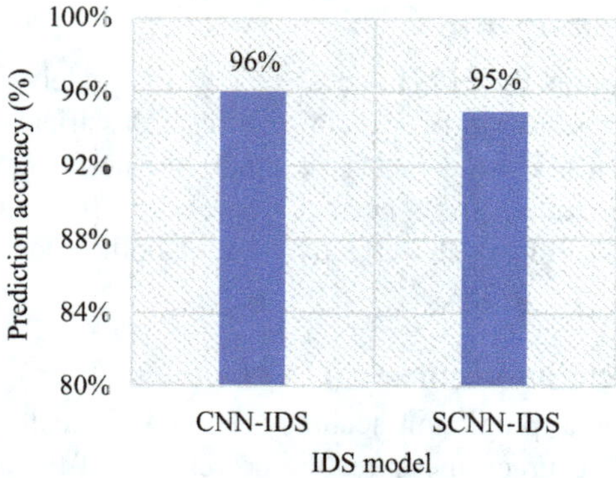

Figure 17.6 CNN vs. SCNN IDS prediction accuracy

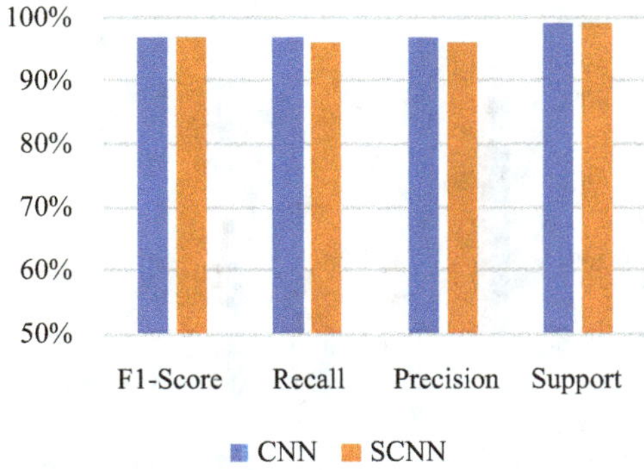

Figure 17.7 CNN vs. SCNN IDS classification accuracy

other factors such as false positives and negatives which are considered crucial metrics in intrusion detection. Our results show that both the CNN-IDS and the SCNN-IDS have comparable results across all four classification metrics as shown in Figure 17.7. The CNN-IDS classifier is only slightly better than the SNN-IDS based on the classification metrics.

We can conclude from the results here in Section 8.1 that the self-normalizing properties of the SCNN-IDS do not adversely impact the performance of the model.

17.8.2 AR of CNN vs. SCNN for IDSs

In our next experiment, we evaluate the adversarial robustness of the SCNN and CNN models for intrusion detection. Five different adversarial attack methods were used in our experiment to generate adversarial samples from the CTU-13 IDS dataset. The adversarial samples were crafted to mislead the classifiers to classify the botnet traffic as normal traffic. Section 17.4.3 above describes the threat model and explains specific details of the attack attributes. The result from all five attack methods demonstrates that the SCNN-IDS is more resistant to adversarial samples as shown in Figure 17.8. Based on the results, there is no differentiation between either category of attack since all five attack methods yielded consistent results with regard to adversarial resilience through self-normalization (Figure 17.9).

17.8.3 Classification accuracy of CNN vs. SCNN for image classification

Figure 17.10 shows the prediction accuracy of CNN vs. SCNN for image classification. Both models demonstrate satisfactory performance using the MNIST dataset. The SCNN-based image classifier model achieves a prediction accuracy of 95% while the CNN-based image classifier model achieves a 96% prediction accuracy.

We evaluate four different classification metrics namely the F1-score, recall, precision, and support. These classification metrics provide a more holistic report on the performance between the two image classifiers by taking into consideration

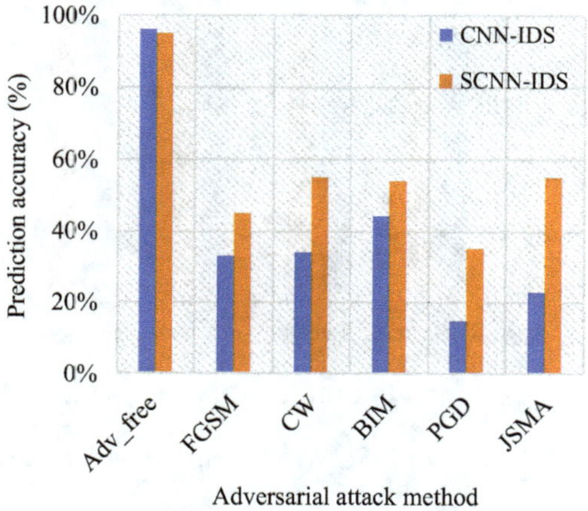

Figure 17.8 CNN vs. SCNN IDS AR

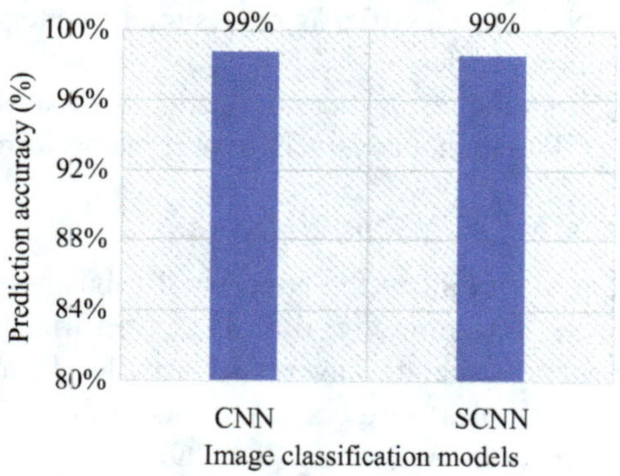

Figure 17.9 CNN vs. SCNN image prediction accuracy

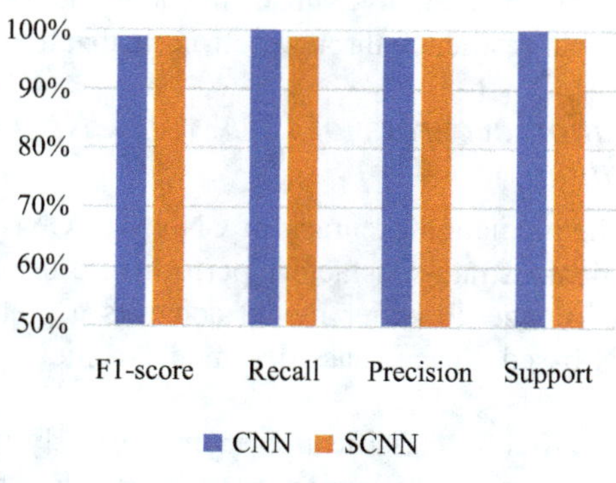

Figure 17.10 CNN vs. SCNN image classification accuracy

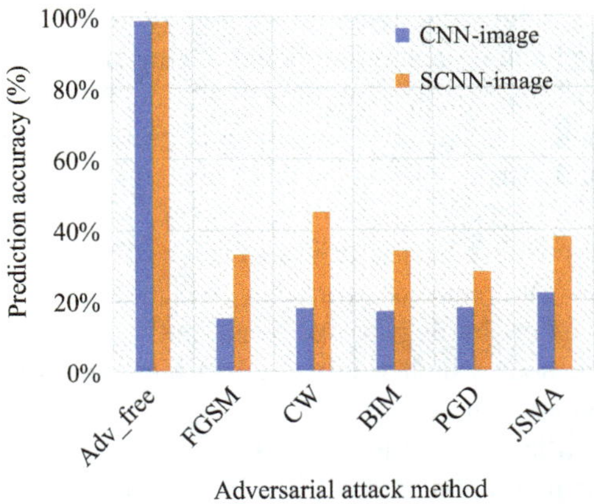

Figure 17.11 CNN vs. SCNN image classification adversarial resilience

other factors such as false positives and negatives which are considered crucial metrics in intrusion detection. Our results show that both the CNN and the SCNN image classifiers have comparable results across all four classification metrics as shown in Figure 17.7. The CNN image classifier is only slightly better than the SCNN image classifier based on the classification metrics.

We can conclude from the results here in Section 8.1 that the self-normalizing properties of the SCNN image classifier do not adversely impact the performance of the model.

17.8.4 AR of CNN vs. SCNN for image classification

In our final experiment, we evaluate the generability of our findings to other domains specifically computer vision by using our SCNN model to create an image classifier for the SocoFing Fingerprint Biometric dataset described in Section 7.3.2 above in terms of prediction accuracy, the SCNN image classifier shows similar results to the SCNN IDS classifier, having comparable results with the CNN models. We further evaluate the adversarial resilience of both CNN and SCNN image classifiers and the results are similar to that of the CNN and SCNN IDS classifiers. The results illustrated in Figure 17.11 show that the SCNN image classifier has comparable results in terms of prediction accuracy and significantly outperforms the CNN image classifier in terms of adversarial resilience.

17.9 Discussion

17.9.1 Comments on CNNs vulnerability to adversarial samples

Adversarial vulnerability in CNNs was previously attributed to non-linearity and over-fitting of the neural network models. However, on the contrary, more recent research has proven that linearity and the ability to generalize very well is a more compelling factors for the adversarial vulnerability in CNNs. Specifically, the authors of [13] argued that by exhibiting linear characteristics in high-dimensional spaces, neural network models become vulnerable to adversarial attacks. The corollary of this theory then is that to overcome adversarial weaknesses, neural network models should have attributes that implement significant effects of nonlinearity.

We attribute the increased adversarial resilience of the SCNN compared to the CNN as a result of the reduced invariance which occurs based on the self-normalizing properties of the SCNN. We importantly note that in this study, the adversarial attack methods which were launched included both model-based and model-agnostic adversarial attack methods.

Li *et al.* [34] utilized spectral analysis to explain adversarial behavior by implementing principal component analysis. Their findings explained the effect of non-linear transformation in the activations on adversarial vulnerability. The authors of [35] suggested that adversarial vulnerability is a consequence of narrow learning. Also, loss surface irregularity of deep neural networks has been shown to be a main cause of adversarial vulnerability [36]. While the loss surface improves the modeling power of the neural network models compared to linear models, it also makes them susceptible to adversarial attacks. This happens because a very minimal movement within the input space, which represents the amount of perturbation generated by the adversarial attack, will correspond to a relatively enormous increase in the loss experienced by the deep learning model. Other factors such as adversarial perturbation time and amount of perturbation were evaluated but there was no direct correlation with the adversarial vulnerability.

17.9.2 Why does self-normalization make SCNN perform better than CNN in the context of adversarial resilience?

Our study has demonstrated that self-normalization in a CNN-based IDS results in greater resilience to adversarial samples. Our proposed version of the CNN-IDS which we term SCNN-IDS uses a SELU in the dense layers as opposed to the ReLU which is commonplace in standard CNN models. The original intent behind self-normalization in CNNs was to enable the capability of high-level abstract representation [7]. This is important, though since the ability of a neural network model to generalize very well depends on its ability to capture high-level abstractions in the dataset [37].

The self-normalizing exponential linear units SELU have been shown to elicit three distinct properties. First, the SELU activation function sets to control the average learning rate (u) of the network using negative and positive values. Second, the SELU maintains a fixed point in the neural network with the aid of a continuous curve. Third, the saturation region dampens the variance alpha (learning rate) and the positive slope augments the alpha (learning rate) [7]. These three properties form the basis of the SELU activation function which sets the mean and variance of the activations at 0 and 1, respectively.

Since CNNs make decisions based on spectral statistical regularities [37], resulting in excessive invariance we argue that adversarial resilience can be reduced by reducing the invariance. We have investigated and found out that all these three properties exhibited by the SELU, including the ability to ensure the invariance of the deep learning model is kept at a minimum, is responsible for reducing the adversarial vulnerability of the SCNN-IDS as demonstrated in our study.

We observed that the adversarial robustness properties of SCNN are more pronounced for image classification than for IDSs. Future work will seek to understand what properties of images make them easier to defend and if this could be transferred to IDSs.

17.10 Conclusion

Self-normalization has been proven to increase the robustness of deep neural networks to vanishing gradient problems. Since adversarial samples take advantage of the irregularity of the loss surface of deep neural networks, we studied the relationship between this irregularity and self-normalization.

Leveraging on a standard CNN model, we proposed a deep learning-based IDS classifier which implements self-normalization within the Dense layers. In our study, we built a self-normalizing CNN IDS classifier (SCNN-IDS) and evaluated its adversarial resilience based on five adversarial attack methods. We compared the performance with a standard CNN IDS classifier (CNN-IDS) and established that the SCNN-IDS is more adversarial resilient. Our results show that self-normalization in the dense layers of a neural network increases the adversarial resilience of CNN models for intrusion detection classifiers. Furthermore, our findings for the deep learning-based IDS classifiers are consistent with image classifiers using the SocoFing fingerprint biometric dataset.

Furthermore, since we evaluated both model-based and model-agnostic adversarial attack methods, our study showed that both categories of adversarial attacks consistently impacted the neural network models which we tested. Hence, in this study, we also conclude that model or model-agnostic methods have no direct relationship with adversarial resilience through self-normalization. Future work would focus on optimizing the detection capabilities of the model to detect and mitigate sophisticated security attacks in beyond 5G and 6G wireless networks.

Acknowledgment

This work was supported by the Natural Sciences and Engineering Research Council of Canada (NSERC) through the NSERC Discovery Grant program. The work of Agbotiname Lucky Imoize is supported in part by the Nigerian Petroleum Technology Development Fund (PTDF) and in part by the German Academic Exchange Service (DAAD) through the Nigerian-German Postgraduate Program under Grant 57473408.

References

[1] Javaid A, Niyaz Q, Sun W, *et al.* A deep learning approach for network intrusion detection system. In: *Proceedings of the 9th EAI International Conference on Bio-inspired Information and Communications Technologies (formerly BIONETICS)*. ICST; 2016. p. 21–26.

[2] Guo W, Mu D, Xu J, *et al.* Lemna: Explaining deep learning based security applications. In: *Proceedings of the 2018 ACM SIGSAC Conference on Computer and Communications Security*. ACM; 2018. p. 364–379.

[3] Szegedy C, Zaremba W, Sutskever I, *et al.* Intriguing properties of neural networks. In *ICLR*; 2014. abs/1312.6199.

[4] Papernot N, McDaniel P, Jha S, *et al.* The limitations of deep learning in adversarial settings. In: *2016 IEEE European Symposium on Security and Privacy (EuroS&P)*. IEEE; 2016. p. 372–387.

[5] Wang Z. Deep learning-based intrusion detection with adversaries. *IEEE Access.* 2018;6:38367–38384.

[6] Ioffe S and Szegedy C. Batch normalization: accelerating deep network training by reducing internal covariate shift. In: *International Conference on Machine Learning*, 2015. p. 448–456.

[7] Klambauer G, Unterthiner T, Mayr A, *et al.* Self-normalizing neural networks. In: *Advances in Neural Information Processing Systems*, 2017. p. 971–980.

[8] Ibitoye O, Shafiq O, and Matrawy A. Analyzing adversarial attacks against deep learning for intrusion detection in IoT networks. In: *2019 IEEE Global Communications Conference (GLOBECOM)*; 2019. p. 1–6.

[9] Madasu A and Rao VA. Effectiveness of Self Normalizing Neural Networks for Text Classification. arXiv preprint arXiv:190501338. 2019.

[10] Zhang C, Costa-Pérez X, and Patras P. Tiki-taka: attacking and defending deep learning-based intrusion detection systems. In: *Proceedings of the 2020 ACM SIGSAC Conference on Cloud Computing Security Workshop*, 2020. p. 27–39.

[11] Abou Khamis R and Matrawy A. Evaluation of adversarial training on different types of neural networks in deep learning-based IDSs. In: *2020 International Symposium on Networks, Computers and Communications (ISNCC)*. IEEE; 2020. p. 1–6.

[12] Bottou L. Large-scale machine learning with stochastic gradient descent. In: *Proceedings of COMPSTAT'2010*. Springer; 2010. p. 177–186.

[13] Goodfellow IJ, Shlens J, and Szegedy C. Explaining and harnessing adversarial examples. In: *CoRR*. 2015; abs/1412.6572.

[14] Su J, Vargas DV, and Sakurai K. One pixel attack for fooling deep neural networks. *IEEE Transactions on Evolutionary Computation.* 2019;23(5):828–841.

[15] Carlini N and Wagner D. Towards evaluating the robustness of neural networks. In: *2017 IEEE Symposium on Security and Privacy (SP)*. IEEE; 2017. p. 39–57.

[16] Papernot N, McDaniel P, Wu X, *et al.* Distillation as a defense to adversarial perturbations against deep neural networks. In: *2016 IEEE Symposium on Security and Privacy (SP)*. IEEE; 2016. p. 582–597.

[17] Carlini N and Wagner D. Defensive Distillation is Not Robust to Adversarial Examples. arXiv preprint arXiv:160704311. 2016.

[18] Kurakin A, Goodfellow I, and Bengio S. Adversarial machine learning at scale. In: *ICLR*, 2017.

[19] Madry A, Makelov A, Schmidt L, *et al.* Towards deep learning models resistant to adversarial attacks. In: *International Conference on Learning Representations*, 2018. Available from: https://openreview.net/forum?id=rJzIBfZAb.

[20] Yuan X, He P, Zhu Q, *et al.* Adversarial examples: attacks and defenses for deep learning. *IEEE Transactions on Neural Networks and Learning Systems*. 2019;30:2805–2824.

[21] Tramèr F, Kurakin A, Papernot N, *et al.* Ensemble adversarial training: attacks and defenses. In: *6th International Conference on Learning Representations*, ICLR2018, 2018.

[22] Ross AS and Doshi-Velez F. Improving the adversarial robustness and interpretability of deep neural networks by regularizing their input gradients. In: *Thirty-Second AAAI Conference on Artificial Intelligence*, 2018.

[23] Farnia F, Zhang JM, and Tse D. Generalizable Adversarial Training Via Spectral Normalization. arXiv preprint arXiv:181107457. 2018.

[24] Garcia S, Grill M, Stiborek J, *et al.* An empirical comparison of botnet detection methods. *Computers & Security*. 2014;45:100–123.

[25] Kuo CCJ. Understanding convolutional neural networks with a mathematical model. *Journal of Visual Communication and Image Representation.* 2016;41:406–413.

[26] Chollet F, Zhu QS, Rahman F, *et al.* Keras. GitHub. Retrieved from: https://github.com/fchollet/keras.

[27] Glorot X and Bengio Y. Understanding the difficulty of training deep feed-forward neural networks. In: *Proceedings of the Thirteenth International Conference on Artificial Intelligence and Statistics*; 2010. p. 249–256.

[28] Srivastava N, Hinton G, Krizhevsky A, *et al.* Dropout: a simple way to prevent neural networks from overfitting. *The Journal of Machine Learning Research.* 2014;15(1):1929–1958.

[29] Kingma DP and Ba J. Adam: A Method for Stochastic Optimization. arXiv preprint arXiv:14126980. 2014;.

[30] LeCun YA, Bottou L, Orr GB, *et al.* Efficient backprop. In: *Neural Networks: Tricks of the Trade.* Springer; 2012. p. 9–48.

[31] Abadi M, Agarwal A, Barham P, *et al.* *TensorFlow: Large-Scale Machine Learning on Heterogeneous Systems*; 2015. Software available from tensorflow.org. Available from: https://www.tensorflow.org/.

[32] Nicolae MI, Sinn M, Tran MN, *et al.* Adversarial Robustness Toolbox v0. 2.2. arXiv preprint arXiv:180701069. 2018.

[33] Gong Z, Wang W, and Ku WS. Adversarial and Clean Data Are Not Twins. arXiv preprint arXiv:170404960. 2017.
[34] Li X and Li F. Adversarial examples detection in deep networks with convolutional filter statistics. In: *Proceedings of the IEEE International Conference on Computer Vision*; 2017. p. 5764–5772.
[35] Jacobsen JH, Behrmann J, Zemel R, *et al.* Excessive invariance causes adversarial vulnerability. In: *International Conference on Learning Representations*; 2018.
[36] Carlini N, Athalye A, Papernot N, *et al.* On Evaluating Adversarial Robustness. arXiv preprint arXiv:190206705. 2019.
[37] Jo J and Bengio Y. Measuring the Tendency of CNNs to Learn Surface Statistical Regularities. arXiv preprint arXiv:171111561. 2017.

Chapter 18

Legal frameworks for security schemes in wireless communication systems

Abdulwaheed Musa[1,2]

Abstract

Wireless communication is one of the very successful technologies that have found applications in our daily lives. It has drawn the attention of researchers, standard bodies, and organizations who have continuously proposed and developed different standards and regulations to further advance the communication system. However, one of the major issues and concerns that have been raised in wireless communication is security with its legal frameworks. Security as well as privacy are very crucial schemes in wireless communication because of the transmission of signals over unprotected media, thus exposing signals to security and privacy attacks such as eavesdropping, modification, and data theft, among others. Depending on the type of data being transmitted, security and privacy and the legal frameworks become even more critical, especially with the adoption of new technology such as cloud computing in the healthcare and other sectors. While framework is a structure that combines, in the form of a single hybrid conceptual solution, different but relevant areas together, the legal frameworks are simply a set of standards which can be utilized to deal with a challenge or to decide what to do. Some legal frameworks have been developed, over the years, particularly in the area of healthcare, Artificial Intelligence (AI) and Internet of Things (IoT). However, there are no sufficient legal frameworks for the security and privacy of wireless network, particularly for the envisioned 6G network. Thus, this chapter presents the fundamental factors of legal frameworks for the security and privacy in a wireless communication network with a focus on the 6G network.

Keywords: 6G networks; Wireless communication; Security and privacy; Legal frameworks; Ethics; Compliance; Conflict resolution

[1]Department of Electrical and Computer Engineering, Kwara State University, Nigeria
[2]Institute for Intelligent Systems, University of Johannesburg, South Africa

18.1 Introduction

Wireless communication has a plethora of benefits over its counterpart in terms of flexibility, scalability, increased data transmission rates, and low cost of deployment, among others. As a result, there is a significant number of end-users and applications of the wireless communication systems, like the cordless mobile phones, wireless local area network (WLAN), satellite communication, etc., which have in turn generated more revenue for the service providers [1]. Similarly, the numerous users of wireless systems have drawn the attention of various standard bodies and organizations who have continuously proposed and developed different standards and regulations as well as defined frequency ranges to further advance wireless communication systems. Over the years, the wireless communications have evolved from the pre-cellular era to the first generation (1G) down to the recently developed fifth generation (5G) network [2]. Researchers have started looking into the future which presents the sixth-generation (6G) network that has been envisioned to facilitate a higher data transmission rate as well as support several novel technologies and applications. However, because of the broadcast nature of wireless communication, concerns about security and privacy, as well as legal frameworks, have been raised.

Security and privacy are the two most crucial issues in wireless communication due to the transmission of signals over unprotected media, thus exposing signals to security and privacy attacks such as eavesdropping, modification, and data theft among others [3]. Depending on the type of data being transmitted, these issues of security and privacy become even more critical, especially with the adoption of new technology, such as cloud computing in the healthcare and other sectors [4]. Legal frameworks are simply a set of standards or rules which can be utilized to deal with a challenge or to decide what to do. On the other hand, ethics are a set of moral principles and standards that guide people's actions. Legal frameworks have been developed, over the years, particularly in the area of healthcare [4], Artificial Intelligence (AI), and the Internet of Things (IoT) [5], however, there are no sufficient legal frameworks for the security and privacy of wireless network, particularly for the envisioned 6G network.

Thus, this chapter aims to present legal frameworks for privacy and security schemes in a wireless communication network including the 6G network. The legal frameworks and innovative solutions to address the privacy and security challenges in 6G wireless networks becomes necessary because of increasing scale and complexity of the systems. Key emerging security and privacy legal frameworks to support massive devices and enabling technologies in the context of 6G wireless networks are established. The chapter presents security legal frameworks and principles, including compliance, data protection, quality of service, and conflict resolution, in 6G networks. The chapter underscores the importance of adherence to laws, regulations, and standards to maintain the privacy and security of 6G networks. Also, ethical and moral principles of 6G wireless network security are highlighted in this chapter, with focus, essentially, on protecting the 6G networks and systems from unauthorized access and malicious attacks, while emphasizing the importance of secure transmission, protecting user privacy, data integrity, accountability, monitoring and auditing, and adherence to best-practice security guidelines and standards.

18.1.1 Contributions

In this subsection, the contributions of the chapter are summarized and highlighted. These are in addition to recommendations for future work in the area of study. The contributions include the following:

- Comprehensive reviews of the security and privacy schemes in wireless communication systems from 1G network up to 6G network.
- Discussion of the foundational background on the evolution of various wireless communication systems.
- Presentation of the security schemes and framework requirements.
- New conceptual security legal frameworks in wireless communication systems, including 6G networks are presented.
- Ethics and moral principles for security and privacy in wireless networks, including the 6G network have been treated.

18.1.2 Chapter organization

Aside Section 18.1 which introduces the chapter, the rest of the chapter is structured such that Section 18.2 presents the evolution of wireless networks, including an overview of the 6G networks; while Sections 18.3 and 18.4 review the security and privacy schemes in wireless communication systems from 1G up to 6G networks. Section 18.5 presents the security framework requirements. In Section 18.6, the wireless network security legal frameworks are presented. Sections 18.7 and 18.8 present the legal as well as ethics and moral principles for security and privacy in the 6G network. Limitations of the study are provided in Section 18.9. Section 18.10 concludes the chapter and provides recommendations for further research directions.

18.2 The evolution of wireless networks

In this section, the previous generations of wireless networks are discussed. The section also presents an overview of the envisaged 6G network, the key enabling technology and some applications of the 6G network. Since the development of wireless communication in 1980, it has continually experienced evolution from 1G up to 5G. These generations have been discussed extensively in the literature where it was noted that the 1G network brought about the advent of mobile communication, having to convey users' voices from the source to the required destination. However, it had many challenges such as being very expensive, little or no roaming, no security and privacy, incompatibility among different technologies, limited to voice communication only, very large equipment, among others [2]. These challenges led to the development of second generation (2G) which was digital and supported both voice and data. As a result, numerous numbers of users started utilizing cellular communication networks. Among other advantages, the 2G provided the use of Short Message Signal (SMS), a feature that was not available in the 1G network. Also, the 2G network came with little roaming and compatibility with other technologies, as well as reduced cost compared to the 1G network. However, it had its challenges such as the inability to support video traffic, lower data rate, weak security, etc. [6,7].

The third-generation (3G) network was then launched to improve the 2G network by supporting voice, data, and video communication. The 3G network introduced technology, such as Massive Input Massive Output (MIMO), is not found in the previous generations. However, there were problems with expensive terminals and high-power consumption, among others [2,6]. The fourth-generation (4G) network was then launched to provide a higher data rate than the 3G network and support voice and video over IP networks as well as online streaming. The 4G network introduced new features such as the harmonious utilization of both Frequency Division Duplex (FDD) and Time Division Duplex (TDD) which increased the capacity of the network. There are other advantages of the 4G network such as higher throughput and simpler architecture, but there are still some disadvantages faced, such as the cost, and battery duration of devices, among others [8].

The recently developed 5G network utilizes new technologies, which include Visible Light Communication (VLC), Network Function Virtualization (NFV), Device-to-Device (D2D) communication, Software-Defined Network (SDN), Heterogenous Network (HetNet), massive MIMO (mMIMO), higher frequencies, network slicing, among others, to significantly improve data throughput, energy efficiency, connection density and reliability [9]. The key 5G use cases are ultra-reliable low latency communications (URLLC), enhanced mobile broadband (eMBB) and massive machine type communication (MTC). eMBB may provide high data speeds of 1G bits/s to mobile users, but URLLC deals more with communication reliability (99.999%) and latency (in milliseconds), particularly for applications such as IoT and vehicle-to-everything (V2X). Massive Machine-Type Communications (mMTC) emphasizes the number of interconnected devices in IoT deployment up to a million connections/km^2 [1,9]. This demonstrated the vast potential of the 5G network.

Figure 18.1 presents a summary of the evolution of wireless communication systems. Some essential features of different wireless generations, 1G to 5G, are shown in Table 18.1.

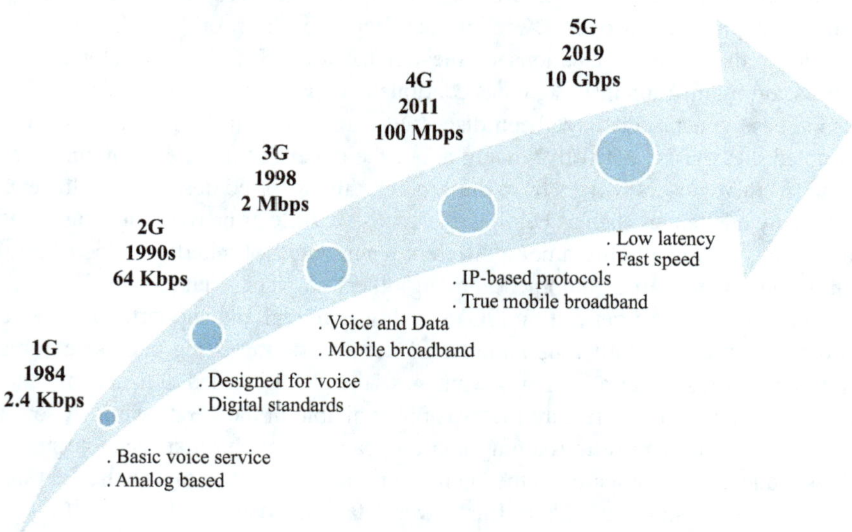

Figure 18.1 Evolution of wireless communication systems [10]

Table 18.1 A comparative of cellular generations

Generations	1G	2G	3G	4G	5G
Deployment	1979	1991	2001	2009	2018
Switching techniques	Circuit switching	Circuit switching	Circuit and packet switching	Packet switching	Packet switching
Access techniques	FDMA	GSM, TDMA, CDMA	WCDMA, UMTS, CDMA 2000, HSUPA/HSDPA	LTEA, OFDMA, SC-FDMA, WiMAX	BDMA, NOMA, FBMC
Application	Voice	Voice and data	Voice, data, and video calls	Voice, data, video calls, HD TV, etc.	Voice, data, video calls, Ultra HD video, VR application
Core network	PSTN	PSTN & packet	Internet	Internet	Internet
Data rate	2.4 kbps	14.4 kbps	384 kbps to 2 Mbps	100 Mbps to 500 Mbps	10 Gbps to 50 Gbps
Frequency band	800 MHz	800 MHz, 900 MHz, 1,800 MHz, 1,900 MHz	800 MHz, 900 MHz, 1,800 MHz, 1,900 MHz, 2,100 MHz	2–8 GHz	1.8 GHz, 2.6 GHz, 30–300 GHz
Merits	Mobility	Digital voice call, low power consumption	Better Internet experience, supports both voice and video calls	High data rate, body area network	Low latency, higher data rate, wider coverage
Demerits	Poor security, low capacity, low coverage	Low data rate, inability to support video transmission	Expensive terminals and high-power consumption	Expensive and high power consumption	Requires many small cell antennas

Recently, researchers have started envisioning the 6G network to improve the traditional wireless communications systems, increase service quality, and support massive data traffic demands. 6G networks seek to maximally increase data rates, minimize energy consumption, improve coverage and broadband connectivity, increase link dependability, reduce latency, and attain cognitive communication. 6G may offer exceptionally high data rates of over 100 Gbps with an end-to-end delay of less than 1 ms [11,12]. 6G is also projected to attain extraordinarily high communication dependability levels. The 6G criteria can be met by employing new advanced and intelligent communication systems. For instance, reconfigurable intelligent surfaces, new spectrum, extra-large MIMO, holographic radio communications, full-duplex wireless communications, modulation and multiple access are all crucial strategies for improving data rates. Furthermore, backscatter and energy harvesting techniques are both essential and required to improve energy efficiency. The merging of terrestrial and non-terrestrial communications and cell-free massive MIMO systems are viable methods for increasing connectivity and providing full coverage. AI and machine learning (ML) are critical strategies for attaining intelligence. Edge computing and holographic teleportation (telepresence) effectively create low-latency and ultra-reliable communications. Finally, quantum communication and blockchain are efficient strategies for increasing communication security, secrecy, and privacy [12–14].

18.3 Privacy and security schemes in wireless communication systems

This section presents the various security and privacy schemes in wireless communication systems from 1G to the envisioned 6G. Due to the broadcast nature of the wireless communication network, it is vulnerable to different security and privacy threats and attacks. These have drawn the attention of researchers, standard organizations and other stakeholders to continuously proffer countermeasures to these challenges. With the advancement of technology, there is a corresponding advancement in the security and privacy threats and attacks as summarized as follows.

One of the major limitations of the 1G wireless network, apart from being analog-based, was the lack of security as well as privacy. The cellular systems lacked authentication and cryptography which made call interception and impersonation attacks easy to conduct [15]. The 2G wireless network, being digital-based, addressed the security and privacy challenges witnessed in the 1G. However, it was susceptible to threats and attacks such as spamming, session hijacking, and fake base stations [16]. Other vulnerabilities witnessed in the 2G include the inability of the operators to be verified by the mobile stations, i.e. one-way authentication [15]. The 3G wireless network advanced from the 2G network and incorporated voice, video, and data, but it was susceptible to attacks and threats, such as denial-of-service, eavesdropping, location update spoofing, identity theft, man-in-the-middle attacks which infringe on the privacy of the users as the attacker accesses confidential information [17]. The 4G was IP-based and was developed to facilitate a higher data transmission rate than the 3G as well as other novel features, such as supporting high-dimension television (HD TV). However, this advancement of technology allowed for even greater security and privacy risks and challenges as the 4G network suffered from eavesdropping, transmission control protocol (TCP) flooding, denial-of-service, intrusion, address spoofing, man-in-the-middle, user identity theft, and malware spreading and authentication delay attacks [18].

The recently developed 5G wireless network facilitated a very high data transmission rate as well as other novel applications and technologies that have significantly advanced the wireless communication system as an entity. The network has a plethora of advantages and has found several applications in various sectors of our daily lives. However, it is also vulnerable to attacks and threats due to numerous applications and technologies it supports. Such threats include eavesdropping, hijacking, denial-of-service, network slice theft, configuration, man-in-the-middle, malware spreading, jamming, and hacking attacks [16,19–21]. The 6G wireless network is expected to be an AI-based autonomous network that would not only facilitate ultra-high data rate transmission with very low latency but also incorporate both space and underwater communication to form a ubiquitous network. However, similar to the 5G network, the 6G network would be susceptible to several threats and attacks such as eavesdropping attacks, jamming attacks, access control attacks, data modification attacks, 51% attacks, and Sybil attacks [22–25]. There are also some identified privacy challenges such as data theft and the misuse of users' data. Table 18.2 presents the taxonomy of some of the attacks and threats in cellular communication systems.

Table 18.2 Taxonomy of threats and attacks in cellular communication systems

Threats	**1G**	**2G**	**3G**	**4G**	**5G**
Eavesdropping	✓	✓	✓	✓	✓
Jamming	×	×	×	✓	✓
Impersonation or identity theft	✓	×	✓	✓	×
Spamming	×	✓	×	×	×
Fabricated BS	×	✓	✓	×	×
Spoofing	×	×	✓	×	✓
TCP flooding	×	×	×	✓	✓
Malware	×	×	×	✓	✓
Hijacking	×	×	×	×	✓
Configuration	×	×	×	×	✓
Network Slice theft	×	×	×	×	✓
Penetration	×	×	×	×	✓
Injection	×	×	×	×	✓
Sniffing	×	×	×	×	✓

Several schemes have been proposed to defend against these threats in all the cellular generations. Schemes, such as voice scrambling, were utilized to provide some level of protection against eavesdropping attacks in the 1G network. However, it does not have very strong encryption, making it possible for attackers to penetrate the network and perform call interception and other malicious activity [17]. Since the 2G network was digital, cryptography schemes, such as COMP128 and A5 cryptographic, were supported. They were used to provide radio-path encryption and user authentication, respectively, but they both have shown weaknesses [25]. For the 3G network, the 3G Partnership Project (3GPP) designed an Authentication and Key Agreement (AKA) scheme which was utilized for mutual authentication of a mobile station (MS) with the network, thus addressing the one-way authentication experienced in the 2G network [26].

Similarly, in the 4G network, the AKA protocol was also utilized to mutually authenticate users and the network as well as the combination of Internet Key Exchange (IKE) and Internet Protocol Security (IP-Sec) was utilized to protect the transmission between the core network and the base station (BS) [27,28]. As mentioned earlier, the overall 5G network vulnerabilities are linked to the vulnerabilities of the key enabling technologies. Thus, some privacy and security schemes have been proposed to secure these enabling technologies, and improve the privacy and security of the 5G network. Some of the schemes include enhanced version of the AKA scheme commonly known as the 5G-AKA, Physical Layer Security (PLS), Transport Layer Security (TLS), and Session Layer Security (SLS) which are used to control channel communication [29]. Furthermore, encryption schemes that were developed for the 5G network include the Elliptic Curve Integrated Encryption Scheme (ECIES) to protect users' identities and International Mobile Subscriber Identity (IMSI) encryption to defend against IMSI catchers [29].

The brief review has shown that security and privacy challenges have been the major concern in wireless communication systems which has attracted the attention of many researchers and stakeholders who have proposed schemes to defend against the threats and attacks. However, a wireless network cannot be completely secured as advancement in technology presents further vulnerabilities, thus there is a need for continuous development and evaluation of security and privacy schemes for wireless communication systems.

18.4 6G wireless network security schemes

The merging use cases for the envisioned 6G wireless network are shown in Figure 18.2. Although the deployment of the 5G radio networks with assured low latency, extraordinary dependability, and mass connectivity, is almost completed [30,31], the network will not be able to satisfy most needs, especially after 2030.

Thus, the impacts of the 6G network include but not limited to the following: an integrated air-ground-space-sea network (SAGIN) through implementation of both terrestrial and non-terrestrial networks [32,33]; increased data speed and network traffic capacity as a result of new radio bands, such as millimeter wave,

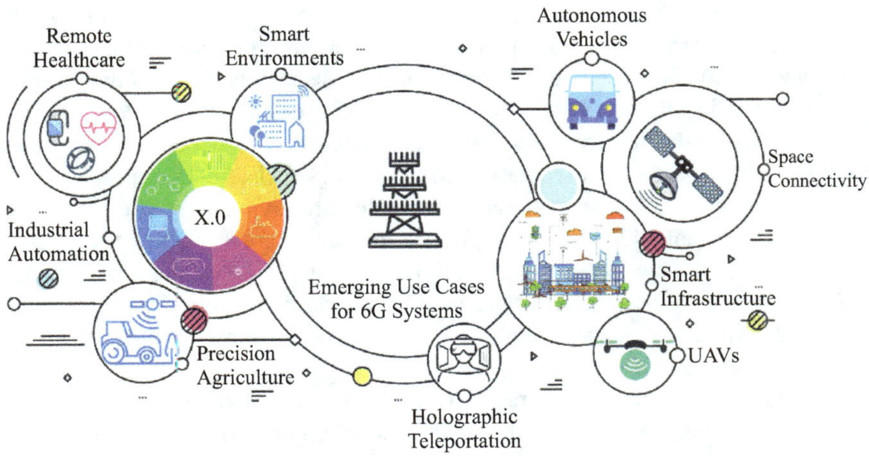

Figure 18.2 6G emerging use cases for 6G [1]

sub-6 GHz, and terahertz (THz); novel intelligent services and applications using AI and big data technologies [34–37]; and improved network security and privacy [38].

Although the privacy and security threats or attacks of the envisioned 6G network are somewhat similar to the 5G network, the 6G network technology is expected to come with better security and more spectrum coverage while using less energy. Using novel technologies, such as waveform design, multiple accesses, channel coding methods, network slicing, cloud edge computing, and various antenna technologies, 6G networks will be able to meet these demands [31].

To maintain secure communications, new security protocols must be developed to protect sensitive data and communications within these technologies. The 6G wireless network security scheme is a critical aspect of the development of 6G technology, which aims to provide a secure, reliable and fast wireless communication infrastructure for various applications. The security of a wireless network is essential to prevent unauthorized access, protect user data and privacy, and ensure the integrity of communication. In this context, several conventional security schemes such as encryption, authentication, and access control have been proposed for 6G networks. Other schemes that have been proposed to secure most of the technologies and applications in the 6G network include PLS-based schemes, quantum cryptographic schemes, Distributed Ledger Technology (DLT), and distributed AI-based security schemes. These schemes have been proposed as potential solutions to the privacy and security challenges in the 6G network and beyond [25,39,40].

The most crucial problems in 6G networks are thought to be threat detection, data processing, data encryption and traffic analysis. Security requirements for 6G use cases are more stringent than those for 5G use cases. It will be more difficult to operate and implement distributed AI security and privacy solutions as a result of the Internet of Everything's (IoE) large range of capabilities and services.

An important challenge in the development of a 6G security scheme is the integration of various security mechanisms that can provide comprehensive security solutions. For instance, encryption is an essential mechanism that ensures the confidentiality of user data, while authentication ensures that only authorized users can access the network. Similarly, access control is necessary to prevent unauthorized access to sensitive network resources [41]. The 6G wireless network security scheme should also consider the complex and dynamic nature of the network's environment [42]. For example, 6G networks are expected to be more diverse than previous generations, including different devices, users, and services. Also, the mobility of users and devices in 6G networks poses significant challenges to the security scheme, as it requires effective mechanisms to protect against unauthorized access [43].

To address these challenges, recent research has proposed the use of advanced security techniques, such as AI, ML, and blockchain. For example, ML algorithms can be used to detect and prevent security threats in real-time, while blockchain technology can provide secure and transparent data sharing between multiple parties. Additionally, the use of network function virtualization (NFV) and software defined networks (SDN) can give more efficient and flexible security solutions for 6G networks.

In summary, a secure and reliable 6G network security scheme is a critical challenge that needs a multi-disciplinary approach. The integration of various security mechanisms and the use of advanced technologies can provide comprehensive security solutions for 6G networks. However, more research is needed to tackle the specific security requirements and challenges of 6G networks, including the protection of user data and privacy, the integration of different security mechanisms, and the mitigation of security threats.

18.5 Security framework requirements

To protect sensitive information and critical systems from unauthorized access, theft, and damage, there is a need for set of guidelines, policies, and standards to ensure wireless networks security. The security framework requirements for wireless networks are crucial for ensuring the integrity, confidentiality and availability of information transmitted over the networks. The requirements include the implementation of various technical and organizational measures to prevent unauthorized access, modification, or destruction of data. The author of [44] provided an overview of the security framework requirements for wireless networks. The study highlights the need for risk assessment, threat modeling, and continuous security monitoring to ensure the security of wireless networks. It also highlights the need for organizations to adopt a proactive approach in identifying potential security threats and implementing appropriate security measures to mitigate the risks.

The security framework requirements aim at protecting the privacy and confidentiality of data transmitted over wireless networks and maintain the reliability and availability of the network services. These requirements also safeguard against unauthorized access and protect against potential security threats to networks. The framework requirements are a mandatory collection of policies and procedures that must be put in place to secure the confidential information of customers, subscribers, network operators, service providers, and public authorities. Subscribers and network service providers who require security to protect their devices, operation and business interests all need networks and asset protection. Some of the criteria that are important and required to protect data and information from threats are as highlighted.

18.5.1 Customers and subscribers

Customers and subscribers need to comply with certain security framework requirements to protect their data and devices. The author of [45] discussed the importance of customer education and awareness in securing wireless networks, and posited that customers should be aware of the security risks associated with wireless networks and should be provided with the necessary tools and information for security. This includes regular security updates, secure password management, and the use of anti-virus software. However, the requirements also include the implementation of secure authentication and access control measures to safeguard customer information and privacy, comprehensive encryption and safe storage of customer data, secured protocols for transmitting customer data, implementation of industry standard authentication procedures for secure customer account access, as well as monitoring and blocking malicious activity. Furthermore, subscribers can defend themselves against security risks by implementing multi-factor authentication on their accounts to greatly reduce the likelihood hacking. Also, subscribers must be cautious before clicking any links and use secure passwords since significant cases of successful cyber-attacks start with phishing.

18.5.2 Network service providers

To ensure the security of wireless networks, network operators and service providers have a significant responsibility in the design, deployment, and maintenance of the networks. They must ensure that the networks are secure and meet standards. The author of [46] highlighted security framework requirements for network operators and service providers to include the use of secure protocols, encryption, and secure authentication mechanisms. In addition, it is important to conduct regular security audits, monitor network activities, and respond promptly to security incidence. The following requirements are also necessary for wireless security: implementing secure authentication and access controls, data encryption and secure data storage for operator information, adopting secure protocols for transmitting data during transit, implementing industry standard authentication procedures for secure network access, monitoring and blocking malicious activities, as well as detecting and responding to network threats.

18.5.3 Public authorities

Public authorities play a crucial role to ensure the security of wireless communication networks by developing and enforcing legal frameworks that promote security. The author of [47] analyzed the security framework requirements for public authorities, including the development of data protection laws and regulations, the establishment of security standards, and the enforcement of penalties for security breaches. It also emphasized the importance of collaboration between public authorities and network operators and service providers to ensure the effective implementation of security framework requirements. As highlighted under subscribers and service providers' requirements, implementing secure authentication and access controls of public information, data encryption and secure data storage of public data, adopting secure protocols for transmitting public data, implementing industry standard authentication procedures for secure network access to public records, monitoring and blocking malicious activities, as well as robust incident response capabilities to effectively address security breaches are also necessary to promote the security legal frameworks for public authorities.

18.6 Legal frameworks for wireless network security

The extensive adoption of wireless networks has made the wireless network security to become a critical aspect. The widespread use of the network could be hampered by the lack of integrity, confidentiality, and security of data [48]. The security of wireless networks presents legal as well as ethics and moral issues. To address these issues and the increasing complexity of the wireless networks, such as the increasing threat of cyber-attacks, it is important for various stakeholders, including government, industry, and academia to collaborate for wireless networks security. Also, the implementation of standards, such as those established by the Institute of Electrical and Electronics Engineers (IEEE), and the management of third-party risk can play a crucial role to ensure the security of wireless networks. Nevertheless, legal frameworks play a key role in ensuring the security of wireless networks by providing guidelines on how to manage security risks and protect their networks and data.

Governments around the world, including the United States, United Kingdom, and the EU, have enacted laws and regulations to protect wireless networks [49] and safeguard data and networks [50,51]. These regulations focus on technical and administrative measures such as encryption, authentication protocols, and the responsibilities of the various parties involved in the network. In the United States, the Federal Communications Commission (FCC) implemented specific rules and regulations for wireless network security. These regulations aim to secure users from unauthorized access and interference, protect user data, and ensure the overall security of the network.

The FCC's rules and regulations for wireless network security mandate that all providers of radio frequency (RF) communication services ensure the security of their networks and protect user privacy. This act requires providers to secure the confidentiality, security, and integrity of user data, and to implement authentication and authorization measures such that only authorized users have access to the network [52].

It is also crucial for companies to be aware of the legal regulations governing their wireless networks and to have established policies to secure their networks. Various companies have thus established their own in-house regulations and processes to guarantee the safety of their wireless networks. These guidelines may touch on subjects such as access authorization protocols, user access control, network surveillance, and antivirus software. Thus, it can be concluded that the security of wireless networks is a multi-faceted and dynamic challenge.

The challenge of securing wireless networks is increased by the constrain of resources, thereby making it difficult to easily implement methods for privacy protection and attack mitigation in traditional networks for emerging wireless networks, such as 6G network. Therefore, there is a need for innovative response techniques of wireless network security and the legal frameworks.

18.7 Security legal principles

The concept of legal security underscores the systematization, stability, rule of law, and protection of human rights through the administration of justice. Security legal principles are the foundational laws and regulations that govern the protection of information, communication networks, and other related systems. These principles are designed to ensure the protection of individuals, organizations, and government entities from malicious cyber threats, unauthorized access, data breaches, and other security risks.

Until recently, wireless network security was only talked about to reduce the risks in data and the technical and organizational infrastructure that supported it. But technology advancement and awareness of the concerns over network security have changed that. The advancement has made it possible for individuals, groups, organizations and government to generate, collect, process, transmit, and store data from various sources and for different purposes. Thus, concerns over how secured what is collected are and the possible harm from privacy violations emanating from their unethical uses have also significantly increased. In recent years, the legal landscape for network security has rapidly evolved, and various organizations have published guidance and best practices to help ensure that security standards are up-to-date, consistent and relevant [53].

The IEEE addresses the importance of security legal principles that govern network security and in the development of 6G networks. A study in [46] highlighted the need for security legal principles in the design of 6G networks. It emphasizes the importance of including privacy and security considerations in the design of 6G networks, and of adhering to international privacy and security standards, such as the ISO 27001 and the NIST SP 800-53. Also, the authors of [54] emphasized the need for privacy policies and security measures that are compliant with international privacy laws and regulations, such as the General Data Protection Regulation (GDPR).

The General Data Protection Regulation (GDPR) provides the basic principles that form the framework for data protection. The fundamental tenets have a direct and indirect effect on the different requirements and restrictions found in the law. Some of the principles contained in the GDPR include the following: Lawful, fair, and transparent processing of personal data. There should be awareness on how personal information is collected, used, or processed. The principle of transparency entails that all information on personal data processing are clear and simple. Second, personal data should only be collected for explicit, specified, and legitimate purposes; this is as determined at the time of gathering of the personal data. Thirdly, processing of personal data must be pertinent and restricted to what is required. Only when there are no other reasonable options for achieving the processing goal can personal data be further processed. Restricting storage of personal data is another (i.e. fourth) principle; a time limitation should be set for deletion or reviews to ensure that data are not stored longer than necessary. Furthermore, processing of personal data should be in such a way that confidentiality and security are ensured, including protection against unauthorized and unlawful use or access, and guarding against accidental loss or damage. Lastly, adhering to the security principles and regulations are very vital, and integral parts.

By establishing appropriate legal measures, as well as security and private policies, the 6G network can be designed and operated to protect the security and privacy of user data. Adherence to security legal principles is a crucial aspect of maintaining the safety and security of data, systems, and networks for companies, governments, and other organizations. Compliance with these laws and regulations forms an integral part of any security plan. This section further pays attention to compliance, data protection, quality of service, and conflict resolution as they relate to the legal security of information.

18.7.1 Compliance

The principle of compliance requires organizations to follow applicable laws, regulations, and industry standards to ensure the safety of their systems and data [47]. This is particularly important for organizations operating in multiple jurisdictions that must comply with the laws of each country they operate in. Organizations should have a clear process for ensuring compliance and procedures for regularly monitoring and updating their efforts [47].

In the context of 6G wireless networks, it is important to consider the adherence to laws, regulations, and standards that organizations must follow to protect sensitive data and maintain the privacy and security of their systems and networks. Compliance with security and privacy regulations as well as national and international laws related to data privacy and cybersecurity, such as the California Consumer Privacy Act (CCPA) and the European General Data Protection Regulation (GDPR) is a critical aspect of any security framework.

Organizations must ensure that their security measures align with industry standards and government regulations to protect sensitive information and maintain customer trust. The author of [55] noted that compliance with regulations can be challenging, as the dynamic nature of IoT-enabled smart cities requires frequent updates to security and privacy measures. To address this challenge, it is suggested that organizations adopt a risk-based approach to compliance, regularly assessing and updating their security and privacy measures to ensure that they align with the latest regulations and standards.

18.7.2 Data protection

Data protection is another important principle of security legal frameworks [56]. Organizations must ensure that sensitive information is protected from unauthorized access and breach to secure and protect the integrity, confidentiality, and availability of sensitive data stored and transmitted over wireless networks. This includes implementing encryption and other security technologies, as well as establishing policies and procedures for access control, incident response, and risk management. Data protection laws vary by jurisdiction, but organizations must generally handle and store personal data securely and inform customers about data usage [57]. Organizations must also be transparent in their data handling practices and follow applicable regulations or standards [58].

In the 6G wireless networks, it will be important to consider new data protection requirements, such as privacy-enhancing technologies and the management of large amounts of data generated by IoT devices. The work in [59] underscores the importance of data protection in the 6G era and the need for innovative solutions to address the privacy and security challenges, especially as the traditional data management approaches are no more sufficient for the security and privacy requirements of 6G wireless systems due to increasing scale and complexity of the systems. To address these challenges, a blockchain-based subscriber data management protocol is proposed. This leverages the decentralized and secure nature of blockchain technology to protect the security and privacy of subscribers' data. The scheme involves the creation of a distributed ledger that records subscribers' data and transactions in a secure and tamper-proof manner, preventing unauthorized access or manipulation of the data.

18.7.3 Quality of Service

Quality of Service (QoS) is another crucial legal principle in wireless security, particularly for organizations handling sensitive information. It ensures that the security framework provides reliable and efficient performance, especially in mission-critical applications. It is important to consider how security measures may impact QoS and vice versa in the context of 6G wireless networks, as well as how to balance these considerations to provide both high levels of security and acceptable levels of performance for users. This requires an understanding of the potential trade-offs between performance and security, and the development of network design and management strategies that take these factors into account.

The European Network and Information Security Agency (ENISA) states that transferring data with minimum disruption and at desired speed and quality is critical to data security [60]. Organizations must ensure quick and secure data transfer to maintain information integrity, and ENISA recommends that they should be able to detect and respond to incidents promptly to efficiently address potential breaches.

18.7.4 Conflict resolutions

Conflict resolution is an important security legal principle that can stem from a variety of sources, including resource allocation, network configuration, and interference from other devices. It is important to ensure that conflicts and disputes between organizations and stakeholders are resolved fairly and successfully. One of the most important key components for a successful conflict resolution framework in 6G networks is the ability to quickly detect and identify conflicts as they occur. This can be accomplished through the use of monitoring and analytics tools that provide real-time visibility into the state of the network. The author of [61] discussed the importance of conflict resolution in the context of the IoT in a smart home environment. The study highlighted the importance of conflict resolution in IoT-enabled smart homes and proposed a rule-based conflict resolution framework to address the challenge of managing multiple devices in a smart home.

The EU General Data Protection Regulation (GDPR) approves that data controllers resolve disputes between themselves and data subjects efficiently [62]. The Global Network Initiative (GNI) promotes a set of principles and guidelines for companies to respect freedom of expression and privacy online, with conflict resolution as a key component [63]. The International Standards Organization (ISO) 27001 requires organizations to have a conflict resolution process in place to address disputes effectively [64].

18.8 Ethics and moral principles

Ethics are a set of moral principles and standards that guide people's actions and behavior. Every user in an organization is required to adhere to certain ethical standards and codes. Less disruption is expected in a company that conducts its activities in accordance with ethical code. Similarly, there will be negative effects on a company that does not uphold ethics. Ethics recognize right from wrong [65]. In computer security for instance, cyber-ethics is what separates security personnel from hackers.

The field of technology, including wireless network security, has a growing importance for ethics and moral principles. As technology continues to rapidly advance, it is important to ensure that its development and deployment align with ethical values. The author of [66] argues that it is important for individuals and organizations to consider ethical and moral considerations when making decisions about the development and deployment of technology and believes that ignoring these principles can have serious consequences, both for individuals and society as a whole. Also, policymakers and regulators must be involved in the development of ethical and moral frameworks, in order to ensure that technology is developed and used in ways that are consistent with the societal values.

The ethical and moral principles of the conventional security schemes and that of 6G wireless network security are similar. Essentially, they focus on protecting data, networks, and systems from unauthorized access and malicious attacks. Both emphasize the importance of protecting user privacy, data integrity, secure transmission, responsible use of technology, accountability, monitoring and auditing, user education, and adherence to best-practice security guidelines and standards.

The authors of [67–71] examine the various conventional security schemes used in wireless networks and their respective strengths and weaknesses. Conventional security measures, such as encryption, authentication and access control have been utilized to secure and protect networks against malicious attacks and unauthorized access. While these methods have proven to be effective, they are, however, with their limitations and are vulnerable to new and evolving security threats. Thus, with the rapid advancement of technology, there is a growing need for the development of more innovative and sophisticated security protocols and schemes to stay ahead of evolving threats and ensure security of the wireless network systems.

18.9 Limitations of the study

This study is not without limitations some of which are as presented thus. Only papers and resources written in English language have been consulted, thereby limiting the research scope to English literature. Also, it is possible to miss some relevant information, but this was minimized as much as possible. The topic covered in this study is multidimensional and interdisciplinary, and this was evident in unavailability of very rich text in the research context.

Lastly, security and its legal frameworks are rarely at the top of stakeholders' concerns, except to potentially comply with fundamental norms or regulatory obligations. Besides, security and privacy issues are typically addressed after the wireless system has been developed and deployed for use. It is therefore difficult to understand the true security requirements. Thus, explanations on protection mechanisms as a form of design solution are given instead of declarative statements about the protection required, even when security requirements are to be highlighted.

18.10 Conclusion and recommendations

Legal frameworks for security and privacy schemes in wireless communication systems have become a critical aspect of modern technology. The increasing need for real-time information exchange, complexity of wireless networks, threat of cyber-attacks and more have exacerbated the need for wireless network security, privacy and legal frameworks. To maintain secure wireless communication systems, security schemes and legal frameworks have been proposed to protect data and communications within these technologies. Furthermore, ethical and moral considerations must be incorporated into the design of the security schemes.

The requirements for the network security framework emanated from various sources which include the customers (i.e. subscribers), the service providers (i.e. network operators), and the public authorities. The customers require confidence in the wireless network and reliability of the services being offered. The network service providers require security of their business interests and operations, as well as meeting their customers and the public's obligations. As for the public authorities, they need security by legislation and orders for services availability, fair and free competition, and privacy protection.

It is recommended that the requirements for the security of wireless networks and services should be premised on international security standards. This will increase interoperability and prevent duplication of efforts. There is a need to balance the cost of security measures and the probable financial impacts of breaches in security since the security mechanisms and services are relatively expensive compared to the value of what are being secured and protected. Thus, it is necessary to have a way of customizing the provided security vis-a-vis the protected services.

Also, it is better to consider security profiles covering a range of wireless network services because of the high number of probable security features combinations. Besides, standardization will allow for reuse of solutions and products, as well as faster and low cost security. The economy of scale in component interoperation and product development within wireless networks in terms of security are important benefits of standardization both for the systems dealers and users.

The security services and mechanisms for wireless networks and operators are associated with protection against malicious attacks, such as eavesdropping, denial of service, spoofing, forgery, or tampering with messages in forms of modification, deletion, delay, replay, misrouting, re-routing, or re-ordering. Protection includes prevention, detection and recovery from threats and attacks, as well as management of personal or security-related data and information. Finally, in the framework, provisions to allow lawful and duly authorized interception are also expected.

References

[1] I. F. Akyildiz, A. Kak, and S. Nie, "6G and Beyond: The Future of Wireless Communications Systems," *IEEE Access* 2020, 8, 133995–134030, doi:10.1109/ACCESS.2020.3010896.

[2] M. H. Mahmud, "Cellular Mobile Technologies (1G to 5G) and Massive MIMO," *Int. J. Sci. Res.* 2019, 8, 929–937.

[3] V. L. Nguyen, P. C. Lin, B. C. Cheng, R. H. Hwang, and Y. D. Lin, "Security and Privacy for 6G: A Survey on Prospective Technologies and Challenges," *IEEE Commun. Surv. Tutorials* 2021, 23, 2384–2428, doi:10.1109/COMST.2021.3108618.

[4] Z. Zandesh, M. Ghazisaeedi, M. V. Devarakonda, and M. S. Haghighi, "Legal Framework for Health Cloud: A Systematic Review," *Int. J. Med. Inform.* 2019, 132, 103953, doi:10.1016/j.ijmedinf.2019.103953.

[5] E. Nehme, H. Salloum, J. Bou Abdo, and R. Taylor, "AI, IoT, and Blockchain: Business Models, Ethical Issues, and Legal Perspectives," In *Internet of Things, Artificial Intelligence and Blockchain Technology*; Springer, 2021; pp. 67–88.

[6] S. Patel, V. Shah, and M. Kansara, "Comparative Study of 2G, 3G and 4G," *Int. J. Sci. Res. Comput. Sci. Eng. Inf. Technol.* 2018, 3, 1962–1964.

[7] M. Arshad, A. Farooq, and A. Shah, "Evolution and Development towards 4th Generation (4G) Mobile Communication Systems," *J. Am. Sci.* 2010, 6, 6.

[8] N. Rawat, "Future and Challenges of 4G Wireless Technology," *Int. J. Sci. Eng. Res.* 2012, 3, 1–7.

[9] A. Shahraki, M. Abbasi, M. Piran, and A. Taherkordi, "A Comprehensive Survey on 6G Networks: Applications, Core Services, Enabling Technologies, and Future Challenges," 2021, *arXiv Prepr. arXiv2101.12475*.

[10] M. Attaran, "The Impact of 5G on the Evolution of Intelligent Automation and Industry Digitization," *J. Ambient Intell. Humaniz. Comput.*, 2021, 14, 5977–5993, doi:10.1007/s12652-020-02521-x.

[11] X. You, C. X. Wang, J. Huang, *et al.*, "Towards 6G Wireless Communication Networks: Vision, Enabling Technologies, and New Paradigm Shifts," *Sci. China Inf. Sci.* 2021, 64, doi:10.1007/s11432-020-2955-6.

[12] H. H. H. Mahmoud, A. A. Amer, and T. Ismail, "6G: A Comprehensive Survey on Technologies, Applications, Challenges, and Research Problems," *Trans. Emerg. Telecommun. Technol.* 2021, 32, e4233, doi:10.1002/ett.4233.

[13] M. H. Alsharif, A. H. Kelechi, M. A. Albreem, S. A. Chaudhry, M. Sultan Zia, and S. Kim, "Sixth Generation (6G) Wireless Networks: Vision, Research Activities, Challenges and Potential Solutions," *Symmetry (Basel).* 2020, 12, 676, doi:10.3390/SYM12040676.

[14] FG-NET-2030 Network 2030 – A Blueprint of Technology, Applications and Market Drivers Towards the Year 2030 and Beyond, 2019.

[15] F. Njoroge, and L. Kamau, "A Survey of Cryptographic Methods in Mobile Network Technologies from 1G to 4G," *Jomo Kenyatta Univ. Agric. Technol.* 2018, 10, 1–6.

[16] I. Ahmad, S. Shahabuddin, T. Kumar, J. Okwuibe, A. Gurtov, and M. Ylianttila, "Security for 5G and Beyond," *IEEE Commun. Surv. Tutor*. 2019, 21, 3682–3722.

[17] S. Sullivan, A. Brighente, S.A.P. Kumar, and M. Conti, "5G Security Challenges and Solutions: A Review by OSI Layers," *IEEE Access* 2021, 9, 116294–116314, doi:10.1109/ACCESS.2021.3105396.

[18] S. Mavoungou, G. Kaddoum, M. Taha, and G. Matar, "Survey on Threats and Attacks on Mobile Networks," *IEEE Access* 2016, 4, 4543–4572, doi:10.1109/ACCESS.2016.2601009.

[19] A. Celik, J. Tetzner, K. Sinha, and J. Matta, "5G Device-to-Device Communication Security and Multipath Routing Solutions," *Appl. Netw. Sci.* 2019, 4, 102, doi:10.1007/s41109-019-0220-6.

[20] I. Ahmad, T. Kumar, M. Liyanage, J. Okwuibe, M. Ylianttila, and A. Gurtov, "5G Security: Analysis of Threats and Solutions," in *2017 IEEE Conference on Standards for Communications and Networking, CSCN 2017* 2017, pp. 193–199, doi:10.1109/CSCN.2017.8088621.

[21] A. Sharma, V. Balasubramanian, and A. Jolfaei, "Security Challenges and Solutions for 5G HetNet," in *Proceedings – 2020 IEEE 19th International Conference on Trust, Security and Privacy in Computing and Communications, TrustCom 2020*; 2020; ISBN 9781665403924.

[22] Y. Siriwardhana, P. Porambage, M. Liyanage, and M. Ylianttila, "AI and 6G Security: Opportunities and Challenges," in *Joint 30th European Conference on Networks and Communications and 3rd 6G Summit, EuCNC/6G Summit 2021*; 2021; ISBN 9781665415262.

[23] P. Porambage, G. Gur, M. Osorio, M. Livanage, and M. Ylianttila, "6G Security Challenges and Potential Solutions," in *Proceeding of the of the Joint European Conference on Networks and Communications & 6G Summit (EuCNC/6G Summit)*; 2021; ISBN 9781665415262.

[24] M. R. Islam, M. M. Rahman, M. Mahmud, M. A. Rahman, M. H. S. Mohamad, and A. H. Embong, "A Review on Blockchain Security Issues and Challenges," in *Proceedings of the 2021 IEEE 12th Control and System Graduate Research Colloquium (ICSGRC),* Shah Alam, Malaysia; 2021; ISBN 9781665440110.

[25] M. Wang, T. Zhu, T. Zhang, J. Zhang, S. Yu, and W. Zhou, "Security and Privacy in 6G Networks: New Areas and New Challenges," *Digit. Commun. Networks* 2020, 6, 281–291, doi:10.1016/j.dcan.2020.07.003.

[26] R. Borgaonkar, L. Hirschi, S. Park, and A. Shaik, "New Privacy Threat on 3G, 4G, and Upcoming 5G AKA Protocols," *Proc. Priv. Enhancing Technol.* 2019, 2019, 108–127, doi:10.2478/popets-2019-0039.

[27] H. Omerani, and T. Mazri, "4G and 5G: Security and Privacy Analysis," in *Proceedings of the ACM International Conference Proceeding Series*; ACM: New York, NY, October 23, 2019; pp. 1–4.

[28] K. A. Alezabi, F. Hashim, S. J. Hashim, and B. M. Ali, "An Efficient Authentication and Key Agreement Protocol for 4G (LTE) Networks," in *Proceedings of the IEEE TENSYMP 2014 – 2014 IEEE Region 10 Symposium*; 2014; pp. 502–507.

[29] R. Khan, P. Kumar, D. N. K. Jayakody, and M. A. Liyanage, "Survey on Security and Privacy of 5G Technologies: Potential Solutions, Recent Advancements, and Future Directions," *IEEE Commun. Surv. Tutor.* 2020, 22, 196–248, doi:10.1109/COMST.2019.2933899.

[30] R. Khan, P. Kumar, D. Jayakody, and M. Liyanage, "A Survey on Security and Privacy of 5G Technologies: Potential Solutions, Recent Advancements, and Future Directions," *IEEE Commun. Surv. Tutor.* 2020, 22, 196–248.

[31] S. S. Ranasinghe and U. M. U. Jayawardena, "6G Wireless Networks: Key Drivers and Research Challenges," *IEEE Access* 2020, 8, pp. 154213–154230.

[32] A. Yazar, S. Dogan-Tusha, and H. Arslan, "6G Vision: An Ultra-flexible Perspective," *ITU. J. Future Evol. Technol.* 2020, 1, 121–140.

[33] C. Alwis, A. Kalla, Q. Pham, *et al.*, "Survey on 6G Frontiers: Trends, Applications, Requirements, Technologies and Future Research," *IEEE Open J. Commun. Soc.* 2021, 2, 836–886.

[34] P. Ray, N. Kumar, and M. Guizani, "A Vision on 6G-Enabled NIB: Requirements, Technologies, Deployments, and Prospects," *IEEE Wirel. Commun.* 2021, 28, 120–127.

[35] H. I. Obakhena, A. L. Imoize, F. I. Anyasi, *et al.*, "Application of Cell-Free Massive MIMO in 5G and Beyond 5G Wireless Networks: A Survey," *J. Eng. Appl. Sci.* 2021, 68, 13, https://doi.org/10.1186/s44147-021-00014-y

[36] A. L. Imoize, O. Adedeji, N. Tandiya, and S. Shetty, "6G Enabled Smart Infrastructure for Sustainable Society: Opportunities, Challenges, and Research Roadmap," *Sensors* 2021, 21, 1709, https://doi.org/10.3390/s21051709

[37] K. Sheth, K. Patel, H. Shah, S. Tanwar, R. Gupta, and N. Kumar, "A Taxonomy of AI Techniques for 6G Communication Networks," *Comput. Commun.* 2020, 161, 279–303.

[38] H. Yang, A. Alphones, Z. Xiong, D. Niyato, J. Zhao, and K. Wu, "Artificial-Intelligence-Enabled Intelligent 6G Networks," *IEEE Netw.* 2020, 34, 272–280.

[39] M. B. Gracia, V. Malele, S. P. Ndlovu, T. E. Mathonsi, L. Maaka, and T. Muchenje, "6G Security Challenges and Opportunities," in *IEEE 13th International Conference on Mechanical and Intelligent Manufacturing Technologies (ICMIMT)*; Cape Town, South Africa, 2022; pp. 339–343.

[40] P. Porambage, G. Gur, D. P. M. Osorio, M. Liyanage, A. Gurtov, and M. Ylianttila, "The Roadmap to 6G Security and Privacy," *IEEE Open J. Commun. Soc.* 2021, 2, 1094–1122, doi:10.1109/OJCOMS.2021.3078081.

[41] Y. F. Xiaoli Liu, "Towards 6G Networks: Key Technologies, Applications, and Open Issues," *IEEE Commun. Mag.* 2020, 58(6), 66–72.

[42] Y. Zheng Li, "Security and Privacy Issues in 6G Wireless Networks: Challenges and Solutions," *IEEE Commun. Mag.* 2020, 58(12), 50–57.

[43] W. Yuhang Zhang, "6G Wireless Networks: Vision, Requirements, Challenges, and Solutions," *IEEE Commun. Mag.* 2021, 59(1), 128–135.

[44] R. S. A. K. K. Kumar, "Securing Wireless Networks through Framework Requirements: A Review," *IEEE Trans. Depend. Secure Comput.* 2020, 17(4), 721–736.

[45] Y. L. Yajuan Zhang, "A Comprehensive Overview of Wireless Network Security," *IEEE J. Selected Areas Commun.* 2020, 37(9), 1925–1932.

[46] X. S. Jianfeng Chen, "Security Framework Requirements for Network Operators and Service Providers," *IEEE Trans. Mobile Comp.* 2022, 21(7), 1467–1475.

[47] Y. Liu, Y. Su, and Y. Yang, "Security Legal Principles in Cloud Computing: A Systematic Literature Review," *Int. J. Network Secur. Appl.* 2018, 10(6), 95–108.

[48] A. Kanuparthi, K. Ramesh, and S. Addepalli, "Hardware and Embedded Security in the Context of Internet of Things," in *Proceedings of the 2013 ACM Workshop on Security, Privacy & Dependability for Cyber Vehicles*, Berlin, Germany, November 2013, pp. 61–64.

[49] M. A. S. Zeki, "Wireless Network Security Legal Frameworks: An Overview," *Int. J. Comput. Netw. Commun.* 2018, 10(3), 1–9.

[50] G. M. Anaya, "Wireless Network Security: Legal Frameworks," *IEEE J. Commun. Netw. Secur.* 2017, 2(1), 1–10.

[51] E. I. Chuang and H. S. Hwang, "The Internationalization of Network Security Regulations: An Overview of Related Laws and Regulations," *IEEE Commun. Mag.* 2010, 48(8), 82–88.

[52] Federal Communications Commission, "Communications Act of 1934," § 222, https://www.fcc.gov/General/communications-act-1934.

[53] Y.-C. L. S.-Y. C. Hsien-Cheng Chen, "Security Framework Requirements for Public Authorities," *IEEE Commun. Mag.* 2023, 61(5), 96–103.

[54] J. L. Liu Xiong, "Privacy and Security Considerations in 6G Wireless Networks," *IEEE Commun. Mag.*2021, 59(2), 102–108.
[55] J. L, J. A. Hernandez-Ramos and V. S. M. A. V. N. A. F. S. G. B. Martinez, "Security and Privacy in Internet of Things-Enabled Smart Cities: Challenges and Future Directions," *IEEE Secur. Privacy* 2020, 19(1), 12–23.
[56] M. Khan, *Data Protection: A Practical Guide to UK and EU Law.* London: Kogan Page; 2015.
[57] B. Hutchinson, C. Mitchell, and P. Regan, *Data Protection Law*, London: Sweet & Maxwell, 2017.
[58] H. Zhang, H. Zhang, and Z. Deng, *Data Protection Law and Practice: An International Approach,* Cambridge: Cambridge University Press, 2019.
[59] X. A. W. Y. M. Z. J. W. Xueqiang Yan, "A Blockchain-based Subscriber Data Management Scheme for 6G Mobile Communication System," in *2021 IEEE Globecom Workshops (GC Wkshps)*, Madrid, Spain, 2021.
[60] ENISA, Quality of Service (QoS), 2018. https://www.enisa.europa.eu/topics/network-and-information-security/network-security/quality-of-service-qos.
[61] M. N. S. S. K. D. T. R. C. Y. L. Thinagaran Perumal, "Rule-Based Conflict Resolution framework for Internet of Things Device Management in Smart Home Environment," in *2016 IEEE 5th Global Conference on Consumer Electronics*, Kyoto, Japan, 2016.
[62] A. Khan, *Conflict Resolution in the Age of the GDPR*, 2019, https://www.itgovernance.co.uk/blog/conflict-resolution-in-the-age-of-the-gdpr.
[63] Global Network Initiative, *GNI Principles & Implementation Criteria*, 2018, https://globalnetworkinitiative.org/principles-implementation-criteria/
[64] ISO, ISO/IEC 27001:2013, 2020. https://www.iso.org/standard/57867.html.
[65] S. Rahman, "Ethics and Philosophy of Security," in J. McPherson (Ed.), *Security and Privacy: Foundations, Technologies and Applications*, IGI Global, pp. 24–52, 2020.
[66] W. James Alexander Hughes, "We Are Not Pontius Pilate: Acknowledging Ethics and Policy," in *2020 IEEE Symposium Series on Computational Intelligence (SSCI)*, Canberra, ACT, Australia, 2021.
[67] V. B. Kumaravelu, A. L. Imoize, F. R. C. Soria, *et al.*, "Outage Probability Analysis and Transmit Power Optimization for Blind-Reconfigurable Intelligent Surface-Assisted Non-Orthogonal Multiple Access Uplink," *Sustainability* 2022, 14, 13188, https://doi.org/10.3390/su142013188.
[68] K. B. Letaief, W. Chen, Y. Shi, J. Zhang, and Y. Zhang, "The Roadmap to 6G: AI Empowered Wireless Networks," *IEEE Commun. Mag.* 2019, 57, 84–90.
[69] G. Gui, M. Liu, F. Tang, N. Kato, and F. Adachi, "6G: Opening New Horizons for Integration of Comfort, Security, and Intelligence," *IEEE Wirel. Commun.* 2020, 27, 126–132. "Conventional Security Schemes in IoT: A Survey," 2021.
[70] S. Saxena, "Security Techniques in Wireless Networks," *Wireless Commun. Mobile Comput.*2005, 5(6), 563–576.
[71] M. Alsabah, M. A. Naser, B. M. Mahmmod, *et al.*, "6G Wireless Communications Networks: A Comprehensive Survey," *IEEE Access* 2021, 9, 148191–148243, doi:10.1109/ACCESS.2021.3124812.

Chapter 19

Design of a quantum true random number generator using quantum gates and benchmarking its performance on an IBM quantum-computer

Vaishnavi Kumar[1] *and Padmapriya Pravinkumar*[1]

Abstract

Random numbers are used in various domains, including quantum communication and cryptography applications such as key generation and authentication. Quantum mechanics has the intrinsic ability to generate truly random numbers, making it an ideal alternative for scientific applications that require randomness. Quantum wireless communication is proposed as a new means of communication that is safe, secure, and efficient. The simplicity of our source rotation gate, combined with its independently verifiable solitary unpredictability, is essential to obtain quantum random number generators without spending any money. In this chapter, we explore and study the 24-qubit random number. The worst-case entropy value for such randomly produced integers is 0.999445, with the min-entropy of such numbers being 0.0008. Additionally, steering restart tests were used to validate later verification of true randomness, and the statistical characteristics of the True Random Number Generation (TRNG) were assessed using an autocorrelation study and the statistical test suites.

Keywords: Quantum random number; Key generation; Entropy; Qubit; IBM-QISKIT

19.1 Background

The idea of randomness has intrigued and puzzled people [1] in every task they handled. We have placed ourselves in the hands of chance throughout history, from reading the inner workings of birds and practising divination in antiquity to rolling dice in casino games to conducting risk assessments and stock market investments

[1]SREE SASTRA Deemed University, India

in the present. The idea of chance and randomness was initially discussed and connected to divinity by ancient Greek thinkers [2]. Since the development of probability theory, randomness has been a topic of ongoing discussion. In the interim, its utility was gradually discovered in several scientific and industrial domains [3].

The study of randomness became even more crucial as modern cryptography advanced since it became evident that producing reliable randomness is essential in almost all information security techniques. However, this ostensibly simple task is challenging. Numerous mathematics and physical approaches have been put forth and used, but only a few numbers are thought to be entirely fulfilling. Moreover, while the quality of randomness creation is a problem, most current communication protocols also call for higher generating speeds. Therefore, one of the foundational elements of creating a secure and effective communication system is having a quick, dependable random number generator that generates truly random numbers. This numerous motivated researchers in the arena [4].

19.1.1 Random numbers

Everybody can comprehend the idea behind a random number: a coin toss generates one random bit with two equally likely outcomes. Since there is no way to foresee the outcome in advance, we refer to it as random [5].

Nevertheless, despite the straightforward concept, providing a good description of random numbers is challenging. D.H. Lehmer pointed out two key characteristics that a collection of random numbers should have in 1951: changeability and a lack of organization in its statistics. First, a random number series should be impossible to anticipate because knowing the previous numbers should not reveal anything about the following number. The string should, nevertheless, appear random and have statistics that are on par with those of truly random series. It is possible to make even stricter arguments that the first feature implied by the second is true: identifying a specific statistical arrangement in the numbers would create the sequence more predictable. Still, a random number sequence with these characteristics should at least be appropriate for most applications found in daily life [6]. It is important to note that additional information theory and computer science initiatives exist. Kolmogorov, Martin-Lof, and Chaitin have all produced studies that explain randomness in terms of computational complexity [7,8].

19.1.2 Importance of randomness

Randomness connotes unpredictability, disorder, and lack of direction. A sequence that lacks any regularities and patterns is said to be statistically random [9]. A number is said to be random when selected randomly from a particular distribution. These numbers almost always have no relationships with preceding numbers and are most likely independent. Random processes have a long, illustrious history. Ancient peoples often used dice to predict their fate. The majority of prehistoric tribes used a variety of divination techniques to try to avoid uncertainty and randomness [10].

Random events are known to occur in nature. For instance, it is impossible to accurately anticipate the physical outcome of an electron being scattered by an atom. Additionally, random events can be produced artificially, and there is no distinction between randomness occurring naturally and not. When artificial random generators are used in place of natural ones, it appears that a succession of random events can be produced. The result of an unexpected event can frequently be translated into a numerical value [11]. As an illustration, a photon that meets with an atom may scatter or trigger other changes inside the bit. There is no practical method to give the alternative alterations a numerical value corresponding to them. Random numbers are uniform distribution random variates. More complicated distributions can be produced with uniform variates, rejection, or inversion techniques. A standard and fundamental definition of a random number series is that each number is independent of the numbers generated before it and is uniformly dispersed among all possible values.

19.1.3 Applications of random numbers

Even in epics, we encounter games with random results based on specific gadgets, proving that the demand for unpredictability is eternal. In reality, lottery games and gambling of some kind were practised by all the ancient civilizations. The use of random numbers has expanded throughout time, with new applications appearing in fields as diverse as encryption, statistical sampling [3,9], and Monte Carlo simulation [12,13]. Unknowingly or not, we frequently employ the results of random number generators in our daily lives. For instance, passwords, pins issued by banks, and CAPTCHAs [14,15] displayed on your screen when you attempt to log in to a website are all supposed to be the output of random number generators [16].

19.1.3.1 Application in cryptography

Cryptography is conceivably the one area where random numbers are used the most. The notion of random numbers is well reflected in the use of random numbers in cryptography, which also offers some guidelines for creating random number generators.

19.1.3.2 One-time pad

One-time pad (OTP) is a cryptographic scheme created in the early twentieth century by Gilbert Vernam and Joseph Mauborgne.

The protocol is straightforward:

1. Binary-code the plaintext before encoding it.
2. Create a key using random bits with a length equal to the plaintext.
3. By bitwise XORing the plaintext and key, you can get the ciphertext.
4. The receiver receives both the key and the ciphertext separately.
5. By accomplishing another bitwise XOR between the ciphertext and the key on the receiving side, the ciphertext is decrypted.

It should go without saying that the OTP is a mechanism of encryption that heavily relies on creating random integers. In that instance, the OTP is regarded as resistant to ciphertext assaults because the encrypted message will look entirely unexpected, provide no information about the plaintext, and take a lifetime to decipher by an observer. The entire plan bottleneck is the generation rate.

19.1.3.3 Key generation

A somewhat extreme protocol created for maximum security is the OTP. Although safe, the protocol is not helpful in most real-world situations. Only those who possess the proper pair of keys may effectively encrypt or decode messages. Eavesdroppers will have to guess the contents of the keys if they do not have access to them to decipher the ciphertext. The intention of the algorithms and, more significantly, the selection of keys determine the protocol's security.

19.1.4 Quantum randomness in cryptography

It is considered secure if someone other than the intended recipient cannot understand the information. The unintended person, called the hacker, is attempting to steal the data. Such data is known as encrypted data in the field of cryptography, which is a branch of cryptology. The confidentiality of the information is the main consideration in the design of the cryptosystem, which also includes security measures to prevent unwanted access. A method for creating encryption and decryption keys is the KSA. The original or user-defined key is a group of bits depending on processing steps or rounds. The encryption algorithm used 40, 128, 192, or 256. The KSA's goal is to make the key so robust that it cannot be broken and prevents hackers from discovering the original key [17].

The process of converting data into an unintelligible format and encrypting it using a key is known as encryption. On the other hand, a decryption algorithm entails utilizing the same or a different key to decode the cypher and turn it back into the original data.

Encryption converts plaintext into ciphertext and secure cyphers by combining several cryptographic qualities like nonlinearity, propagation criteria, correlation, and algebraic immunity. The security of a cipher is determined by the degree of the key susceptible to a cryptanalysis attack. Random numbers can be employed to increase the strength and complexity of the key in addition to KSA, which breaks the key down into subkeys. The random numbers or bits required to achieve randomness in a cryptosystem are produced by random number generators. By using a random number generator-based cryptosystem, encryption transforms plain text into ciphertext; the KSA key also uses random number bits created by a random number generator [18].

19.1.4.1 Method of quantum computing

1. Most approaches to building a quantum computer are based on one of three techniques for controlling quantum particles.
2. Manipulating a nucleus or subatomic particles spin. The superposition of up and down spins yields a qubit.
3. Altering the electrical charge. A qubit is created by superimposing two or more distinct charge sites.

4. Altering a photon polarization or phase. Phase or polarization angle superposition is used to create qubits.
5. Nuclear magnetic resonance

Nuclear spin control is the foundation of a nuclear magnetic resonance quantum computer. This has been accomplished by adjusting the nuclear spins of atoms within a molecule. Spin is affected by magnetic pulses within an NMR chamber magnetic field. The chemical bonds between nearby atoms form the entanglement of spins needed to establish a qubit. Scalability is the primary flaw.

19.1.4.1.1 Ion trap

Nuclear spin control is also the foundation of an ion trap quantum computer. An electromagnetic field can trap a single ion or even small groups. Theoretically, this method is scalable, but it needs a cryogenic setting. Only single qubit systems have been demonstrated up to this point.

19.1.4.1.2 Quantum dot

Electrical charge, spin, or energy state manipulation may occur in a quantum dot computer. A quantum dot is often a tiny molecular mound on a silicon substrate that is 180 nm or smaller and contains a small number of electrons or potentially only one electron. Such dots would be arranged in a predictable pattern to form a computer. This approach appears to have the best chance of being commercially saleable.

19.1.4.1.3 Josephson junction

In a Josephson junction, electron cooper pairs are tunneled via a thin semiconducting layer that is sandwiched between two layers of superconducting material. Voltages applied across the junction can change the likelihood of which side of the sandwich the electron pairs will occupy. There have been demonstrated two-qubit operations. The production of these junctions is seen as scalable. This idea is the foundation of the most promising system now in use.

19.1.4.1.4 Optical

The polarization or phase of each photon is the foundation of an optical quantum computer. By simultaneously creating identical photons, entanglement is made possible. Polarizing lenses, phase shifters, and beam splitters are used to manage the superposition of polarization phase states. Theoretically, this technique can handle several qubytes.

19.1.5 Quantum information processing

The analogous data unit in quantum computing systems is the qubit. It is capable of simultaneous computations since it can exist in both states (0 and 1) at any given time. Superposition is the idea that a particle can exist in two states simultaneously. Quantum entanglement is a crucial idea in quantum computing [19]. In correlation, photon particles interact at some time and can be entangled in pairs (similar to a couple of magnets interacting with each other). For example, we may determine the spin state of the second entangled photon particle by determining the spin state of one of the photons. As a result, quantum entanglement enables instantaneous interaction between qubits separated by vast distances.

The combined effects of superposition and quantum entanglement boost computing capability. When qubits are expanded, capacity increases. As a result, quantum computers are better at solving problems than conventional computers. As a result, enormous volumes of information may be stored and processed in very few storage units.

By using the idea of polarization, it is feasible to exclude a portion of the light waves oriented geometrically. An essential part of quantum key distribution is photon and light polarization. So far as processing and storage speeds are concerned, quantum computing is a game changer. Quantum computing can provide better answers more quickly, helping to maximize profit.

19.1.6 Highlights of the proposed work

- Utilized the rotation gates Ry and Rz for the quantum true random number generator.
- The random number generator is made out of 24 qubits.
- Non-repeatability is ensured through the restart analysis and the statistical autocorrelation analysis.
- Min entropy and worst-case entropy are calculated.
- Post-processing is not employed in the generation process.
- Randomness verified through NIST 800-22 test.
- The entropy source is verified through NIST 800-90B test.

In this chapter, Part 2 reviews the literature survey for the proposed system, Part 3 adds the necessary information regarding the gates and system imparted in the proposed system, Part 4 gives a detailed description of the method of generation of the proposed quantum true random number generator, Part 5 gives the result analysis of the designed random number generator, and Part 6 gives the conclusion and the future scope.

19.2 Literature survey

19.2.1 Methods of generating random numbers

Naturally, creating random number generators is of great interest to scientists and technologists, and in recent years, a variety of RNGs have been developed and suggested [20,21]. The criteria of homogeneity and independence should ideally be followed by the output of the random number generator. The strength of the cryptographic algorithms employed in modern computers depends on the randomness of the seed, which is used to seed the key. Since the encryption process is deterministic, attackers who can predict or restrict the random seed's range will have an easier time decrypting the data.

Therefore, random number generation is the cornerstone of contemporary cryptographic security. Unfortunately, many encrypted systems have been rendered useless by poor RNGs [22]. A random number generator should produce bits that are independent of one another and have an equal chance of being 0 or 1. Additionally, a random number generator must be unpredictable, with no way to predict the results. Currently, there are two primary methods for producing random numbers. Both hardware and pseudorandom number generators. Usually, the random seed is created by combining true random number generator (TRNG) and pseudorandom number generator (PRNG):

1. Pseudorandom number generator
2. True random number generator

19.2.2 Survey of pseudorandom number generators

PRNGs use a sequence of mathematical processes to produce random numbers. However, PRNGs are deterministic and can be anticipated if the PRNG's state is known [23]. PRNGs need inputs known as seeds to increase predictability. Because the seed influences the sequences produced by PRNGs, PRNGs alone cannot make genuinely random data [24]. The requirement for randomness and unpredictability in the seed itself leads us back to the original issue. In a procedure known as entropy input, PRNGs are frequently seeded with an HRNG to address this issue. It would seem that utilizing PRNGs defeats the purpose of using them since one already needs a random number to start the process because PRNGs need to be seeded with a random number [25].

However, PRNGs are advantageous because they can generate numbers faster than HRNGs and produce superior statistical features for unpredictability. The Blum–Blum–Shub (BBS) generator, Mersenne twister, and linear congruential generator are a few well-known pseudorandom number generation algorithms that are entirely based on arithmetic techniques. The most typical applications of these techniques are in generic programming and Monte Carlo simulations [26]. The pseudorandom number generator's deterministic nature raises the possibility that results may be foreseen, which is not suitable for cryptography. Pseudorandom numbers are not random, strictly speaking.

By using the information from a physical process, HRNGs generate random numbers. Typically, sensors gather chaotic signals like background noise from the environment to generate a random number. HRNGs frequently rely on methods like camera data pointing towards an entropic scene that is challenging to simulate and model. It is unavoidable that random data has a significant level of entropy, but this does not imply that the data is truly random; various criteria other than entropy are advised for cryptographic RNGs [27]. Other HRNG techniques draw information from sources like disc drive timing data, mouse movement, and keyboard delays. Since these techniques lack true randomness, they must be sent through a randomness extractor before being accepted for use in cryptographic standards. HRNG also generates quantum random numbers, but many quantum random number-generating techniques frequently need randomness extraction. The issue of creating an HRNG that is impossible to manage, compute, or predict persists.

With sufficient knowledge of the beginning state, it is theoretically possible to identify HRNGs that depend on processes prescribed by classical physics.

Unfortunately, fundamentally unpredictable non-deterministic quantum processes exist. Quantum processes in identical stacked states will not necessarily yield the same result from repeated measurements. To put it another way, quantum systems are essentially random. Therefore, the final version of HRNGs and potentially the answer to the RNG puzzle will be quantum random number generators.

19.2.2.1 Middle square method

The middle square technique was the initial idea Von Neumann and Metropolis offered for an arithmetic generator. Using the middle digits of the square of the preceding number, the following number in the sequence is generated using this procedure. The middle digits of earlier numbers serve as the seed value for the next number. Numerous researchers have explored this approach and discovered that it has not always yielded good results.

19.2.2.2 Linear congruential generator

It is one of the earliest methods, and it is quick at the same time as being incredibly simple to learn and put into practice. An easy linear equation based on modulo arithmetic can be used to generate a pseudorandom sequence. Mathematicians often use modulo arithmetic, in which integers wrap around after they reach a specific value. A recurrence relation is used to define the linear congruential generator:

19.2.3 Physical random number generator

A hardware random number generator, commonly referred to as a physical random number generator, produces random numbers via physical operations. They take measurements of noisy physical systems and transform the data into random bits rather than utilizing deterministic techniques. Randomness has been attributed to various physical processes, including ambient noise and radioactive decay events. In addition, events involving human activity may potentially be beneficial sources of randomness. For example, a unique device file called /dev/ random, included in many Unix-like operating systems, acts as a random number generator by gathering unpredictability from mouse movements, keyboard timings, and other potential environmental disturbances [28].

19.2.4 Survey of true random number generators

Bruce Schneier claims that a truly random sequence is one that not only meets the criteria for a pseudorandom sequence but also cannot be reliably duplicated. A real random sequence generator will produce two completely distinct random sequences if you run it twice with the same input. True random number generators generate randomness by sampling and digitizing real-world physical phenomena and entropy sources. Physical principles ensure the generated values' unpredictability [29].

Such physical phenomena include background radiation, thermal noise, jitter, photonic emission in semiconductors, radioactive decay and quantum vacuum. Entropy here refers to the degree of uncertainty surrounding a result. Entropy, which makes a real random number generator unpredictable, is produced by physical processes. However, there would be a limit to how quickly random events could be recorded and generated [30]. Most real random number generators struggle significantly with their poor generation rates. The fact that these generators depend on hardware is another significant drawback. They require physical instruments to record the event because they utilize real-world phenomena. Because of this, they are creating real random generators can be very expensive, especially if the required hardware is not widely available. These systems often rely on microscopic events like thermal noise and the photoelectric effect, which uses a beam splitter to produce low-level, statistically surprising noise signals. These stochastic processes are completely unpredictable in theory. A transducer is frequently used in hardware random number generators to transform physical phenomena into electrical signals. An amplifier and other electronic gear boost the random fluctuations' amplitude to a quantifiable level. Some analogue to digital converter turns the output into a digital number, often only the binary digits 0 or 1 be represented mathematically. Random numbers can be collected by periodically sampling the randomly fluctuating signal [31].

19.2.5 Unpredictable random number generators

Unpredictable random number generators (URNGs) are one of the practical TRNG approximations. The intricacy of the underlying phenomenon causes unpredictability, which is the foundation of URNGs. These generators typically take randomness from readily accessible sources, such as computer parts, unstable internal processor states, human–computer interaction, race situations and so on. As a result, they may offer a high level of randomness. A detailed understanding of the underlying phenomenon is fundamental because the components are deterministic, suppressing the ease of prediction of internal states of the features and, consequently, upcoming values.

19.2.6 Quantum random number generator

Quantum random number generators (QRNGs), a specific type of physical random number generator, get their randomness from the unpredictable outcome of a quantum measurement [27]. The fundamental assumptions of quantum mechanics contain the idea of randomness. A measured quantum system will randomly enter one of its potential eigenstates and produce the matching result. These measurement results can be truly non-deterministic, the authors of [32,33] confirmed by studies, making them excellent candidates for random number generation.

Semi self-testing

1. Trusted
2. Self-testing

19.2.6.1 Trusted device QRNGs

Trusted device QRNG assumes that the sources and devices utilized to extract randomness are accurately described. They are suitable for practical applications because of their simplicity in theory and implementation. However, they can only be applied if the performance closely matches the model. A trusted device is a complicated experimental setup because it is challenging to manage quantum systems. These tools cannot confirm if the output bits are truly random or under the control of an enemy. It is referred regarded be a trustworthy gadget because of this. You could select suitable quantum protocols to produce random numbers if you are confident that your gadget is operating as intended. The categories of non-optical and optical trusted device QRNGs will be used to categorize trusted device QRNGs in the following sections. Several possible implementations for each sub-class are briefly detailed below regarding their advantages and disadvantages. Let us first talk about other non-optical reliable QRNG solutions that might be used.

19.2.6.2 Non-optical trusted device QRNGs

The categories of non-optical and optical trusted device QRNGs will be used to categorize trusted device QRNGs in the following sections. Several possible implementations for each sub-class are briefly detailed below regarding their advantages and disadvantages. Let us first talk about other non-optical reliable QRNG solutions that might be used.

19.2.6.3 Radioactive decay

One of the earliest quantum processes employed to produce random numbers was the radioactive decay of particles [18]. Only the uncertainty principle of quantum mechanics can explain radioactivity. The radioactivity-based random number generators used the sensitive Geiger–Muller (GM) tubes and well-known radioactive sources of α, β, and γ radiations. Radiation detectors are much more extensively utilized since they are much simpler [34]. GM tubes use a Townsend avalanche to generate a pulse for each detected particle.

The special rate depends on several variables, including the half-life of the sample, its position, and the gas condition in the GM tube. The pulses have a Poisson distribution. There are two approaches to producing random numbers from the arrival of pulses at random [35]: (1) the Fast clock method [36], where the clock frequency is higher than the mean detection rate; and (2) the Slow clock method, where the counts occur more frequently than a clock cycle. The unpredictability in the arrival time is transformed into bits by counting the number of clock cycles between two consecutive clicks. The fast clock reads and resets itself to zero after each detection, and the resulting time is utilized to produce a random number [37]. By using the count parity (odd/even), we may more uniformly distribute the results. The least significant digit or the coefficients of 2^0 in binary form should be used if the counts are taken in binary. For instance, if there are two counts, the binary representation of that number is "10," and the resulting random number is "0." In binary, the generated number for an odd number, such as "3" or "11," would be "0." The slow clock method produces random numbers by counting how many times a fixed amount of time passes. A modulo counter should be used to limit the number of counts to an integer N. Furthermore, distributions with tiny arbitrary bias [18] can be obtained by adding various counts modulo. Some contemporary radioactive decay-based random number generators employ semiconductor devices rather than GM tubes because they are more practical and work at lower voltages. As a result, the output signals can still be amplified further, even though they are weaker. In addition, they streamline the generators' design. Interestingly, the proposals produce uniform distribution using an RC circuit that converts an exponential or Poissonian random variable to one with a uniform distribution.

The usage of QRNGs based on radioactive sources is severely constrained by these disadvantages. The unpredictability source must be handled with extreme caution and expertise because it is radioactive. The radioactivity can also impact the detectors, reducing their effectiveness over time. Another restriction is the detector dead time brought on by an accumulation of ions inside the detector. After a successful detection event, GM tubes and semiconductor tubes require some time to regain their full detection capabilities due to the detector's dead time. This must be considered when producing random bits and the necessary post-processing processes.

19.2.6.4 Electronic noise

Entropy can also be obtained from the noise in electronic circuits to extract unpredictability. Typical electronic noise-based random number generators derive their entropy from circuit components like resistors or diodes. The charge carrier's quantum nature is often responsible for the noise. Once more, this is a result of the quantum mechanical uncertainty principle. Once it has been created, the noise can be amplified and used for extraction. A straightforward technique would be used to test the noisy element voltage output against a threshold value to produce random bits. Utilizing time of arrival techniques is an additional strategy [36].

Shot noise and thermal noise are two general noise categories in such systems. Shot noise develops because current carriers are quantum objects, making it a probabilistic phenomenon. Shot noise is produced when carriers tunnel through p–n junction-created quantum barriers to create reverse or leakage current. Under a sufficiently low current, the voltage peaks across a diode represent the hopping of quantum carriers. The motion of the carriers in reaction to the surrounding temperature, however, produces thermal noise. Since shot-noise is an actual quantum process, it would be ideal for extracting randomness from it. However, extraction of randomness from shot noise suffers from this flaw since it is challenging to disentangle these effects in practice.

Furthermore, as voltage spikes depend on past charge flow in the diode and therefore affect the quality of the extracted randomness, they also have memory effects. Nevertheless, under the right circumstances, shot-noise effects have been known to manifest predominately in Zener diodes and transistors. Moreover, commercial QRNGs exploit shot noise as a source of entropy [38].

19.2.6.5 Atomic systems

Apart from the approaches mentioned above for quantum random number generation, there have been proposals for random number generation using atomic systems. For example, trapped ions were used for random number generation [39].

However, the experimental setups needed for random number generation using trapped ions are much more complex, and such arrangements produce random numbers with low generation rates. Still, various proposals for generating random numbers using atomic systems have been reported [40]. Specifically, a QRNG based on the spin noise of an alkali metal vapor was proposed. The spin noise arises due to inherent quantum uncertainty due to the interaction between different systems of atoms. The signal acquired from such a process has been proved to arise from quantum noise, not optical pumping effects.

19.2.6.6 Optical trusted device QRNG

Optical QRNGs take advantage of the quantum nature of photons for generating random bits. Optics-based protocols are more accessible to implement than the methods mentioned above due to the ease of equipment availability and extensive research that has already been performed for various other purposes. Sources of entropy in this class of QRNGs are the light emitted from LASERs, LEDs, and single-photon sources. The light is then manipulated using optical elements and eventually measured. Different protocols utilize different aspects of the quantum nature of light to extract randomness. We classify the different optical QRNGs based on the type of detector used as QRNGs using single-photon and QRNGs using macroscopic detectors [41].

19.2.6.7 QRNGs using single photon detectors

The devices belonging to this class utilize single-photon detectors for their functioning. Various techniques can be used to construct single-photon sources [42,43].

19.2.6.7.1 Qubit state

These devices generate randomness by measuring qubits in superposition. The principle behind these devices lies in quantum theory axioms, including collapse upon measurement. Specifically, the quantum state of a photon can be in a superposition of possible paths. For example, if a photon is passed through a beam splitter. It exists in the superposition state of the state on the reflected side and the state on the transmitted side [44]. Similar is the case when a linearly polarized photon inserts on a polarizing beam splitter, a device that transmits horizontally polarized light and reflects vertically polarized light. At the output, the photon exists in a horizontal and vertical polarization superposition. Any quantum system can live in a superposition of the basis states. One can easily design a QRNG using this quantum property of photon along with the collapse on measurement postulate that states that a quantum state would collapse to one of the basis states upon measurement. In the computational basis representing the absence or presence of the photon, we can define a state which means one photon in the first path and no photon in the second path and a state with the photon in the second path and no photon in the first path.

19.2.6.8 Self-testing QRNGs

The methods for creating random numbers that have been detailed thus far rely entirely on the reliability of the tools being utilized. However, realistic scenarios do not support this. In actual use, the gadgets might not work properly or might even be under enemy control [45]. Therefore, some testing system must be in place to confirm the random number generator output sequence. They can be modeled to keep track of the generator's internal state and alert the user if an unexpected failure or skewed output occurs. However, the same issue arises because the user is now expected to believe the checking device. However, trusted device QRNGs cannot even distinguish the impacts of classical noise from the generated numbers without an appropriate checking mechanism; thus, it is better to incorporate at least some self-testing methods in terms of security.

19.2.6.9 Semi-self-testing QRNGs

Some components of the random number generator may be better characterized than others. In reality, one might believe certain aspects of the device while ignoring information regarding other factors. This is especially important if our only issues are with loud channels and faulty devices, not adversarial situations. Such device security is between fully self-tested and trusted devices. The choice is between self-testing QRNGs with high credibility but low performance and practical, trusted devices that offer excellent performance at low cost and low credibility [46].

Noisy intermediate-scale quantum (NISQ) computers, like the IBM quantum processor based on superconducting technology, have been the focus of current scientific research efforts [47]. This research shows how to use quantum rotation gates to create a 24-qubit quantum true random number generator. We demonstrate the random number generation using Python scripts in Qiskit [48]. The quantum circuits are executed on actual quantum computers with reasonable agreement. The incorrect performance of the superconducting devices offered by IBM is linked to the experimental departures from the initial findings.

19.3 Preliminaries

The following parts describe how quantum computing is relevant to business and services. In this chapter, we lay down the foundations of quantum information processing. First, we present the most common notation used in quantum physics. Then, we define a qubit, how to compose larger systems from qubits, how the state of a system can be changed, and how we can learn something about the state using a measurement.

19.3.1 Dirac notation

Quantum mechanics uses the Dirac notation for vectors, also called the bra-ket notation. The vector v is denoted by v>. The inner product between u> and |v> is denoted by <u|v>. We sometimes denote the inner product by (|u>, |v>) because it can make expressions easier to read. The outer product between |u> and |v> is symbolized by |u><v|.

19.3.2 Quantum system

The state can be applied to calculations in the manner listed below. The system starting state is encoded in the input. The state is then evolved, and a measurement obtains output. The quantum system used most often in quantum computing is called a qubit.

19.3.3 Qubit

A qubit is a quantum system with a two-dimensional Hilbert space associated with it called H. We work only with this mathematical abstraction. The physical qubit can then be realized in various ways. We define {|0>, |1>} as an orthonormal basis of H. This basis is called the computational basis. The state of a qubit is a vector |v> C H |v> = a|0> + b|1>, Where $| a |^2 + | b |^2 = 1$. Vectors where represent the same state as vector |v>. They differ only in their global phase, which is physically insignificant. It means that vectors yield the same measurement statistics and cannot be distinguished. Two often used states with their unique label are states represented in (19.1):

$$|+>=\frac{1}{\sqrt{2}}|0>+\frac{1}{\sqrt{2}}|1>,|->=\frac{1}{\sqrt{2}}|0>-\frac{1}{\sqrt{2}}|1> \tag{19.1}$$

States |+>, |-> form an orthonormal basis.

A qubit is the fundamental unit of computing in a quantum computer. A quantum gate, also known as a quantum logic gate, is a quantum circuit that can operate on multiple qubits. A qubit function $|q_0\rangle$ can be represented as in (19.2) and (19.3):

$$|q_0\rangle=\frac{1}{\sqrt{2}}|0\rangle+\frac{i}{\sqrt{2}}|1\rangle \tag{19.2}$$

$$|0\rangle=\begin{bmatrix}1\\0\end{bmatrix}|1\rangle=\begin{bmatrix}0\\1\end{bmatrix} \tag{19.3}$$

A qubit can be symbolized in a normal state as in (19.4):

$$|\Psi\rangle=\cos\left(\frac{\theta}{2}\right)|0\rangle+e^{i\Psi}\sin\left(\frac{\theta}{2}\right)|1\rangle \tag{19.4}$$

19.3.4 Bloch sphere

The Bloch sphere may be employed to depict the state of a qubit. We can always rewrite the vector |v> up to the global phase as in (19.5):

$$\left|v>=\cos\frac{\theta}{2}\right|0>+e^{i\phi}\sin\frac{\theta}{2}\left|1>\right. \tag{19.5}$$

Angles θ, ϕ with the term $e^{i\phi}$ is called the relative phase. Contrary to the global phase, the relative phase can be determined by measurements. The Bloch sphere representation does not have any extension to multi-qubit systems. The tensor product combines vector spaces into a larger vector space. Suppose we have n qubits with their two-dimensional Hilbert spaces H. The vector space of the composite scheme is then H^n = H .... H.

The state of a 2-qubit system with both qubits in states |0> is |0>|0>. To make the notation more compact, we write tensor products also as |0>|1> |0>|1> |01>. To make this notation even more concise, we often write labels in decimal base, e.g. |101> |5>.

The computational basis of an n-qubit system can be written as in (19.6):

$$\{|\mathrm{i} > \mathrm{iC}\{0, \ldots.., 2^n - 1\}\} \tag{19.6}$$

Product states may be expressed as the tensor product of two 1-qubit states. Entangled states are those that cannot be written in this manner. The state is

$$\frac{1}{\sqrt{2}}|00> + \frac{1}{\sqrt{2}}|11> \tag{19.7}$$

Equation (19.7) is an example of an entangled state. The state

$$\frac{1}{\sqrt{2}}|00> + \frac{1}{\sqrt{2}}|01> = |0> \otimes \left(\frac{1}{\sqrt{2}}|0> + \frac{1}{\sqrt{2}}|1>\right) = |0> \otimes |+> \tag{19.8}$$

Equation (19.8) is an example of a product state.

19.3.5 Evolution of a quantum system

A closed quantum system is a system that does not interact with its surroundings. If we assume that qubits form a secure quantum system, then the development of their state is labeled by a linear transformation. Since the resulting state must be normalized, this transformation must be unitary. If we start in a state |v>, then after the state evolves according to unitary U, it will end up in the state U|v>. Often we want to apply a unitary U only to one qubit and leave other qubits unchanged, and can be achieved by using the unitary $I^{\otimes i_1} \otimes \mathrm{U} \otimes I^{\otimes i_2}$ to the state of the system, where $i_1 + i_2 + 1$ is the number of qubits, and U is applied to $(i_1 + 1)$th qubit. Unitary operators can be composed of more straightforward unitary operators. In quantum computing, we are interested in sets of unitary operators that can decompose arbitrary unitary operators.

19.4 Proposed method

The IBM quantum lab is the primary source for generating genuine quantum random numbers. The suggested context aims to quantify the unpredictable and difficult-to-reproduce random number sequence. The Rotation gate Ry and Rz is used in this technique to produce random numbers. There are two sections to the suggested method. While segment 2 covers the design of the random number generator, segment 1 describes quantum programming with Qiskit. In this study, the Rotation gates will be used for measurement in the ibmq qasm Simulator.

Segment 1

19.4.1 Qiskit quantum programming

IBM Qiskit Python unit ropes quantum programming for simulators and actual devices. The built-in functions of Qiskit can be used to design quantum registers and circuits. Quantum programme run via joining to the IBM hardware in the real world. Invariably, a regular computer coexists with a quantum gadget.

19.4.1.1 Role of Qiskit

It converts conventional programming language into quantum machine language and provides tools for creating and modifying quantum programmes. It combines the physical quantum device and quantum algorithms to execute quantum processing. The actual quantum gadget is programmed using quantum languages in Qiskit, known as open quantum assembly language (OpenQASM). Even though software varies, minor syntactical variations with new version announcements do not drastically modify the product's functionality.

Segment 2

19.4.2 Scheme of random number generator

In our decorum, the single-qubit gate $RY\left(\frac{\pi}{2}\right), RZ\left(\frac{\pi}{2}\right)$ *and* $P\left(\frac{\pi}{2}\right)$.

The RY $\left(\frac{\pi}{2}\right)$) gate can be articulated by square matrices in (19.9),

$$\text{i.e., RY}\left(\frac{\pi}{2}\right) = \begin{pmatrix} \cos\left(\frac{\pi}{4}\right) & -\sin\left(\frac{\pi}{4}\right) \\ \sin\left(\frac{\pi}{4}\right) & \cos\left(\frac{\pi}{4}\right) \end{pmatrix} \tag{19.9}$$

The RZ $\left(\frac{\pi}{2}\right)$ is stated with square matrices in (19.10):

$$\text{RZ}\left(\frac{\pi}{2}\right) = \begin{pmatrix} e^{-i\frac{\pi}{4}} & 0 \\ 0 & e^{i\frac{\pi}{4}} \end{pmatrix} \tag{19.10}$$

The P $\left(\frac{\pi}{2}\right)$ gate can be uttered by square matrices in (19.11):

$$\text{P}\left(\frac{\pi}{2}\right) = e^{i\frac{\pi}{2}}\begin{pmatrix} 1 & 0 \\ 0 & 1 \end{pmatrix} \tag{19.11}$$

In the quantum computer, the initial state of a qubit is usually organized in $|0>$ and can be represented with a state vector as $[1\ 0]^T$. By applying RY $\left(\frac{\pi}{2}\right)$, RZ $\left(\frac{\pi}{2}\right)$, P $\left(\frac{\pi}{2}\right)$ gate, the superposition state $|+>=\frac{(|0>+|1>)}{\sqrt{2}}$ can be acquired as in Figure. 19.1. Based on the mathematical hypothesis of quantum mechanics, if the quantum computer is noiseless and the quantum operations are flawless, the measurement results in the computational basis $|0>$ and $|1>$ should be equally random. In contrast to traditional and pseudorandom number generators, which need random seeds to generate separate sequences from the same root. Because all possible outputs

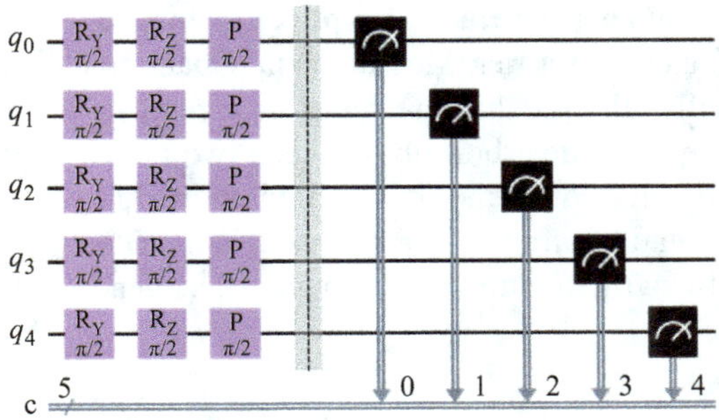

Figure 19.1 Random number generator of 5 qubits

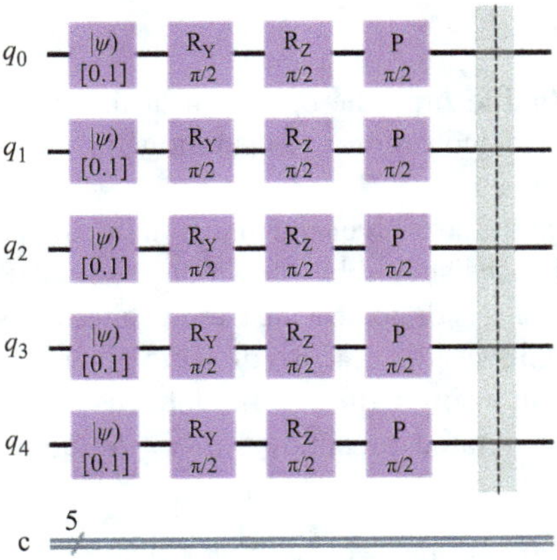

Figure 19.2 State vectors for 5 qubits random number generation

originate from the same source, there is less chance that a random number generator's production can be predicted.

Unpredictability is one of the most fundamental features of quantum mechanics that underlie reality. A qubit must be put into superposition to generate a random bit, after which its measured value will likely be either zero or one. Linear algebra and Dirac notation can explain the quantum state and the use of quantum gates in a circuit.

Statevector ([0.125−0.125j, 0.125−0.125j, 0.125−0.125j, 0.125−0.125j, 0.125−0.125j, 0.125−0.125j, 0.125−0.125j, 0.125−0.125j, 0.125−0.125j, 0.125−0.125j, 0.125−0.125j, 0.125−0.125j, 0.125−0.125j, 0.125−0.125j, 0.125−0.125j, 0.125−0.125j, 0.125−0.125j, 0.125−0.125j, 0.125−0.125j, 0.125−0.125j,0.125−0.125j, 0.125−0.125j, 0.125−0.125j, 0.125−0.125j, 0.125−0.125j, 0.125−0.125j, 0.125−0.125j, 0.125−0.125j, 0.125−0.125j, 0.125−0.125j, 0.125−0.125j, 0.125−0.125j],dims=(2, 2, 2, 2, 2))

A state vector simulator is used for the measurement of the state vectors for the random number generators for 5 qubits as a sample which is indicated in Figure 19.2.

1. A true random number generator comprises Qubits set at state "0," Rotating operators *Y*, *Z*, and Ph, Barrier, Measurement, Quantum registers, and Classical registers. Initially, all 24 qubits are set to state zero.
2. Transform the qubit state arbitrarily between two points on the Bloch sphere.
   This is accomplished by the unitary transformation of gates which correspond to state rotations around different axes in the Bloch sphere. The rotation operator $R_x\left(\frac{\pi}{2}\right)$ is written with the SU(2) gate. The general SU(2) gate is given in (19.12):

$$U(\theta, \phi, \delta) = \begin{pmatrix} \cos\left(\frac{\theta}{2}\right) & -ie^{i\delta}\sin\left(\frac{\theta}{2}\right) \\ -ie^{i\phi}\sin\left(\frac{\theta}{2}\right) & e^{i(\delta+\phi)}\cos\left(\frac{\theta}{2}\right) \end{pmatrix} \text{ with } \phi = 0, \delta = 0, \theta = \frac{\pi}{2} \tag{19.12}$$

3. The barrier is applied to limit the operation of the superposed qubit state; the barrier is used as the indication of separating the functions on the quantum registers.
4. Since all the 24 qubits are given on the individual quantum register wires, 24 measurement devices should be shown on the registers.
5. Once the measurement is done, the measured states are stored on the classical registers. No of qubits used = 24, No of shots given = 65,536 ("shots" is the no of times it runs in a single time), Simulator used = qasm Simulator (IBM lab), No of bits generated at single run = 24 qubits × 65,536 shots = 1,572,864.

Because of cloud computing, we can now collaborate with pioneers in quantum computing. One of these innovators is IBM Quantum Experience, which allows clients to design, polish, and construct their circuits using actual quantum computers and simulators. However, the physical realizations of the circuits are constrained by the designs of these devices [49]. Furthermore, IBM QX comes with features that enable users to project and use various trials for various claims. In this scenario, cloud quantum computing has been used to create random numbers. This study uses the IBM QX to design a new QRNG circuit. As a result, both the IBM QX simulator (also recognized as the Qasm simulator) and the actual quantum processor are incorporated.

The measurement outputs of the superposition state paired with the computational base are supposed to create random number sequences, according to quantum physics. Accordingly, the designed arrangements would satisfy the statistical requirements for RNGs indicated earlier. In this instance, the two-dimensional Hilbert space is covered by the computational framework. A quantum computer can create the needed state by applying the rotation gate to one quantum bit. It is crucial to remember that this process always begins in the same place. Since every possible output comes from the same seed, there is less chance that a random number generator output can be predicted.

19.5 Testing random number generators statistically

It is crucial to authorize the random number generator statistical tests for encryption techniques. In this case, the random number generators used must be unpredictable. This test discovers whether the random number generators produce substantial bias or correlation sequences.

A random number generator is seen as a "black box" when employed in statistical testing. One would anticipate the generator output to have particular characteristics under the null hypothesis that it is impartial and independent. The test statistic, whose probability distribution is infamous, counts the features of the output. The test statistic tells us how likely a genuinely random number generator will provide a result with a test statistic value below one. The *p*-value refers to this probability. Statistical tests are often designed as test suites to be comprehensive. Well-known test suites include NIST SP 800-22 and NIST SP 800-90B, for instance.

If a biased random number generator were used, statistical tests would fail because they are designed to look for statistical irregularities below the assumption that the generator is unbiased. It can be hard to test quantum random number generators since these machines can be biased and unexpected. Since statistically faulty generators can provide incredible results, the structure of statistical tests falls short of expressing the core of randomness: unpredictability. Quantum inequalities have been used to ensure that unpredictability exists, but they have not yet advanced to the point where they can fully replace statistical tests.

19.5.1 Restart experiment

A renowned appropriate analysis is used to differentiate between actual randomness and pseudo-randomness. The result of the TRNG is that it will consistently produce different sequences instead of the PRNG, which always generates identical sequences for a given seed. This experiment was conducted to demonstrate the proposed QTRNG's genuine unpredictability. For the analysis, 128 starting bit sequences are recorded.

S1 01001110001011011111000000010000010010100010010011100100010 001010111100000010101001111101100011000101011101

S2 10010111001110111110011101000010010100001110111111010011000 110011000010111110100010100111100011011110001011

S3 01011100101100100001111111011010011100100000111010100100100 010101111000011011111110100011001000100100111001

S4 11010110111100101010110110111001100011110000010000100110000 001111001010111101110010010001101110000111110001

S5 01101110010101110110001000000001001010100000000111110000111 101001001010011010010011100101001110100101001100

19.5.2 Statistical test suite – autocorrelation analysis

Autocorrelation assesses the similarity between data points. When measurements close to one another have comparable values, there is no repeating pattern relation

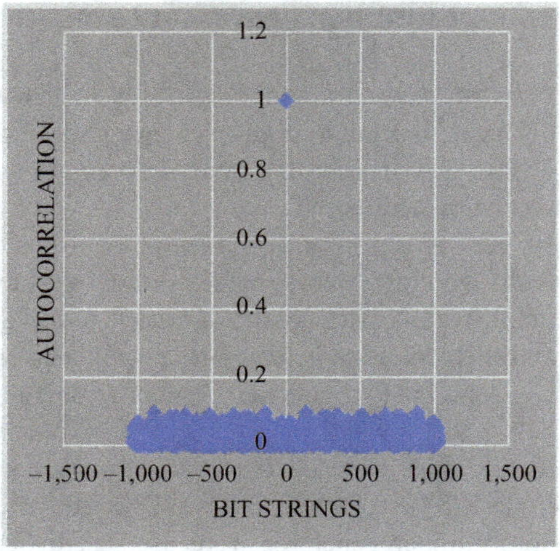

Figure 19.3 Autocorrelation analysis

in an orderly process. The autocorrelation values represented in (19.13) are calculated using a mathematical method:

$$\text{ATC} = \frac{|I - NI|}{S} \tag{19.13}$$

S is the size of the sequence; I is the number of identical bit positions between shifted bits; NI is the number of non-identical bit positions between shifted bits.

From Figure 19.3, we can clearly state that the random numbers are highly autocorrelated. The impulse is indicated on point 1, and from this, we can clearly state that the statistical analysis for the generated random numbers is passed.

19.5.3 National Institute of Standards and Technology (NIST) SP 800-22

This test suite is an established statistical test for testing cryptographic random numbers. Random number generators must be unexpected for cryptography applications, even though this unpredictability property is not required in other contexts like simulation and modeling. In the test suites 16 tests as in Table 19.1, the amount that binary sequences deviate from randomness is expressed using a separate test statistic. Three steps make up the NIST SP 800-22 testing procedure. The 16 tests must be performed on each sample first. Next, the likelihood that a fair, independent RNG is generated in each sample is reported for each test. Then, the p-value, a measure of the significance of 0.01, is applied to this likelihood. If the p-value is less than the level of importance, the illustration bombs the test [50].

Table 19.1 NIST SP 800-22

S. no.	Type of test	*p*-value	Conclusion		
1	T1	0.0155	Random		
2	T2	0.6104	Random		
3	T3	0.1052	Random		
4	T4	0.1310	Random		
5	T5	0.9166	Random		
6	T6	0.9634	Random		
7	T7	0.3011	Random		
8	T8	0.4043	Random		
9	T9	0.0734	Random		
10	T10	0.4624	Random		
11	T11	0.8675	Random		
		0.8005	Random		
12	T12	0.6540	Random		
13	T13	0.0174	Random		
14	T14	0.0096	Non-Random		
15	T15	State	Chi-squared	*p*-value	Conclusion
		−4	7.3511	0.1958	Random
		−3	5.8948	0.3165	Random
		−2	4.1968	0.5214	Random
		−1	3.4653	0.6286	Random
		+1	2.2520	0.8132	Random
		+2	5.3737	0.3719	Random
		+3	2.6757	0.7498	Random
		+4	1.3310	0.9317	Random
16	T16	State	Counts	*p*-value	Conclusion
		−9.0	291	0.5274	Random
		−8.0	301	0.5642	Random
		−7.0	289	0.4573	Random
		−6.0	299	0.4866	Random
		−5.0	308	0.5108	Random
		−4.0	340	0.7676	Random
		−3.0	364	0.9601	Random
		−2.0	360	0.9828	Random
		−1.0	363	0.9406	Random
		+1.0	340	0.4344	Random
		+2.0	342	0.6830	Random
		+3.0	357	0.9469	Random
		+4.0	367	0.9327	Random
		+5.0	402	0.6110	Random
		+6.0	434	0.4127	Random
		+7.0	417	0.5632	Random
		+8.0	352	0.9310	Random
		+9.0	290	0.5216	Random

T1–T16: Represents the various tests like Frequency Test, Frequency check within a Block, Run Test, Longest Run of ones in a Block, Binary Matrix Rank Test, Discrete Fourier Transform (spectral) Test, Non-overlapping Template Matching Test, Overlapping Template Matching Test, Maurer's Universal Statistical Test, Linear Complexity Test, Serial Test, Approximate Entropy Test, Cumulative Sums (forward) Test, Cumulative Sums (reverse) Test, Random Excursion Test, and Random Excursions Variant Test, respectively.

According to the National Institute of Standards and Technology (NIST), these methods can spot deviancies from randomness in binary sequences. The NIST test suites are 16 tests that evaluate the unpredictable nature of binary sequences produced by hardware or software-based random number generators. The value of the next element in the series cannot be anticipated, regardless of what has been created. If the *p*-value is lesser than 0.01, the series is considered non-random; otherwise, if it is greater than or equal to 0.01, it is called random. A total of 2 crore bits is admitted in this test and verified for randomness.

19.5.4 NIST 800-90B statistical test

Estimate the entropy of the given raw random number to refine quantum randomness. It is necessary to enumerate the extractable randomness by the worst-case min-entropy from the quantum entropy source [45]. The minimum entropy of a probability distribution is well-defined through $H_{original}$(minimum entropy/sample) and $H_{bitstring}$(additional entropy estimation/bit). We used NIST statistical test suite SP800-90B for entropy assessment. It is used for deterministic and non-deterministic design and testing entropy sources. According to the NIST recommendation, ten entropy estimators are employed to authorize an entropy source. For accuracy, a consecutive raw dataset of 100 Mb sample values was acquired straight from the proposed random number generator. By following the test, tracking of IID or non-IID is possible. By referring to Table 19.2, the min-entropy for all raw sequences is determined through the T-Tuple test estimate as 0.0001/1 bit.

Tests 1–10: Represents the estimate of most common value, Collision test, Markov test, Compression test, T-Tuple test, LRS test, Multi most common in window prediction test, Lag prediction test, Multi Markov model with counting prediction test, LZ78Y prediction test respectively.

Table 19.2 NIST 800-90B statistical test

S. no.	Type of test		*p*-value/1 bit
1	Entropic statistics	1	0.8295
		2	1.0000
		3	0.8474
		4	0.5681
2	Tuple	5	0.0001
		6	0.0003
3	Predictor	7	0.8295
		8	0.8228
		9	0.0701
		10	0.8295
4	Original_H		0.0008
5	Bitstring_H		0.0001
6	Minimum of (4,5)		0.0008
7	(2^{-m})Entropy_Worst case		0.999445

19.6 Conclusion and future scope

Researchers were able to reckon the qubits in a cloud quantum computer and use arithmetic tests for random number generators to assess the device's potential health indicators. Device samples were statistically examined to check for bias and patterns. The min-entropy was computed to determine each sample's degree of homogeneity. Statistical testing for random number generators is a worthwhile technique aimed at assessing the stability and condition of qubits inside a cloud quantum computer. The outcomes using IBM Quantum Device quantum gates are more encouraging. Assuming more qubits become available, the contemporary era of quantum computing will emerge. The new set of quantum gates will be used to create genuine quantum random number generators, and the random numbers they produce will be compared. In particular, the generated prime random numbers will be applied to the quantum key distribution for security purposes as part of the examination of the structural arrangement of the gates. In our future work, we intend to concentrate on various areas, including applying generated real quantum random numbers in symmetric cryptography. We will generate real quantum random numbers using a set of quantum gates, compare the results, and analyze the non-identical arrangement in the gate structure to derive prime numbers from it. The quantum key distribution will use the random prime numbers generated for security reasons. Prime numbers analysis will be carried out from the different random numbers obtained through various random number generators using quantum gates.

References

[1] P. L'Ecuyer, "History of uniform random number generation," in *Proceedings of the Winter Simulation Conference*, pp. 202–230, 2017.

[2] D. J. Bennett, *Randomness*, Cambridge, MA: Harvard University Press, p. 238, 1998.

[3] C. E. Shannon, "Communication theory of secrecy systems," *Bell Syst. Tech. J.*, vol. 28, no. 4, pp. 656–715, 1949.

[4] R. Stevanović, G. Topić, K. Skala, M. Stipčević, and B. M. Rogina, "Quantum random bit generator service for Monte Carlo and other stochastic simulations," *Lecture Notes in Computer Science*, vol. 4818, pp. 508–515, 2007.

[5] S. Chari, *Randomness as a Computational Resource: Issues in Efficient Computation*, Cornell University, 1994.

[6] M. Pivoluska, M. Plesch, M. Farkas, *et al.*, "Semi-device-independent random number generation with flexible assumptions," *npj Quantum Inf. 2021 71*, vol. 7, no. 1, pp. 1–12, 2021.

[7] P. Martin-Löf, "The definition of random sequences," *Inf. Control*, vol. 9, no. 6, pp. 602–619, 1966.

[8] G. J. Chaitin, "Information, randomness and incompleteness," in *Information, Randomness and Incompleteness*, Singapore: World Scientific, 1990.

[9] R. Gennaro, "Randomness in cryptography," *IEEE Secur. Priv.*, vol. 4, no. 02, pp. 64–67, 2006.

[10] Y. Zheng and T. Matsumoto, "Breaking real-world implementations of cryptosystems by manipulating their random number generation 3," in *Pre-proceedings of the 1997 Symposium on Cryptography and Information Security*, 29 January–1 February 1997, Fukuoka, pp. 6–7.

[11] M. Iavich, T. Kuchukhidze, S. Gnatyuk, and A. Fesenko, "Computer network and information security," *Comput. Netw. Inf. Secur.*, vol. 3, pp. 28–38, 2021.

[12] N. Metropolis and S. Ulam, "The Monte Carlo method," *J. Am. Stat. Assoc.*, vol. 44, no. 247, pp. 335, 1949.

[13] N. Metropolis, "*The Beginning of the Monte Carlo Method*," Los Alamos Science Special Issue, Vol. 15. Scientific Research Publishing, pp. 125–130, 1987.

[14] R. M. Karp, "An introduction to randomized algorithms," *Discret. Appl. Math.*, vol. 34, no. 1–3, pp. 165–201, 1991.

[15] R. Motwani and P. Raghavan, "Randomized algorithms," in *The Computer Science and Engineering Handbook*, London: CRC Press, 1996.

[16] R. Motwani and P. Raghavan, *Randomized Algorithms*, Cambridge: Cambridge University Press, 1995.

[17] C. H. Bennett and G. Brassard, "Quantum cryptography: public key distribution and coin tossing," *Theor. Comput. Sci.*, vol. 560, no. P1, pp. 7–11, 2020.

[18] H. Schmidt, "Quantum-mechanical random-number generator," *J. Appl. Phys.*, vol. 41, no. 2, pp. 462, 2003.

[19] N. Zettili, *Quantum Mechanics Concepts and Applications*, 2nd ed., Jacksonville: Jacksonville State University, 2009.

[20] "Francis Galton – 'Dice for statistical experiments.' Nature(42 1890):13-4." Available: https://galton.org/bib/JournalItem.aspx_action=view_id=193. [Accessed: 08-Sep-2022].

[21] J. von Neumann, "Various techniques used in connection with random digits," *National Bureau of Standards Applied Mathematics Series* (Notes by G. E. Forsythe), vol. 12, pp. 36–38, 1951.

[22] T. E. Hull and A. R. Dobell, "Random number generators," *SIAM*, vol. 4, no. 3, pp. 230–254, 2006. http://dx.doi.org/10.1137/1004061.

[23] W. Hörmann, J. Leydold, and G. Derflinger, "Automatic nonuniform random variate generation," *Computational Statistics*, vol. 19, pp. 659–660, 2004.

[24] P. L'Ecuyer, "Random number generation," in *Handbook of Computational Statistics*, New York, NY: Springer, 2012, pp. 35–71.

[25] F. James, "A review of pseudorandom number generators," *Comput. Phys. Commun.*, vol. 60, no. 3, pp. 329–344, 1990.

[26] Q. Zhou, R. Valivarthi, C. John, and W. Tittel, "Practical quantum random number generator based on sampling vacuum fluctuations," *Quantum Eng.*, vol. 1, no. 1, pp. e8, 2017.

[27] M. Herrero-Collantes and J. C. Garcia-Escartin, "Quantum random number generators," *Rev. Mod. Phys.*, vol. 89, no. 1, p. 015004, 2017.

[28] K. Tamura and Y. Shikano, "Quantum random number generation with the superconducting quantum computer IBM 20Q Tokyo," *Cryptol. ePrint Arch.*, 2020.

[29] F. Steinlechner, M. Jofre, M. Curty, *et al.*, "True random numbers from amplified quantum vacuum," *Opt. Express*, vol. 19, no. 21, pp. 20665–20672, 2011.

[30] M. Soucarros, C. Canovas-Dumas, J. Clédière, P. Elbaz-Vincent, and D. Réal, "Influence of the temperature on true random number generators," in *IEEE International Symposium on Hardware Oriented Security and Trust. HOST 2011*, pp. 24–27, 2011.

[31] H. Guo and W. Wei, "A bias free, quantum random number generator," *Front. Opt. 2009/Laser Sci. XXV/Fall 2009 OSA Opt. Photonics Tech. Dig. (2009), Pap. FWL4*, p. FWL4, Oct. 2009.

[32] A. Aspect, P. Grangier, and G. Roger, "Experimental tests of realistic local theories via Bell's theorem," *Phys. Rev. Lett.*, vol. 47, no. 7, pp. 460, 1981.

[33] A. Aspect, P. Grangier, and G. Roger, "Experimental realization of Einstein-Podolsky-Rosen-Bohm *Gedankenexperiment*: a new violation of Bell's inequalities," *Phys. Rev. Lett.*, vol. 49, no. 2, pp. 91, 1982.

[34] H. Friedman, "Geiger counter tubes," *Proc. IRE*, vol. 37, no. 7, pp. 791–808, 1949.

[35] M. P. Silverman, W. Strange, C. R. Silverman, and T. C. Lipscombe, "Tests of alpha-, beta-, and electron capture decays for randomness," *Phys. Lett. A*, vol. 262, no. 4–5, pp. 265–273, 1999.

[36] C. H. Vincent, "The generation of truly random binary numbers," *J. Phys. E.*, vol. 3, no. 8, pp. 594, 1970.

[37] Y. Yoshizawa, H. Kimura, H. Inoue, K. Fujita, M. Toyama, and O. Miyatake, "Physical random numbers generated by radioactivity," *J. Japanese Soc. Comput. Stat.*, vol. 12, no. 1, pp. 67–81, 1999.

[38] Y. Shen, L. Tian, and H. Zou, "Practical quantum random number generator based on measuring the shot noise of vacuum states," *Phys. Rev. A*, vol. 81, no. 6, pp. 063814, 2010.

[39] M. Gude, "Concept for a high performance random number generator based on physical random phenomena," *Frequenz*, vol. 39, no. 7–8, pp. 187–190, 1985.

[40] C. S. Pétrie and J. Alvin Connelly, "A noise-based ic random number generator for applications in cryptography," *IEEE Trans. Circuits Syst. I Fundam. Theory Appl.*, vol. 47, no. 5, pp. 615–621, 2000.

[41] M. J. Applegate, O. Thomas, J. F. Dynes, Z. L. Yuan, D. A. Ritchie, and A. J. Shields, "Efficient and robust quantum random number generation by photon number detection," *Appl. Phys. Lett.*, vol. 107, no. 7, p. 071106, 2015.

[42] M. D. Eisaman, J. Fan, A. Migdall, and S. V. Polyakov, "Invited review article: single-photon sources and detectors," *Rev. Sci. Instrum.*, vol. 82, no. 7, p. 071101, 2011.

[43] G. S. Buller and R. J. Collins, "Single-photon generation and detection," *Meas. Sci. Technol.*, vol. 21, no. 1, p. 012002, 2009.

[44] G. Weihs, T. Jennewein, C. Simon, H. Weinfurter, and A. Zeilinger, "Violation of Bell's inequality under strict Einstein locality conditions," *Phys. Rev. Lett.*, vol. 81, no. 23, pp. 5039–5043, 1998.

[45] M. S. Turan, E. Barker, J. Kelsey, K. A. McKay, M. L. Baish, and M. Boyle, "Recommendation for the entropy sources used for random bit generation," Tech. rep., NIST, 2018.

[46] X. Ma, X. Yuan, Z. Cao, B. Qi, and Z. Zhang, "Quantum random number generation," *npj Quantum Inf. 2016 21*, vol. 2, no. 1, pp. 1–9, 2016.

[47] "IBM – India | IBM." Available: https://www.ibm.com/in-en [Accessed: 12-May-2022].

[48] "Qiskit." Available: https://qiskit.org/ [Accessed: 12-May-2022].

[49] V. Kumar, J. B. B. Rayappan, R. Amirtharajan, and P. Praveenkumar, "Quantum true random number generation on IBM's cloud platform," *J. King Saud Univ. – Comput. Inf. Sci.*, vol. 34, no. 8, pp. 6453–6465, 2022.

[50] "A Statistical Test Suite for Random and Pseudorandom Number Generators for Cryptographic Applications | NIST." Available: https://www.nist.gov/publications/statistical-test-suite-random-and-pseudorandom-number-generators-cryptographic [Accessed: 18-Oct-2022].

Chapter 20

Security challenges and prospects of 6G network in cloud environments

Peace Busola Falola[1], *Emmanuel Abidemi Adeniyi*[1], *Joseph Bamidele Awotunde*[2], *Rasheed Gbenga Jimoh*[2] and *Agbotiname Lucky Imoize*[3,4]

Abstract

Global technological and industrial development is advancing in the current world. The development of the 6G communication system has been facilitated by the pervasive adoption of the latest generation of information and communication technologies (ICTs), such as artificial intelligence (AI), virtual reality (VR), augmented reality (AR), extended reality (XR), the Internet of Things (IoT), and blockchain technology. Exploratory research into 6G networks as the next wave of services is likely motivated by the limits of 5G networks that have been discovered as additional 5G networks are installed. These analyses cover the fundamental privacy and security concerns raised by 6G technology. This chapter discusses 6G security problems as a foundation for future research directions. Last, the security problems of cloud computing were explored.

Keywords: 6G networks; Cloud computing; Security; Technologies; Internet of Medical Things; Artificial intelligence; Communication

20.1 Introduction

6G will investigate novel communication modes without regard for current network concepts or technologies. It includes compatible novel ideas, frameworks, protocols, and approaches that assist the present and future. Intelligent, deep, holographic, and

[1]Department of Computer Science, Precious Cornerstone University, Nigeria
[2]Department of Computer Science, Faculty of Communication and Information Sciences, University of Ilorin, Nigeria
[3]Department of Electrical & Electronics Engineering, Faculty of Engineering, University of Lagos, Nigeria
[4]Department of Electrical Engineering and Information Technology, Institute of Digital Communication, Ruhr University, Germany

pervasive connectivity are characteristics of the 6G system. Every part of the data transmission, such as the competence of the net components and net design, the smartness of the integrated entity (the node sensor), then the smartness of the information transferred to facilitate the smart support, reflects the intelligence of the data transmission. Profound connection implies deep sensing, learning, and cognition. Holographic connectivity has the following characteristics: high-quality, seamless coverage across augmented reality (AR) and virtual reality (VR), and holographic communication anywhere (and at any time). Ubiquitous connectedness is a multi-dimensional coverage link that spans all terrains and spaces [1,2].

Essentially, 6G is expanding and upgrading the previous generation, 5G. Regarding network access, the 6G standard will cover mobile cellular, satellite, airborne, undersea, acoustic, and visible light communications (VLCs). To deliver genuinely worldwide seamless coverage of integrated information, a space-spanning interconnected network, the air, the sea, and land, will be created by 6G [3,4]. 6G will considerably enhance data transmission rate, point-to-point lag, dependability, communication concentration, bandwidth proficiency, and net performance to satisfy the diverse network specifications of several sectors of the economy. The 6G network and its users will become a cohesive matter regarding network intelligence. Each user will be able to enhance their experience with AI assistants (AIAs) since AI will look into consumer desires in more detail. Connecting persons, machinery, and tangible items in the real globe to those in the digital globe will be the aim of 6G services, going beyond network service borders.

A 6G mobile network system must provide very fast speeds, expanded capacity, and the need for distance to enable potential new applications like virtual reality, vigorous medicine, and computer disaster predicting. Given the earlier control on cellular networks, the premiere 6G links will be mainly based on the current 5G structure, taking profit off the benefits obtained in 5G. For instance, the upsurge in permitted bandwidths also enhanced the architecture of a distributed network and modified how we work and play [5]. Moreover, data will likely influence humans by 2030, providing quick and limitless cellular access. 6G should therefore enhance existing wireless technology and enable system prosecution. Regarding speed, 6G is designed to utilize a more sophisticated frequency band than earlier generations to boost the data transmission rate, which is predicted to be 100–1,000 times quicker than 5G [6]. To be more precise, 6G systems will enable broadband spectrum to connect at hundreds of GB per second, for instance, by combining the utilization of spectrum from one to three GHz, a terahertz band, and a millimeter wave band.

Fundamental societal and personal needs are what motivated the projected improvement in 2030. Global Tesla predictions are possible, and 6G will play an exceptional part in this transition by providing an ICT infrastructure that will allow end users to be absorbed by a sizable artificial brain. It delivers virtual storage, infinite storage, and collective intelligence. The crucial goal of 6G is to offer mobile phone users access to several services, including network identifiers in diverse areas, multimedia apps, and web connection for cellular users with high-speed data speeds without damaging the network. Figure 20.1 illustrates the prospect of a 6G network in the cloud ecosystem.

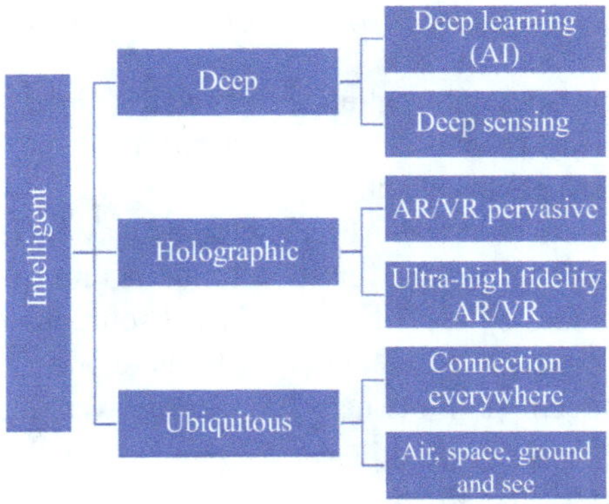

Figure 20.1 Prospect of 6G network in the cloud ecosystem

Artificial Intelligence (AI) technology is anticipated to provide the foundation for constructing the 6G network. 6G will be essentially "intelligent" due to the so-called "Intelligent connection." Larger, more complex networks, a greater range of endpoints and networking devices, and more complex and diverse company kinds will all provide challenges for 6G networks. "Intelligent Connectivity" will concurrently meet two requirements: first, all linked devices on a network that are connected are smart, and the services that are connected to them have been smart; second, the intricate and massive network itself demands intelligent management. Deep Connectivity, Holographic Connectivity, and Ubiquitous Connectivity, the 6G network's other three essential components, will all be supported by "Intelligent Connectivity," which will be its essential characteristic. According to our predictions, deep coverage will give way to "Deep connectivity" in the following 10 years (2030) of 6G systems. In 10 years (2030), planar multimedia will predominate in media communication, along with advanced value augmented reality/virtual reality (AR/VR) interactions, hologram data interactions, and wireless hologram information exchange becoming a reality.

High-quality augmented AR/VR will be readily available, together with holographic communication and projection that can be utilized anytime, anywhere, to achieve the information exchange goal of supposed "holographic connectedness." As a result, individuals can participate in completely immersive, interactive holographic experiences at any time or location [7]. A vast world will be more and more available if the "Anytime, Anywhere" connection condition is ultimately reached a decade ahead (2030), that is, to attain true "Ubiquitous connectivity." The 6G network trunk is created by the three extra attributes of "Deep Connectivity," "Holographic Connectivity," and "Ubiquitous Connectivity," while "Intelligent Connectivity" will eventually act as the network brain and nerve. Together, these four features will create a fully organic, soul-filled 6G network. Based on the current 5G, the communication system will be improved and developed further. As a result, the network will reduce the distance between everything, information will overcome the barriers of time and space, and seamless integration of people and everything else will be possible [8].

20.1.1 The primary contribution of this chapter is as follows

(i) Presents various issues and challenges related to 6G network communication.
(ii) Describe the many privacy and security concerns the 6G technology may cause in the cloud environment.
(iii) Analyze how blockchain and AI technologies may be leveraged to treat the privacy and security concerns linked with the 6G network in the cloud.
(iv) The study concludes by identifying AI-based solutions regarding safety and confidentiality concerns of 6G networks in the cloud environment.
(v) The potential directions in 6G network security were also discussed.

20.1.2 Chapter organization

Section 20.2 introduces the 6G network issues and solutions. Section 20.3 discusses the application of AI in 6G networks. Section 20.4 presents the application of blockchain in the 6G network. Section 20.5 presents the security challenges of 6G networks and cloud environments. Section 20.6 discusses security challenges in cloud environment. Section 2.7 covers security requirements for 6G networks in cloud environment. Finally, Section 20.8 discuss the AI solution to 6G privacy and security issues in cloud environment, while Section 20.9 concludes the chapter.

20.2 6G network issues and solutions

(A) Network security issue

Security is a main issue for 6G wireless networks, especially while utilizing the Terrestrial Space Integrated Network (STIN) method. With the traditional physical series security, 6G must look at other sorts of confidentiality, such as coordinated network security. An innovative security approach that depends on low complexity and high levels of security must thus be strengthened. In light of this, several physical layer security mechanisms created for 5G systems may be expanded to 6G systems, including a secure multiple input and multiple output (MIMO) with low-density parity control (LDPC); mm-Wave (millimeter wave) safe techniques can also be used for "UM-MIMO (ultramassive multiple input and multiple output) and Terahertz (THz) spectrum applications." Integral network security depends on the availability of a suitable management objective for multiple function approaches for various security segments. For example, a centralized distribution management solution, which might be used for STIN, considers the administration of multilingual and information exchange keys that do not require a certificate. This comprehensive solution, which successfully protects secret data and then secrecy on 6G networks, may be created by integrating physical and network-level safety measures with efficient administration and application [9].

With novel approaches, 6G communications should considerably increase safety, confidentiality, and secrecy. However, the safety and confidentiality of transmissions are still ensured in 5G networks by using conventional encryption techniques built on the public cryptosystem. Rivest–Shamir–Adleman (RSA) crypto-spores are less concerned about the demand from big data and AI technologies than they are about the lack of privacy protections in the 5G future.

(B) Resource as a Service (RaaS)
Software-defined networking (SDN) and network functionality verification (NFV) have created RaaS, an integrated and resource-oriented resource allocation strategy that has been made easier. As a result, the physical infrastructure seems to be divided into many virtual networks. It allows cellular providers or telecommunications companies to deploy virtual network resources according to the needs of a certain service. Software specificities and customizable metro conditions are likely to be among the resources of the network. As a result, network screening using proprietary software and metro-determined surfaces, ranging from a machine-activated cloud access network (C-RAN) to open-form wireless, will only be network functionality verification (NFV) development patterns throughout the 6G cycle [10].

(C) Heterogeneous high-frequency bands
The utilization of mm-Wave and THz in 6G results in several new open problems. A significant outstanding question for mm-Wave will be the support for big motions at those frequencies. New architectural designs and propagation models are necessary for THz [11]. The transmitter's high power, high sensitivity, and low noise levels are essential characteristics necessary to offset for THz loss on a high route. It is imperative to fully comprehend these physical series elements before creating new connections and network layer procedures that better use cross-frequency resources. The knowledge of how microwaves, mm-Waves, and THz cells interact in each series is an additional crucial method.

20.2.1 Secure and privacy issue in 6G network transmission technology

Streaming content is turned into encrypted information blocks for transmission and caching using network coding technology in accordance with a detailed grasp of the needs for 6G mobile services that concentrate on huge, low-latency streaming media services [12]. This improves the effectiveness of data transfer while preserving user privacy and content security. Additionally, feature data is retrieved from the service content using edge computing and AI technologies for high-security user services. They are subsequently routed back to the cloud technology facility. This guarantees user information confidentiality while decreasing the burden on the 6G backhaul network and enhancing customer service quality. They are subsequently routed back to the cloud technology facility. This guarantees client information confidentiality while decreasing the burden on the 6G backhaul network and enhancing customer service value.

20.3 Application of AI in 6G network

Although applying AI to 6G is a current trend, it is erroneous to see AI as a 6G technology that is simply tacked onto mobile communications. We can only understand the technological trend of 6G mobile communications by delving into user demands and examining the connections between intelligence, communication, and the future of mankind. Three stages of the interaction between AI and humans were anticipated by Israeli historian Yuval Harari in "A Brief History of the Future": (1) AI is the super oracle of humans, capable of comprehending and mastering all of a personal psychological and physiological quality and providing timely and correct advice on how to live and work; nonetheless, it is up to the person receiving the advice to decide whether to follow it; (2) AI will evolve into a human super-agent and replace some of the human capacity for decision-making. It is fully empowered to manage matters on behalf of people; (3) as AI develops, it eventually becomes the ruler of humanity, the master of humanity, and all human behavior is governed by its plans. According to the forecast above, 6G should continue the trajectory of the AI–human connection and enter the oracle stage, the first stage of relationship evolution. Figure 20.2 shows AI application in 6G network framework requirement.

The capabilities of 6G will further into different systems, the actual world and the online world, serving as a crucial implementation base for the oracle stage. The actual-world system services are backward compatible with standard situations like enhanced mobile broadband (eMBB), massive machine-type communications (mMTC), and ultra-reliable low latency communications (uRLLC) in 5G to meet the fundamental requirements of the interconnectedness of everything in the real world. The simulated world system services are an expansion of real-world services and are tailored to meet the diverse demands of the virtual world. In the simulated world in which the 6G has developed, users may have their own AIA, which can

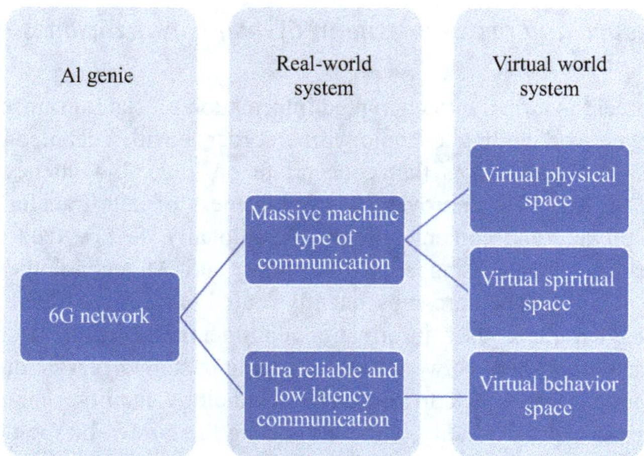

Figure 20.2 AI application in 6G network framework requirement

record, save, and interact with everything they say, see, and think. The simulated world system creates a complete three-dimensional simulation of each user and allows for the digital abstraction and representation of the varied unique demands of human users. The three spaces that make up the simulated world system are virtual physical space, VBS, and virtual spiritual space.

The virtual physical space (VPS) produces a virtual environment that mirrors the actual world (including geographic environment, houses, highways, vehicles, internal structures, etc.) and offers a simulated digital space for massive users' AIA information interaction. It operates on real-time huge data transfer of common situations that is 6G compatible. For example, a large sporting event, a huge celebration, a disaster aid and rescue operation, a military operation, simulated e-commerce, an online factory, etc. may all be supported by the data in VPS due to its real-time updating and high-precision simulation capabilities.

The 5G mMTC scenario is expanded by virtual behavior space (VBS). VBS can gather and track the bodily movements as well as biological functions of people in live time, convey diagnostic and treatment information to AIA promptly, and do all of this by utilizing the 6G human–machine interface and biosensor network. AIA forecasts the user's health state according to the data interpretation supplied by VBS and offers prompt and efficient treatment options. The widespread implementation of precision medicine is an example application that VBS supports.

Virtual spiritual space may be built based on the analysis of virtual physical space, VBS, and service scenario data and the huge information interaction. The advancement of differential demands perception and the development of semantic information theory have allowed AIA better to recognize the distinct psychological and users' religious demands. These perceived demands range from virtual needs like games and hobbies to actual needs like social connection and job searching. AIA offers comprehensive recommendations as well as services for consumers' health lives and fun according to the perceived needs captured by VSS. For instance, with the help of 6G, data engagement and cooperation, AIA of various users can give users comprehensive advice on choosing a partner and getting married, accurately evaluate consumers' job searching and promotion, and assist users in establishing, maintaining, and developing better social relationships.

In order to back up AI Genie's semantic perception and analysis, 6G must process semantic information in addition to digital information. This necessitates overcoming the constraints of classical information theory and developing generalized information theory to create a thorough processing system for syntax and linguistic information. This is also the conceptual underpinning for the implementation of human–computer intelligent communication.

20.4 Application of blockchain security in 6G network

A decentralized ledger, a blockchain, stores information in a series of data chunks [13]. Blockchain aggregates many network, consensus, and automated management technologies [13]. All of these technologies must be appropriately blended and chosen to provide the needed security characteristics necessary for the underlying application situation [13]. The future wave of cellular networks, notably 6G, which will gradually become softwarized, decentralized, and a network of free devices, is thought to require blockchain technology's safe and complete automation [14]. In particular, it is anticipated that the use of blockchain technology will enhance both the 6G technology (safety, confidentiality, broadband service management, resources usage, and bandwidth planning) as well as 6G applications [healthcare, power web, autonomous aerial vehicles (UAVs), self-driving cars (CAVs), and augmented reality] [14]. Blockchain technology has much more to offer than only security measures for 6G networks. Blockchain-based resource optimization can achieve the extremely high data speeds needed for 6G [14]. Blockchain adoption is very advantageous for networks with decentralized control. Additionally, blockchain can support the 6G RAN need for cooperation [15].

Blockchain is seen as a crucial technology for 6G applications [13]. Future wireless networks' cutting-edge applications may undergo a revolution thanks to a combination of 6G and blockchain [16]. Several 6G applications need high safety and confidentiality, which blockchain can deliver [16]. The backing of developments like reconfigurable intelligent surfaces (RIS), Terahertz (THz) communication, AI, and tiny cell networks would be necessary to meet the demanding network performance needs of 6G applications [14]. Coordination and cooperation in an unreliable and open area are required to make these technologies work together effectively to supply resources and meet performance criteria [13]. These technologies also call for the implementation of dense networks, which will require additional infrastructure and network deployment that is more complex [13].

Since 51% of systems are to be specifically designed to aim at a blockchain network, a major advantage of blockchain is its potential to produce an attack-proof network. Consequently, it is a decentralized, immutable ledger where all transactions take place [16]. Decentralization of the network will be required to streamline network deployment. Blockchain will give the decentralized network the required transparency and trustlessness. Blockchain technology's key features are decentralized ledger, decentralized networks, openness, immutability, and irreversibility [16]. It is therefore anticipated to offer greater efficiency and cheaper cost. Because it is immutable, blockchain is known to be trustworthy to support a lot of users across several sectors. Blockchain will also fulfill the strict security requirements of emerging communications networks owing to its designed security features [13]. The function of blockchain in 6G networks is depicted in Figure 20.3.

Massive connection in future systems in terms of scalability, real-time communication with minimal delay, better efficiency, and synchronization are some of the problems of 6G [17]. It would be difficult to modify the structure of 6G networks to accommodate an unforeseen traffic request, given the billions of connected devices that would operate in future industrial ecosystems [17]. For correct functioning in real-time with minimal latency, device-to-device and machine-to-machine interactions need strong accuracy with almost zero delays. For use cases like automated driving and augmented reality (AR)-supported medical systems, it could be required to have a consistently low latency communication capacity in

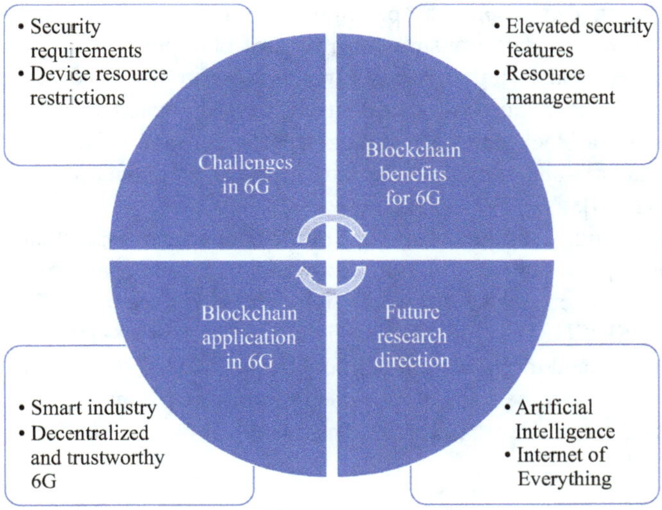

Figure 20.3 Role of blockchain in 6G networks security

large-scale data transfer [17]. Billion-device connections are required for higher throughput, and ground stations and other network elements must be able to manage the big transaction volume in real-time [17]. Synchronization is a critical need in industrial applications that require timing. Security requirements like as confidentiality, integrity, availability, authentication, access control, and auditing are problematic in 6G computing environments. This section explores the potential and advantages of utilizing blockchain technology to resolve any potential 6G issues.

20.4.1 Intelligent resource management

The spectrum, processing power, and other available resources and infrastructure would all be in high demand for 6G applications. Spectrum sharing, orchestration, and decentralized computation are resource management procedures that must be compatible with extremely large infrastructures [17]. In Beyond 5G, Zhang *et al.* [18] provided a paradigm for edge intelligence and IoT with safe and flexible service management. In order to control the interaction between owners and users, Maksymyuk *et al.* [19] suggested an efficient network structure that uses blockchain technology. The researchers created an unlicensed bands allocation mechanism using game theory. Dai *et al.* have applied blockchain and deep reinforcement learning [20] for effective resource allocation solutions, such as bandwidth distribution and energy management. Mafakheri1 *et al.* [21] used blockchain technology for sharing resources and showed how smart contracts might be used to give functionality for self-organizing networks.

The most crucial resource in the network is the spectrum. Due to the introduction of the 6G latest services, this resource is scarce and needs to be utilized carefully [16]. As the data rate is closely proportional to the provided broadband, spectrum sharing can help with spectrum management or spectrum sharing to support the increased data transmission needs of 6G applications [13]. Licensed spectrum owners and unauthorized bandwidth users in any band can collaborate and coordinate under various general guidelines automated by smart contracts implemented on a blockchain to maximize spectrum usage [13]. Selecting a consortium blockchain and suitable consensus methods may make this system more secure for 6G [13].

The application of blockchain in infrastructure and asset management is demonstrated by the sample of a user who wants to reduce communication lag by locating the finest channels of transmission offered by specialized asset service providers (SASPs) [13]. A verified and authorized user in a system for managing blockchain assets and infrastructure will look for the closest communication relays [13]. One of the major benefits of blockchain in resource control is its capacity to store data, such as location-based data. This historical data can be used to promote access equity [16].

20.4.2 Elevated security features

Blockchain integration in 6G helps to manage the need for privacy and secrecy, data integrity, non-reliability, and auditability. Blockchain can offer the needed degrees of data integrity, non-repudiation, and auditability by choosing the right network, consensus, and automated management algorithms [13]. Considering security from the standpoint of privacy is important. The varied ways that data confidentiality is used in the intricate safety needs of the upcoming 6G network environment [17]. For improved data privacy and confidentiality, blockchain enables asymmetric PKI-based encryption and the addition of confidentiality protection mechanisms [13]. The blockchain accepts fresh blocks only after verification by several P2P nodes using a consensus process [13]. Blockchain technology complies with confidentiality, integrity, and availability (CIA) criteria since it is a new technology with access control to ensure privacy, authenticity, and accessibility. Because blockchain data is distributed, secrecy cannot be guaranteed; yet, because of its distributed ledger and tamper-proof architecture, it retains integrity and availability [22].

The large amount of information that will be generated in future computer settings poses a significant difficulty with regard to data integrity. A blockchain-based approach to stop assaults that compromise data integrity was described by Adat *et al.* [23]. A blockchain-based architecture was suggested by Ortega *et al.* [24] to guarantee the accuracy of data shared through the network. Any block data integrity may be easily checked by looking at the hash trees [13]. Additionally, because of the connections between the chained blocks, data tampering becomes much more challenging as the blockchain grows [13]. Blockchain uses two layers of security, with the data being encrypted for data verification. The block hash is cascaded into the following block for data storage [13]. By creating a unique key pair for each transaction on the blockchain to prevent transaction linkage, privacy is guaranteed [13].

Only upon confirmation by a consensus process among several P2P nodes are new blocks accepted by the blockchain. Each unit is connected to its parent unit using a cryptography hash mechanism, which is the preceding block in the chain. As a result, data can be audited and verified back to the beginning block [13]. Future communication ecosystems will demand additional elements like service availability [17]. Rodrigues *et al.* [25] created a distributed denial of service (DDoS) protection system using blockchain technology. For the execution of crucial security services like DDoS attack prevention, data protection, and access control, Sharma *et al.* [26] advocated the use of blockchain with SDN.

Another important aspect of the 6G network ecosystem is its accountability. Overall, blockchain and decentralized ledger technologies may be used to achieve network security, surveillance, and governance [17]. The distributed ledger continues to serve as an unchangeable, transparent log of every event that may be used for event auditing [17]. In 6G systems, scalability is a crucial necessity. To meet the anticipated massive connection demand in the future, blockchain and smart contracts can eliminate the scalability restrictions of centralized systems [17]. In addition, the edge and fog computing nodes' decentralized nature and ease of integration will boost such networks' capacity for service [17].

20.5 Security challenges of 6G networks and cloud environment

Security has become a critical concern in all industries where major efforts are made regularly [27]. Technological advancements, new gadgets, and usage habits have evolved, resulting in many vulnerabilities. These vulnerabilities may be exploited maliciously. Furthermore, the increased usage of IoT devices with new innovations, such as multi-access edge computing (MEC), may significantly expand security holes [27]. As a result, trust in new breakthroughs always begins at a very low level. Trust difficulties will occur with an even more mysterious technology, such as 6G. This may be regarded as a substantial problem that must be conquered [27]. Safety and confidentiality concerns in 6G technology are addressed in this section.

20.5.1 The 6G technologies: security and privacy issues

Certain critical innovations are now intriguing in key sections of the 6G network. They provide 6G networks with strong dependability, minimal delay, and secure and effective communication solutions. However, most of these technologies pose serious privacy and security hazards.

(A) AI

AI is frequently seen as one of the most crucial elements of the upcoming network design in relation to every other technology anticipated to be used in 6G networks [28]. It would be an understatement to say that network specialists are intrigued by AI. A greater number of new safety and confidentiality matters result from this concentration [28]. Although AI on the 5G network seems to be run in remote areas with lots of learning information and sophisticated yet secure computing hubs, AI will have a larger impact on the 6G network [28]. AI and machine learning are used to secure different 6G security, defense, and protection frameworks [29]. Security systems have become more autonomous and correct, thanks to the usage of AI and machine learning, and security analytics is given predictive powers [29]. The architectural layers that physical and computing technologies serve (physical layers, which include elements like information links and Internet infrastructure, as well as computational stacks consist of, among other things, software-defined networks, network function virtualization, and cloud/edge/fog computing) are used to classify AI technologies [29].

The physical layer may use various AI-based technologies, including deep neural networks, K-means, and supervised/unsupervised learning [28]. By enhancing connections, these techniques not only raise physical layer performance but also forecast traffic and enhance security [28]. A machine learning-based antenna design was created by Hong *et al.* [30] with the potential to stop information leakage in physical layer communication. Sattiraju *et al.* [31] created an unsupervised learning technique that might be used to the verification method to improve physical layer security. Nawaz *et al.* also suggested using quantum encryption techniques and machine learning to safeguard 6G network communication lines.

According to Loven *et al.* [32], AI might augment edge security through safety procedures and smooth network architecture controls. In their investigation of AI technologies, Zhou *et al.* [33] suggested that more accurate deep learning may be used to identify risks in edge computing. But more research has to be done on this proposition.

AI is helpful for large-scale data processing, dispersed AI, asset optimization, network optimization, and the physical layers and network design. In order to increase the safety of 6G networks, Dang *et al.* [34] claim that AI may help identify network problems and provide early warning systems. According to Tomkos *et al.* [35], the deployment of decentralized and federated AI in a 6G network reduces the need for edge systems for transmitting data, enhancing network safety. Furthermore, Wang *et al.* [14] provide numerous differential confidentiality strategies that would be perfect for resolving some of 6G confidentiality difficulties. In contrast, Zhang *et al.* [36] claimed that the effect of data correlation in particular machine learning algorithms might exacerbate confidentiality breaches.

(B) Quantum communication

Quantum transmission is yet one potential solution for the 6G network. It has the capability to enhance data transfer security and reliability considerably [37]. Security and reliability are seen to be two major concerns for quantum transmission that is greatly enhanced [29]. The quantum state is affected by any negative eavesdropping [37]. The quantum state will change if an attacker edits or copies information in the quantum transmission [29]. In principle, quantum transmission enables absolute reliability and, with the right innovation, is well-suited for long-distance communication. It provides a range of creative solutions that raise the standardization of communication [29]. Security and essential improvements are both provided by quantum communication. It could improve communication and offer solutions in ways that conventional communication tactics cannot. It does not, however, address every security concern. Despite efforts to build quantum cryptography, fiber attenuation is a challenge in big gap quantum information exchange [37]. Nawaz *et al.* and Zhang *et al.* [36,38] have developed quantum approaches for private information exchange using quantum key allocation systems. Research on quantum computing is still in its early stages, although it is moving quickly [39].

(C) Blockchain technology

Blockchain is yet another crucial 6G network technology. It may be used for a variety of things, including bandwidth sharing, decentralized ledger technology, and network decentralization [37]. Network decentralization was employed by Dang *et al.* [34] to improve network performance. Through the use of distributed ledger technology, Strinati *et al.* [40] increased authentication security. Additionally, spectrum monopolies and unused spectrum may be resolved with the aid of blockchain technology [41]. The three key privacy concerns are transmission, verification, and access management. Blockchain technology was used by Ling *et al.* [42] to demonstrate verification and safe network access.

(D) VLC

In order to address the continually growing needs for wireless communication, VLC is a potential solution [43]. Ad hoc networks in cars and indoor navigation devices have both employed VLC. An indoor locating device based on VLC was shown by Luo *et al.* [44]. VLC has been shown to lessen EM interference. LEDs have been used to show high-speed data transmission in a number of research activities. Resilience to electromagnetic interference and increased bandwidths are features of VLC [45]. However, VLC communication suffers from a number of flaws. For example, VLC technology is frequently used to assist indoor environments adversely affected by natural light. Among the safety and confidentiality issues are criminal activities and communication methods with VLC [46]. An attacker must be visible to the victim in order to assault a VLC operation that is already in progress, according to Pathak *et al.* [47]. This would undoubtedly make it simpler for attackers to be found. Aggressive behaviors are among the other VLC technology security risks. The SecVLC protocol [29] is created for information data transfer in a vehicle network. A thorough summary of optical wireless communication (OWC) and 6G was provided by Mehedi *et al.* [46] in their most recent systematic review. It is anticipated that 6G will transform underwater optical wireless communication (UWOC) and underwater wireless power transfer (UWPT) [37].

(E) Terahertz technology (THz)

Future 6G technology cannot use the current RF band [48]. As a result, THz technology has drawn more interest. The frequency spectrum from the range of 0.1 and 10 THz was utilized by THz communication technology. Additionally, electromagnetic waves and optical signals are used. According to authors in [3], the THz spectrum provides a number of benefits, including a 100 Gbps data throughput, strong security, and less listening in the intercell effect. THz can lessen the intercell effect significantly [41]. Strianti *et al.* [40] looked at the issue of energy consumption in THz communication. Security flaws in THz authentication and malicious attack systems exist.

(F) Molecular communication

A particularly promising technique for 6G communications is molecular communication technology. However, it is an emerging multidisciplinary method [28]. Utilizing biochemical signals to communicate information is the main principle of molecular communication [28]. Bionano machines communicate utilizing organic signals or particles in aquatic environments, making molecular communication a potential 6G method in various health applications [49]. However, there are a number of safety and confidentiality concerns related to communication, verification, and encryption procedures since molecular communication deal with highly sensitive information. As a result, it is crucial to provide safe molecular communication [49]. In order to ensure information integrity, the concept of biochemical cryptography was first proposed in [50], where a biological macro-composition molecule and structure might be used as a means [49].

Nakano *et al.* [51] demonstrated how the sender, receiver, and connected nodes might communicate while moving using a mobile molecular communication method. However, the communication, verification, and encryption processes have already been linked to several security and privacy concerns. According to Farsad *et al.* [52], only a small number of studies have even considered the safety of molecular communication linkages, and an enemy might break this communication channel. In order to increase the security of the transmitted data, Lu *et al.* [53] suggested a coding method. Furthermore, Loscri *et al.* [54] suggested numerous conceivably useful key molecular communication routes that would encourage the creation of fresh validation systems to safeguard data safety and confidentiality. These researchers also go through a number of attack methods at various molecular communication levels, including flooding attacks, jamming, and desynchronization [44]. Even though it is clear that additional time is needed to create workable routes for molecular transfer in the 6G network, this technology should perform what conventional communication approaches cannot do [44].

20.6 Security challenges in cloud environment

There has recently been a lot of interest in cloud computing due to the ever-growing demand [55]. While cloud computing has provided many useful services, it has also opened up a wide range of opportunities and risks [55,56]. Due to the volume of data being sent over the network and stored in specialized cloud services, there are various weaknesses that malicious actors may take advantage of [55]. Security is one of the issues that is seen to be a major barrier to cloud computing success [57,58]. Therefore, many cloud computing security issues are active [56].

Regardless of the infrastructure being used, data security risks are at the top of the list [58,59]. Cloud computing is no different; its distributed architecture and multi-tenant design emphasize security issues more [59]. The generation, storage, usage, distribution, and disposal of data are all included in the data life cycle. Therefore, each cloud service provider (CSP) should offer appropriate security features to all data life cycle phases [58,60]. A customer could be able to look into another customer's data without that customer's knowledge and remove or edit it, for instance if the software application (shared application) is not constructed securely [58,61]. To avoid this, adequate safety measures must be put in place. The point of erasing information is significant in the cloud once more. The CSP should handle it carefully to ensure that data is completely demolished at a customer's request [59]. In addition, the customer should have access to and be able to audit the data backups being used to prevent data losses (scope, saving intervals, saving timings, storage duration, etc.). Utilizing a cloud service will provide all of these issues as well as several more [59].

Another crucial aspect of cloud computing security is data location [58,62]. Site transparency is among the most notable benefits of cloud computing. Still, there is a potential threat as well: without knowing the exact location of information storage, privacy laws for a particular region may be seriously affected and violated [56]. Thus, privacy of private user information is critical in cloud computing [56]. Since technology protection alone is inadequate to address the issue, cloud providers' strategic strategies are crucial for maintaining the security of their clients' personal or business data [56,58,63].

Because cloud computing sometimes necessitates accessing open connections and thus exposes information being sent to the rest of the world, cyberattacks of any kind are to be expected. Security flaws have been discovered in popular modern cloud-based systems that an attacker might possibly use. Cloud computing considers safety and confidentiality due to the features of its computational architecture [64]. Both network and information protection flaws can affect cloud computing due to how it is implemented [65,66].

Third-party interactions may become a problem for cloud systems and other security issues brought on by infrastructure and virtual machine (VM) features [67]. Cloud security is dynamically difficult because of things like software flaws, social engineering, and human mistakes [68]. Intrusion detection is the most crucial element of smooth network monitoring to reduce security threats. If the current generation of intrusion detection systems (IDSs) is inadequate, a security breach in the cloud environment may result [69]. Operating systems, load balancing, memory management, and concurrency control are just a few components that might provide a security risk in a cloud environment, in addition to databases, virtual servers, and networks [70].

Security on cloud servers may be breached through data loss and other botnets. In terms of security, the multi-tenancy model is also crucial to consider [58].

Secrecy, accessibility, and integrity are the main categories into which cloud environment security issues can be divided. There are difficulties with cloud infrastructure for both data and equipment [58].

When using cloud services, trust, which is closely linked to the reliability and legitimacy of CSPs, raises security concerns. Building trust could be vital to developing a productive cloud computing ecosystem. Whether a provider has overcome all obstacles, including data protection, VM protection, and other legal and regulatory requirements, forms the basis of trust. The three factors considered in this examination of cloud system security are confidentiality, integrity, and availability (CIA) [71].

Cloud-based services are susceptible to a number of assaults, including man-in-the-middle (MITM) attacks and phishing schemes, intercepting, spying, and other related attacks that can attack a computer network and data in transit. Attacks by DDoS botnets are a common but severe hazard to the cloud computing infrastructure [58]. If there is no way to prevent the well-known DDoS attack, cloud computing may be in danger. The consistency and level of protection of a cloud environment will be mainly determined by the security of VMs [58]. When considering the protection of cloud computing, accounting, authentication, and the use of encryption are all components of safe computing [58].

20.6.1 Important concepts in cloud security

Cloud security encompasses a wide variety of subjects. It is necessary to present the essential ideas that may be utilized to pinpoint the origin of threats and vulnerabilities in order to understand them. This part covers these subjects, starting with an overview of virtualization components and moving on to multi-tenancy. The idea of outsourcing data and cloud software is also discussed. Before concluding with a discussion on trust, the section examines data storage security and standards.

20.6.2 Virtualization elements

Virtualization is heavily integrated as a fundamental technology in cloud computing [72]. Virtualization describes how computational resources, including hardware, software, and storage, are shared or aggregated between instances [72]. Virtualization is a necessary technology inextricably linked to the cloud computing theory. It is the virtualization technology that facilitates cloud services, particularly platform as a service (PaaS) and software as a service (SaaS), where a single physical infrastructure comprises services or platforms to offer to numerous cloud users [72]. This adds the entire security features of virtualization technology to the current cloud computing security problems and challenges [72]. In addition, virtualization is a crucial technique for attaining cost-effectiveness and availability in cloud infrastructure, which dramatically cuts costs and helps CSPs [72].

Furthermore, the reliance on virtualization implies that security is equally important; yet, it has been identified as a weak target [72]. Cloud privacy and security are allegedly directly threatened by virtualization and its multi-tenancy [72,73]. For instance, virtualization and equipment characteristics make attacks through the side and backdoor channels possible.

20.6.3 Trust

Trust can be a subjectively measurable scale that influences decisions depending on the underlying assumptions. Various criteria are used to evaluate trust, a complex and multi-phased phenomenon with many dimensions. Because of this, it is very varied and depends on the underlying circumstance [72,74]. For example, trust difficulties develop when a customer infrastructure is kept on an external basis and is handled by an external party company in cloud settings. These two factors imply that a human element that clients are unaware of and interact with the infrastructure exists [72,74].

The cloud provider's duties include configuring the underlying SaaS, PaaS, or IaaS infrastructure. The inclusion of security management is very crucial. The infrastructure, including bare metal, hardware, and data centers, is also related to trust [72,74]. When potentially important data is entrusted to someone else almost entirely, problems might develop with the smallest asset and the most comprehensive security picture [72,74].

20.7 Security requirements for 6G network in cloud environment

The 6G network security structure was created with flexibility in mind. Considering 6G is designed to provide an enormous advantage over 5G, the distinction between within and beyond the system will become increasingly hazy [29]. Consequently, conventional network security methods such as IPsec and firewalls will be ineffective in protecting the network from attackers. The 6G safety model should include the crucial safety principle of zero trust (ZT) in the cloud environment network to overcome this problem. ZT is a safety idea that prioritizes memory space preservation. ZT assumes that an intrusive party might be present throughout the network, as well as the network infrastructure is vulnerable to intrusion or dishonesty from outsiders. Regularly conducting such an evaluation and taking precautions will minimize the risk of internal resource harm [75]. The ZT design (ZTD) is a framework to enhance that includes interactions between networked devices (NDs), protocols procedures, and access controls. As a result, ZTD should serve as the core of 6G security architecture. Using the ZT paradigm, some security needs may be met to allow secure 6G networks. The safety needs that must be addressed and addressed by the safety infrastructure in 6G networks are explained in the following paragraphs.

- Virtualization security solution: To address VMs security risks, a scheme with a reliable virtualization layer is required, as well as security mechanisms that detect hidden malicious programs, such as rootkits. Furthermore, the hypervisor must provide a complete division of compute, capacity, as well as network capabilities utilizing hypertext transfer protocol (HTTPS) protocol such as transport layer security (TLS), secure shell (SSH), a virtual private network (VPN), and so on. VM inspection (VMI) is a hypervisor capability that analyses and detects safety threats by monitoring each VM virtual central processing unit (vCPU) register metadata, file IO, and communications packets to avoid invasion. When adopting containerization, the operating system should suitably establish the rights of the various containers and prohibit the installation of critical system domains and immediate access to the host machine storage package.

- Autonomous management solution: The essential thing to undertake when dealing with open source concerns is to control flaws produced by using, updating, and removing public documents. As a result, timely identification of attacks needs an intelligent decision-making system capable of detecting loopholes and applying updates. A further step is required to guarantee that the modified software is administered promptly and correctly using the secure over-the-air (OTA) approach. Moreover, a safety governance model must be built to deal with (1) long-term open source weaknesses, (2) shifts in developer perspective, and (3) the implementation of safety mechanisms.
- AI-powered data security: AI systems must be transparent about how they defend users and the cloud ecosystem from attackers in order to guarantee that they are secure against threats. The first step in the procedure is to create reliable AI models. Additionally, it is necessary to validate if the AI algorithms operating on user equipment (UE), radio access networks (RAN), and the kernel have been unethically modified or changed due to an unfriendly assault by using a method like digital signatures. A system must perform self-healing or recovery methods when a hostile AI model has been discovered. Additionally, the system must restrict data collection for AI training to strong network elements.
- Users' privacy: To protect their safety, users' private details should be maintained and utilized in line with established procedures between the service provider, the cloud users, the client, and the cloud ecosystem. The 6G system secures private information in a secured execution environment (SEE) and trustworthy software while reducing or anonymizing the volume of publicly available data. Before cloud users publish private information, its integrity and authorization must be confirmed. Another alternative when engaging with customer data is to employ homomorphic encryption (HE) to ensure that the content may be retrieved in an encoded format. AI-based approaches, like a learning-based confidentiality offloading system, might also be utilized to protect the geolocation and use behavior.
- Post-quantum encryption scheme: Because supercomputers will render present asymmetric key authentication schemes unsafe, the 6G system must abandon them. Many academics have focused on post-quantum encryption (PQE) methods such as lattice-based encryption, code-based cryptology, multimodal polynomial encryption technology, and hash-based signatures. The US National Institute of Standards and Technology (NIST) plans to choose the best PQE techniques between 2022 and 2024 as part of its PQC research. The

key length for PQE is predicted to be several orders of magnitude greater than Rivest–Shamir–Adleman (RSA). In addition, PQEs are anticipated to be more computationally expensive than the present RSA technique. As a result, it is critical that PQE be properly incorporated into the hardware and software performance and service requirements of the 6G network.

20.8 AI solution to 6G privacy and security issues in cloud environment

Intelligence network administration configurations have several security problems. First, automated closed-loop networks may be vulnerable to security threats, including MITM attacks, deception, and denial of service (DoS) [75]. DoS assaults can be carried out by gradually developing fictitious high demands on virtual network functions (VNFs), which would upsurge the size of VMs. MITM attacks can reroute traffic through malicious devices by generating bogus fault events and collecting domain control signals. Attacks using deceit may be made by altering the sent data. Second, if 6G networks use information-disclosure-prone Intent-Based Interfaces like Zero touch network & Service Management (ZSM), attacks including malicious configuration and aberrant behavior may occur. Information that unauthorized parties intercept with harmful intent might jeopardize system security objectives (such as confidentiality and privacy) and inspire other attacks. The entire security of the management system may be compromised by undesired settings in intent-based interfaces, such as modifying the mapping from intent to action or lowering the security level. A faulty intention might potentially lead to similar effects.

In addition to smart surfaces, zero-energy IoT devices, improved AI methodologies, potential quantum computing systems, AI-powered automated devices, AI-driven air interfaces, humanoid robots, and autonomous networks, 6G is predicted to establish the way for the development of a number of new innovations [76]. Future developments in digital societies will also call for new applications, including the huge availability of little data, the rise in the old population, the meeting of communication, sensing, and computation, and gadget-free communication. AI technologies demonstrate a stronger influence on privacy issues related to 6G security issues in two ways [77]. First, while the proper use of AI may improve privacy in 6G, there is also a chance that AI attacks may violate privacy [78,79]. Second, AI models may be the target of privacy attacks during the testing and training phases (e.g., a poisoning assault, reverse, membership interference, adversarial attacks) [79].

With the utilization of AI and a 6G network, networks may be able to self-configure, self-organize, self-heal, optimize, self-segment, and recover from faults. As a result, it will boost feasibility, shorten recovery times, and improve service quality for the customer. Figure 20.4 illustrates how AI is used in 6G networks.

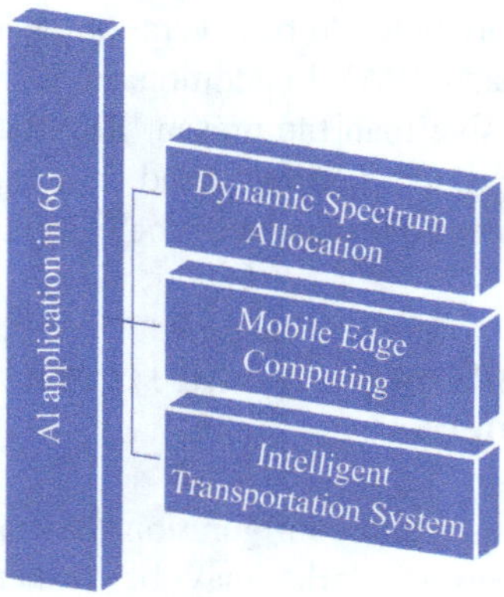

Figure 20.4 A solution catalogue for the selected applications of AI

20.9 Conclusions

The influence of AI is pervasive across all industries, and when 6G is progressively implemented, it is anticipated to fully utilize AI approaches. A 6G architecture that is AI-enabled and makes use of a variety of AI technologies, from the physical layer up to the application layer. This chapter introduces to research on the 6G network security issues and requirements. It depicts the progression of privacy in older wireless networks, beginning with the 1G network and progressing to the eventual 6G network. This chapter introduces critical 6G network needs as well as 6G architecture. Security issues in the 6G network and cloud environment were explored. The advantages and challenges of incorporating cloud computing/mobile edge computing into 6G were also discussed. A more in-depth study on various assaults on the 6G network connected to cloud computing is still required. This chapter delves into the essential issues and challenges of 6G network security and privacy in the cloud environment. Various technology and security problems are also covered. This chapter provides an overview of 6G applications and safety needs. Furthermore, the use of AI in reducing the safety problems of 6G in the cloud context was explored. Finally, this chapter covers the security of 6G networks as well as AI-based solutions. Finding a way to defend 6G critical security challenges in the cloud environment is a significant topic that could be explored in the future to ensure the security of 6G technologies.

Acknowledgment

The work of Agbotiname Lucky Imoize is supported in part by the Nigerian Petroleum Technology Development Fund (PTDF) and, in part, by the German Academic Exchange Service (DAAD) through the Nigerian–German Postgraduate Program under Grant 57473408.

References

[1] G. Marco, M. Polese, M. Mezzavilla, S. Rangan, and M. Zorzi, Toward 6G networks: use cases and technologies, *IEEE Commun. Mag.* 58(3) (2020) 55–61.

[2] A. L. Imoize, H. I. Obakhena, F. I. Anyasi, and S. N. Sur, A review of energy efficiency and power control schemes in ultra-dense cell-free massive MIMO systems for sustainable 6G wireless communication, *Sustainability* 14(17) (2022) 11100.

[3] T. Huang, W. Yang, J. Wu, J. Ma, X. Zhang, and D. Zhang, A survey on green 6G network: architecture and technologies, *IEEE Access* 7 (2019) 175758–175768.

[4] L. D. Xu, Industrial information integration – an emerging subject in industrialization and informatization process, *J. Ind. Inf. Integr.* 17 (2020) 100128.

[5] K. David and H. Berndt, 6G vision and requirements: is there any need for beyond 5G? *IEEE Veh. Technol. Mag.* 13(3) (2018) 72–80.

[6] A. L. Imoize, O. Adedeji, N. Tandiya, and S. Shetty, 6G enabled smart infrastructure for sustainable society: opportunities, challenges, and research roadmap, *Sensors* 21(5) (2021) 1709.

[7] E. Bastug, M. Bennis, M. Médard, and M. Debbah, Toward interconnected virtual reality: opportunities, challenges, and enablers, *IEEE Commun. Mag.* 55(6) (2017) 110–117.

[8] Z. Yajun, Y. Guanghui, and X. U. Hanqing, 6G mobile communication networks: vision, challenges, and key technologies, *Sci. Sin. Inf.* 49(8) (2019) 963–987.

[9] P. Yang, Y. Xiao, M. Xiao, and S. Li, 6G wireless communications: vision and potential techniques, *IEEE Netw.* 33(4) (2019) 70–75.

[10] A. E. Adeniyi, K. M. Abiodun, J. B. Awotunde, M. Olagunju, O. S. Ojo, and N. P. Edet, Implementation of a block cipher algorithm for medical information security on cloud environment: using modified advanced encryption standard approach, *Multimed. Tools Appl.* 82 (2023) 20537–20551.

[11] Y. Xing and T. S. Rappaport, Propagation measurement system and approach at 140 GHz-moving to 6G and above 100 GHz, In *2018 IEEE Global Communications Conference (GLOBECOM)* (pp. 1–6). IEEE, 2018.

[12] S. Fouladi, J. Emmons, E. Orbay, *et al.* Salsify: low-latency network video through tighter integration between a video codec and a transport protocol, In *Networked Systems Design and Implementation*, 2018, pp. 267–282.

[13] J. B. Awotunde, S. Misra, O. B. Ayoade, R. O. Ogundokun, and M. K. Abiodun, Blockchain-based framework for secure medical information in internet of things system, In *Blockchain Applications in the Smart Era* (pp. 147–169). Cham: Springer International Publishing, 2022.

[14] J. B. Awotunde, C. Chakraborty, and S. O. Folorunso, A secured transaction based on blockchain architecture in mobile banking platform, *Int. J. Internet Technol. Secured Trans.* 12(4) (2022) 287–303.

[15] E. A. Adeniyi, R. O. Ogundokun, S. Misra, J. B. Awotunde, and K. M. Abiodun, Enhanced security and privacy issue in multi-tenant environment of green computing using blockchain technology, In *Blockchain Applications in the Smart Era* (pp. 65–83). Cham: Springer International Publishing, 2022.

[16] H. Taherdoost, The role of smart contract blockchain in 6G wireless communication system, *Appl. Proc. Comput. Sci*. 215 (2022) 44–50.

[17] T. Hewa, G. Gur, A. Kalla, M. Ylianttila, An Braeken, and M. Liyanagek, The role of blockchain in 6G: challenges, opportunities and research directions, In *Proceedings of the 2020 2nd 6G Wireless Summit 2020: Gain Edge for the 6G Era, (6G SUMMIT)*, Levi, Finland, 17–20 March 2020.

[18] K. Zhang, Y. Zhu, S. Maharjan, and Y. Zhang, Edge intelligence and blockchain empowered 5G beyond for the industrial Internet of Things, *IEEE Netw.* 33(5) (2019) 12–19.

[19] T. Maksymyuk, J. Gazda, L. Han, and M. Jo, Blockchain-based intelligent network management for 5G and beyond, In *2019 3rd International Conference on Advanced Information and Communications Technologies (AICT)*, 2019, pp. 36–39.

[20] Y. Dai, D. Xu, S. Maharjan, Z. Chen, Q. He, and Y. Zhang, Blockchain and deep reinforcement learning empowered intelligent 5G beyond, *IEEE Netw.* 33(3) (2019) 10–17.

[21] B. Mafakheri, T. Subramanya, L. Goratti, and R. Riggio, Blockchain based infrastructure sharing in 5G small cell networks, In *2018 14th International Conference on Network and Service Management (CNSM)*, IEEE, 2018, pp. 313–317.

[22] K. Shahzad, A. O. Aseeri, and M. A. Shah, A blockchain-based authentication solution for 6G communication security in tactile networks, *Electronics* 11 (2022) 1374. https://doi.org/10.3390/ electronics11091374

[23] V. Adat, I. Politis, C. Tselios, and S. Kotsopoulos, "Blockchain enhanced SECRET small cells for the 5G environment, In *2019 IEEE 24th International Workshop on Computer Aided Modeling and Design of Communication Links and Networks (CAMAD)*, 2019, pp. 1–6.

[24] L.K. Ramasamy, Firoz Khan K.P., A. L. Imoize, *et al.*, Blockchain-based wireless sensor networks for malicious node detection: a survey, *IEEE Access* 9 (2021) 128765–128785.

[25] J. B. Awotunde, S. Misra, and Q. T. Pham, A secure framework for internet of medical things security based system using lightweight cryptography enabled blockchain, In *Future Data and Security Engineering. Big Data, Security and Privacy, Smart City and Industry 4.0 Applications: 9th International Conference, FDSE 2022, Ho Chi Minh City, Vietnam, November 23–25, 2022, Proceedings,* Singapore: Springer Nature Singapore, 2022, pp. 258–272.

[26] P. K. Sharma, S. Singh, Y.-S. Jeong, and J. H. Park, Distblocknet: a distributed blockchains-based secure SDN architecture for IoT networks, *IEEE Commun. Mag.* 55(9) (2017) pp. 78–85.

[27] M. Banafaa, I. Shayea, J. Din, *et al.*, 6G mobile communication technology: requirements, targets, applications, challenges, advantages, and opportunities, *Alexandria Eng. J.* 64 (2023) 245–274, https://doi.org/10.1016/j.aej.2022.08.017

[28] M. Wang, T. Zhu, T. Zhang, J. Zhang, S. Yu, and W. Zhou, Security and privacy in 6G networks: new areas and new challenges, *Digit. Commun. Netw.* 6 (2020) 281–291, https://doi.org/10.1016/j.dcan.2020.07.003

[29] S. A. Abdel Hakeem, H. H. Hussein, and H. Kim, Security requirements and challenges of 6G technologies and applications, *Sensors* 22(5) (2022) 1969, https:// doi.org/10.3390/s22051969

[30] C. Meshram, A. L. Imoize, A. Elhassouny, A. Aljaedi, A. R. Alharbi, and S. S. Jamal, IBOOST: a lightweight provably secure identity-based online/offline signature technique based on FCM for massive devices in 5G wireless sensor networks, *IEEE Access* 9 (2021) 131336–131347.

[31] R. Sattiraju, A. Weinand, and H. D. Schotten, Ai-assisted Phy Technologies for 6g and beyond wireless networks, arXiv Preprint arXiv:1908.09523.

[32] L. Loven, T. Leppanen, and E. Peltonen, Edge Ai: a vision for distributed, edge-native artificial intelligence in future 6G networks, In *The 1st 6G Wireless Summit*, 2019, pp. 1–2.

[33] Z. Zhou, H. Liao, B. Gu, K. M. S. Huq, S. Mumtaz, and J. Rodriguez, Robust mobile crowd sensing: when deep learning meets edge computing, *IEEE Netw.* 32 (4) (2018) 54–60.

[34] S. Dang, O. Amin, B. Shihada, and M.-S. Alouini, What should 6G be? *Nat. Electron.* 3(1) (2020) 20–29.

[35] I. Tomkos, D. Klonidis, E. Pikasis, and S. Theodoridis, Toward the 6G network era: opportunities and challenges, *IT Prof.* 22(1) (2020) 34–38.

[36] K. S. Adewole, T. T. Salau-Ibrahim, A. L. Imoize, *et al.*, Empirical analysis of data streaming and batch learning models for network intrusion detection, *Electronics* 11(19) (2022) 3109.

[37] S. A. Hassnain Mohsan, A. Mazinani, W. Malik, *et al.*, 6G: envisioning the key technologies, applications and challenges, *Int. J. Adv. Comput. Sci. Appl.* 11(9) (2020) 14–23, https://doi.org/10.14569/IJACSA.2020.0110903

[38] S. J. Nawaz, S. K. Sharma, S. Wyne, M. N. Patwary, and M. Asaduzzaman, Quantum machine learning for 6G communication networks: state-of-the-art and vision for the future, *IEEE Access* 7 (2019) 46317–46350.

[39] J. B. Awotunde, R. G. Jimoh, S. O. Folorunso, E. A. Adeniyi, K. M. Abiodun, and O. O. Banjo, Privacy and security concerns in IoT-based healthcare systems, In *The Fusion of Internet of Things, Artificial Intelligence, and Cloud Computing in Health Care* (pp. 105–134). Cham: Springer International Publishing, 2021.

[40] E. C. Strinati, S. Barbarossa, J. L. Gonzalez-Jimenez, D. Ktenas, N. Cassiau, and C. Dehos, 6G: The Next Frontier, arXiv_Preprint arXiv:1901.03239.

[41] Z. Zhang, Y. Xiao, Z. Ma, *et al.*, 6G wireless networks: vision, requirements, architecture, and key technologies, *IEEE Veh. Technol. Mag.* 14(3) (2019) 28–41.

[42] X. Ling, J. Wang, T. Bouchoucha, B. C. Levy, and Z. Ding, Blockchain radio access network (b-ran): towards decentralized secure radio access paradigm, *IEEE Access* 7 (2019) 9714–9723.
[43] M. S. Islim, R. X. Ferreira, X. He, *et al.*, Towards 10 gb/s orthogonal frequency division multiplexing-based visible light communication using a GaN violet micro-LED, *Photon. Res.* 5(2) (2017) A35–A43.
[44] J. Luo, L. Fan, and H. Li, Indoor positioning systems based on visible light communication: state of the art, *IEEE Commun. Surv. Tutor.* 19(4) (2017) 2871–2893.
[45] L. U. Khan, Visible light communication: applications, architecture, standardization and research challenges, *Digit. Commun. Netw.* 3(2) (2017) 78–88.
[46] S. Ucar, S. Coleri Ergen, O. Ozkasap, D. Tsonev, and H. Burchardt, Secvlc: secure visible light communication for military vehicular networks, In *Proceedings of the 14th ACM International Symposium on Mobility Management and Wireless Access*, 2016, pp. 123–129.
[47] S. A. H. Mohsan, M. Mehedi, A. Mazinani, *et al.*, A systematic review on practical considerations, recent advances and research challenges in underwater optical wireless communication, 2020, pp. 11–17.
[48] S. Elmeadawy and R. Shubair, 6G future wireless communications: future technologies and research challenges, In *2019 International Conference on Electrical and Computing Technologies and Applications (ICECTA)*, 2019, doi:10.1109/ICECTA48151.2019.8959607
[49] P. Porambage, G. Gür, D. P. Moya Osorio, M. Livanage, and M. Ylianttila, 6G security challenges and potential solutions, In *2021 Joint European Conference on Networks and Communications & 6G Summit (EuCNC/6G Summit)*, Porto, Portugal, 2021, pp. 622–627, doi:10.1109/EuCNC/6GSummit51104.2021.9482609
[50] F. Dressler and F. Kargl, Towards security in nano-communication: challenges and opportunities, *Nano Commun. Netw.* 3(3) (2012) 151–160.
[51] T. Nakano, Y. Okaie, S. Kobayashi, T. Hara, Y. Hiraoka, and T. Haraguchi, Methods and applications of mobile molecular communication, *Proc. IEEE* 107(7) (2019) 1442–1456.
[52] N. Farsad, H. B. Yilmaz, A. Eckford, C.-B. Chae, and W. Guo, A comprehensive survey of recent advancements in molecular communication, *IEEE Commun. Surv. Tutor.* 18(3) (2016) 1887–1919.
[53] Y. Lu, M. D. Higgins, and M. S. Leeson, Comparison of channel coding schemes for molecular communications systems, *IEEE Trans. Commun.* 63 (11) (2015) 3991–4001.
[54] V. Loscri, C. Marchal, N. Mitton, G. Fortino, and A. V. Vasilakos, Security and privacy in molecular communication and networking: opportunities and challenges, *IEEE Trans. Nanobiosci.* 13(3) (2014) 198–207.
[55] H. Tabrizchi and M. Kuchaki Rafsanjani, A survey on security challenges in cloud computing: issues, threats, and solutions, *J. Supercomput.* 76(12) (2020) 9493–9532, https://doi.org/10.1007/s11227-020-03213-1

[56] M. Ahmed and M. Ashraf Hossain, Cloud computing and security issues in the cloud, *Int. J. Netw. Secur. Appl.* 6(1) (2014) 25–36, https://doi.org/10.5121/ijnsa.2014.6103

[57] T. M. Khorshed, A. B. M. S. Ali, and S. A. Wasimi, A survey on gaps, threat remediation challenges and some thoughts for proactive attack detection in cloud computing, *Future Gener. Comput. Syst.* 28(2012) 833–851, doi:10.1016/j.future.2012.01.006

[58] M. Ahmed and M. A. Hossain, Cloud computing and security issues in the cloud, *Int. J. Netw. Secur. Appl. (IJNSA)* 6(1) 2014.

[59] N. Alrehailiand and A. Mutaha, Cloud computing security challenges, *Iarjset* 7(8) (2020) 120–123, https://doi.org/10.17148/iarjset.2020.7817

[60] D. Chen and H. Zhao, Data security and privacy protection issues in cloud computing, In *2012 International Conference on Computer Science and Electronics Engineering*, Hangzhou, 23–25 March 2012, pp. 647–651, https://doi.org/10.1109/ICCSEE.2012.193

[61] T. S. Chou, Security threats on cloud computing vulnerabilities, *Int. J. Comput. Sci. Inf. Technol.* 5(3) (2013) 79.

[62] D. Teneyuca, Internet cloud security: the illusion of inclusion, *Inf. Secur. Tech. Rep.* 16 (2011) 102–107, doi:10.1016/j.istr.2011.08.005

[63] A. Joint and E. Baker, Knowing the past to understand the present 1 – issues in the contracting for Cloud based services, *Comput. Law Secur. Rev.* 27 (2011) 407–415. doi:10.1016/j.clsr.2011.05.002

[64] A. Bisong and S. S. M. Rahman, An overview of the security concerns in enterprise cloud computing, *Int. J. Netw. Secur. Appl.* 3(1) (2011) 30–45, doi:10.5121/ijnsa.2011.3103

[65] M. AbdulRaheem, I. D. Oladipo, A. González-Briones, J. B. Awotunde, A. R. Tomori, and R. G. Jimoh, An efficient lightweight speck technique for edge-IoT-based smart healthcare systems, In *5G IoT and Edge Computing for Smart Healthcare* (pp. 139–162). London: Academic Press.

[66] S. Qaisar and K. F. Khawaja, Cloud computing: network/security threats and countermeasures, *Interdiscip. J. Contemp. Res. Business* 3(9) (2012) 1323–1329.

[67] Hashizume, D. G. Rosado, E. F. -Medina, *et al.*, An analysis of security issues for cloud computing, *J. Internet Serv. Appl.* 4(5) (2013) 1–13.

[68] W. Kim, Cloud computing: today and tomorrow, *J. Object Technol.* 8(1) (2009) 65–72.

[69] C. B. Westphall, C. M. Westphall, F. L. Koch, *et al.*, Management and security for grid, cloud and cognitive networks, *Rev. Sistemas Informação FSMA* 8 (2011) 8–21.

[70] K. Hamlen, M. Kantarcioglu, L. Khan, and V. Thuraisingham, Security issues for cloud computing, *Int. J. Inf. Secur. Privacy* 4(2) (2010) 39–51, doi:10.4018/jisp.2010040103

[71] S. Basu, A. Bardhan, K. Gupta, *et al.*, *Cloud Computing Security Challenges & Solutions—A Survey*, New York, NY: IEEE, 2018, pp. 347–356.

[72] M. Ahmed and A. T. Litchfield, Taxonomy for identification of security issues in cloud computing environments, *J. Comput. Inf. Syst.* 58(1) (2018) 79–88, https://doi.org/10.1080/08874417.2016.1192520
[73] C. Modi, D. Patel, B. Borisaniya, H. Patel, A. Patel, and M. Rajarajan, A survey of intrusion detection techniques in cloud, *J. Netw. Comput. Appl.* 36 (2013) 42–57.
[74] D. A. B. Fernandes, L.F. B. Soares, J. V. Gomes, M. M. Freire, and P. R. M. Inácio, Security issues in cloud environments: a survey, *Int. J. Inf. Secur.* 13 (2014) 113–170.
[75] I. Tomkos, D. Klonidis, E. Pikasis, and S. Theodoridis, Toward the 6G network era: opportunities and challenges, *IT Professional* 22(1) (2020) 34–38.
[76] C. de Alwis, A. Kalla, and Q. V. Pham, Survey on 6G frontiers: trends, applications, requirements, technologies and future research, *IEEE Open J. Commun. Soc.* (2021) 1–1.
[77] Y. Sun, J. Liu, J. Wang, Y. Cao, and N. Kato, When machine learning meets privacy in 6G: a survey, *IEEE Commun. Surv. Tutor.* 22(4) (2020) 526–565.
[78] P. H. Pathak, X. Feng, P. Hu, and P. Mohapatra, Visible light communication, networking, and sensing: a survey, potential and challenges, *IEEE Commun. Surv. Tutor.* 17(4) (2015) 2047–2077.
[79] C. Benzaid and T. Taleb, ZSM security: threat surface and best practices, *IEEE Netw.* 34(3) (2020) 124–133.

Index

www.ingramcontent.com/pod-product-compliance
Lightning Source LLC
LaVergne TN
LVHW011547200926
848484LV00001B/292
* 9 7 8 1 8 3 9 5 3 6 6 3 2 *